CWNA: Certified Wireless Network Administrator Official Study Guide

CWNA Exam (PW0-104) Objectives

OBJECTIVE	CHAPTER
Radio Frequency (RF) Technologies	
1.1. RF Fundamentals	2
1.1.1. Define and explain the basic concepts of RF behavior: gain; loss; reflection; refraction; diffraction; scatterings; VSWR; return loss; amplification; attenuation; absorption; wave propagation; free space path loss; delay spread.	
1.2. RF Mathematics	3
1.2.1. Understand and apply the basic components of RF mathematics: watt; milliwatt; decibel (dB); dBm; dBi; dBd; SNR; RSSI; system operating margin (SOM); fade margin; link budget; intentional radiator; equivalent isotropically radiated power (EIRP).	
1.3. RF Signal and Antenna Concepts	4, 18
1.3.1. Identify RF signal characteristics, the applications of basic RF antenna concepts, and the implementation of solutions that require RF antennas: visual LOS; RF LOS; the Fresnel zone; beamwidths; azimuth and elevation; passive gain; isotropic radiator; polarization; simple antenna diversity; MIMO diversity; radio chains; spatial multiplexing (SM); transmit beamforming (TxBF); maximal ratio combining (MRC); wavelength; frequency; amplitude; phase.	
1.3.2. Explain the applications of physical RF antenna and antenna system types and identify their basic attributes, purpose, and function: omnidirectional/dipole antennas; semidirectional antennas; highly directional antennas; sectorized antennas.	
1.3.3. Describe the proper locations and methods for installing RF antennas: pole/mast mount; ceiling mount; wall mount.	
1.4. RF Antenna Accessories	4
1.4.1. Identify the use of the following WLAN accessories and explain how to select and install them for optimal performance and regulatory domain compliance: amplifiers; attenuators; lightning arrestors; mounting systems; grounding rods/wires; towers, safety equipment, and concerns; RF cables; RF connectors; RF signal splitters.	
IEEE 802.11 Regulations and Standards	
2.1. Spread Spectrum Technologies	6, 7, 8, 18
2.1.1. Identify some of the uses for spread spectrum technologies: wireless LANs; wireless PANs; wireless MANs; wireless WANs.	
2.1.2. Comprehend the differences between, and explain the different types of, spread spectrum technologies and how they relate to the IEEE 802.11-2007 standards (as amended and including 802.11n draft 2.0) PHY clauses: DSSS; HR-DSSS; ERP; OFDM; HT (MIMO).	
2.1.3. Identify the underlying concepts of how spread spectrum technology works: modulation; coding.	
2.1.4. Identify and apply the concepts that make up the functionality of spread spectrum technology: colocation; channel centers and widths (all PHYs); primary and secondary channels; overlapping and nonoverlapping channels; carrier frequencies; throughput vs. data rate; bandwidth; communication resilience; physical carrier-sense (CSMA/CA); virtual carrier-sense (NAV).	
2.2. IEEE 802.11-2007 Standard (as Amended and Including 802.11n Draft 2.0)	5, 6, 9
2.2.1. Identify, explain, and apply the frame types and frame exchange sequences covered by the IEEE 802.11-2007 standard.	

Sybex®
An Imprint of
WILEY

OBJECTIVE	CHAPTER
2.2.2. Identify and apply regulatory domain requirements: dynamic frequency selection (DFS); transmit power control (TPC); available channels; output power.	
2.2.3. OSI model layers affected by the 802.11-2007 standard and amendments	
2.2.4. Use of ISM and UNII bands in Wi-Fi networks	
2.2.5. Supported data rates for each IEEE 802.11-2007 PHY	
2.3. IEEE 802.11 Industry Organizations and Their Roles	1
2.3.1. Define the roles of the following organizations in providing direction, cohesion, and accountability within the WLAN industry: regulatory domain governing bodies; IEEE; Wi-Fi Alliance.	
IEEE 802.11 Protocols and Devices	
3.1. IEEE 802.11 Protocol Architecture	7, 8, 9
3.1.1. Summarize the processes involved in authentication and association: the IEEE 802.11 state machine; Open System authentication, Shared Key authentication, and deauthentication; association, reassociation, and disassociation.	
3.1.2. Define, describe, and apply the following concepts associated with WLAN service sets: stations and BSSs; starting and joining a BSS; BSSID and SSID; Ad Hoc mode and IBSS; Infrastructure mode and ESS; distribution system (DS); distribution system media; layer 2 and layer 3 roaming.	
3.1.3. Explain and apply the following power-management features of WLANs: Active mode; Power Save mode; unscheduled automatic power save delivery (U-APSD); WMM Power Save (WMM-PS); Power Save Multi Poll (PSMP); spatial multiplexing power save (SMPS); TIM/DTIM/ATIM.	
3.2. IEEE 802.11 MAC and PHY Layer Technologies	8, 9, 18
3.2.1. Describe and apply the following concepts surrounding WLAN frames: IEEE 802.11 frame format vs. IEEE 802.3 frame format; layer 3 protocol support by IEEE 802.11 frames; terminology review: frames, packets, and datagrams; terminology review: bits, bytes, and octets; terminology: MAC and PHY; guard interval (GI); PSDU; PPDU; PPDU formats; MSDU; MPDU; A-MPDU; A-MSDU; 802.11 frame format; 802.11 frame types; interframe spaces (RIFS, SIFS, PIFS, DIFS, AIFS, EIFS); block acknowledgments; jumbo frame support (layer 2); MTU discovery and functionality (layer 3).	
3.2.2. Identify methods described in the IEEE 802.11-2007 standard for locating, joining, and maintaining connectivity with an IEEE 802.11 WLAN: active scanning (probes); passive scanning (beacons); dynamic rate switching.	
3.2.3. Define, describe, and apply IEEE 802.11 coordination functions and channel access methods and features available for optimizing data flow across the RF medium: DCF and HCF coordination functions; EDCA channel access method; RTS/CTS and CTS-to-Self protocols; HT Dual-CTS protection; HT L-SIG protection; HT channel width operation (20 MHz, 20/40 MHz, PCO); HT operation modes (0, 1, 2, 3); fragmentation.	
3.3. WLAN Infrastructure and Client Devices	10
3.3.1. Identify the purpose of the following WLAN infrastructure devices and describe how to install, configure, secure, and manage them: autonomous access points; lightweight access points; mesh access points / routers; enterprise WLAN controllers; remote office WLAN controllers; PoE injectors (single and multiport) and PoE-enabled Ethernet switches; WLAN bridges; residential WLAN gateways; enterprise encryption gateways.	
3.3.2. Describe the purpose of the following WLAN client devices and explain how to install, configure, secure, and manage them: PC Cards (ExpressCard, CardBus, and PCMCIA); USB2, CF, and SD devices; PCI, Mini PCI, and Mini PCIe cards; workgroup bridges.	

Exam specifications and content are subject to change at any time without prior notice and at the CWNP's sole discretion. Please visit CWNP's website (`www.cwnp.com`) for the most current information on their exam content.

Sybex®
An Imprint of
WILEY

CWNA®

Certified Wireless Network Administrator

Official Study Guide

David Coleman

David Westcott

Wiley Publishing, Inc.

Acquisitions Editor: Jeff Kellum
Development Editor: Sara Barry
Technical Editor: Sam Coyl
Production Editor: Rachel McConlogue
Copy Editor: Sharon Wilkey
Production Manager: Tim Tate
Vice President and Executive Group Publisher: Richard Swadley
Vice President and Publisher: Neil Edde
Assistant Project Manager: Jenny Swisher
Associate Producer: Kit Malone
Quality Assurance: Angie Denny
Book Designers: Judy Fung and Bill Gibson
Compositor: Craig Woods, Happenstance Type-O-Rama
Proofreader: Nancy Bell
Indexer: Ted Laux
Project Coordinator, Cover: Lynsey Stanford
Cover Designer: Ryan Sneed

Published simultaneously in Canada

ISBN: 978-0-470-43890-9

For general information on our other products and services or to obtain technical support, please contact our Customer Care Department within the U.S. at (877) 762-2974, outside the U.S. at (317) 572-3993, or fax (317) 572-4002.

Wiley also publishes its books in a variety of electronic formats. Some content that appears in print may not be available in electronic books.

Library of Congress Cataloging-in-Publication Data is available from the publisher.

10 9 8 7 6 5

Dear Reader,

Thank you for choosing *CWNA: Certified Wireless Network Administrator Official Study Guide*. This book is part of a family of premium-quality Sybex books, all of which are written by outstanding authors who combine practical experience with a gift for teaching.

Sybex was founded in 1976. More than thirty years later, we're still committed to producing consistently exceptional books. With each of our titles we're working hard to set a new standard for the industry. From the paper we print on, to the authors we work with, our goal is to bring you the best books available.

I hope you see all that reflected in these pages. I'd be very interested to hear your comments and get your feedback on how we're doing. Feel free to let me know what you think about this or any other Sybex book by sending me an email at `nedde@wiley.com`, or if you think you've found a technical error in this book, please visit `http://sybex.custhelp.com`. Customer feedback is critical to our efforts at Sybex.

Best regards,

Neil Edde
Vice President and Publisher
Sybex, an Imprint of Wiley

We dedicate this book to all the men and women of the United States Armed Forces for putting their private lives aside to preserve and protect freedom. Thank you for your service and your sacrifice.

Acknowledgments

David Coleman would once again like to thank his children, Brantley and Carolina, for their patience and understanding of their father throughout the writing of this book. I love you kids very much. David would also like to thank his mother, Marjorie Barnes, and his stepfather, William Barnes, for many years of support and encouragement.

David Westcott would like to thank his parents, Kathy and George, who have provided so much support and love and from whom he has learned so much. He would also like to thank Janie, Jennifer, and Samantha for their patience and understanding of life on the road and for their support throughout the writing of this book.

Writing this edition of the CWNA Study Guide has been a rapid-fire version of writing the original book. Like writing the original book, it was an adventure from the start. We would like to thank all of the following individuals for their support and contributions during the entire process.

We must first thank the acquisitions editor of Sybex, Jeff Kellum, for initially finding us and bringing us on to this project four years ago. Jeff is an extremely patient and understanding editor who occasionally sends a nasty email message. We would also like to thank our development editor, Sara Barry and her special assistant Kathleen Avery Barry, who was born December 11, 2008—8 lbs. and 4 oz. We also need to send special thanks to our editorial manager Pete Gaughan and our production editor Rachel McConlogue, and Sharon Wilkey, our copyeditor.

We also need to give a big shout-out to our technical editor, Sam Coyl. Sam is a member of the IEEE with many years of practical experience in wireless communications. His contributions to the book were nothing short of invaluable. When Sam is not providing awesome technical editing, he is vice president of business development for Netrepid (www.netrepid.com), a wireless solutions provider.

We would also like to thank Devin Akin, Cary Chandler, Kevin Sandlin, Scott Turner, and Scott Williams of the CWNP program (www.cwnp.com). You gentlemen should be proud of the international renowned wireless certification program that you have developed. It has been a pleasure working with all of you the past seven years.

Thanks to Proxim and to Ken Ruppel (kenruppel@gmail.com) for allowing us to include the video *Beam Patterns and Polarization of Directional Antennas* on the CD-ROM. Many thanks to Andrew Potter for making himself available for our photography needs.

Special thanks goes to Andras Szilagyi for not only creating the EMANIM software program but for all the extra assistance he provided by creating a customized version of the program for the CD-ROM.

David Coleman would like to thank Wayne McAllister for his support and friendship over the past several years and for his “canary in the coal mine” analogy. David Coleman would also like to send a shout out to his *homies* in South Korea: Gene Connor, Rick McConnell, Ken Ring, and Joe Vowell.

We would also like to thank the following individuals and companies for their support and contributions to the book:

Aerohive Networks (www.aerohive.com)—Adam Conway

AeroScout (www.aeroscout.com)—Steffan Haithcox and Amit Larom

AirDefense (www.airdefense.net)—Nico Darrow, Ralf Deltrap, Bryan Harkins, and David Thomas

AirMagnet (www.airmagnet.com)—Dilip Advani, Chia Chee Kuan, and Joey Kuo

Aruba Networks (www.arubanetworks.com)—Chris Leach, Kevin Hamilton, Carolyn Cutler, and Susan Wells

Berkeley Varitronics Systems (www.bvsystems.com)—Carmine Caferra

By Light (www.by-light.com)—Steve Hurdle

CACE Technologies (www.cacetech.com)—Janice Spampinato

Caster Tray (www.castertray.com)—Joel Baldevarona

Cisco Systems (www.cisco.com)—Joel Barrett, Neil Diener, and Paul Shiffer

Cushcraft (www.cushcraft.com)—Mark Miller

Fluke Networks (www.flukenetworks.com)—Carolyn Carter, Thomas Doumas, and Lori Whitmer

General Dynamics (www.generaldynamics.com)—Kevin Terrell

Honeywell (www.honeywell.com)—Ray Durham and Nik Wong

Juniper Networks (www.juniper.net)—Curt Hooper

Meru Networks (www.merunetworks.com)—Kamal Anand

Motorola (www.motorola.com)—Debra McDonald and Socrates Sakellaropoulos

NetStumbler (www.netstumbler.com)—Marius Milner

PacStar (www.www.pacstar.com)—Bob Forman

Polycom (www.polycom.com)—Tricia Allen, Justin Borthwick, Geri Mitchell-Brown, Jonathan Cherry, Michelle Chessler, David Gibbs, Keith Hayden, Kate Lepore, David Mangham, Barbara McVicker, Wylee Post, Ken Rains, Robin Raulf-Sager, Josh Redmore, Steve Rolapp, Tom Thumser, Robert Tomer, Sue Yingling, and many others.

Potter Images (www.potterimages.com)—Andrew Potter

Proxim Wireless Corporation (www.proxim.com)—Ken Day, Pamela Valentine, and Amit Malhotra

Siemens (www.siemens.com)—Karl-Heinz Marks

Times Microwave Systems (www.timesmicrowave.com)—Joe Lanoue

Vocera (www.vocera.com)—Victoria Holl and Christopher O'Donnell

Wi-Fi Alliance (www.wifi.org)—Kelly Davis-Felner and Krista Ford

About the Authors

David D. Coleman is a WLAN security consultant and trainer. He teaches the CWNP classes that are recognized throughout the world as the industry standard for wireless networking certification, and he also conducts vendor-specific Wi-Fi training. David has instructed IT professionals from around the globe in wireless networking administration, wireless security, and wireless frame analysis. The company he founded, AirSpy Networks (www.airspy.com), specializes in corporate training and has worked in the past with Avaya, Nortel, Polycom, and Siemens. AirSpy Networks also specializes in government classes, and it has trained numerous computer security employees from various law enforcement agencies, the U.S. Marines, the U.S. Army, the U.S. Navy, the U.S. Air Force, and other federal and state government agencies. David has written many books and white papers about wireless networking, and he is considered an authority on 802.11 technology. Readers of this book are encouraged to follow David on Twitter at http://twitter.com/mistermultipath.

David is also a member of the Certified Wireless Network Expert (CWNE) Roundtable, a selected group of individuals who work with the CWNP program to provide direction for the CWNP exams and certifications. David resides in Atlanta, Georgia, where he shares a home with his two children, Carolina and Brantley. David Coleman is CWNE #4, and he can be reached via email at david@airspy.com.

David Westcott is an independent consultant and technical trainer with more than 20 years of experience in information technology, specializing in wireless networking and security. In addition to providing advice and direction to corporate clients, David has been a certified trainer for more than 16 years, providing training to government agencies, corporations, and universities around the world. David was an adjunct faculty member for Boston University's Corporate Education Center for more than 10 years and has developed courseware on wireless networking, wireless mesh networking, wireless packet analysis, wired networking, and security for Boston University and many other clients.

Since installing his first wireless network in 1999, David has become a Certified Wireless Network Trainer, Administrator, Security Professional, and Analysis Professional. David is also a member of the CWNE Roundtable, a selected group of individuals who work with the CWNP Program to provide direction for the CWNE exam and certification. David has earned certifications from Cisco, Microsoft, EC-Council, CompTIA, and Novell. David lives in Concord, Massachusetts. A licensed pilot, he enjoys flying his Piper Cherokee 180 around New England when he is not flying around the world commercially. David is CWNE #7 and can be reached via email at david@westcott-consulting.com.

Contents at a Glance

Contents

Table of Exercises

Foreword

Wireless LANs seem to be everywhere these days. The technology is advancing so rapidly that it seems almost impossible to stay abreast of all of the changes. The SOHO sector is adopting new WLAN technologies well before ratified amendments or interoperability certifications are in place for each technology. The SMB sector is slightly more cautious but often serves as a test bed for many leading-edge technologies. The enterprise has adopted 802.11 technology at an increasing pace over the past few years, and adoption has often been due to driving factors such as saving money, as with VoWiFi and device-tracking technologies, or being able to accomplish new business goals that could not be achieved without wireless technology.

Now that new Wi-Fi vendors aren't popping up every other day and market consolidation has begun, we're seeing some stabilization. With the reduced stream of new technologies and products, the focus is on educating and certifying people. CWNP is focused strictly on staying abreast of all facets of 802.11 technology: standards, products, and new technologies. The CWNA certification is the first step in the CWNP line of certifications and is focused on administering an enterprise 802.11 WLAN. CWNA includes topics such as 802.11 standards, security, management, protocol analysis, QoS, site surveying, and radio frequency. Additional certifications focus more intensely on security, protocol analysis, QoS, design, advanced surveying, VoWiFi, location tracking, and RF spectrum management.

David Coleman and David Westcott have worked as Certified Wireless Network Trainers (CWNTs) for as long as the CWNT certification has been available, and each was quick to pursue all CWNP certifications as they were released. Each has years of experience with a breadth of WLAN technologies and leading-edge products, which is obvious to their students and anyone working alongside them in the field. Having worked with each of these gentlemen for years, I can confidently say there could be no finer pair of seasoned trainers collaborating on a CWNA book. These WLAN veterans have devoted hundreds of hours to pouring their experience into this book, and the reader is assured to acquire a plethora of 802.11 knowledge. Mr. Coleman and Mr. Westcott have participated in the shaping of CWNP as a whole since its earliest days and have each added tremendous value to the CWNA certification specifically. I would like to thank each of these fine gentlemen for their unwavering support of CWNP, and I would like to congratulate them on their diverse accomplishments as engineers, trainers, and authors.

Devin Akin
Chief Technology Officer
The CWNP program

Introduction

If you have purchased this book or if you are thinking about purchasing this book, you probably have some interest in taking the CWNA® (Certified Wireless Network Administrator) certification exam or in learning more about what the CWNA certification exam is about. We would like to congratulate you on this first step, and we hope that our book can help you on your journey. Wireless networking is one of the hottest technologies on the market. As with many fast-growing technologies, the demand for knowledgeable people is often greater than the supply. The CWNA certification is one way to prove that you have the knowledge and skills to support this growing industry. This Study Guide was written with that goal in mind.

This book was written to help teach you about wireless networking so that you have the knowledge needed not only to pass the CWNA certification test, but also to be able to design, install, and support wireless networks. We have included review questions at the end of each chapter to help you test your knowledge and prepare for the test. We have also included labs, white papers, videos, and presentations on the CD to further facilitate your learning.

Before we tell you about the certification process and requirements, we must mention that this information may have changed by the time you are taking your test. We recommend that you visit `www.cwnp.com` as you prepare to study for your test to determine what the current objectives and requirements are.

Do not just study the questions and answers! The practice questions in this book are designed to test your knowledge of a concept or objective that is likely to be on the CWNA exam. The practice questions will be different from the actual certification questions. If you learn and understand the topics and objectives, you will be better prepared for the test.

About CWNA® and CWNP®

If you have ever prepared to take a certification test for a technology that you are unfamiliar with, you know that you are not only studying to learn a different technology, but probably also learning about an industry that you are unfamiliar with. Read on and we will tell you about CWNP.

CWNP is an abbreviation for *Certified Wireless Network Professional*. There is no CWNP test. The CWNP program develops courseware and certification exams for wireless LAN technologies in the computer networking industry. The CWNP certification program is a vendor-neutral program.

The objective of CWNP is to certify people on wireless networking, not on a specific vendor's product. Yes, at times the authors of this book and the creators of the certification will talk about, demonstrate, or even teach how to use a specific product; however, the goal is the overall understanding of wireless, not the product itself. If you learned to drive a car, you had to physically sit and practice in one. When you think back and reminisce, you probably do not tell someone you learned to drive a Ford; you probably say you learned to drive using a Ford.

There are five wireless certifications offered by the CWNP program:

CWTS™: Certified Wireless Technology Specialist The CWTS certification is the latest certification from the CWNP program. CWTS is an entry-level enterprise WLAN certification, and a recommended prerequisite for the CWNA certification. This certification is geared specifically toward both WLAN sales and support staff for the enterprise WLAN industry. The CWTS certification verifies that sales and support staff are specialists in WLAN technology and have all the fundamental knowledge, tools, and terminology to more effectively sell and support WLAN technologies.

CWNA®: Certified Wireless Network Administrator The CWNA certification is a foundation-level Wi-Fi certification; however, it is not considered an entry-level technology certification. Individuals taking this exam (exam PW0-104) typically have a solid grasp on network basics such as the OSI model, IP addressing, PC hardware, and network operating systems. Many candidates already hold other industry-recognized certifications, such as the CompTIA Network+ or Cisco CCNA, and are looking for the CWNA certification to enhance or complement existing skills.

CWSP®: Certified Wireless Security Professional The CWSP certification exam (PW0-200) is focused on standards-based wireless security protocols, security policy, and secure wireless network design. This certification introduces candidates to many of the technologies and techniques that intruders use to compromise wireless networks and that administrators use to protect wireless networks. With recent advances in wireless security, WLANs can be secured beyond their wired counterparts.

CWNE®: Certified Wireless Network Expert The CWNE certification is the highest-level certification in the CWNP program. By successfully completing the CWNE requirements, you will have demonstrated that you have the most advanced skills available in today's wireless LAN market. The CWNE exam (PW0-300) focuses on advanced WLAN analysis, design, troubleshooting, quality-of-service (QoS) mechanisms, spectrum management, and extensive knowledge of the IEEE 802.11 standard as amended.

CWNT®: Certified Wireless Network Trainer Certified Wireless Network Trainers are qualified instructors certified by the CWNP program to deliver CWNP training courses to IT professionals. CWNTs are technical and instructional experts in wireless technologies, products, and solutions. To ensure a superior learning experience for our customers, CWNP Education Partners are required to use CWNTs when delivering training using official CWNP courseware.

How to Become a CWNA

To become a CWNA, you must do the following two things: agree that you have read and will abide by the terms and conditions of the CWNP Confidentiality Agreement and pass the CWNA certification test.

A copy of the CWNP Confidentiality Agreement can be found online at the CWNP website.

When you sit to take the test, you will be required to accept this confidentiality agreement before you can continue with the test. After you have agreed, you will be able to continue with the test, and if you pass the test, you are then a CWNA.

The information for the exam is as follows:

- Exam name: Wireless LAN Administrator
- Exam number: PW0-104
- Cost: $175 (in U.S. dollars)
- Duration: 90 minutes
- Questions: 60
- Question types: Multiple choice/multiple answer
- Passing score: 70 percent (80 percent for instructors)
- Available languages: English
- Availability: Register at Pearson VUE (`www.vue.com/cwnp`)

When you schedule the exam, you will receive instructions regarding appointment and cancellation procedures, ID requirements, and information about the testing center location. In addition, you will receive a registration and payment confirmation letter. Exams can be scheduled weeks in advance or, in some cases, even as late as the same day.

After you have successfully passed the CWNA exam, the CWNP program will award you a certification that is good for three years. To recertify, you will need to pass the current PW0-104 exam, the CWSP exam, or the CWNE exam. If the information you provided the testing center is correct, you will receive an email from CWNP recognizing your accomplishment and providing you with a CWNP certification number. After you earn any CWNP certification, you can request a certification kit. The kit includes a congratulatory letter, a certificate, and a wallet-sized personalized ID card. You will need to log in to the CWNP tracking system, verify your contact information, and request your certification kit.

Who Should Buy This Book?

If you want to acquire a solid foundation in wireless networking and your goal is to prepare for the exam, this book is for you. You will find clear explanations of the concepts you need to grasp and plenty of help to achieve the high level of professional competency you need in order to succeed.

If you want to become certified as a CWNA, this book is definitely what you need. However, if you just want to attempt to pass the exam without really understanding wireless, this Study Guide is not for you. It is written for people who want to acquire hands-on skills and in-depth knowledge of wireless networking.

How to Use This Book and the CD

We have included several testing features in the book and on the CD-ROM. These tools will help you retain vital exam content as well as prepare you to sit for the actual exam.

Before you begin At the beginning of the book (right after this introduction) is an assessment test that you can use to check your readiness for the exam. Take this test before you start reading the book; it will help you determine the areas that you may need to brush up on. The answers to the assessment test appear on a separate page after the last question of the test. Each answer includes an explanation and a note telling you the chapter in which the material appears.

Chapter review questions To test your knowledge as you progress through the book, there are review questions at the end of each chapter. As you finish each chapter, answer the review questions and then check your answers—the correct answers appear on the page following the last review question. You can go back and reread the section that deals with each question you answered wrong to ensure that you answer correctly the next time you are tested on the material.

Electronic flashcards You will find flashcard questions on the CD for on-the-go review. These are short questions and answers, just like the flashcards you probably used in school. You can answer them on your PC or download them onto a Palm device for quick and convenient reviewing.

Test engine The CD also contains the Sybex Test Engine. With this custom test engine, you can identify weak areas up front and then develop a solid studying strategy that includes each of the robust testing features described previously. Our thorough readme file will walk you through the quick, easy installation process.

In addition to the assessment test and the chapter review questions, you will find three bonus exams. Use the test engine to take these practice exams just as if you were taking the actual exam (without any reference material). When you have finished the first exam, move on to the next one to solidify your test-taking skills. If you get more than 95 percent of the answers correct, you are ready to take the certification exam.

Labs and exercises Several chapters in this book have labs that use software, spreadsheets, and videos that are also provided on the CD-ROM that is included with this book. These labs and exercises will provide you with a broader learning experience by providing hands-on experience and step-by-step problem solving.

White papers Several chapters in this book reference wireless networking white papers that are also provided on the CD-ROM included with this book. These white papers serve as additional reference material for preparing for the CWNA exam.

Exam Objectives

The CWNA exam measures your understanding of the fundamentals of RF behavior, your ability to describe the features and functions of wireless LAN components, and your knowledge of the skills needed to install, configure, and troubleshoot wireless LAN hardware peripherals and protocols.

The skills and knowledge measured by this examination were derived from a survey of wireless networking experts and professionals. The results of this survey were used in weighing the subject areas and ensuring that the weighting is representative of the relative importance of the content.

The following chart provides the breakdown of the exam, showing you the weight of each section:

Subject Area	% of Exam
Radio frequency (RF) technologies	21%
802.11 regulations and standards	17%
802.11 protocols and devices	17%
802.11 network implementation	17%
802.11 network security	10%
802.11 RF site surveying	18%
Total	**100%**

Radio Frequency (RF) Technologies—21%

1.1. RF Fundamentals

1.1.1. Define and explain the basic concepts of RF behavior.

- Gain
- Loss
- Reflection
- Refraction
- Diffraction
- Scattering
- VSWR
- Return loss
- Amplification
- Attenuation
- Absorption

- Wave propagation
- Free space path loss
- Delay spread

1.2. RF Mathematics

1.2.1. Understand and apply the basic components of RF mathematics.

- Watt
- Milliwatt
- Decibel (dB)
- dBm
- dBi
- dBd
- SNR
- RSSI
- System operating margin (SOM)
- Fade margin
- Link budget
- Intentional radiator
- Equivalent isotropically radiated power (EIRP)

1.3. RF Signal and Antenna Concepts

1.3.1. Identify RF signal characteristics, the applications of basic RF antenna concepts, and the implementation of solutions that require RF antennas.

- Visual LOS
- RF LOS
- Fresnel zone
- Beamwidths
- Azimuth and elevation
- Passive gain
- Isotropic radiator
- Polarization
- Simple antenna diversity
- MIMO diversity
- Radio chains
- Spatial multiplexing (SM)

- Transmit beam forming (TxBF)
- Maximal ratio combining (MRC)
- Wavelength
- Frequency
- Amplitude
- Phase

1.3.2. Explain the applications of physical RF antenna and antenna system types and identify their basic attributes, purpose, and function.

- Omnidirectional/dipole antennas
- Semidirectional antennas
- Highly directional antennas
- Sectorized antennas

1.3.3. Describe the proper locations and methods for installing RF antennas.

- Pole/mast mount
- Ceiling mount
- Wall mount

1.4. RF Antenna Accessories

1.4.1. Identify the use of the following WLAN accessories and explain how to select and install them for optimal performance and regulatory domain compliance.

- Amplifiers
- Attenuators
- Lightning arrestors
- Mounting systems
- Grounding rods/wires
- Towers, safety equipment, and concerns
- RF cables
- RF connectors
- RF signal splitters

IEEE 802.11 Regulations and Standards—17%

2.1. Spread Spectrum Technologies

2.1.1. Identify some of the uses for spread spectrum technologies.

- Wireless LANs
- Wireless PANs

- Wireless MANs
- Wireless WANs

2.1.2. Comprehend the differences between, and explain the different types of, spread spectrum technologies and how they relate to the IEEE 802.11-2007 standard's (as amended and including IEEE 802.11n draft 2.0) PHY clauses.

- DSSS
- HR-DSSS
- ERP
- OFDM
- HT (MIMO)

2.1.3. Identify the underlying concepts of how spread spectrum technology works.

- Modulation
- Coding

2.1.4. Identify and apply the concepts that make up the functionality of spread spectrum technology.

- Colocation
- Channel centers and widths (all PHYs)
- Primary and secondary channels
- Overlapping and nonoverlapping channels
- Carrier frequencies
- Throughput vs. data rate
- Bandwidth
- Communication resilience
- Physical carrier sense (CSMA/CA)
- Virtual carrier sense (NAV)

2.2. IEEE 802.11-2007 Standard (as amended and including 802.11n draft 2.0)

2.2.1. Identify, explain, and apply the frame types and frame exchange sequences covered by the IEEE 802.11-2007 standard.

2.2.2 Identify and apply regulatory domain requirements.

- Dynamic frequency selection (DFS)
- Transmit power control (TPC)
- Available channels
- Output power

2.2.3 OSI model layers affected by the 802.11-2007 standard and amendments

2.2.4 Use of ISM and UNII bands in Wi-Fi networks

2.2.5 Supported data rates for each IEEE 802.11-2007 PHY

2.3. 802.11 Industry Organizations and Their Roles

2.3.1. Define the roles of the following organizations in providing direction, cohesion, and accountability within the WLAN industry.

- Regulatory domain governing bodies
- IEEE
- Wi-Fi Alliance

802.11 Protocols and Devices—17%

3.1. 802.11 Protocol Architecture

3.1.1. Summarize the processes involved in authentication and association.

- The 802.11 state machine
- Open System authentication, Shared Key authentication, and deauthentication
- Association, reassociation, and disassociation
- Deauthentication

3.1.2. Define, describe, and apply the following concepts associated with WLAN service sets.

- Stations and BSSs
- Starting and joining a BSS
- BSSID and SSID
- Ad Hoc mode and IBSS
- Infrastructure mode and ESS
- Distribution system (DS)
- Distribution system medium (DSM)
- Layer 2 and layer 3 roaming

3.1.3. Explain and apply the following power-management features of WLANs.

- Active mode
- Power Save mode
- Unscheduled automatic power save delivery (U-APSD)
- WMM Power Save (WMM-PS)
- Power Save Multi Poll (PSMP)

- Spatial multiplexing power save (SMPS)
- TIM/DTIM/ATIM

3.2. 802.11 MAC and PHY Layer Technologies

3.2.1. Describe and apply the following concepts surrounding WLAN frames.

- IEEE 802.11 frame format vs. IEEE 802.3 frame format
- Layer 3 protocol support by IEEE 802.11 frames
- Terminology review: frames, packets, and datagrams
- Terminology review: bits, bytes, and octets
- Terminology: MAC and PHY
    - Guard interval (GI)
    - PSDU
    - PPDU
    - PPDU formats
    - MSDU
    - MPDU
    - A-MPDU
    - A-MSDU
    - 802.11 frame format
    - 802.11 frame types
    - Interframe spaces (RIFS, SIFS, PIFS, DIFS, AIFS, EIFS)
    - Block acknowledgments
- Jumbo frame support (layer 2)
- MTU discovery and functionality (layer 3)

3.2.2. Identify methods described in the IEEE 802.11-2007 standard for locating, joining, and maintaining connectivity with an 802.11 WLAN.

- Active scanning (probes)
- Passive scanning (beacons)
- Dynamic rate switching

3.2.3. Define, describe, and apply 802.11 coordination functions and channel access methods and features available for optimizing data flow across the RF medium.

- DCF and HCF coordination functions
- EDCA channel access method
- RTS/CTS and CTS-to-Self protocols

- HT Dual-CTS protection
- HT L-SIG protection
- HT channel width operation (20 MHz, 20/40 MHz, PCO)
- HT operation modes (0, 1, 2, 3)
- Fragmentation

3.3. WLAN Infrastructure and Client Devices

3.3.1. Identify the purpose of the following WLAN infrastructure devices and describe how to install, configure, secure, and manage them.

- Autonomous access points
- Lightweight access points
- Mesh access points/routers
- Enterprise WLAN controllers
- Remote office WLAN controllers
- PoE injectors (single and multiport) and PoE-enabled Ethernet switches
- WLAN bridges
- Residential WLAN gateways
- Enterprise encryption gateways

3.3.2. Describe the purpose of the following WLAN client devices and explain how to install, configure, secure, and manage them.

- PC Cards (ExpressCard, CardBus, and PCMCIA)
- USB2, CF, and SD devices
- PCI, Mini PCI, and Mini PCIe cards
- Workgroup bridges

802.11 Network Implementation—17%

4.1. 802.11 Network Design, Implementation, and Management

4.1.1. Identify technology roles for which WLAN technology is appropriate and describe implementation of WLAN technology in those roles.

- Corporate data access and end-user mobility
- Network extension to remote areas
- Building-to-building connectivity (bridging)
- Last-mile data delivery (wireless ISP)
- Small office/home office (SOHO) use

- Mobile office networking
- Educational/classroom use
- Industrial (warehousing and manufacturing)
- Healthcare (hospitals and offices)
- Hotspots (public network access)
- Municipal networks
- Transportation networks (trains, planes, automobiles)
- Law enforcement networks

4.2. 802.11 Network Troubleshooting

4.2.1. Identify and explain how to solve the following WLAN implementation challenges by using features available in enterprise-class WLAN equipment.

- System throughput
- Co-channel and adjacent-channel interference
- RF noise and noise floor
- Narrowband and wideband RF interference
- Multipath (in SISO and MIMO environments)
- Hidden nodes
- Near/far
- Weather

4.3. Power over Ethernet (PoE)

4.3.1. IEEE 802.3-2005, clause 33 (formerly IEEE 802.3af)

4.3.2. Powering HT (IEEE 802.11n) devices

- Proprietary midspan and endpoint PSEs
- IEEE 802.3at draft midspan and endpoint PSEs

4.4. WLAN Architectures

4.4.1. Define, describe, and implement autonomous APs.

- Network connectivity
- Common feature sets
- Configuration, installation, and management
- Advantages and limitations
- QoS and VLANs

4.4.2. Define, describe, and implement WLAN controllers that use centralized and/or distributed forwarding.

- Network connectivity
- Core, distribution, and access layer forwarding
- Lightweight, mesh, and portal APs
- WLAN profiles
- Multiple BSSIDs per radio
- Scalability
- Intra- and inter-controller station handoffs
- Configuration, installation, and management
- Advantages and limitations
- Tunneling, QoS, and VLANs

4.4.3. Define, describe and implement a WNMS that manages autonomous APs, WLAN controllers, and mesh nodes.

- Network connectivity
- Common feature sets
- Configuration, installation, and management
- Advantages and limitations

4.4.4. Define, describe, and implement a multiple channel architecture (MCA) network model.

- BSSID/ESSID configuration
- Site surveying methodology
- Network throughput capacity
- Co-channel and adjacent-channel interference
- Cell sizing (including micro-cell)

4.4.5. Define, describe, and implement a single channel architecture (SCA) network model.

- BSSID/ESSID configuration (including virtual BSSIDs)
- Site surveying methodology
- Network throughput capacity
- Co-channel and adjacent-channel interference
- Cell sizing
- Transmission coordination
- Channel stacking

4.4.6. Define and describe alternative WLAN architectures.

- WLAN arrays
- Cooperative control
- Mesh networks

802.11 Network Security—10%

5.1. 802.11 Network Security Architecture

5.1.1. Identify and describe the strengths, weaknesses, appropriate uses, and implementation of the following IEEE 802.11 security-related items.

- Legacy security mechanisms
  - WEP cipher suite
  - Open System authentication
  - Shared Key authentication
  - MAC filtering
  - SSID hiding
- Modern security mechanisms
  - WPA-/WPA2-Enterprise
  - WPA-/WPA2-Personal
  - Wi-Fi protected setup (WPS)
  - TKIP and CCMP cipher suites
  - 802.1X/EAP framework
  - Preshared key (PSK)/passphrase authentication
- Additional mechanisms
  - Secure device management protocols (HTTPS, SNMPv3, SSH2)
  - Role-based access control (RBAC)

5.2. 802.11 Network Security Analysis, Performance Analysis, and Troubleshooting

5.2.1. Describe, explain, and illustrate the appropriate applications for the following wireless security solutions.

- Wireless intrusion prevention system (WIPS)
  - Security monitoring, containment, and reporting
  - Performance monitoring and reporting
  - Troubleshooting and analysis

- Protocol analyzers
    - Security and performance monitoring
    - Troubleshooting and analysis

5.3. 802.11 Network Security Policy Basics

5.3.1. Describe the following general security policy elements.

- Applicable audience
- Risk assessment
- Impact analysis
- Security auditing
- Policy enforcement
- Monitoring, response, and reporting
- Asset management

5.3.2. Describe the following functional security policy elements.

- Design and implementation best practices
    - Small office/home office (SOHO)
    - Small and medium business (SMB)
    - Enterprise
- Password policy
- Acceptable use and abuse policy
- Training requirements
- Physical security
- Social engineering

802.11 RF Site Surveying—18%

6.1. 802.11 Network Site Survey Fundamentals

6.1.1. Explain the importance of and the processes involved in information collection for manual and predictive RF site surveys. (These happen in preparation for an RF site survey.)

- Gathering business requirements
- Interviewing managers and users
- Defining physical and data security requirements
- Gathering site-specific documentation
- Documenting existing network characteristics

- Gathering permits and zoning requirements
- Indoor- or outdoor-specific information
- Identifying infrastructure connectivity and power requirements
- Understanding RF coverage requirements
- Understanding data capacity and client density requirements
- VoWiFi considerations for delay and jitter
- Client connectivity requirements
- Antenna use considerations
- Aesthetics requirements
- Tracking system considerations
- WIPS sensor considerations

6.1.2. Explain the technical aspects involved in performing manual and predictive RF site surveys. (These happen as part of the RF site survey.)

- Locating and identifying RF interference sources
- Defining AP and antenna types to be used
- Defining AP and antenna placement locations
- Defining AP output power and channel assignments
- Defining co-channel and adjacent-channel interference
- Testing applications for proper operation

6.1.3. Describe site survey reporting and follow-up procedures for manual and predictive RF site surveys. (These happen after the RF site survey.)

- Reporting methodology
- Customer reporting requirements
- Hardware recommendations and bills of material
- Application analysis for capacity and coverage verification

6.2. 802.11 Network Site Survey Systems and Devices

6.2.1. Identify the equipment, applications, and system features involved in performing predictive site surveys.

- Predictive analysis/simulation applications (also called RF planning and management tools)
- Integrated predictive site survey features of WLAN controllers
- Site survey verification tools and/or applications
- Indoor site surveys vs. outdoor site surveys

6.2.2. Identify the equipment, applications, and methodologies involved in performing manual site surveys.

- Site survey hardware kits
- Spectrum analyzers
- Protocol analyzers
- Active site survey tools and/or applications
- Passive site survey tools and/or applications
- VoWiFi site survey best practices (dB boundaries, antenna use, balanced links)
- Manufacturers' client utilities

6.2.3. Identify the equipment, applications, and methodologies involved in self-managing RF technologies.

- Automated RF resource management

CWNA Exam Terminology

The CWNP program uses specific terminology when phrasing the questions on any of the CWNP exams. The terminology used most often mirrors the same language that is used in the IEEE 802.11-2007 standard. While technically correct, the terminology used in the exam questions often is not the same as the marketing terminology that is used by the Wi-Fi Alliance. The most current IEEE version of the 802.11 standard is the IEEE 802.11-2007 document, which includes all the amendments that have been ratified prior to the document's publication. Standards bodies such as the IEEE often create several amendments to a standard before "rolling up" the ratified amendments (finalized or approved versions) into a new standard.

For example, you might already be familiar with the term *802.11g*, which is a ratified amendment that has now been integrated into the IEEE 802.11-2007 standard. The technology that was originally defined by the 802.11g amendment is called Extended Rate Physical (ERP). Although the name 802.11g effectively remains the more commonly used marketing terminology, any exam questions will use the technical term ERP instead of 802.11g.

To properly prepare for the CWNA exam, any test candidate should become 100 percent familiar with the terminology used by the CWNP program. This book defines and covers all terminology; however, the CWNP program maintains an updated current list of exam terms that can be downloaded from the following URL: `www.cwnp.com/exams/exam_terms.html`.

Tips for Taking the CWNA Exam

Here are some general tips for taking your exam successfully:

- Bring two forms of ID with you. One must be a photo ID, such as a driver's license. The other can be a major credit card or a passport. Both forms must include a signature.

- Arrive early at the exam center so you can relax and review your study materials, particularly tables and lists of exam-related information.
- Read the questions carefully. Do not be tempted to jump to an early conclusion. Make sure you know exactly what the question is asking.
- There will be questions with multiple correct responses. When there is more than one correct answer, a message at the bottom of the screen will prompt you to either "choose two" or "choose all that apply." Be sure to read the messages displayed to know how many correct answers you must choose.
- When answering multiple-choice questions you are not sure about, use a process of elimination to get rid of the obviously incorrect answers first. Doing so will improve your odds if you need to make an educated guess.
- Do not spend too much time on one question. This is a form-based test; however, you cannot move backward through the exam. You must answer the current question before you can move to the next question, and after you have moved to the next question, you cannot go back and change your answer on a previous question.
- Keep track of your time. Because this is a 90-minute test consisting of 60 questions, you have an average of 90 seconds to answer each question. You can spend as much or as little time on any one question, but when 90 minutes is up, the test is over. Check your progress. After 45 minutes, you should have answered at least 30 questions. If you have not, do not panic. You will simply need to answer the remaining questions at a faster pace. If on average you can answer each of the remaining 30 questions 4 seconds quicker, you will recover 2 minutes. Again, do not panic; just pace yourself.
- For the latest pricing on the exams and updates to the registration procedures, visit CWNP's website at `www.cwnp.com`.

Assessment Test

1. At which layers of the OSI model does 802.11 technology operate? (Choose all that apply.)
   A. Data-Link
   B. Network
   C. Physical
   D. Presentation
   E. Transport

2. Which Wi-Fi Alliance certification defines the mechanism for conserving battery life that is critical for handheld devices such as barcode scanners and VoWiFi phones?
   A. WPA2-Enterprise
   B. WPA2-Personal
   C. WMM-PS
   D. WMM-SA
   E. CWG-RF

3. Which of these frequencies has the longest wavelength?
   A. 750 KHz
   B. 2.4 GHz
   C. 252 GHz
   D. 2.4 MHz

4. Which of these terms can best be used to compare the relationship between two radio waves that share the same frequency?
   A. Multipath
   B. Multiplexing
   C. Phase
   D. Spread spectrum

5. A bridge transmits at 10 mW. The cable to the antenna produces a loss of 3 dB, and the antenna produces a gain of 20 dBi. What is the EIRP?
   A. 25 mW
   B. 27 mW
   C. 4 mW
   D. 1,300 mW
   E. 500 mW

6. What are some possible effects of voltage standing wave ratio (VSWR)? (Choose all that apply.)
   A. Increased amplitude
   B. Decreased signal strength
   C. Transmitter failure
   D. Erratic amplitude
   E. Out-of-phase signals

7. When installing a higher-gain omnidirectional antenna, which of the following occurs? (Choose two.)
   A. The horizontal coverage increases.
   B. The horizontal coverage decreases.
   C. The vertical coverage increases.
   D. The vertical coverage decreases.

8. OFDM radio cards are backward compatible with which IEEE 802.11 radios?
   A. FHSS radios
   B. ERP radios
   C. DSSS radios
   D. HR-DSSS radios
   E. HT radios
   F. None of the above

9. Which IEEE 802.11 draft amendment specifies the use of the 5.850 to 5.925 GHz frequency band?
   A. IEEE 802.11a
   B. IEEE 802.11h
   C. IEEE 802.11p
   D. IEEE 802.11g
   E. IEEE 802.11u

10. Which of the following are valid ISM bands? (Choose all that apply.)
    A. 902–928 MHz
    B. 2.4–2.4835 MHz
    C. 5.725–5.825 GHz
    D. 5.725–5.875 GHz

11. Choose two spread spectrum signal characteristics.
    A. Narrow bandwidth
    B. Low power
    C. High power
    D. Wide bandwidth

12. A service set identifier is often synonymous with which of the following?
    A. IBSS
    B. ESSID
    C. BSSID
    D. Basic service set identifier
    E. BSS

13. Which ESS design scenario is defined by the IEEE 802.11-2007 standard?
    A. Two or more access points with overlapping coverage cells
    B. Two or more access points with overlapping disjointed coverage cells
    C. One access point with a single BSA
    D. Two basic service sets connected by a DS with colocated coverage cells
    E. None of the above

14. What CSMA/CA conditions must be met before an 802.11 radio card can transmit? (Choose all that apply.)
    A. The NAV timer must be equal to zero.
    B. The random back-off timer must have expired.
    C. The CCA must be positive.
    D. The proper interframe space must have occurred.
    E. The access point must be in PCF mode.

15. Beacon management frames contain which of the following information? (Choose all that apply.)
    A. Channel information
    B. Destination IP address
    C. Basic data rate
    D. Traffic indication map (TIM)
    E. Vendor proprietary information
    F. Time stamp
    G. Spread spectrum parameter sets

16. Anthony Dean was hired to perform a wireless packet analysis of your network. While performing the analysis, he noticed that many of the data frames were preceded by an RTS frame followed by a CTS frame. What could cause this phenomenon to occur? (Choose all that apply.)
    A. Because of high RF noise levels, some of the stations have automatically enabled RTS/CTS.
    B. Some stations were manually configured for RTS/CTS.
    C. A nearby OFDM radio is causing some of the nodes to enable a protection mechanism.
    D. The network is a mixed-mode environment.

17. What is another name for an 802.11 data frame?
    A. PPDU
    B. PSDU
    C. MSDU
    D. MPDU
    E. BPDU

18. Which WLAN device uses self-healing and self-forming mechanisms and layer 2 routing protocols?
    A. WLAN switch
    B. WLAN controller
    C. WLAN VPN router
    D. WLAN mesh access point

19. Which WLAN device offers AP management, user management, intrusion detection, and spectrum management?
    A. Sectorized array
    B. Autonomous AP
    C. WLAN controller
    D. Enterprise wireless gateway (EWG)
    E. All of the above

20. Wi-Fi technology is used in many different vertical markets. Which of these verticals markets are you most likely to still find legacy 802.11 FHSS technology?
    A. Healthcare
    B. Manufacturing
    C. Education
    D. Law enforcement
    E. Hotspots

21. Wireless mesh routers often have two radio cards. One radio is used for client connectivity, and the other is used for backhaul. Which of these statements bests meets this model? (Choose all that apply.)
    A. A 2.4 GHz radio is used for distribution, while a 5 GHz radio is used for access.
    B. An ERP radio is used for client connectivity, while an HR-DSSS radio is used for backhaul.
    C. An ERP radio is used for client connectivity, while an OFDM radio is used for backhaul.
    D. A 2.4 GHz radio is used for access, while a 5 GHz radio is used for distribution.

22. If IEEE 802.1X/EAP security is in place, what type of roaming solution is needed for time-sensitive applications such as VoWiFi?
    A. Nomadic roaming solution
    B. Proprietary layer 3 roaming solution
    C. Seamless roaming solution
    D. Mobile IP solution
    E. Fast secure roaming solution

23. The hidden node problem occurs when one client station's transmissions are not heard by all the other client stations in the coverage area of a basic service set (BSS). What are some of the consequences of the hidden node problem? (Choose all that apply.)
    A. Retransmissions
    B. Intersymbol interference (ISI)
    C. Collisions
    D. Increased throughput
    E. Decreased throughput

24. What are some potential causes of layer 2 retransmissions? (Choose all that apply.)
    A. Multipath
    B. Mismatched client and AP power settings
    C. Dual-frequency transmissions
    D. Fade margin
    E. Multiplexing

25. Which of these solutions would be considered strong WLAN security?
    A. SSID cloaking
    B. MAC filtering
    C. WEP
    D. Shared Key authentication
    E. CCMP/AES

26. Which security standard defines port-based access control?
    A. IEEE 802.11x
    B. IEEE 802.3b
    C. IEEE 802.11i
    D. IEEE 802.1X
    E. IEEE 802.11s

27. Which is the best tool for detecting an RF jamming denial-of-service attack? (Choose all that apply.)
    A. Time-domain analysis software
    B. Layer 2 distributed WIPS
    C. Spectrum analyzer
    D. Layer 1 distributed WIPS
    E. Oscilloscope

28. Which of these attacks can be detected by a wireless intrusion detection system (WIDS)? (Choose all that apply.)
    A. Deauthentication spoofing
    B. MAC spoofing
    C. Rogue ad hoc network
    D. Association flood
    E. Rogue AP

29. You have been hired by the XYZ Company based in the United States for a wireless site survey. What government agencies need to be informed before a tower is installed of a height that exceeds 200 feet above ground level? (Choose all that apply.)
    A. RF regulatory authority
    B. Local municipality
    C. Fire department
    D. Tax authority
    E. Aviation authority

30. You have been hired by the ABC Corporation to conduct an indoor site survey. What information will be in the final site survey report that is delivered? (Choose two.)
    A. Security analysis
    B. Coverage analysis
    C. Spectrum analysis
    D. Routing analysis
    E. Switching analysis

31. Name potential sources of interference in the 5 GHz UNII band. (Choose all that apply.)
    A. Perimeter sensors
    B. Nearby OFDM (802.11a) WLAN
    C. Cellular phone
    D. DSSS access point
    E. Bluetooth
    F. Nearby HT (802.11n) WLAN

32. Which of these measurements are taken for indoor coverage analysis? (Choose all that apply.)

A. Received signal strength

B. Signal-to-noise ratio

C. Noise level

D. Path loss

E. Packet loss

33. What problems may result due to access points with too much transmission amplitude? (Choose all that apply.)

A. Multipath

B. Poor client capacity

C. Co-channel interference

D. Layer 3 roaming failure

E. Access point buffer overflow

34. What must a powered device (PD) do to be considered IEEE 802.3-2005 clause 33 (PoE) compliant? (Choose all that apply.)

A. Be able to accept power in either of two ways (through the data lines or unused pairs).

B. Reply with a classification signature.

C. Reply with a 35 ohm detection signature.

D. Reply with a 25 ohm detection signature.

E. Receive 30 watts of power from the power sourcing equipment.

35. An HT network can operate on which frequency bands? (Choose all that apply.)

A. 902–928 MHz

B. 2.4–2.4835 GHz

C. 5.15–5.25 GHz

D. 5.47–5.725 GHz

36. What are some of the methods used to reduce MAC layer overhead as defined by the 802.11n draft amendment? (Choose all that apply.)

A. A-MSDU

B. A-MPDU

C. MRC

D. MCS

E. PPDU

Answers to Assessment Test

1. A, C. The IEEE 802.11-2007 standard defines communication mechanisms at only the Physical layer and MAC sublayer of the Data-Link layer of the OSI model. For more information, see Chapter 1.

2. C. WMM-PS helps conserve battery power for devices using Wi-Fi radios by managing the time the client device spends in sleep mode. Conserving battery life is critical for handheld devices such as barcode scanners and VoWiFi phones. To take advantage of power-saving capabilities, both the device and the access point must support WMM Power Save. For more information, see Chapter 1.

3. A. A 750 KHz signal has an approximate wavelength of 1,312 feet, or 400 meters. A 252 GHz signal has an approximate wavelength of less than 0.05 inches, or 1.2 millimeters. Remember, the higher the frequency of a signal, the smaller the wavelength property of an electromagnetic signal. To calculate the wavelength, use the formula $\lambda = c/f$. For more information, see Chapter 2.

4. C. Phase involves the positioning of the amplitude crests and troughs of two waveforms. For more information, see Chapter 2.

5. E. The 10 mW of power is decreased by 3 dB, or divided by 2, giving 5 mW. This is then increased by 20 dBi, or multiplied by 10 twice, giving 500 mW. For more information, see Chapter 3.

6. B, C, D. Reflected voltage caused by an impedance mismatch may cause a degradation of amplitude, erratic signal strength, or even the worst-case scenario of transmitter burnout. See Chapter 4 for more information.

7. A, D. When the gain of an omnidirectional antenna is increased, the vertical coverage area decreases while the horizontal coverage area is increased. See Chapter 4 for more information.

8. F. OFDM (802.11a) clause 17 radios transmit in the 5 GHz UNII bands and are not compatible with FHSS (802.11 legacy) clause 14 radios, DSSS (802.11 legacy) clause 15 radios, HR-DSSS (802.11b) clause 18 radios, or ERP (802.11g) clause 19 radios, which transmit in the 2.4 GHz ISM frequency band. OFDM radios are forward compatible but not backward compatible with HT radios. HT (802.11n) clause 20 radios, which can transmit on either frequency band, are backward compatible with OFDM radios as well as ERP, HR-DSSS, and DSSS radios. For more information, see Chapter 5.

9. C. The IEEE 802.11p draft amendment defines enhancements to the IEEE 802.11-2007 standard to support communications between high-speed vehicles and roadside infrastructure in the licensed ITS band of 5.9 GHz. For more information, see Chapter 5.

10. A, D. The ISM bands are 902–928 MHz, 2.4–2.4835 GHz, and 5.725–5.875 GHz. 5.725–5.825 is the upper UNII band. There is no ISM band that operates in a 2.4 MHz range. See Chapter 6 for more information.

11. B, D. A spread spectrum signal utilizes bandwidth that is wider than what is required to carry the data and has low transmission power requirements. See Chapter 6 for more information.

12. B. The logical network name of a wireless LAN is often called an ESSID (extended service set identifier) and is essentially synonymous with SSID (service set identifier), which is another term for a logical network name in the most common deployments of a WLAN. For more information, see Chapter 7.

13. E. The scenarios described in options A, B, C, and D are all examples of how an extended service set may be deployed. The IEEE 802.11-2007 standard defines an extended service set (ESS) as "a set of one or more interconnected basic service sets." However, the IEEE 802.11-2007 standard does not mandate any of the correct given examples. For more information, see Chapter 7.

14. A, B, C, D. Carrier Sense Multiple Access with Collision Avoidance (CSMA/CA) is a medium access method that utilizes multiple checks and balances to try to minimize collisions. These checks and balances can also be thought of as several lines of defense. The various lines of defense are put in place to hopefully ensure that only one radio is transmitting while all other radios are listening. The four lines of defense include the network allocation vector, the random back-off timer, the clear channel assessment, and interframe spacing. For more information, see Chapter 8.

15. A, C, D, E, F, G. The only information not contained in the beacon management frame is the destination IP address. The body of all 802.11 management frames contain only layer 2 information; therefore, IP information is not included in the frame. For more information, see Chapter 9.

16. B, D. Stations can be manually configured to use RTS/CTS for all transmissions. This is usually done to diagnose hidden node problems. This network could also be a mixed-mode HR-DSSS (802.11b) and ERP (802.11g) network. The ERP (802.11g) nodes have enabled RTS/CTS as their protection mechanism. For more information, see Chapter 9.

17. D. The technical name for an 802.11 data frame is a MAC Protocol Data Unit (MPDU). An MPDU contains a layer 2 header, a frame body, and a trailer that is a 32-bit CRC known as the frame check sequence (FCS). Inside the frame body of an MPDU is a MAC Service Data Unit (MSDU), which contains data from the LLC and layers 3–7. For more information, see Chapter 9.

18. D. WLAN mesh access points create a self-forming WLAN mesh network that automatically connects access points at installation and dynamically updates routes as more clients are added. Because interference may occur, a self-healing WLAN mesh network will automatically reroute data traffic in a Wi-Fi mesh cell by using proprietary layer 2 routing protocols. For more information, see Chapter 10.

19. C. WLAN controllers, also known as WLAN switches, use centralized management and configuration of thin access points. User management capabilities are available through the use of role-based access control (RBAC). Most WLAN switches also have internal wireless intrusion detection systems (WIDS) and offer spectrum management capabilities. For more information, see Chapter 10.

20. B. Warehouse and manufacturing environments often deploy wireless handheld devices such as bar code scanners, which are often used for inventory control. Most of the original deployments of wireless handheld scanners used 802.11 frequency hopping technology. Some legacy 802.11 FHSS wireless LANs still exist today in manufacturing and warehouse environments. For more information, see Chapter 11.

21. C, D. When installing a mesh router, it is best to use two or more radio cards. The 2.4 GHz HR-DSSS and ERP radios are often used for client connectivity, while the 5 GHz OFDM radios are used for mesh connectivity or backhaul. For more information, see Chapter 11.

22. E. When using an IEEE 802.1X/EAP security solution in the enterprise, the average time involved during the authentication process can be 700 milliseconds or longer. Voice over Wi-Fi (VoWiFi) requires a handoff of 150 milliseconds or less when roaming. A fast secure roaming (FSR) solution is needed if IEEE 802.1X/EAP security and time-sensitive applications are used together in a wireless network. In the past, FSR solutions have been proprietary; however, the IEEE 802.11r amendment defines fast secure roaming standard mechanisms. For more information, see Chapter 12.

23. A, C, E. The stations that cannot hear the hidden node will transmit at the same time that the hidden node is transmitting. This will result in continuous transmission collisions in a half-duplex medium. Collisions will corrupt the frames and they will need to be retransmitted. Anytime retransmissions are necessary, more overhead is added to the medium, resulting in decreased throughput. Intersymbol interference is a result of multipath and not the hidden node problem. For more information, see Chapter 12.

24. A, B. Layer 2 retransmissions can be caused by many different variables in a WLAN environment. Multipath, mismatched client and AP transmission power, RF interference, hidden nodes, adjacent cell interference and low signal-to-noise ratio (SNR) are all possible causes of layer 2 retransmissions. For more information, see Chapter 12.

25. E. Although you can hide your SSID to cloak the identity of your wireless network from script kiddies and nonhackers, it should be clearly understood that SSID cloaking is by no means an end-all wireless security solution. Because of spoofing and because of all the administrative work that is involved, MAC filtering is not considered a reliable means of security for wireless enterprise networks. WEP and Shared Key authentication are legacy 802.11 security solutions. CCMP/AES is defined as the default encryption type by the IEEE 802.11i security amendment. Cracking the AES cipher would take the lifetime of the sun using the tools that are available today. For more information, see Chapter 13.

26. D. The IEEE 802.1X standard is not specifically a wireless standard and often is mistakenly referred to as IEEE 802.11x. The IEEE 802.1X standard is a port-based access control standard. IEEE 802.1X provides an authorization framework that allows or disallows traffic to pass through a port and thereby access network resources. For more information, see Chapter 13.

27. C, D. Although the layer 2 wireless intrusion detection and prevention products might be able to detect some RF jamming attacks, the only tool that will absolutely identify an interfering signal is a spectrum analyzer. A spectrum analyzer is a frequency domain tool that can detect any RF signal in the frequency range that is being scanned. Layer 1 distributed spectrum analysis is now available in some WIPS enterprise solutions. For more information, see Chapter 14.

28. A, B, C, D, E. 802.11 wireless intrusion detection systems may be able to monitor for as many as 100 or more attacks. Any layer 2 DoS attack and spoofing attack and most rogue devices can be detected. For more information, see Chapter 14.

29. A, B, E. In the United States, if any tower exceeds a height of 200 feet above ground level (AGL), you must contact both the FCC and FAA, which are communications and aviation regulatory authorities. Other countries will have similar height restrictions, and the proper RF regulatory authority and aviation authority must be contacted to find out the details. Local municipalities may have construction regulations, and a permit may be required. For more information, see Chapter 15.

30. B, C. The final site survey report, known as the deliverable, will contain spectrum analysis information identifying potential sources of interference. Coverage analysis will also define RF cell boundaries. The final report also contains recommended access point placement, configuration settings, and antenna orientation. Application throughput testing is often an optional analysis report included in the final survey report. Security, switching, and routing analysis are not included in a site survey report. For more information, see Chapter 15.

31. A, B, F. Nearby OFDM (802.11a) WLAN and perimeter sensors both transmit in the 5 GHz UNII bands. HT (802.11n) WLAN radios can transmit at either 2.4 or 5 GHz. A nearby HT WLAN operating at 5 GHz can potentially be a source of interference. DSSS access points and Bluetooth devices transmit in the 2.4 GHz frequency space. Cellular phones transmit in licensed frequencies. For more information, see Chapter 16.

32. A, B, C, E. RF coverage cell measurements that are taken during an indoor passive site survey include received signal strength, noise levels, signal-to-noise ratio (SNR), and data rates. Packet loss can be an additional measurement recorded during an active manual site survey. Packet loss is a calculation needed for an outdoor wireless bridging survey. For more information, see Chapter 16.

33. B, C. A common mistake often made when deploying a WLAN is configuring the access points with too much transmission amplitude. Problems that can result from overpowered access points include hidden nodes, co-channel interference, and capacity problems. For more information, see Chapter 12.

34. A, D. For a powered device (PD) such as an access point to be considered compliant with the IEEE 802.3-2005 clause 33 PoE standard, the device must be able to receive power through the data lines or the unused twisted pairs of an Ethernet cable. The PD must also reply to the power sourcing equipment (PSE) with a 25-ohm detection signature. The PD may reply with a classification signature, but it is optional. The current PoE standard allows for a maximum draw of 12.95 watts by the PD from the power-sourcing equipment. For more information, see Chapter 17.

35. B, C, D. High Throughput (HT) technology is defined by the IEEE 802.11n draft amendment and is not frequency dependent. HT can operate in the 2.4 GHz ISM band as well as all of the 5 GHz UNII frequency bands. For more information, see Chapter 18.

36. A, B. The 802.11n amendment introduces two new methods of frame aggregation to help reduce the overhead. Frame aggregation is a method of combining multiple frames into a single frame transmission. The first method of frame aggregation is known as an Aggregate MAC Service Data Unit (A-MSDU). The second method of frame aggregation is known as an Aggregate MAC Protocol Data Unit (A-MPDU).

Chapter 1

Overview of Wireless Standards, Organizations, and Fundamentals

IN THIS CHAPTER, YOU WILL LEARN ABOUT THE FOLLOWING:

- ✓ **History of WLAN**
- ✓ **Standards organizations**
  - Federal Communications Commission
  - International Telecommunication Union Radiocommunication Sector
  - Institute of Electrical and Electronics Engineers
  - Wi-Fi Alliance
  - International Organization for Standardization
- ✓ **Core, distribution, and access**
- ✓ **Communications fundamentals**

Wireless local area network (WLAN) technology has a long history that dates back to the 1970s with roots as far back as the 19th century. In this chapter, you will learn a brief history of WLAN technology. Learning a new technology can seem like a daunting task. There are so many new acronyms, abbreviations, terms, and ideas to become familiar with. One of the keys to learning any subject is to learn the basics. Whether you are learning to drive a car, fly an airplane, or install a wireless computer network, there are basic rules, principles, and concepts that, once learned, provide the building blocks for the rest of your education.

IEEE 802.11 technology, more commonly referred to as Wi-Fi, is a standard technology for providing local area network (LAN) communications using radio frequencies (RFs). The IEEE designated the 802.11-2007 standard as a guideline to provide operational parameters for WLANs. There are numerous standards organizations and regulatory bodies that help govern and direct wireless technologies and the related industry. Having some knowledge of these different organizations can provide you with insight as to how IEEE 802.11 functions, and sometimes even how and why the standards have evolved the way they have.

As you become more knowledgeable about wireless networking, you may want or need to read some of the standards that are created by the different organizations. Along with the information about the standards bodies, this chapter includes a brief overview of their documents.

In addition to reviewing the different standards organizations that guide and regulate Wi-Fi, this chapter discusses where WLAN technology fits in with basic networking design fundamentals. Finally, this chapter reviews some fundamentals of communications and data keying that are not part of the CWNA exam but that may help you better understand wireless communications.

History of WLAN

In the 19th century, numerous inventors and scientists, including Michael Faraday, James Clerk Maxwell, Heinrich Rudolf Hertz, Nikola Tesla, David Edward Hughes, Thomas Edison, and Guglielmo Marconi, began to experiment with wireless communications. These innovators discovered and created many theories about the concepts of electrical magnetic *radio frequency (RF)*.

Wireless networking technology was first used by the U.S. military during World War II to transmit data over an RF medium using classified encryption technology, to send battle plans across enemy lines. The *spread spectrum* radio technologies often used in today's WLANs were also originally patented during the era of World War II, although they were not implemented until almost two decades later.

In 1970, the University of Hawaii developed the first wireless network, called ALOHAnet, to wirelessly communicate data between the Hawaiian Islands. The network used a LAN communication Open Systems Interconnection layer 2 protocol called ALOHA on a wireless shared medium in the 400 MHz frequency range. The technology used in ALOHAnet is often credited as a building block for the Medium Access Control technologies of Carrier Sense Multiple Access with Collision Detection (CSMA/CD) used in Ethernet and Carrier Sense Multiple Access with Collision Avoidance (CSMA/CA) used in 802.11 radios. You will learn more about CSMA/CA in Chapter 8, "802.11 Medium Access."

In the 1990s, commercial networking vendors began to produce low-speed wireless data networking products, most of which operated in the 900 MHz frequency band. The Institute of Electrical and Electronics Engineers (IEEE) began to discuss standardizing WLAN technologies in 1991. In 1997, the IEEE ratified the original 802.11 standard that is the foundation of the WLAN technologies that you will be learning about in this book.

Legacy 802.11 technology was deployed between 1997 and 1999 mostly in warehousing and manufacturing environments for the use of low-speed data collection with wireless barcode scanners. In 1999, the IEEE defined higher data speeds with the 802.11b amendment. The introduction of data rates as high as 11 Mbps, along with price decreases, ignited the sales of wireless home networking routers in the small office, home office (SOHO) marketplace. Home users soon became accustomed to wireless networking in their homes and began to demand that their employers also provide wireless networking capabilities in the workplace. After initial resistance to 802.11 technology, small companies, medium-sized businesses, and corporations began to realize the value of deploying 802.11 wireless technology in their enterprises.

If you ask the average user about their 802.11 wireless network, they will probably give you a very strange look. The name that most people recognize for the technology is *Wi-Fi*. Wi-Fi is a marketing term, recognized worldwide by millions of people as referring to 802.11 wireless networking.

What Does the Term *Wi-Fi* Mean?

Many people mistakenly assume that *Wi-Fi* is an acronym for the phrase *wireless fidelity* (much like *hi-fi* is short for *high fidelity*), but Wi-Fi is simply a brand name used to market 802.11 WLAN technology. Ambiguity in IEEE framework standards for wireless communications allowed manufacturers to interpret the 802.11 standard in different ways. As a result, multiple vendors could have IEEE 802.11–compliant devices that did not interoperate with each other. The group Wireless Ethernet Compatibility Alliance (WECA) was created to further define the IEEE standard in such a way as to force interoperability between vendors. WECA, now the Wi-Fi Alliance, chose the term *Wi-Fi* as a marketing brand. The Wi-Fi Alliance champions enforcing interoperability among wireless devices. To be Wi-Fi compliant, vendors must send their products to a Wi-Fi Alliance test lab that thoroughly tests compliance to the Wi-Fi certification. More information about the origins of the term Wi-Fi can be found online at Wi-Fi Net News: `www.wifinetnews.com/archives/006029.html`.

Wi-Fi radios are used for numerous enterprise applications and can also be found in laptops, cellular phones, cameras, televisions, printers, and many other consumer devices. More than 300 million Wi-Fi chipsets were shipped in 2007, with current estimates of annual sales of over one billion Wi-Fi chipsets by the year 2011.

According to the Wi-Fi Alliance, there are currently more than 450 million Wi-Fi users worldwide. In a survey that they conducted, 68 percent of those users would rather give up chocolate than do without Wi-Fi. Since the original standard was created in 1997, 802.11 technology has grown to enormous proportions; Wi-Fi has now become part of our worldwide culture.

Standards Organizations

Each of the standards organizations discussed in this chapter help to guide a different aspect of the wireless networking industry.

The International Telecommunication Union Radiocommunication Sector (ITU-R) and local entities such as the Federal Communications Commission (FCC) set the rules for what the user can do with a radio transmitter. These organizations manage and regulate frequencies, power levels, and transmission methods. They also work together to help guide the growth and expansion that is being demanded by wireless users.

The Institute of Electrical and Electronics Engineers (IEEE) creates standards for compatibility and coexistence between networking equipment. The IEEE standards must adhere to the rules of the communications organizations, such as the FCC.

The Wi-Fi Alliance performs certification testing to make sure wireless networking equipment conforms to the 802.11 WLAN communication guidelines, similar to the IEEE 802.11-2007 standard.

The International Organization for Standardization (ISO) created the Open Systems Interconnection (OSI) model, which is an architectural model for data communications.

You will look at each of these organizations in the following sections.

Federal Communications Commission (FCC)

To put it simply, the *Federal Communications Commission (FCC)* regulates communications within the United States as well as communications to and from the United States. Established by the Communications Act of 1934, the FCC is responsible for regulating interstate and international communications by radio, television, wire, satellite, and cable. The task of the FCC in wireless networking is to regulate the radio signals that are used for wireless networking. The FCC has jurisdiction over the 50 states, the District of Columbia, and U.S. possessions. Most countries have governing bodies that function similarly to the FCC.

The FCC and the respective controlling agencies in the other countries typically regulate two categories of wireless communications: licensed and unlicensed. The difference is that unlicensed users do not have to go through the license application procedures before they can install a wireless system. Both licensed and unlicensed communications are typically regulated in the following five areas:

- Frequency
- Bandwidth
- Maximum power of the intentional radiator (IR)
- Maximum equivalent isotropically radiated power (EIRP)
- Use (indoor and/or outdoor)

Real World Scenario

What Are the Advantages and Disadvantages of Using an Unlicensed Frequency?

As stated earlier, licensed frequencies require an approved license application, and the financial costs are very high. One main advantage of an unlicensed frequency is that permission to transmit on the frequency is free. Although there are no financial costs, you still must abide by transmission regulations and other restrictions. In other words, transmitting in an unlicensed frequency may be free, but there still are rules.

The main disadvantage to transmitting in an unlicensed frequency band is that anyone else can also transmit in that same frequency space. Unlicensed frequency bands are often very crowded; therefore, transmissions from other individuals can cause interference with your transmissions. If someone else is interfering with your transmissions, you have no legal recourse as long as the other individual is abiding by the rules and regulations of the unlicensed frequency.

Essentially, the FCC and other regulatory bodies set the rules for what the user can do regarding RF transmissions. From there, the standards organizations create the standards to work within these guidelines. These organizations work together to help meet the demands of the fast-growing wireless industry.

The FCC rules are published in the Code of Federal Regulations (CFR). The CFR is divided into 50 titles that are updated yearly. The title that is relevant to wireless networking is Title 47, *Telecommunications*. Title 47 is divided into many parts; Part 15, "Radio Frequency Devices," is where you will find the rules and regulations regarding wireless networking related to 802.11. Part 15 is further broken down into subparts and sections. A complete reference will look like this example: 47CFR15.3.

The FCC transmit power regulations for the 2.4 GHz ISM frequency band and the 5 GHz UNII bands can be found in the appendix of this book. More information can be found at `www.fcc.gov` and `http://wireless.fcc.gov`.

International Telecommunication Union Radiocommunication Sector (ITU-R)

A global hierarchy exists for management of the RF spectrum worldwide. The United Nations has tasked the *International Telecommunication Union Radiocommunication Sector (ITU-R)* with global spectrum management. The ITU-R maintains a database of worldwide frequency assignments and coordinates spectrum management through five administrative regions.

The five regions are broken down as follows:

Region A: North and South America Inter-American Telecommunication Commission (CITEL)

`www.citel.oas.org`

Region B: Western Europe European Conference of Postal and Telecommunications Administrations (CEPT)

`www.cept.org`

Region C: Eastern Europe and Northern Asia Regional Commonwealth in the field of Communications (RCC)

`www.rcc.org.ru/en/`

Region D: Africa African Telecommunications Union (ATU)

`www.atu-uat.org`

Region E: Asia and Australasia Asia-Pacific Telecommunity (APT)

`www.aptsec.org`

Within each region, local government RF regulatory bodies such as the following manage the RF spectrum for their respective countries:

- Australia: Australian Communications and Media Authority (ACMA)
- Japan: Association of Radio Industries and Businesses (ARIB)
- New Zealand: Ministry of Economic Development
- United States: Federal Communications Commission (FCC)

It is important to understand that communications are regulated differently in many regions and countries. For example, European RF regulations are very different from the regulations used in North America. When deploying a WLAN, please take the time to learn about rules and policies of the local *regulatory domain authority*.

More information about the ITU-R can be found at `www.itu.int/ITU-R/`.

Institute of Electrical and Electronics Engineers (IEEE)

The *Institute of Electrical and Electronics Engineers*, commonly known as the *IEEE*, is a global professional society with more than 350,000 members. The IEEE's mission is to "foster technological innovation and excellence for the benefit of humanity." To networking professionals, that means creating the standards that we use to communicate.

The IEEE is probably best known for its LAN standards, the IEEE 802 project.

The 802 project is one of many IEEE projects; however, it is the only IEEE project addressed in this book.

IEEE projects are subdivided into working groups to develop standards that address specific problems or needs. For instance, the IEEE 802.3 working group was responsible for the creation of a standard for Ethernet, and the IEEE 802.11 working group was responsible for creating the WLAN standard. The numbers are assigned as the groups are formed, so the 11 assigned to the wireless group indicates that it was the 11th working group formed under the IEEE 802 project.

As the need arises to revise existing standards created by the working groups, task groups are formed. These task groups are assigned a sequential single letter (multiple letters are assigned if all single letters have been used) that is added to the end of the standard number (for example, 802.11a, 802.11g, and 802.3af). Some letters are not assigned. For example *o* and *l* are not assigned to prevent confusion with the numbers 0 and 1. Other letters may not be assigned to task groups to prevent confusion with other standards. For example, 802.11x has not been assigned because it can be easily confused with the 802.1X standard and because 802.11*x* has become a common casual reference to the 802.11 family of standards.

More information about the IEEE can be found at `www.ieee.org`.

It is important to remember that the IEEE standards, like many other standards, are written documents describing how technical processes and equipment should function. Unfortunately, this often allows for different interpretations when the standard is being implemented, so it is common for early products to be incompatible between vendors, as was the case with the early 802.11 products.

The history of the 802.11 standard and amendments is covered extensively in Chapter 5, "IEEE 802.11 Standards." The CWNA exam (PW0-104) is based on the most recently published version of the standard, 802.11-2007. The 802.11-2007 standard can be downloaded from `http://standards.ieee.org/getieee802/802.11.html`.

Wi-Fi Alliance

The *Wi-Fi Alliance* is a global, nonprofit industry association of more than 300 member companies devoted to promoting the growth of WLANs. One of the primary tasks of the Wi-Fi Alliance is to market the Wi-Fi brand and raise consumer awareness of new 802.11 technologies as they become available. Because of the Wi-Fi Alliance's overwhelming marketing success, the majority of the worldwide 450 million Wi-Fi users immediately recognize the Wi-Fi logo seen in Figure 1.1.

FIGURE 1.1 Wi-Fi logo

The Wi-Fi Alliance's main task is to ensure the interoperability of WLAN products by providing certification testing. During the early days of the 802.11 standard, the Wi-Fi Alliance further defined some of the ambiguous standards requirements and provided a set of guidelines to assure compatibility between different vendors. As seen in Figure 1.2, products that pass the Wi-Fi certification process receive a Wi-Fi Interoperability Certificate that provides detailed information about the individual product's Wi-Fi certifications.

FIGURE 1.2 Wi-Fi Interoperability Certificate

Wi-Fi® Interoperability Certificate

Certification ID: WFAxxxx

a b g Wi Fi n DRAFT CERTIFIED®

This certificate indicates the capabilities and features that successfully completed interoperability testing by the Wi-Fi Alliance. You may find detailed descriptions of these features at www.wi-fi.org/certification_programs.php.

Certificate Date: date_of_product_certification
Category: primary_product_category
Company: company_name
Product: product_name
Model/SKU #: model_number/sku

This product has the following Wi-Fi Certifications:

IEEE Standard	Security	Multimedia	Convergence
IEEE 802.11a IEEE 802.11b IEEE 802.11g IEEE 802.11n draft 2.0 IEEE 802.11d IEEE 802.11h	WPA™ - Enterprise, Personal WPA2™ - Enterprise, Personal **EAP Type(s)** EAP-TLS EAP-TTLS/MSCHAPv2 PEAPv0/EAP-MSCHAPv2 PEAPv1/EAP-GTC EAP-SIM **Vendor EAP Type(s)** EAP-TLS EAP-TTLS/MSCHAPv2 PEAPv0/EAP-MSCHAPv2 PEAPv1/EAP-GTC EAP-SIM	WMM ® WMM Power Save **Special Features** Wi-Fi Protected Setup™ PIN PBC NFC	Voice – Personal CWG-RF Profile (contact manufacturer for results)

For more information: www.wi-fi.org/certification_programs.php

The Wi-Fi Alliance, originally named the Wireless Ethernet Compatibility Alliance (WECA), was founded in August 1999. The name was changed to the Wi-Fi Alliance in October 2002.

The Wi-Fi Alliance has certified more than 4,600 Wi-Fi products for interoperability since testing began in April 2000. Multiple Wi-Fi CERTIFIED™ programs exist that cover basic connectivity, security, quality of service (QoS), and more. Testing of vendor Wi-Fi products is performed in 12 independent authorized test laboratories worldwide. The guidelines for interoperability for each Wi-Fi CERTIFIED™ program are usually based on key components and functions that are defined in the IEEE 802.11-2007 standard and various 802.11 amendments. In fact, many of the same engineers who belong to 802.11 task groups are also contributing members of the Wi-Fi Alliance. However, it is important to understand that the IEEE and the Wi-Fi Alliance are two separate organizations. The IEEE 802.11 task group defines the WLAN standards, and the Wi-Fi Alliance defines interoperability certification programs. The Wi-Fi CERTIFIED™ programs include the following:

802.11a, b, or g—IEEE 802.11 baseline The baseline program certifies 802.11a, b, and/or g interoperability to ensure that the essential wireless data transmission works as expected. 802.11b and g utilize spectrum in the 2.4 GHz band. 802.11g has a higher data rate (54 Mbps) than 802.11b (11 Mbps). 802.11a utilizes frequencies in the 5 GHz band and has a maximum data rate of 54 Mbps. Each certified product is required to support one frequency band as a minimum, but it can support both. The CWNA exam will not use the terms 802.11 a/b/g; however, the a/b/g terminology is commonplace within the industry because of the Wi-Fi Alliance baseline certifications.

Wi-Fi Protected Access 2 (WPA2)—security WPA2 is based on the security mechanisms that were originally defined in the IEEE 802.11i amendment that defines a *robust security network (RSN)*. Two versions of WPA2 exist: WPA2-Personal defines security for a SOHO environment, and WPA2-Enterprise defines stronger security for enterprise corporate networks. Each certified product is required to support WPA2-Personal or WPA2-Enterprise. More-detailed discussion of WPA2 security can be found in Chapter 13, "802.11 Network Security Architecture."

802.11n draft 2.0—IEEE 802.11 baseline This certification program is based on the 802.11n draft amendment that defines a High Throughput (HT) wireless network utilizing multiple-input multiple-output (MIMO) technology. *High Throughput (HT)* provides PHY and MAC enhancements to support throughput of 100 Mbps and greater. This technology is discussed in detail in Chapter 18, "High Throughput (HT) and 802.11n." Please note that Wi-Fi Multimedia (WMM) certification is required for the 802.11n draft 2.0 certification.

Wi-Fi Multimedia (WMM)—multimedia WMM is based on the QoS mechanisms that were originally defined in the IEEE 802.11e amendment. WMM enables Wi-Fi networks to prioritize traffic generated by different applications. In a network where WMM is supported by both the access point and the client device, traffic generated by time-sensitive applications such as voice or video can be prioritized for transmission on the half-duplex RF medium. WMM mechanisms are discussed in greater detail in Chapter 9, "802.11 MAC Architecture."

WMM Power Save (WMM-PS)—multimedia WMM-PS helps conserve battery power for devices using Wi-Fi radios by managing the time the client device spends in sleep mode.

Conserving battery life is critical for handheld devices such as barcode scanners and VoWiFi phones. To take advantage of power-saving capabilities, both the device and the access point must support WMM Power Save. WMM-PS and legacy power-saving mechanisms are discussed in greater detail in Chapter 9.

Wi-Fi Protected Setup—security Wi-Fi Protected Setup defines simplified and automatic WPA and WPA2 security configurations for home and small-business users. Users can easily configure a network with security protection by using a personal identification number (PIN) or a button located on the access point and the client device.

CWG-RF—multimedia Converged Wireless Group-RF Profile (CWG-RF) was developed jointly by the Wi-Fi Alliance and the Cellular Telecommunications and Internet Association (CTIA), now known as The Wireless Association. CWG-RF defines performance metrics for Wi-Fi and cellular radios in a converged handset to help ensure that both technologies perform well in the presence of the other. All CTIA-certified handsets now include this certification.

Voice Personal—application Voice Personal offers enhanced support for voice applications in residential and small-business Wi-Fi networks. These networks include one access point, mixed voice and data traffic from multiple devices (such as phones, PCs, printers, and other consumer electronic devices), and support for up to four concurrent phone calls. Both the access point and the client device must be certified to achieve performance matching the certification metrics.

As 802.11 technologies evolve, new Wi-Fi CERTIFIED™ programs will be detailed by the Wi-Fi Alliance. The next certification will probably be Voice Enterprise, which will define enhanced support for voice applications in the enterprise environment. Some aspects of the 802.11r (secure roaming) and 802.11k (resource management) amendments will probably be tested in Voice Enterprise.

Wi-Fi Alliance and Wi-Fi CERTIFIED

More information about the Wi-Fi Alliance can be found at `www.wi-fi.org`. The following five white papers from the Wi-Fi Alliance are also included on the CD that accompanies this book:

- *Wi-Fi CERTIFIED™ for WMM™: Support for Multimedia Applications with Quality of Service in Wi-Fi Networks*
- *WMM™ Power Save for Mobile and Portable Wi-Fi CERTIFIED Devices*
- *Wi-Fi Protected Access: Strong, Standards-Based, Interoperable Security for Today's Wi-Fi Networks*
- *Wi-Fi CERTIFIED™ 802.11n Draft 2.0: Longer-Range, Faster-Throughput, Multimedia-Grade Wi-Fi Networks*
- *Wi-Fi CERTIFIED™ Voice-Personal: Delivering the Best End-User Experience for Voice over Wi-Fi*

International Organization for Standardization (ISO)

The *International Organization for Standardization*, commonly known as the *ISO*, is a global, nongovernmental organization that identifies business, government, and society needs and develops standards in partnership with the sectors that will put them to use. The ISO is responsible for the creation of the Open Systems Interconnection (OSI) model, which has been a standard reference for data communications between computers since the late 1970s.

Why Is It ISO and Not IOS?

ISO is not a mistyped acronym. It is a word derived from the Greek word *isos*, meaning *equal*. Because acronyms can be different from country to country, based on varying translations, the ISO decided to use a word instead of an acronym for its name. With this in mind, it is easy to see why a standards organization would give itself a name that means *equal*.

The OSI model is the cornerstone of data communications, and learning to understand it is one of the most important and fundamental tasks a person in the networking industry can undertake.

The layers of the OSI model are as follows:

- Layer 7, Application
- Layer 6, Presentation
- Layer 5, Session
- Layer 4, Transport
- Layer 3, Network
- Layer 2, Data-Link
  - LLC sublayer
  - MAC sublayer
- Layer 1, Physical

The IEEE 802.11-2007 standard defines communication mechanisms only at the Physical layer and MAC sublayer of the Data-Link layer of the OSI model. How 802.11 technology is used at these two OSI layers is discussed in detail throughout this book.

You should have a working knowledge of the OSI model for both this book and the CWNA exam. Make sure you understand the seven layers of the OSI model and how communications take place at the different layers. If you are not comfortable with the concepts of the OSI model, spend some time reviewing it on the Internet or from a good networking fundamentals book prior to taking the CWNA test.

More information about the ISO can be found at `www.iso.org`.

Core, Distribution, and Access

If you have ever taken a networking class or read a book about network design, you have probably heard the terms *core*, *distribution*, and *access* when referring to networking architecture. Proper network design is imperative no matter what type of network topology is used. The core of the network is the high-speed backbone or the superhighway of the network. The goal of the core is to carry large amounts of information between key data centers or distribution areas, just as superhighways connect cities and metropolitan areas.

The core layer does not route traffic nor manipulate packets but rather performs high-speed switching. Redundant solutions are usually designed at the core layer to ensure the fast and reliable delivery of packets. The distribution layer of the network routes or directs traffic toward the smaller clusters of nodes or neighborhoods of the network.

The distribution layer routes traffic between virtual LANs (VLANs) and subnets. The distribution layer is akin to the state and county roads that provide medium travel speeds and distribute the traffic within the city or metropolitan area.

The access layer of the network is responsible for slower delivery of the traffic directly to the end user or end node. The access layer mimics the local roads and neighborhood streets that are used to reach your final address. The access layer ensures the final delivery of packets to the end user. Remember that speed is a relative concept.

Because of traffic load and throughput demands, speed and throughput capabilities increase as data moves from the access layer to the core layer. The additional speed and throughput tends to also mean higher cost.

Just as it would not be practical to build a superhighway so that traffic could travel between your neighborhood and the local school, it would not be practical or efficient to build a two-lane road as the main thoroughfare to connect two large cities such as New York and Boston. These same principles apply to network design. Each of the network layers—core, distribution, and access—are designed to provide a specific function and capability to the network. It is important to understand how wireless networking fits into this network design model.

Wireless networking can be implemented as either point-to-point or point-to-multipoint solutions. Most wireless networks are used to provide network access to the individual client stations and are designed as point-to-multipoint networks. This type of implementation is designed and installed on the access layer, providing connectivity to the end user. 802.11 wireless networking is most often implemented at the access layer. In Chapter 10, "Wireless Devices," you will learn about the difference between *autonomous access points* and the centralized *WLAN controller* solutions that utilize *lightweight access points*. All access points are deployed at the access layer; however, lightweight access points tunnel 802.11 wireless traffic to WLAN controllers that are typically deployed at the distribution or core layer.

Wireless bridge links are typically used to provide connectivity between buildings in the same way that county or state roads provide distribution of traffic between neighborhoods. The purpose of wireless bridging is to connect two separate, wired networks wirelessly. Routing data traffic between networks is usually associated with the distribution layer. Wireless bridge links cannot typically meet the speed or distance requirements of the core layer, but they can be very effective at the distribution layer. An 802.11 bridge link is an example of wireless technology being implemented at the distribution layer.

Although wireless is not typically associated with the core layer, you must remember that speed and distance requirements vary greatly between large and small companies and that one person's distribution layer could be another person's core layer. Very small companies may even implement wireless for all networking, forgoing any wired devices. Higher-bandwidth proprietary wireless bridges and some 802.11 mesh network deployments could be considered an implementation of wireless at the core layer.

Communications Fundamentals

Although the CWNA certification is considered one of the entry-level certifications in the Certified Wireless Network Professional (CWNP) wireless certification program, it is by no means an entry-level certification in the computing industry. Most of the candidates for the CWNA certificate have experience in other areas of information technology. However, the background and experience of these candidates varies greatly.

Unlike professions for which knowledge and expertise is learned through years of structured training, most computer professionals have followed their own path of education and training.

When people are responsible for their own education, they typically will gain the skills and knowledge that are directly related to their interests or their job. The more-fundamental knowledge is often ignored because it is not directly relevant to the tasks at hand. Later, as their knowledge increases and they become more technically proficient, people realize that they need to learn about some of the fundamentals.

Many people in the computer industry understand that in data communications, bits are transmitted across wires or waves. They even understand that some type of voltage change or wave fluctuation is used to distinguish the bits. When pressed, however, many of these same people have no idea what is actually happening with the electrical signals or the waves.

In the following sections, you will review some fundamental communications principles that directly and indirectly relate to wireless communications. Understanding these concepts will help you to better understand what is happening with wireless communications and to more easily recognize and identify the terms used in this profession.

Understanding Carrier Signals

Because data ultimately consists of bits, the transmitter needs a way of sending both 0s and 1s to transmit data from one location to another. An AC or DC signal by itself does not perform this task. However, if a signal fluctuates or is altered, even slightly, the signal can be interpreted so that data can be properly sent and received. This modified signal is now capable of distinguishing between 0s and 1s and is referred to as a *carrier signal*. The method of adjusting the signal to create the carrier signal is called *modulation*.

Three components of a wave that can fluctuate or be modified to create a carrier signal are amplitude, frequency, and phase.

This chapter reviews the basics of waves as they relate to the principles of data transmission. Chapter 2, "Radio Frequency Fundamentals," covers radio waves in much greater detail.

All radio-based communications use some form of modulation to transmit data. To encode the data in a signal sent by AM/FM radios, cellular telephones, and satellite television, some type of modulation is performed on the radio signal that is being transmitted. The average person typically is not concerned with how the signal is modulated, only that the device functions as expected. However, to become a better wireless network administrator, it is useful to have a better understanding of what is actually happening when two stations communicate. The rest of this chapter provides an introduction to waves as a basis for understanding carrier signals and data encoding and introduces you to the fundamentals of encoding data.

Amplitude and Wavelength

RF communication starts when radio waves are generated from an RF transmitter and picked up or "heard" by a receiver at another location. RF waves are similar to the waves that you see in an ocean or lake. Waves are made up of two main components: wavelength and amplitude (see Figure 1.3).

Amplitude is the height, force, or power of the wave. If you were standing in the ocean as the waves came to shore, you would feel the force of a larger wave much more than you would a smaller wave. Transmitters do the same thing, but with radio waves. Smaller waves are not as noticeable as bigger waves. A bigger wave generates a much larger electrical signal picked up by the receiving antenna. The receiver can then distinguish between highs and lows.

FIGURE 1.3 This drawing shows the wavelength and amplitude of a wave.

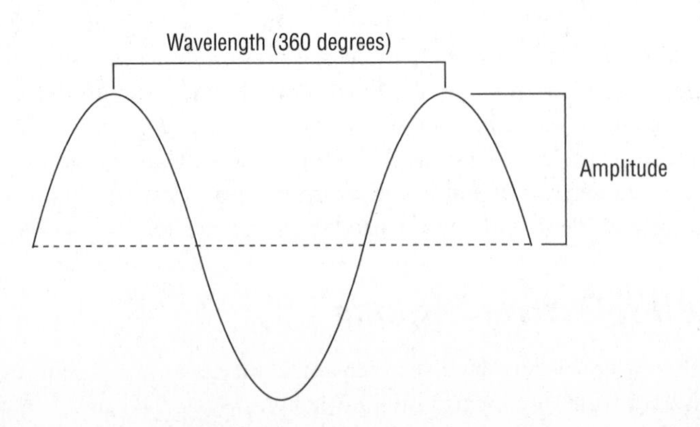

Wavelength is the distance between similar points on two back-to-back waves. When measuring a wave, the wavelength is typically measured from the peak of a wave to the peak of the next wave. Amplitude and wavelength are both properties of waves.

Frequency

Frequency describes a behavior of waves. Waves travel away from the source that generates them. How fast the waves travel, or more specifically, how many waves are generated over a 1-second period of time, is known as frequency. If you were to sit on a pier and count how often a wave hits it, you could tell someone how frequently the waves were coming to shore. Think of radio waves in the same way; however, radio waves travel much faster than the waves in the ocean. If you were to try to count the radio waves that are used in wireless networking, in the time it would take for one wave of water to hit the pier, several billion radio waves would have also hit the pier.

Phase

Phase is a relative term. It is the relationship between two waves with the same frequency. To determine phase, a wavelength is divided into 360 pieces referred to as *degrees* (see Figure 1.4). If you think of these degrees as starting times, then if one wave begins at the 0 degree point and another wave begins at the 90 degree point, these waves are considered to be 90 degrees out of phase.

In an ideal world, waves are created and transmitted from one station and received perfectly intact at another station. Unfortunately, RF communications do not occur in an ideal world. There are many sources of interference and many obstacles that will affect the wave in its travels to the receiving station. In Chapter 2, we'll introduce you to some of the outside influences that can affect the integrity of a wave and your ability to communicate between two stations.

FIGURE 1.4 This drawing shows two waves that are identical; however, they are 90 degrees out of phase with each other.

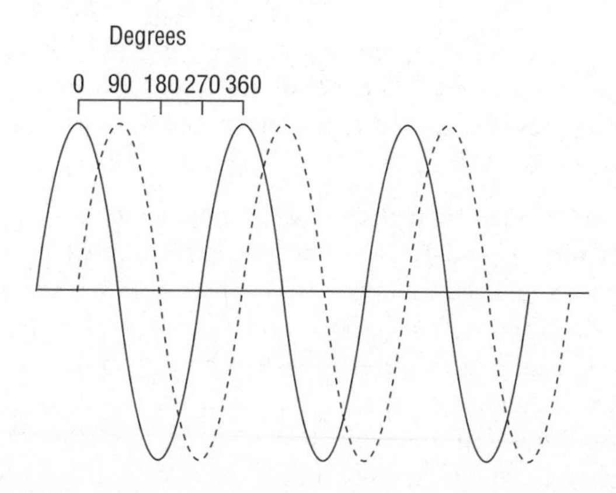

Time and Phase

Suppose you have two stopped watches and both are set to noon. At noon you start your first watch, and then you start your second watch 1 hour later. The second watch is 1 hour behind the first watch. As time goes by, your second watch will continue to be 1 hour behind. Both watches will maintain a 24-hour day, but they are out of synch with each other. Waves that are out of phase behave similarly. Two waves that are out of phase are essentially two waves that have been started at two different times. Both waves will complete full 360-degree cycles, but they will do it out of phase, or out of synch with each other.

Understanding Keying Methods

When data is sent, a signal is transmitted from the transceiver. In order for the data to be transmitted, the signal must be manipulated so that the receiving station has a way of distinguishing 0s and 1s. This method of manipulating a signal so that it can represent multiple pieces of data is known as a *keying method*. A keying method is what changes a signal into a carrier signal. It provides the signal with the ability to encode data so that it can be communicated or transported.

There are three types of keying methods that are reviewed in the following sections: amplitude-shift keying (ASK), frequency-shift keying (FSK), and phase-shift keying (PSK). These keying methods are also referred to as *modulation techniques*. Keying methods use two different techniques to represent data:

Current state With current state techniques, the current value (the current state) of the signal is used to distinguish between 0s and 1s. The use of the word *current* in this context does not refer to current as in voltage but rather to current as in the present time. Current state techniques will designate a specific or current value to indicate a binary 0, and another value to indicate a binary 1. At a specific point in time, it is the value of the signal that determines the binary value. For example, you can represent 0s and 1s by using an ordinary door. Once a minute you can check to see whether the door is open or closed. If the door is open, it represents a 0, and if the door is closed, it represents a 1. The current state of the door, open or closed, is what determines 0s or 1s.

State transition With state transition techniques, the change (or transition) of the signal is used to distinguish between 0s and 1s. State transition techniques may represent a 0 by a change in a wave's phase at a specific time, whereas a 1 would be represented by no change in wave's phase at a specific time. At a specific point in time, it is the presence of a change or the lack of presence of a change that determines the binary value. The upcoming "phase-shift keying" section provides examples of this in detail, but a door can be used again to provide a simple example. Once a minute you check the door. In this case, if the door is moving (opening or closing), it represents a 0, and if the door is still (either open or closed), it represents a 1. In this example, the state of transition (moving or not moving) is what determines 0s or 1s.

Amplitude-shift keying

Amplitude-shift keying (ASK) varies the amplitude, or height, of a signal to represent the binary data. ASK is a current state technique, where one level of amplitude can represent a 0 bit and another level of amplitude can represent a 1 bit. Figure 1.5 shows how a wave can modulate an ASCII letter *K* by using amplitude-shift keying. The larger amplitude wave is interpreted as a binary 1, and the smaller amplitude wave is interpreted as a binary 0.

FIGURE 1.5 An example of amplitude-shift keying (ASCII code of an uppercase *K*)

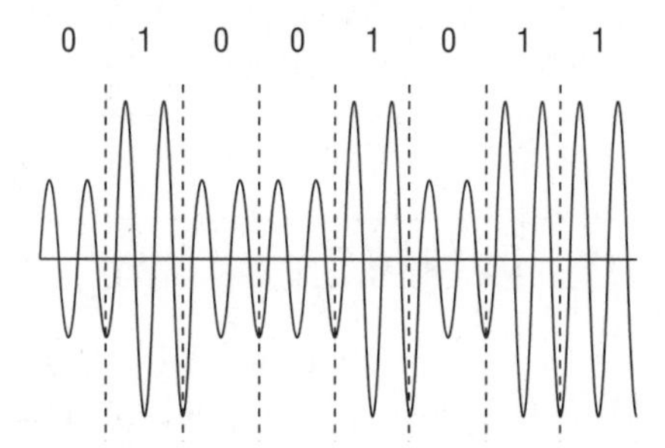

This shifting of amplitude determines the data that is being transmitted. The way the receiving station performs this task is to first divide the signal being received into periods of time known as *symbol periods*. The receiving station then samples or examines the wave during this symbol period to determine the amplitude of the wave. Depending on the value of the wave's amplitude, the receiving station can determine the binary value.

As you will learn later in this book, wireless signals can be unpredictable and also subject to interference from many sources. When noise or interference occurs, it usually affects the amplitude of a signal. Because a change in amplitude due to noise could cause the receiving station to misinterpret the value of the data, ASK has to be used cautiously.

Frequency-shift keying

Frequency-shift keying (FSK) varies the frequency of the signal to represent the binary data. FSK is a current state technique, where one frequency can represent a 0 bit and another frequency can represent a 1 bit (Figure 1.6). This shifting of frequency determines the data that is being transmitted. When the receiving station samples the signal during the symbol period, it determines the frequency of the wave, and depending on the value of the frequency, the station can determine the binary value.

Figure 1.6 shows how a wave can modulate an ASCII letter *K* by using frequency-shift keying. The faster frequency wave is interpreted as a binary 1, and the slower frequency wave is interpreted as a binary 0.

FSK is used in some of the legacy deployments of 802.11 wireless networks. With the demand for faster communications, FSK techniques would require more-expensive technology to support faster speeds, making it less practical.

FIGURE 1.6 An example of frequency-shift keying (ASCII code of an uppercase *K)*

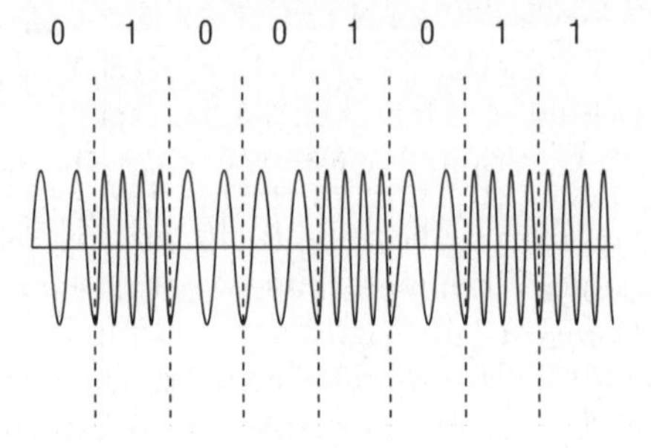

> **Why Haven't I Heard about Keying Methods Before?**
>
> You might not realize it, but you *have* heard about keying methods before. AM/FM radio uses amplitude modulation (AM) and frequency modulation (FM) to transmit the radio stations that you listen to at home or in your automobile. The radio station modulates the voice and music into its transmission signal, and your home or car radio demodulates it.

Phase-shift keying

Phase-shift keying (PSK) varies the phase of the signal to represent the binary data. PSK is a state transition technique, where one phase can represent a 0 bit and another phase can represent a 1 bit. This shifting of phase determines the data that is being transmitted. When the receiving station samples the signal during the symbol period, it determines the phase of the wave and the status of the bit.

Figure 1.7 shows how a wave can modulate an ASCII letter *K* by using phase-shift keying. A phase change at the beginning of the symbol period is interpreted as a binary 1, and the lack of a phase change at the beginning of the symbol period is interpreted as a binary 0.

FIGURE 1.7 An example of phase-shift keying (ASCII code of an uppercase *K*)

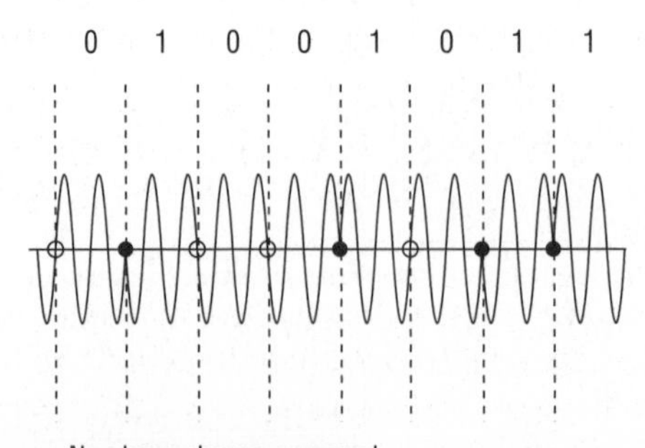

PSK technology is used extensively for radio transmissions as defined by the 802.11-2007 standard. Typically, the receiving station samples the signal during the symbol period and compares the phase of the current sample with the previous sample and determines the difference. This degree difference, or *differential*, is used to determine the bit value.

More-advanced versions of PSK can encode multiple bits per symbol. Instead of using two phases to represent the binary values, four phases can be used. Each of the four phases is capable of representing two binary values (00, 01, 10, or 11) instead of one (0 or 1), thus shortening the transmission time. When more than two phases are used, this is referred to as *multiple phase-shift keying (MPSK)*. Figure 1.8 shows how a wave can modulate an ASCII letter *K* by using a multiple phase-shift keying method. Four possible phase changes can be monitored, with each phase change now able to be interpreted as 2 bits of data instead of just 1. Notice that there are fewer symbol times in this drawing than there are in the drawing in Figure 1.5.

FIGURE 1.8 An example of multiple phase-shift keying (ASCII code of an uppercase *K*)

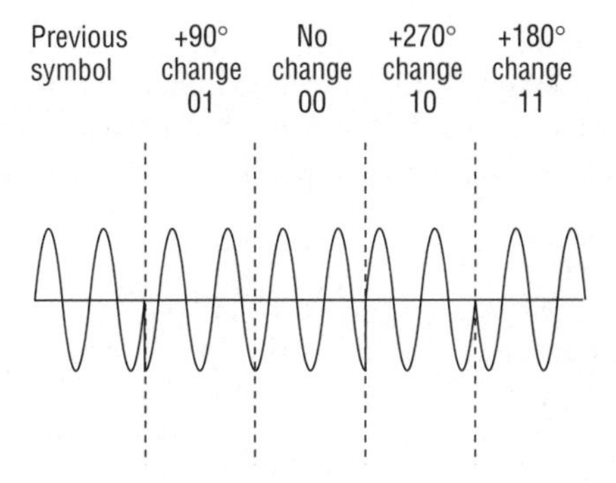

> **Where Else Can I Learn More about 802.11 Technology and the Wi-Fi Industry?**
>
> Reading this book from cover to cover is a great way to start understanding Wi-Fi technology. In addition, because of the rapidly changing nature of 802.11 WLAN technologies, the authors of this book would like to recommend these additional resources:
>
> **WNN** Wi-Fi Net News is a highly respected blog and daily newsletter about all the latest events and happenings in the Wi-Fi industry. Wi-Fi Net News (WNN) has more than 100,000 subscribers and is maintained by blogger Glenn Fleishman. Do yourself a favor and subscribe to Wi-Fi Net News at `www.wifinetnews.com`.
>
> **Wi-Fi Alliance** As mentioned earlier in this chapter, the Wi-Fi Alliance is the marketing voice of the Wi-Fi industry and maintains all the industry's certifications. The knowledge center section of the Wi-Fi Alliance website, `www.wi-fi.org`, is an excellent resource.

CWNP The Certified Wireless Networking Professional program maintains learning resources such as user forums and a WLAN white paper database. The website www.cwnp.com is also the best source of information about all the vendor-neutral CWNP wireless networking certifications.

WLAN vendor websites Although the CWNA exam and this book take a vendor-neutral approach about 802.11 education, the various WLAN vendor websites are often excellent resources for information about specific Wi-Fi networking solutions. Many of the major WLAN vendors are mentioned throughout this book, and a complete listing of most of the major WLAN vendor websites can be found in Chapter 11, "WLAN Deployment and Vertical Markets."

Summary

This chapter explained the history of wireless networking and the roles and responsibilities of the three key organizations involved with the wireless networking industry:

- FCC and other regulatory domain authorities
- IEEE
- Wi-Fi Alliance

To provide a basic understanding of the relationship between networking fundamentals and 802.11 technologies, we discussed these concepts:

- OSI model
- Core, distribution, and access

To provide a basic knowledge of how wireless stations transmit and receive data, we introduced some of the components of waves and modulation:

- Carrier signals
- Amplitude
- Wavelength
- Frequency
- Phase
- Keying methods, including ASK, FSK, and PSK

When troubleshooting RF communications, having a solid knowledge of waves and modulation techniques can help you understand the fundamental issues behind communications problems and hopefully help lead you to a solution.

Exam Essentials

Know the three industry organizations. Understand the roles and responsibilities of the regulatory domain authorities, the IEEE, and the Wi-Fi Alliance.

Understand core, distribution, and access. Know where 802.11 technology is deployed in fundamental network design.

Understand wavelength, frequency, amplitude, and phase. Know the definitions of each RF characteristic.

Understand the concepts of modulation. ASK, FSK, and PSK are three carrier signal modulation techniques.

Key Terms

Before you take the exam, be certain you are familiar with the following terms:

access
amplitude
amplitude-shift keying (ASK)
autonomous access points
carrier signal
core
distribution
Federal Communications Commission (FCC)
frequency
frequency-shift keying (FSK)
High Throughput (HT)
Institute of Electrical and Electronics Engineers (IEEE)
International Organization for Standardization (ISO)
International Telecommunication Union Radiocommunication Sector (ITU-R)
keying method
lightweight access points
modulation
phase
phase-shift keying (PSK)
radio frequency (RF)
regulatory domain authority
robust security network (RSN)
spread spectrum
wavelength
Wi-Fi
Wi-Fi Alliance
wireless local area network (WLAN)
WLAN controllers

Review Questions

1. 802.11 technology is typically deployed at which fundamental layer of network architecture?
   - A. Core
   - B. Distribution
   - C. Access
   - D. Network
2. Which organization is responsible for enforcing maximum transmit power rules in an unlicensed frequency band?
   - A. IEEE
   - B. Wi-Fi Alliance
   - C. ISO
   - D. IETF
   - E. None of the above
3. 802.11 wireless bridge links are typically associated with which network architecture layer?
   - A. Core
   - B. Distribution
   - C. Access
   - D. Network
4. The 802.11-2007 standard was created by which organization?
   - A. IEEE
   - B. OSI
   - C. ISO
   - D. Wi-Fi Alliance
   - E. FCC
5. What organization ensures interoperability of WLAN products?
   - A. IEEE
   - B. ITU-R
   - C. ISO
   - D. Wi-Fi Alliance
   - E. FCC

6. What type of signal is required to carry data?
   A. Communications signal
   B. Data signal
   C. Carrier signal
   D. Binary signal
   E. Digital signal

7. Which keying method is most susceptible to interference from noise?
   A. FSK
   B. ASK
   C. PSK
   D. DSK

8. Which sublayer of the OSI model's Data-Link layer is used for communication between 802.11 radios?
   A. LLC
   B. WPA
   C. MAC
   D. FSK

9. The term *Wi-Fi* is an acronym for which of these phrases?
   A. Wireless fundamentals
   B. Wireless hi-fidelity
   C. Wireless fidelity
   D. Wireless functionality
   E. None of the above

10. The Wi-Fi Alliance is responsible for which of the following certification programs? (Choose all that apply.)
    A. WPA2
    B. WEP
    C. 802.11-2007
    D. WMM
    E. PSK

11. Which wave properties can be modulated to encode data? (Choose all that apply.)
    A. Amplitude
    B. Frequency
    C. Phase
    D. Wavelength

12. The IEEE 802.11-2007 standard defines communication mechanisms at which layers of the OSI model? (Choose all that apply.)

    A. Network

    B. Physical

    C. Transport

    D. Application

    E. Data-Link

    F. Session

13. The height or power of a wave is known as what?

    A. Phase

    B. Frequency

    C. Amplitude

    D. Wavelength

14. Global spectrum management is tasked to what organization?

    A. FCC

    B. Wi-Fi Alliance

    C. ITU-R

    D. IEEE

15. A modulated signal capable of carrying data is known as what?

    A. Data transmission

    B. Communications channel

    C. Data path

    D. Carrier signal

16. Which of the following wireless communications parameters and usage are typically governed by a local regulatory authority? (Choose all that apply.)

    A. Frequency

    B. Bandwidth

    C. Maximum transmit power

    D. Maximum EIRP

    E. Indoor/outdoor usage

17. The Wi-Fi Alliance is responsible for which of the following certification programs? (Choose all that apply.)

    A. WECA

    B. Voice Personal

    C. 802.11v

    D. WAVE

    E. WMM-PS

18. A wave is divided into degrees. How many degrees make up a complete wave?

A. 100

B. 180

C. 212

D. 360

19. What are the advantages of using unlicensed frequency bands for RF transmissions? (Choose all that apply.)

A. There are no government regulations.

B. There is no financial cost.

C. Anyone can use the frequency band.

D. There are no rules.

20. The OSI model consists of how many layers?

A. Four

B. Six

C. Seven

D. Nine

Answers to Review Questions

1. C. 802.11 wireless networking is typically used to connect client stations to the network via an access point. Autonomous and lightweight access points are deployed at the access layer, not the core or distribution layer. The Physical layer is a layer of the OSI model, not a network architecture layer.

2. E. RF communications are regulated differently in many regions and countries. The local regulatory domain authorities of individual countries or regions define the spectrum policies and transmit power rules.

3. B. 802.11 wireless bridge links are typically used to perform distribution layer services. Core layer devices are usually much faster than 802.11 wireless devices, and bridges are not used to provide access layer services. The Network layer is a layer of the OSI model, not a network architecture layer.

4. A. The Institute of Electrical and Electronics Engineers (IEEE) is responsible for the creation of all of the 802 standards.

5. D. The Wi-Fi Alliance provides certification testing, and when a product passes the test, it receives a Wi-Fi Interoperability Certificate.

6. C. A carrier signal is a modulated signal that is used to transmit binary data.

7. B. Because of the effects of noise on the amplitude of a signal, amplitude-shift keying (ASK) has to be used cautiously.

8. C. The IEEE 802.11-2007 standard defines communication mechanisms at only the Physical layer and MAC sublayer of the Data-Link layer of the OSI model. The Logical Link Control (LLC) sublayer of the Data-Link layer is not defined by the 802.11-2007 standard. WPA is a security certification. FSK is a modulation method.

9. E. The most common assumption is that that *Wi-Fi* is an acronym for *wireless fidelity*. The problem is that there is no such thing as wireless fidelity, which is in fact a meaningless term that is similar to high fidelity. Wi-Fi is simply a brand marketing name that is used by the Wi-Fi Alliance to promote 802.11 WLAN technology.

10. A, D. 802.11-2007 is the IEEE standard, and WEP (Wired Equivalent Privacy) is defined as part of the IEEE 802.11-2007 standard. PSK is not a standard; it is an encoding technique. Wi-Fi Multimedia (WMM) is a Wi-Fi Alliance certification program that enables Wi-Fi networks to prioritize traffic generated by different applications. WPA2 is a certification program that defines Wi-Fi security mechanisms.

11. A, B, C. The three keying methods that can be used to encode data are amplitude-shift keying (ASK), frequency-shift keying (FSK), and phase-shift keying (PSK).

12. B, E. The IEEE 802.11-2007 standard defines communication mechanisms at only the Physical layer and MAC sublayer of the Data-Link layer of the OSI model.

13. C. Height or power are two terms that describe the amplitude of a wave. Frequency is how often a wave repeats itself. Wavelength is the actual length of the wave, typically measured from peak to peak. Phase refers to the starting point of a wave in relation to another wave.

14. C. The International Telecommunication Union Radiocommunication Sector (ITU-R) has been tasked with global spectrum management.

15. D. A carrier signal is a signal that has been modulated to carry data.

16. A, B, C, D, E. All of these are typically regulated by the local or regional RF regulatory authority.

17. B, E. The Wi-Fi Alliance maintains certification programs to ensure vendor interoperability. Voice Personal is a certification program that defines enhanced support for voice applications in residential and small-business Wi-Fi networks. WMM-PS is a certification program that defines methods to conserve battery power for devices using Wi-Fi radios by managing the time the client device spends in sleep mode.

18. D. A wave is divided into 360 degrees.

19. B, C. The main advantage of an unlicensed frequency is that permission to transmit on the frequency is free and anyone can use the unlicensed frequency. Although there are no financial costs, you still must abide by transmission regulations and other restrictions. The fact that anyone can use the frequency band is also a disadvantage because of overcrowding.

20. C. The OSI model is sometimes referred to as the seven-layer model.

Chapter 2

Radio Frequency Fundamentals

IN THIS CHAPTER, YOU WILL LEARN ABOUT THE FOLLOWING:

✓ **Definition of radio frequency signal**

✓ **Radio frequency characteristics**

- Wavelength
- Frequency
- Amplitude
- Phase

✓ **Radio frequency behaviors**

- Wave propagation
- Absorption
- Reflection
- Scattering
- Refraction
- Diffraction
- Loss (attenuation)
- Free space path loss
- Multipath
- Gain (amplification)

In addition to understanding the OSI model and basic networking concepts, you must broaden your understanding of many other networking technologies in order to properly design, deploy, and administer an 802.11 wireless network. For instance, when administering an Ethernet network, you typically need a comprehension of TCP/IP, bridging, switching, and routing. The skills to manage an Ethernet network will also aid you as a WLAN administer because most 802.11 wireless networks act as "portals" into wired networks. The IEEE defines the 802.11 communications at the Physical layer and the MAC sublayer of the Data-Link layer.

To fully understand the 802.11 technology, you need to have a clear concept of how wireless works at the first layer of the OSI model, and at the heart of the Physical layer is *radio frequency (RF)* communications.

In a wired LAN, the signal is confined neatly inside the wire, and the resulting behaviors are anticipated. However, just the opposite is true for a wireless LAN. Although the laws of physics apply, RF signals move through the air in a sometimes unpredictable manner. Because RF signals are not saddled inside an Ethernet wire, you should always try to envision a wireless LAN as an "ever changing" network.

Does this mean that you must be an RF engineer from Georgia Tech to perform a WLAN site survey or monitor a Wi-Fi network? Of course not, but if you have a good grasp of the RF characteristics and behaviors defined in this chapter, your skills as a wireless network administrator will be ahead of the curve. Why does a wireless network perform differently in an auditorium full of people than it does inside an empty auditorium? Why does the performance of a wireless LAN seem to degrade in a storage area with metal racks? Why does the range of a 5 GHz radio transmitter seem shorter than the range of a 2.4 GHz radio card? These are the types of questions that can be answered with some basic knowledge of how RF signals work and perform.

Wired communications travel across what is known as *bounded medium*. Bounded medium contains or confines the signal (small amounts of signal leakage can occur). Wireless communications travel across what is known as *unbounded medium*. Unbounded medium does not contain the signal, which is free to radiate into the atmosphere in all directions (unless restricted or redirected by some outside influence).

In this chapter, we first define what an RF signal does. Then we will discuss both the properties and the behaviors of RF.

What Is a Radio Frequency (RF) Signal?

This book is by no means intended to be a comprehensive guide on the laws of physics, which is the science of motion and matter. However, a basic understanding of some of the concepts of physics as they relate to radio frequency (RF) is important for even an entry-level wireless networking professional.

The *electromagnetic (EM) spectrum*, which is usually simply referred to as *spectrum*, is the range of all possible electromagnetic radiation. This radiation exists as self-propagating electromagnetic waves that can move through matter or space. Examples of electromagnetic waves include gamma rays, X-rays, visible light, and radio waves. Radio waves are electromagnetic waves occurring on the radio frequency portion of the electromagnetic spectrum, as pictured in Figure 2.1.

FIGURE 2.1 Electromagnetic spectrum

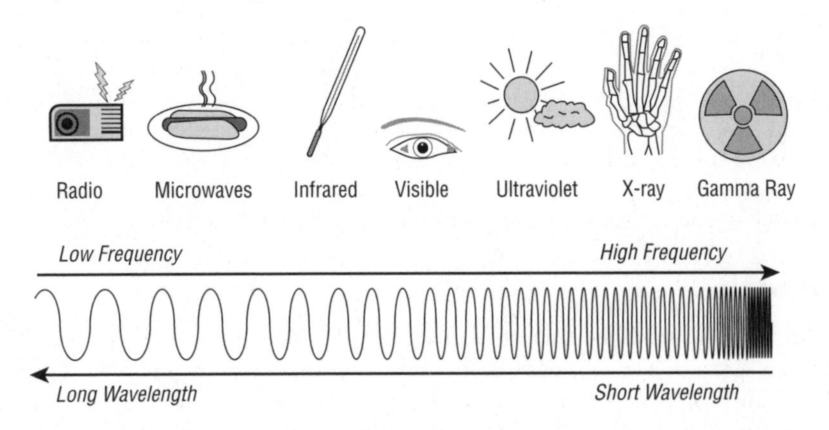

An RF signal starts out as an electrical *alternating current (AC)* signal that is originally generated by a transmitter. This AC signal is sent through a copper conductor (typically a coaxial cable) and radiated out of an antenna element in the form of an electromagnetic wave. This electromagnetic wave is the wireless signal. Changes of electron flow in an antenna, otherwise known as *current*, produce changes in the electromagnetic fields around the antenna.

An alternating current is an electrical current with a magnitude and direction that varies cyclically, as opposed to direct current, the direction of which stays in a constant form. The shape and form of the AC signal—defined as the *waveform*—is what is known as a sine wave, as shown in Figure 2.2. Sine wave patterns can also be seen in light, sound, and the ocean. The fluctuation of voltage in an AC current is known as cycling, or *oscillation*.

An RF electromagnetic signal radiates away from the antenna in a continuous pattern that is governed by certain properties such as wavelength, frequency, amplitude, phase, and polarity. Additionally, electromagnetic signals can travel through mediums of different materials or travel in a perfect vacuum. When an RF signal travels through a vacuum, it moves at the speed of light, which is approximately 300,000,000 meters per second, or 186,000 miles per second.

FIGURE 2.2 A sine wave

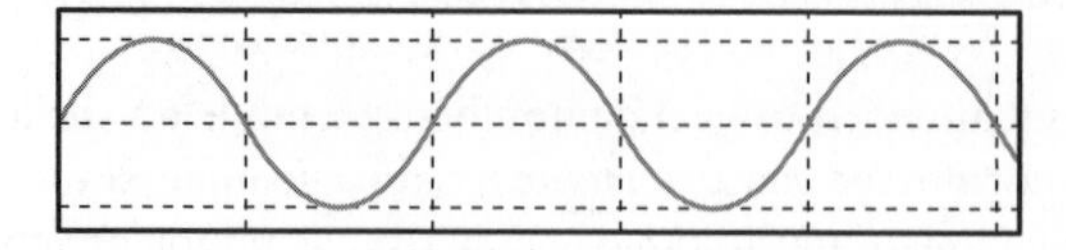

RF electromagnetic signals travel using a variety or combination of movement behaviors. These movement behaviors are referred to as *propagation behaviors*. We discuss some of these propagation behaviors later in this chapter, including absorption, reflection, scattering, refraction, diffraction, amplification, and attenuation.

Radio Frequency Characteristics

These characteristics, defined by the laws of physics, exist in every RF signal:

- Wavelength
- Frequency
- Amplitude
- Phase

You will look at each of these in more detail in the following sections.

Wavelength

As stated earlier, an RF signal is an alternating current (AC) that continuously changes between a positive and negative voltage. An oscillation, or cycle, of this alternating current is defined as a single change from up to down to up, or as a change from positive to negative to positive.

A *wavelength* is the distance between the two successive crests (peaks) or two successive troughs (valleys) of a wave pattern, as pictured in Figure 2.3. In simpler words, a wavelength is the distance that a single cycle of an RF signal actually travels.

FIGURE 2.3 Wavelength

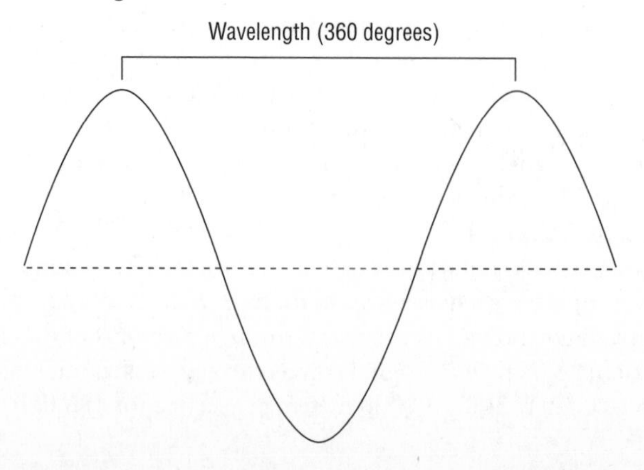

The Greek symbol λ (lambda) represents wavelength. Frequency is usually denoted by the Latin letter *f*. The Latin letter *c* represents the speed of light in a vacuum. This is derived from *celeritas*, the Latin word meaning speed.

It is very important to understand that there is an inverse relationship between wavelength and frequency. The three components of this inverse relationship are frequency (f, measured in hertz, or Hz), wavelength (λ, measured in meters, or m), and the speed of light (c, which is a constant value of 300,000,000 m/sec). The following reference formulas illustrate the relationship: $\lambda = c/f$ and $f = c/\lambda$. A simplified explanation is that the higher the frequency of an RF signal, the smaller the wavelength of that signal. The larger the wavelength of an RF signal, the lower the frequency of that signal.

AM radio stations operate at much lower frequencies than WLAN 802.11 radios, while satellite radio transmissions occur at much higher frequencies than WLAN radios. For instance, radio station WSB-AM in Atlanta broadcasts at 750 KHz and has a wavelength of 1,312 feet, or 400 meters. That is quite a distance for one single cycle of an RF signal to travel. In contrast, some radio navigation satellites operate at a very high frequency, near 252 GHz, and a single cycle of the satellite's signal has a wavelength of less than 0.05 inches, or 1.2 millimeters. Figure 2.4 displays a comparison of these two extremely different types of RF signals.

FIGURE 2.4 750 KHz wavelength and 252 GHz wavelength

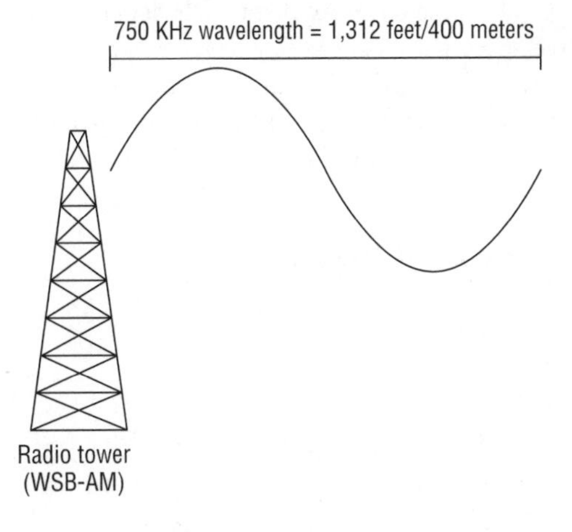

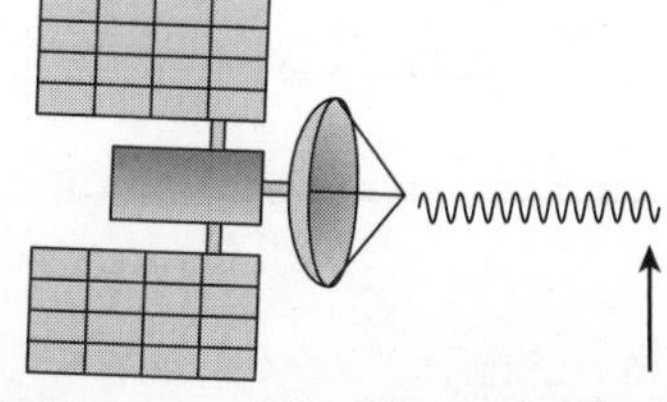

As RF signals travel through space and matter, they lose signal strength (attenuate). An electromagnetic signal with a smaller wavelength will attenuate *faster,* to an amplitude level below the sensitivity of a receiver radio. Theoretically, in a vacuum, electromagnetic signals will travel forever. However, as a signal travels through our atmosphere, the signal will attenuate to amplitudes below the receive sensitivity threshold of a receiving radio. Essentially, the signal will arrive at the receiver, but it will be too weak to be detected. An electromagnetic signal with a larger wavelength will maintain an amplitude level above the sensitivity of a receiver radio over greater distances.

The perception is that the higher frequency signal with smaller wavelength will not travel as far as the lower frequency signal with larger wavelength. The reality is that the higher frequency signal has just become too weak that it is below the receive sensitivity threshold of the receiving radio, whereas the lower frequency signal is still above the receivers sensitivity threshold. A good analogy to a receiving radio would be the human ear. The next time you hear a car coming down the street with loud music, notice that the first thing you hear will be the bass (lower frequencies). This practical example shows that the lower frequency signals with the larger wavelength will be heard from a greater distance than the higher frequency signal with the smaller wavelength.

The majority of wireless LAN (WLAN) radio cards operate in either the 2.4 GHz frequency range or the 5 GHz range. In Figure 2.5, you see a comparison of a single cycle of the two different frequency WLAN radio cards.

The higher frequencies will attenuate faster through space. This is important for a wireless engineer to know for two reasons. First, the coverage distance is dependent on the attenuation through the air (referred to as free space path loss, discussed later in this chapter). Second, the higher the frequency, the less the signal will penetrate through obstructions. For example, a 2.4 GHz signal will pass through walls, windows, and doors with greater strength than a 5 GHz signal. Think of how much farther you can hear an AM station (lower frequency) versus an FM station (higher frequency).

FIGURE 2.5 2.45 GHz wavelength and 5.775 GHz wavelength

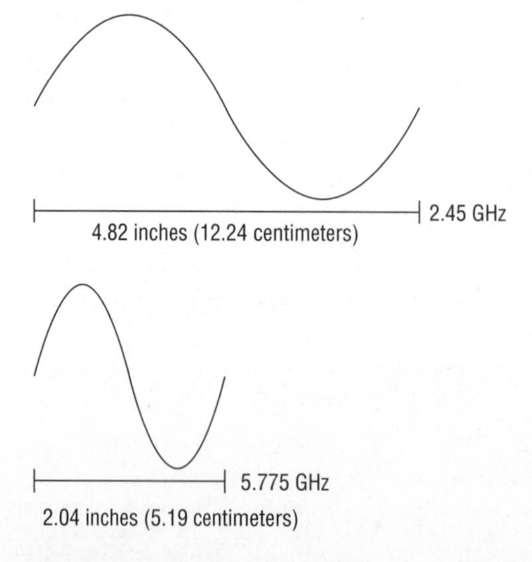

Note that the length of a 2.45 GHz wave is about 4.8 inches, or 12 centimeters. The length of a 5.775 GHz wave is a distance of only about 2 inches, or 5 centimeters.

As you can see in Figures 2.4 and 2.5, the wavelengths of the different frequency signals are different because, although each signal cycles only one time, the waves travel dissimilar distances. In Figure 2.6, you see the formulas for calculating wavelength distance in either inches or centimeters.

FIGURE 2.6 Wavelength formulas

Inches: wavelength = 11.811/frequency (GHz)

Centimeters: wavelength = 30/frequency (GHz)

Throughout this study guide, you will be presented with various formulas. You will not need to know these formulas for the CWNA certification exam. The formulas are in this study guide to demonstrate concepts and to be used as reference material.

Real World Scenario

How Does the Wavelength of a Signal Concern Me?

Because the wavelength property is shorter in the 5 GHz frequency range, Wi-Fi equipment using 5 GHz radio cards will have shorter range and coverage area than Wi-Fi equipment using 2.4 GHz radio cards.

Part of the design of the WLAN includes what is called a *site survey*. Part of the responsibility of the site survey is to determine zones, or cells, of usable received signal coverage in your facilities. The 2.4 GHz access points will provide greater RF footprints (coverage area) for client stations than the higher-frequency equipment simply because of the different wavelengths of the two frequency signals. More 5 GHz access points may have to be installed to meet the same coverage needs that are achieved by a lesser number of 2.4 GHz access points. The penetration of these signals will also reduce coverage for 5 GHz more than it will for 2.4 GHz.

Frequency

As previously mentioned, an RF signal cycles in an alternating current in the form of an electromagnetic wave. You also know that the distance traveled in one signal cycle is the wavelength. But what about how often an RF signal cycles in a certain time period?

Frequency is the number of times a specified event occurs within a specified time interval. A standard measurement of frequency is *hertz (Hz)*, which was named after the German physicist Heinrich Rudolf Hertz. An event that occurs once in 1 second is equal to 1 Hz. An event that occurs 325 times in 1 second is measured as 325 Hz. The frequency at which electromagnetic waves cycle is also measured in hertz. Thus, the number of times an RF signal cycles in 1 second is the frequency of that signal, as pictured in Figure 2.7.

FIGURE 2.7 Frequency

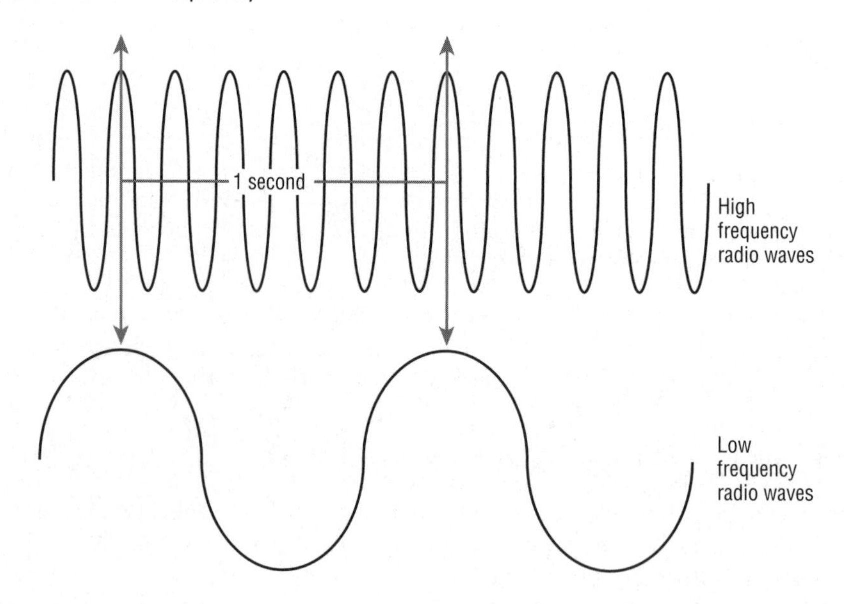

Different metric prefixes can be applied to the hertz (Hz) measurement of radio frequencies:

1 hertz (Hz) = 1 cycle per second

1 kilohertz (KHz) = 1,000 cycles per second

1 megahertz (MHz) = 1,000,000 (million) cycles per second

1 gigahertz (GHz) = 1,000,000,000 (billion) cycles per second

So when we are talking about 2.4 GHz WLAN radio cards, the RF signal is oscillating 2.4 billion times per second!

Inverse Relationship

Remember that there is an inverse relationship between wavelength and frequency. The three components of this inverse relationship are frequency (f, measured in hertz, or Hz), wavelength (λ, measured in meters, or m), and the speed of light (c, which is a constant value of 300,000,000 m/sec). The following reference formulas illustrate the relationship: $\lambda = c/f$ and $f = c/\lambda$. A simplified explanation is that the higher the frequency of an RF signal, the shorter the wavelength will be of that signal. The longer the wavelength of an RF signal, the lower the frequency will be of that signal.

Amplitude

Another very important property of an RF signal is the amplitude, which can be characterized simply as the signal's strength, or power. When speaking about wireless transmissions, this is often referenced as how loud or strong the signal is. *Amplitude* can be defined as the maximum displacement of a continuous wave. With RF signals, the amplitude corresponds to the electrical field of the wave. When you look at an RF signal in an oscilloscope, the amplitude is represented by the positive crests and negative troughs of the sine wave.

In Figure 2.8, you can see that λ represents wavelength and y represents amplitude. The first signal's crests and troughs have more magnitude, thus it has more amplitude. The second signal's crests and troughs have decreased magnitude, and therefore the signal has less amplitude.

FIGURE 2.8 Amplitude

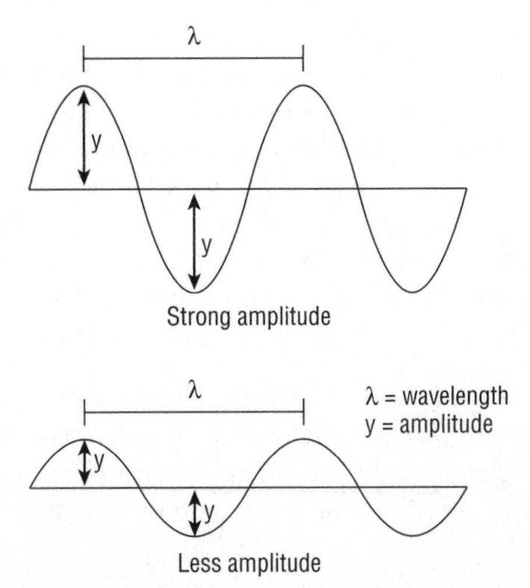

Note that although the signal strength (amplitude) is different, the frequency and wavelength of the signal remains constant. A variety of factors can cause an RF signal to lose amplitude, otherwise known as attenuation, which we discuss later in this chapter, in the section "Loss (Attenuation)."

When discussing signal strength in a WLAN, amplitude is usually referred to as either transmit amplitude or received amplitude. *Transmit amplitude* is typically defined as the amount of initial amplitude that leaves the radio transmitter. For example, if you configure an access point to transmit at 50 milliwatts (mW), that is the transmit amplitude. Cables and connectors will attenuate the transmit amplitude while an antenna will amplify the transmit amplitude. When a radio receives an RF signal, the received signal strength is most often referred to as *received amplitude*. RF signal strength measurements taken during a site survey is an example of received amplitude.

Different types of RF technologies require varying degrees of transmit amplitude. AM radio stations may transmit narrow band signals with as much power as 50,000 watts. The radio cards in most indoor 802.11 access points have a transmit power range between 1 mW and 100 mW. You will learn later that Wi-Fi radio cards can receive signals with amplitudes as low as billionths of a milliwatt.

Phase

Phase is not a property of just one RF signal but instead involves the relationship between two or more signals that share the same frequency. The phase involves the relationship between the position of the amplitude crests and troughs of two waveforms.

Phase can be measured in distance, time, or degrees. If the peaks of two signals with the same frequency are in exact alignment at the same time, they are said to be *in phase*. Conversely, if the peaks of two signals with the same frequency are not in exact alignment at the same time, they are said to be *out of phase*. Figure 2.9 illustrates this concept.

What is important to understand is the effect that phase has on amplitude when radio cards receive multiple signals. Signals that have 0 (zero) degree phase separation actually combine their amplitude, which results in a received signal of much greater signal strength, or twice the amplitude. If two RF signals are 180 degrees out of phase (the peak of one signal is in exact alignment with the trough of the second signal), they cancel each other out and the effective received signal strength is null. Phase separation has a cumulative effect. Depending on the amount of phase separation of two signals, the received signal strength may be either increased or diminished. The phase difference between two signals is very important to understanding the effects of an RF phenomenon known as multipath, which is discussed later in this chapter.

On your CD is a freeware program called EMANIM. Toward the end of this chapter, you will use this program to execute Exercise 2.1, which is a lab that demonstrates the changes in amplitude due to phase relationships of RF signals.

FIGURE 2.9 Phase

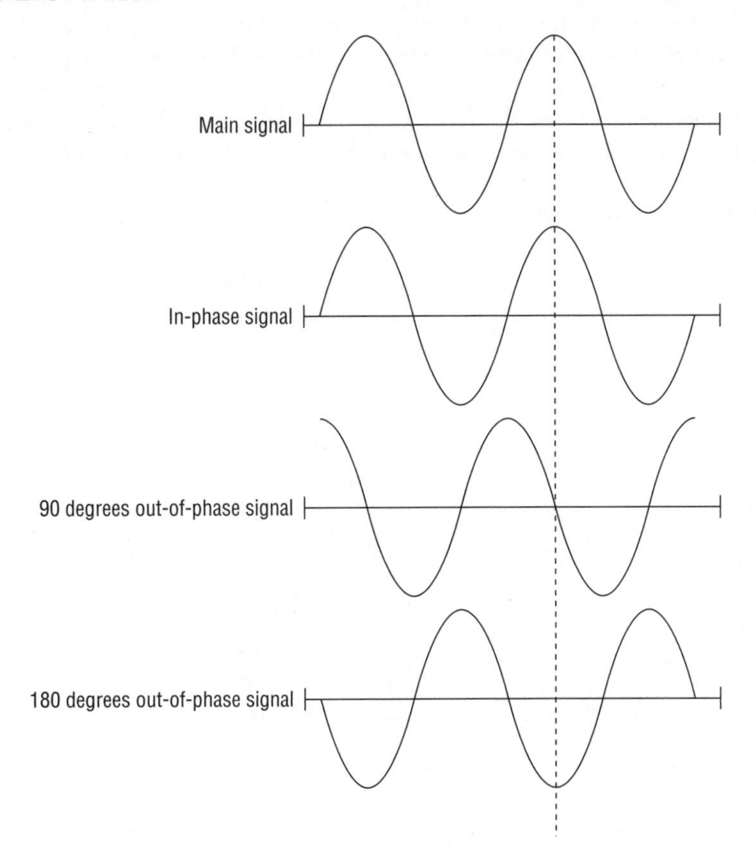

Radio Frequency Behaviors

As an RF signal travels through the air and other mediums, it can move and behave in different manners. RF propagation behaviors include absorption, reflection, scattering, refraction, diffraction, free space path loss, multipath, attenuation, and gain.

Wave Propagation

Now that you have learned about some of the various characteristics of an RF signal, it is important to understand the way an RF signal behaves as it moves away from an antenna. As stated before, electromagnetic waves can move through a perfect vacuum or pass through materials of different mediums. The way in which the RF waves move—known as wave *propagation*—can vary drastically depending on the materials in the signal's path. Drywall will have a much different effect on an RF signal than metal.

What happens to an RF signal between two locations is a direct result of how the signal propagates. When we use the term *propagate*, try to envision an RF signal broadening or spreading as it travels farther away from the antenna. An excellent analogy is shown in Figure 2.10, which depicts an earthquake. Note the concentric seismic rings that propagate away from the epicenter of the earthquake. RF waves behave in much the same fashion. The manner in which a wireless signal moves is often referred to as *propagation behavior*.

FIGURE 2.10 Propagation analogy

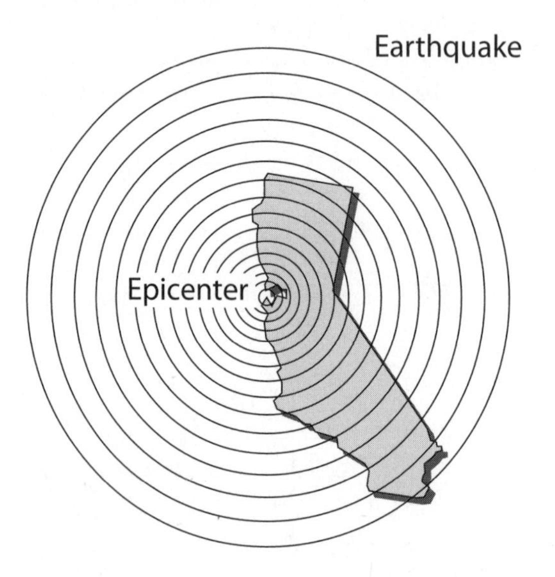

As a WLAN engineer, it is important to understand RF propagation behaviors for making sure that access points are deployed in the proper location, for making sure the proper type of antenna is chosen, and for monitoring the health of the wireless network.

Absorption

The most common RF behavior is *absorption*. If a signal does not bounce off an object, move around an object, or pass through an object, then 100 percent absorption has occurred. Most materials will absorb some amount of an RF signal to varying degrees.

Brick and concrete walls will absorb a signal significantly, whereas drywall will absorb a signal to a lesser degree. Water is another example of a medium that can absorb a signal to a large extent. Absorption can be a leading cause of attenuation, which is discussed later in this chapter. Even objects with large water content such as paper, cardboard, fish tanks, and so forth, can absorb signals. In some cases, specific leaves such as pine needles can absorb WLAN signals because the length of the needle is very close to the wavelength of the 2.4 GHz frequency.

Real World Scenario

User Density

Mr. Akin performed a wireless site survey at a campus lecture hall. He determined how many access points were required and their proper placement so that he would have the necessary RF coverage. Ten days later, Professor Sandlin gave a heavily attended lecture on business economics. During this lecture, the signal strength and quality of the WLAN was less than desirable. What happened? Human bodies!

An average adult body is 50 to 65 percent water. Water causes absorption, which results in attenuation. User density is an important factor when designing a wireless network. One reason is the effects of absorption. Another reason is the amount of available bandwidth, which we discuss in Chapter 15, "Radio Frequency Site Survey Fundamentals."

Reflection

One of the most important RF propagation behaviors to be aware of is reflection. When a wave hits a smooth object that is larger than the wave itself, depending on the media, the wave may bounce in another direction. This behavior is categorized as *reflection*. An analogous situation could be a child bouncing a ball off a sidewalk and the ball changing direction. Figure 2.11 depicts another analogy, a laser beam pointed at a single small mirror. Depending on the angle of the mirror, the laser beam bounces or reflects off in a different direction. RF signals can reflect in the same manner, depending on the objects or materials the signals encounter.

There are two major types of reflections: sky wave reflection and microwave reflection. Sky wave reflection can occur in frequencies below 1 GHz, where the signal has a very large wavelength. The signal bounces off the surface of the charged particles of the ionosphere in the earth's atmosphere. This is why you can be in Charlotte, North Carolina, and listen to radio station WLS-AM from Chicago on a clear night.

Microwave signals, however, exist between 1 GHz and 300 GHz. Because they are higher-frequency signals, they have much smaller wavelengths, thus the term *microwave*. Microwaves can bounce off smaller objects like a metal door. Microwave reflection is what we are concerned about in WLAN environments. In an outdoor environment, microwaves can reflect off large objects and smooth surfaces such as buildings, roads, bodies of water, and even the earth's surface. In an indoor environment, microwaves reflect off smooth surfaces such as doors, walls, and file cabinets. Anything made of metal will absolutely cause reflection. Other materials such as glass and concrete may cause reflection as well.

FIGURE 2.11 Reflection analogy

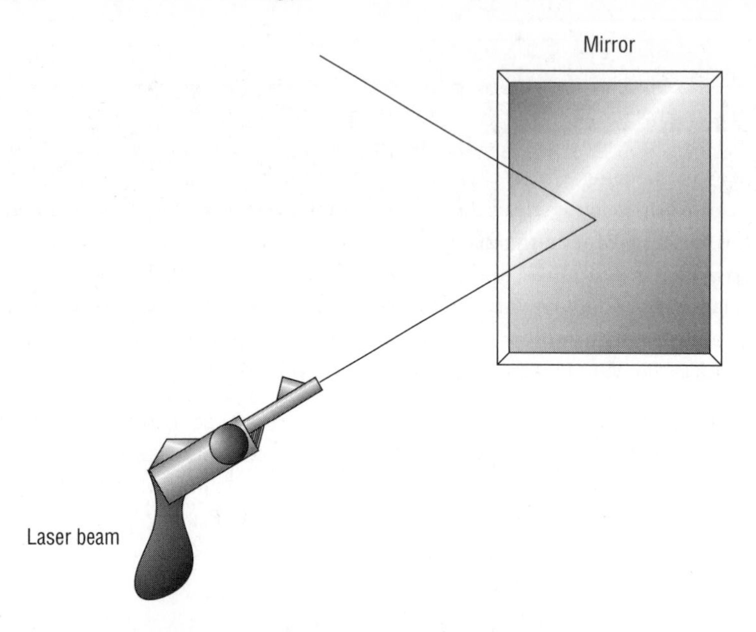

Reflection Is a Major Source of Poor WLAN Performance

Reflection can be the cause of serious performance problems in a WLAN. As a wave radiates from an antenna, it broadens and disperses. If portions of this wave are reflected, new wave fronts will appear from the reflection points. If these multiple waves all reach the receiver, the multiple reflected signals cause an effect called multipath.

Multipath can degrade the strength and quality of the received signal or even cause data corruption or cancelled signals. (Further discussion of multipath occurs later in this chapter. Hardware solutions to compensate for the negative effects of multipath, such as directional antennas and antenna diversity, are discussed in Chapter 4, "Radio Frequency Signal and Antenna Concepts.")

Although reflection and multipath can be your number one enemy, new antenna technologies such as multiple-input multiple-output (MIMO) will become commonplace in the future to actually take advantage of reflected signals.

In some cases, this reflection can be planned. A wireless engineering firm called Netrepid was hired to connect two buildings with a wireless bridge. The site survey revealed that a building was in the way of this path. However, Netrepid was able to engineer a wireless signal to connect the two buildings together. The signal was bounced off a tall office building with a glass exterior that was off to the side. Testing showed that the glass was able to reflect enough of the signal in a specific direction that a reliable signal could be achieved.

Scattering

Did you know that the color of the sky is blue because the wavelength of light is smaller than the molecules of the atmosphere? This blue sky phenomenon is known as Rayleigh scattering. The shorter blue wavelength light is absorbed by the gases in the atmosphere and radiated in all directions. This is another example of an RF propagation behavior called *scattering*, sometimes called *scatter.*

Scattering can most easily be described as multiple reflections. These multiple reflections occur when the electromagnetic signal's wavelength is larger than pieces of whatever medium the signal is passing through.

Scattering can happen in two ways. The first type of scatter is on a smaller level and has a lesser effect on the signal quality and strength. This type of scattering may manifest itself when the RF signal moves through a substance and the individual electromagnetic waves are reflected off the minute particles within the medium. Smog in our atmosphere and sandstorms in the desert can cause this type of scattering.

The second type of scattering occurs when an RF signal encounters some type of uneven surface and is reflected into multiple directions. Chain link fences, tree foliage, and rocky terrain commonly cause this type of scattering. When striking the uneven surface, the main signal dissipates into multiple reflected signals, which can cause substantial signal downgrade and may even cause a loss of the received signal.

Figure 2.12 shows a flashlight being shined against a disco mirror ball. Note how the main signal beam is completely displaced into multiple reflected beams with less amplitude and into many different directions.

FIGURE 2.12 Scattering analogy

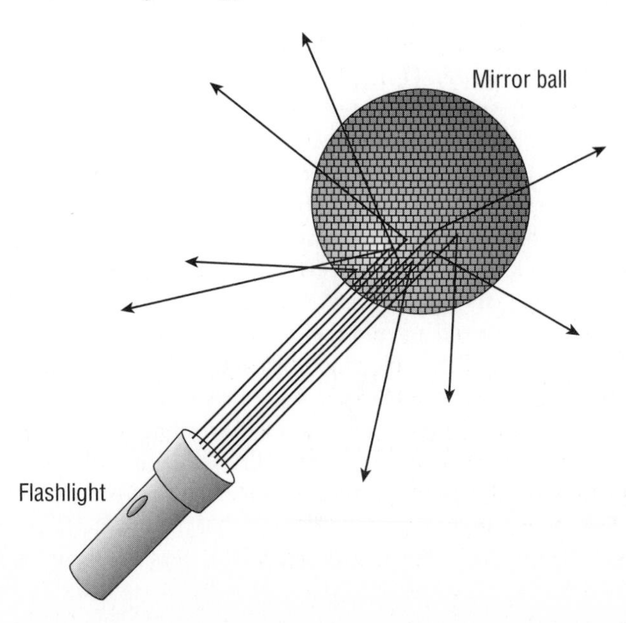

Refraction

In addition to RF signals being absorbed or bounced (via reflection or scattering), if certain conditions exist, an RF signal can actually be bent in a behavior known as *refraction*. A straightforward definition of refraction is the bending of an RF signal as it passes through a medium with a different density, thus causing the direction of the wave to change. RF refraction most commonly occurs as a result of atmospheric conditions.

When you are dealing with long-distance outdoor bridge links, an instance of refractivity change that might be a concern is what is known as the *k-factor*. A k-factor of 1 means there is no bending. A k-factor of less than 1, such as 2/3, represents the signal bending away from the earth. A k-factor of more than 1 represents bending toward the earth. Normal atmospheric conditions have a k-factor of 4/3, which is bending slightly toward the curvature of the earth.

The three most common causes of refraction are water vapor, changes in air temperature, and changes in air pressure. In an outdoor environment, RF signals typically refract slightly back down toward the earth's surface. However, changes in the atmosphere may cause the signal to bend away from the earth. In long-distance outdoor wireless bridge links, refraction can be an issue. An RF signal may also refract through certain types of glass and other materials that are found in an indoor environment. Figure 2.13 shows two examples of refraction.

Diffraction

Not to be confused with refraction, another RF propagation behavior exists that also bends the RF signal; it is called *diffraction*. Diffraction is the bending of an RF signal around an object (whereas refraction, as you recall, is the bending of a signal as it passes through a medium). Diffraction is the bending and the spreading of an RF signal when it encounters an obstruction. The conditions that must be met for diffraction to occur depend entirely on the shape, size, and material of the obstructing object as well as the exact characteristics of the RF signal, such as polarization, phase, and amplitude.

Typically, diffraction is caused by some sort of partial blockage of the RF signal, such as a small hill or a building that sits between a transmitting radio and a receiver. The waves that encounter the obstruction bend around the object, taking a longer and different path. The waves that did not encounter the object do not bend and maintain the shorter and original path. The analogy depicted in Figure 2.14 is a rock sitting in the middle of a river. Most of the current maintains the original flow; however, some of the current that encounters the rock will reflect off the rock and some will diffract around the rock.

Sitting directly behind the obstruction is the receiver radio that is now in an area known as the *RF shadow*. Depending on the change in direction of the diffracted signals, the area of the RF shadow can become a dead zone of coverage or still possibly receive degraded signals.

The concept of RF shadows is important when selecting antenna locations. Mounting to a beam or other wall structure can create a virtual RF blind spot.

FIGURE 2.13 Refraction

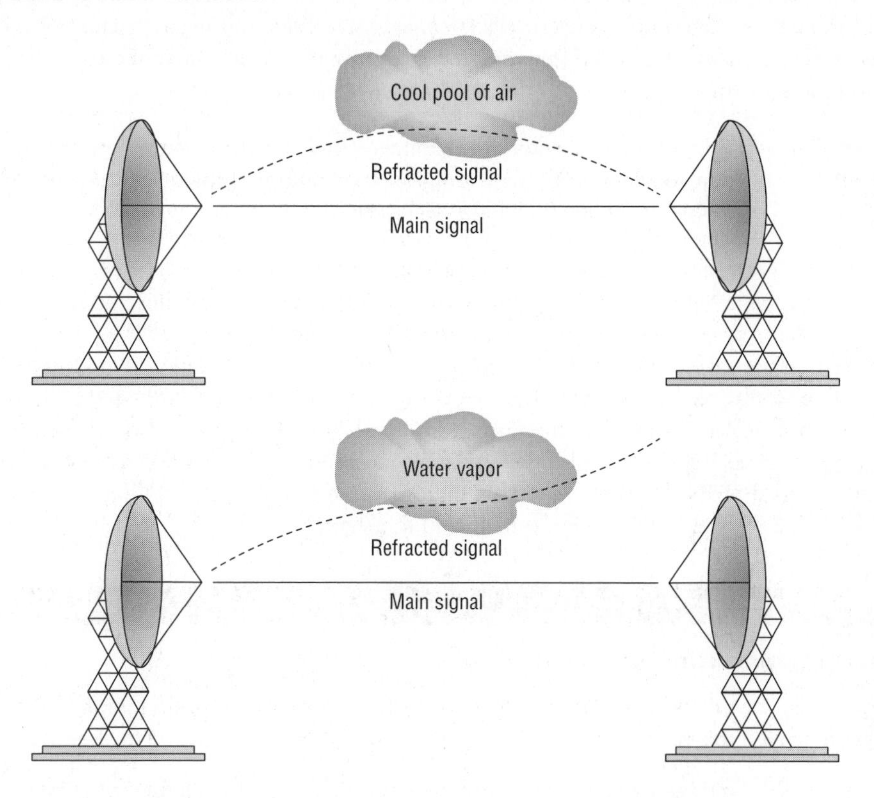

FIGURE 2.14 Diffraction analogy

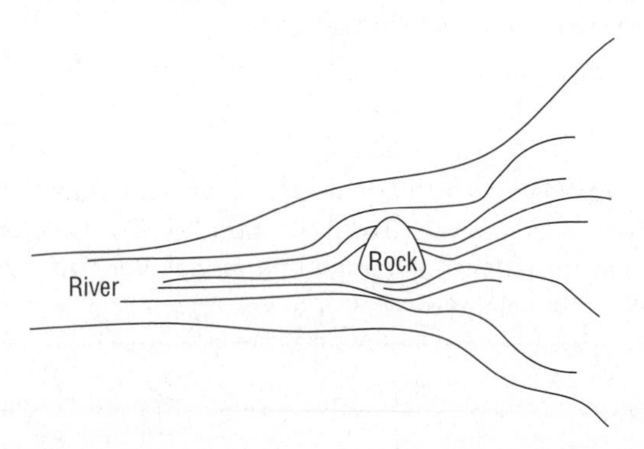

Loss (Attenuation)

Loss, also known as *attenuation*, is best described as the decrease of amplitude, or signal strength. A signal may lose strength while on a wire or in the air. On the wired portion of the communications (RF cable), the AC electrical signal will lose strength because of the electrical impedance of coaxial cabling and other components such as connectors.

In Chapter 4, we discuss impedance, which is the measurement of opposition to the AC current. You will also learn about impedance mismatches, which can create signal loss on the wired side.

Attenuation is typically not desired; however, on rare occasions an RF engineer may add a hardware attenuator device on the wired side of an RF system to introduce attenuation to remain compliant with power regulations or for capacity design purposes.

After the RF signal is radiated into the air via the antenna, the signal will attenuate due to absorption, distance, and the negative effects of multipath. You already know that as an RF signal passes through different mediums, the signal can be absorbed into the medium, which in turn causes a loss of amplitude. Different materials typically yield different attenuation results. As discussed earlier, water is a major source of absorption as well as dense materials such as cinder blocks, all of which lead to attenuation.

EXERCISE 2.1

Visual Demonstration of Absorption

In this exercise, you will use a program called EMANIM, a freeware program found on the CD that comes with this book, to view the attenuation effect of materials due to absorption.

1. Insert the CD included with this book into your computer and install the EMANIM program by double-clicking `emanim_setup.exe`.
2. From the main EMANIM menu, click Phenomenon.
3. Click Sybex CWNA Study Guide.
4. Click Exercise 2.2.
5. When a radio wave crosses matter, the matter absorbs part of the wave. As a result, the amplitude of the wave decreases. The extinction coefficient determines how much of the wave is absorbed by unit length of material. Vary the length of the material and the extinction coefficient for Wave 1 to see how it affects the absorption.

Both loss and gain can be gauged in a relative measurement of change in power called decibels (dB), which is discussed extensively in Chapter 3, "Radio Frequency Components, Measurements, and Mathematics." Table 2.1 shows the different attenuation values for several materials.

TABLE 2.1 Attenuation Comparison of Materials

Material	2.4 GHz
Foundation wall	–15 dB
Brick, concrete, concrete blocks	–15 dB
Elevator or metal obstacle	–10 dB
Metal rack	–6 dB
Drywall or Sheetrock	–3 dB
Nontinted glass windows or door	–3 dB
Wood door	–3 dB
Cubicle wall	–2 dB

Table 2.1 is meant as a reference chart and is not information that will be covered on the CWNA exam. Actual measurements may vary from site to site depending on specific environmental factors.

It is important to understand that an RF signal will also lose amplitude merely as a function of distance in what is known as free space path loss. Also, reflection propagation behaviors can produce the negative effects of multipath, and as a result cause attenuation in signal strength.

Free Space Path Loss

Because of the laws of physics, an electromagnetic signal will attenuate as it travels despite the lack of attenuation caused by obstructions, absorption, reflection, diffraction, and so on. *Free space path loss (FSPL)* is the loss of signal strength caused by the natural broadening of the waves, often referred to as *beam divergence*. RF signal energy spreads over larger areas as the signal travels farther away from an antenna, and as a result, the strength of the signal attenuates.

One way to illustrate free space path loss is to use a balloon analogy. Before a balloon is filled with helium, it remains small but has a dense rubber thickness. After the balloon is inflated and has grown and spread in size, the rubber becomes very thin. RF signals lose strength in much the same manner. Luckily, this loss in signal strength is logarithmic and not linear; thus the amplitude does not decrease as much in a second segment of equal length as it decreases in the first segment. A 2.4 GHz signal will change in power by about 80 dB after 100 meters but will lessen only another 6 dB in the next 100 meters.

Here are the formulas to calculate free space path loss:

$FSPL = 36.6 + (20\log_{10}(f)) + (20\log_{10}(D))$

FSPL = path loss in dB

f = frequency in MHz

D = distance in miles between antennas

$FSPL = 32.44 + (20\log_{10}(f)) + (20\log_{10}(D))$

FSPL = path loss in dB

f = frequency in MHz

D = distance in kilometers between antennas

Free space path loss formulas are provided as a reference and are not included on the CWNA exam. Many of the formulas in this book are provided in the form of spreadsheet calculators on the book's CD. An online calculator for FSPL and other RF calculators can be found at `www.airspy.com/calculators.php`.

An even simpler way to estimate free space path loss is called the 6 dB rule (remember for now that decibels are a measure of gain or loss, and further details of dB are covered extensively in Chapter 3). The 6 dB rule states that doubling the distance will result in a loss of amplitude of 6 dB. Table 2.2 shows estimated path loss and confirms the 6 dB rule. Also notice that the 5 GHz signal attenuates more than the 2.4 GHz signal. It should be noted that higher-frequency signals attenuate faster because of the shorter wavelength.

TABLE 2.2 Attenuation Due to Free Space Path Loss

Distance (km)	Attenuation (dB)	
	2.4 GHz	**5 GHz**
1	100.0	106.4
2	106.1	112.4
4	112.1	118.5
8	118.1	124.5

Real World Scenario

Why Is Free Space Path Loss Important?

All radio devices have what is known as a receive sensitivity level. The radio receiver can properly interpret and receive a signal down to a certain fixed amplitude threshold. If a radio device receives a signal above its amplitude threshold, the card can differentiate between the signal and other RF noise that is in the background. The background noise is typically referred to as the *noise floor.*

After the amplitude of a received signal falls below the radio device's receive sensitivity threshold, the device can no longer make the distinction between the signal and the background noise. The concept of free space path loss also applies to road trips in your car. When you are in a car listening to an AM radio station, eventually you will drive out of range and will no longer be able to hear the music above the static noise.

When designing both indoor WLANs and outdoor wireless bridge links, you must make sure that the RF signal will not attenuate below the receive sensitivity level of your wireless radio card simply because of free space path loss. You typically achieve this goal indoors during a site survey. An outdoor bridge link requires a series of calculations called a *link budget.* (Site surveys are covered in Chapters 15 and 16, and link budgets are covered in Chapter 3.)

Multipath

Multipath is a propagation phenomenon that results in two or more paths of a signal arriving at a receiving antenna at the same time or within nanoseconds of each other. Because of the natural broadening of the waves, the propagation behaviors of reflection, scattering, diffraction, and refraction will occur differently in dissimilar environments. A signal may reflect off an object or scatter, refract, or diffract. These propagation behaviors can all result in multiple paths of the same signal.

In an indoor environment, reflected signals and echoes can be caused by long hallways, walls, desks, floors, file cabinets, and numerous other obstructions. Indoor environments with large amounts of metal surfaces such as airport hangers, warehouses, and factories are notoriously high-multipath environments because of all the reflections. The propagation behavior of reflection is typically the main cause of high-multipath environments. In an outdoor environment, multipath can be caused by a flat road, large body of water, building, or atmospheric conditions. Therefore, we have signals bouncing and bending in many different directions. The principal signal will still travel to the receiving antenna, but many of the bouncing and bent signals may also find their way to the receiving antenna via different paths. In other words, multiple paths of the RF signal arrive at the receiver, as seen in Figure 2.15.

FIGURE 2.15 Multipath

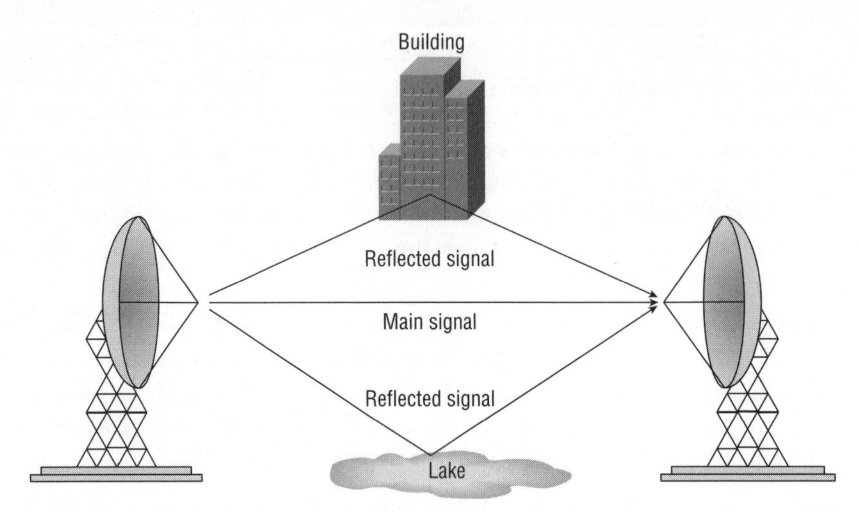

It usually takes a little bit longer for reflected signals to arrive at the receiving antenna because they must travel a longer distance than the principal signal. The time differential between these signals can be measured in billionths of a second (nanoseconds). The time differential between these multiple paths is known as the *delay spread*. You will learn later in the book that certain spread spectrum technologies are more tolerant than others of delay spread.

So what exactly happens when mutipath presents itself? In television signal transmissions, multipath causes a ghost effect with a faded duplicate image to the right of the main image. With RF signals, the effects of multipath can be either constructive or destructive. Quite often they are destructive. Due to the differences in phase of the multiple paths, the combined signal will often attenuate, amplify, or become corrupted. These effects are sometimes called *Rayleigh fading*, a phenomenon named after British physicist Lord Rayleigh.

The four possible results of multipath are as follows:

Downfade This is decreased signal strength. When the multiple RF signal paths arrive at the receiver at the same time and are out of phase with the primary wave, the result is a decrease in signal strength (amplitude). Phase differences of between 121 and 179 degrees will cause *downfade*.

Upfade This is increased signal strength. When the multiple RF signal paths arrive at the receiver at the same time and are in phase or partially out of phase with the primary wave, the result is an increase in signal strength (amplitude). Smaller phase differences of between 0 and 120 degrees will cause *upfade*. Please understand, however, that the final received signal can never be stronger than the original transmitted signal because of free space path loss.

Nulling This is signal cancellation. When the multiple RF signal paths arrive at the receiver at the same time and are 180 degrees out of phase with the primary wave, the result will be *nulling*. Nulling is the complete cancellation of the RF signal.

Data corruption Because of the difference in time between the primary signal and the reflected signals known as the delay spread, along with the fact that there may be multiple reflected signals, the receiver can have problems demodulating the RF signal's information. The delay spread time differential can cause bits to overlap with each other, and the end result is corrupted data, as seen in Figure 2.16. This type of multipath interference is often known as *intersymbol interference (ISI).*

FIGURE 2.16 Data corruption ISI

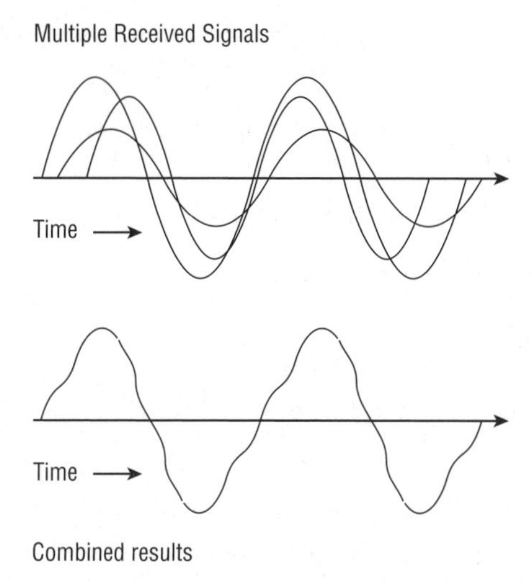

The bad news is that high-multipath environments usually will result in data corruption because of intersymbol interference caused by the delay spread. The good news is that the receiving station will detect the errors through an 802.11-defined cyclic redundancy check (CRC) because the checksum will not calculate accurately. The 802.11 standard requires that all unicast frames must be acknowledged by the receiving station with an acknowledgment (ACK) frame; otherwise, the transmitting station will have to retransmit the frame. The receiver will *not* acknowledge a frame that has failed the CRC. Therefore, unfortunately, the frame must be retransmitted, but this is better than it being misinterpreted.

Mutipath can have a very negative effect on the performance or throughput of your WLAN because of layer 2 retransmissions that are a direct result of intersymbol interference. Layer 2 retransmissions affect the overall throughput of an 802.11 WLAN and can also affect the delivery of time-sensitive packets of applications such as VoIP. In Chapter 12, "WLAN Troubleshooting," we discuss the multiple causes of layer 2 retransmissions and how to troubleshoot and minimize them. Multipath is one of the main causes of layer 2 retransmissions that negatively affect the throughput and latency of a WLAN.

So how is a hapless WLAN engineer supposed to deal with all these multipath issues? The use of directional antennas will often reduce the number of reflections, and antenna diversity can also be used to compensate for the negative effects of multipath. Sometimes, reducing transmit power or using a lower-gain antenna can solve the problem as long as there is enough signal to provide connectivity to the remote end.

On your CD is a freeware program called EMANIM. Use this program for Exercise 2.2, which demonstrates the effects of phase and multipath fading.

EXERCISE 2.2

Visual Demonstration of Multipath and Phase

In this exercise, you will use a program called EMANIM to view the effect on amplitude due to various phases of two signals arriving at the same time.

1. Insert the CD included with this book into your computer and install the EMANIM program by double-clicking `emanim_setup.exe`.
2. From the main EMANIM menu, click Phenomenon.
3. Click Sybex CWNA Study Guide.
4. Click Exercise 2.1a.
5. Two identical, vertically polarized waves are superposed (you might not see both of them because they cover each other). The result is a wave having double the amplitude of the component waves.
6. Click Exercise 2.1b.
7. Two identical, 70-degree out-of-phase waves are superposed. The result is a wave with an increased amplitude over the component waves.
8. Click Exercise 2.1c.
9. Two identical, 140-degree out-of-phase waves are superposed. The result is a wave with a decreased amplitude over the component waves.
10. Click on Exercise 2.1d.
11. Two identical, vertically polarized waves are superposed. The result is a cancellation of the two waves.

Gain (Amplification)

Gain, also known as *amplification*, can best be described as the increase of amplitude, or signal strength. The two types of gain are known as active gain and passive gain. A signal's amplitude can be boosted by the use of external devices.

Active gain is usually caused by the use of an amplifier on the wire that connects the transceiver to the antenna. The amplifier is usually bidirectional, meaning that it increases the AC voltage both inbound and outbound. Active gain devices require the use of an external power source.

Passive gain is accomplished by focusing the RF signal with the use of an antenna. Antennas are passive devices that do not require an external power source. Instead, the internal workings of an antenna focus the signal more powerfully in one direction than another.

The proper use of RF amplifiers and antennas is covered extensively in Chapter 4.

Despite the usual negative effects of multipath, it should be reiterated that when multiple RF signals arrive at the receiver at the same time, and are in phase or partially out of phase with the primary wave, the result can be an increase, or gain, in amplitude. However, an increase in signal amplitude is usually a result of passive or active gain before the signal radiates from the antenna.

Two very different tools can be used to measure the amplitude of a signal at a given point. The first, a frequency domain tool, can be used to measure amplitude in a finite frequency spectrum. The frequency domain tool used by WLAN engineers is called a *spectrum analyzer*. The second tool, a time domain tool, can be used to measure how a signal's amplitude changes over time. The conventional name for a time domain tool is an *oscilloscope*. Figure 2.17 shows how both these tools can be used to measure amplitude. It should be noted that spectrum analyzers are often used by WLAN engineers during site surveys. An oscilloscope is rarely if ever used when deploying a WLAN; however, oscilloscopes are used by RF engineers in laboratory test environments.

FIGURE 2.17 RF signal measurement tools

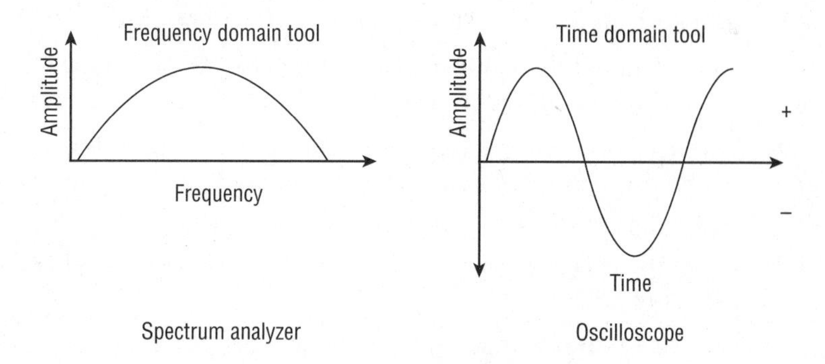

Summary

This chapter covered the meat and potatoes, the basics, of radio frequency signals. To properly design and administer a WLAN network, it is essential to have a thorough understanding of the following principles of RF properties and RF behaviors:

- Electromagnetic waves and how they are generated

- The relationship between wavelength, frequency, and the speed of light
- Signal strength and the various ways in which a signal can either attenuate or amplify
- The importance of the relationship between two or more signals
- How a signal moves by bending, bouncing, or absorbing in some manner

When troubleshooting an Ethernet network, the best place to start is always at layer 1, the Physical layer. WLAN troubleshooting should also begin at the Physical layer. Learning the RF fundamentals that exist at layer 1 is an essential step in proper wireless network administration.

Exam Essentials

Understand wavelength, frequency, amplitude, and phase. Know the definition of each RF characteristic and how it can affect wireless LAN design.

Remember all the RF propagation behaviors. Be able to explain the differences between each RF behavior (such as reflection, diffraction, scattering, and so on) and the various mediums that are associated with each behavior.

Understand what causes attenuation. Loss can occur either on the wire or in the air. Absorption, free space path loss, and multipath downfade are all causes of attenuation.

Define free space path loss. Despite the lack of any obstructions, electromagnetic waves attenuate in a logarithmic manner as they travel away from the transmitter.

Remember the four possible results of multipath and their relationship to phase. Multipath may cause downfade, upfade, nulling, and data corruption.

Know the results of intersymbol interference and delay spread. The time differential between a primary signal and reflected signals may cause corrupted bits and affect throughput and latency due to layer 2 retransmissions.

Explain the difference between active and passive gain. RF amplifiers are active devices, whereas antennas are passive devices.

Explain the difference between transmit and received amplitude. Transmit amplitude is typically defined as the amount of initial amplitude that leaves the radio transmitter. When a radio receives an RF signal, the received signal strength is most often referred to as received amplitude.

Key Terms

Before you take the exam, be certain you are familiar with the following terms:

absorption
active gain
alternating current (AC)
amplification
amplitude
attenuation
delay spread
diffraction
downfade
electromagnetic (EM) spectrum
free space path loss (FSPL)
frequency
gain
hertz (Hz)
intersymbol interference (ISI)
link budget
loss
microwave
multipath
noise floor
nulling
oscillation
oscilloscope
passive gain
phase
propagation
propagation behavior
radio frequency (RF)
Rayleigh fading
received amplitude
reflection
refraction
RF shadow
scattering
site survey
spectrum analyzer
transmit amplitude
upfade
wavelength

Review Questions

1. What are some results of multipath interference? (Choose all that apply.)
   A. Cross polarization
   B. Upfade
   C. Excessive retransmissions
   D. Absorption
2. What term best defines the linear distance traveled in one positive-to-negative-to-positive oscillation of an electromagnetic signal?
   A. Crest
   B. Frequency
   C. Trough
   D. Wavelength
3. Which of the following statements are true about amplification? (Choose all that apply.)
   A. Some antennas require an outside power source.
   B. RF amplifiers require an outside power source.
   C. Antennas are passive gain amplifiers that focus the energy of a signal in one direction.
   D. RF amplifiers passively increase signal strength by focusing the AC current of the signal.
   E. Signal strength may passively increase because of multipath.
4. A standard measurement of frequency is called what?
   A. Hertz
   B. Milliwatt
   C. Nanosecond
   D. Decibel
   E. K-factor
5. When an RF signal bends around an object, this propagation behavior is known as what?
   A. Stratification
   B. Refraction
   C. Scattering
   D. Diffraction
   E. Attenuation

6. When the multiple RF signals arrive at a receiver at the same time and are ________ with the primary wave, the result can be ________ of the primary signal.

   A. out of phase, scattering

   B. in phase, intersymbol interference

   C. in phase, attenuation

   D. 180 degrees out of phase, amplification

   E. in phase, cancellation

   F. 180 degrees out of phase, cancellation

7. Which of the following statements are true? (Choose all that apply.)

   A. As a result of upfade, a final received signal will be stronger than the original transmitted signal.

   B. As a result of downfade, a final received signal will never be stronger than the original transmitted signal.

   C. As a result of upfade, a final received signal will never be stronger than the original transmitted signal.

   D. As a result of downfade, a final received signal will be stronger than the original transmitted signal.

8. What is the frequency of an RF signal that cycles 2.4 million times per second?

   A. 2.4 hertz

   B. 2.4 MHz

   C. 2.4 GHz

   D. 2.4 kilohertz

   E. 2.4 KHz

9. What is an example of a time domain tool that could be used by an RF engineer?

   A. Oscilloscope

   B. Spectroscope

   C. Spectrum analyzer

   D. Refractivity gastroscope

10. What are some objects or materials that are common causes of reflection? (Choose all that apply.)

   A. Metal

   B. Trees

   C. Asphalt road

   D. Lake

   E. Carpet floors

11. Which of these propagation behaviors can result in multipath? (Choose all that apply.)
    - A. Refraction
    - B. Diffraction
    - C. Reflection
    - D. Scattering
    - E. None of the above

12. Which behavior can be described as an RF signal encountering a chain link fence, causing the signal to bounce into multiple directions?
    - A. Diffraction
    - B. Scatter
    - C. Reflection
    - D. Refraction
    - E. Multiplexing

13. What is another name for background noise?
    - A. Noise ceiling
    - B. Background interference
    - C. Noise floor
    - D. Background information

14. Which of the following can cause refraction of an RF signal traveling through it? (Choose all that apply.)
    - A. Shift in air temperature
    - B. Change in air pressure
    - C. Humidity
    - D. Smog
    - E. Wind
    - F. Lightning

15. Which of the following statements are true about free space path loss? (Choose all that apply.)
    - A. RF signals will attenuate as they travel despite the lack of attenuation caused by obstructions.
    - B. Path loss occurs at a constant linear rate.
    - C. RF signals will attenuate as they travel because of obstructions.
    - D. Path loss occurs at a logarithmic rate.

16. What term is used to describe the time differential between a primary signal and a reflected signal arriving at a receiver?

A. Path delay

B. Spread spectrum

C. Multipath

D. Delay spread

17. What is an example of a frequency domain tool that could be used by an RF engineer?

A. Oscilloscope

B. Spectroscope

C. Spectrum analyzer

D. Refractivity gastroscope

18. Using knowledge of RF characteristics and behaviors, which two options should a WLAN engineer be most concerned about during an indoor site survey? (Choose all that apply.)

A. Firewall door

B. Indoor temperature

C. User density

D. Drywall

19. Which three properties are interrelated?

A. Frequency, wavelength, and the speed of light

B. Frequency, amplitude, and the speed of light

C. Frequency, phase, and amplitude

D. Amplitude, phase, and the speed of sound

20. Which RF behavior best describes a signal striking a medium and bending in a different direction?

A. Refraction

B. Scattering

C. Diffusion

D. Diffraction

E. Microwave reflection

Answers to Review Questions

1. B, C. Mutipath may result in attenuation, amplification, signal loss, or data corruption. If two signals arrive together in phase, the result is an increase in signal strength called upfade. The delay spread may also be too significant and cause data bits to be corrupted, resulting in excessive layer 2 retransmissions.

2. D. The wavelength is the linear distance between the repeating crests (peaks) or repeating troughs (valleys) of a single cycle of a wave pattern.

3. B, C, E. RF amplifiers introduce active gain with the help of an outside power source. Passive gain is typically created by antennas that focus the energy of a signal without the use of an outside power source. Passive gain may also result in the form of upfade, which is one possible effect of multipath.

4. A. The standard measurement of the number of times a signal cycles per second is hertz (Hz). One Hz is equal to one cycle in 1 second.

5. D. Often confused with refraction, the diffraction propagation is the bending of the wave front around an obstacle. Diffraction is caused by some sort of partial blockage of the RF signal, such as a small hill or a building that sits between a transmitting radio and a receiver.

6. F. Nulling, or cancellation, can occur when multiple RF signals arrive at the receiver at the same time and are 180 degrees out of phase with the primary wave.

7. B, C. When the multiple RF signals arrive at the receiver at the same time and are in phase or partially out of phase with the primary wave, the result is an increase in signal strength (amplitude). However, the final received signal, whether affected by upfade or downfade, will never be stronger than the original transmitted signal because of free space path loss.

8. B. 802.11 wireless LANs operate in the 5 GHz and 2.4 GHz frequency range. However, 2.4 GHz is equal to 2.4 billion cycles per second. The frequency of 2.4 million cycles per second is 2.4 MHz.

9. A. An oscilloscope is a time domain tool that be used to measure how a signal's amplitude changes over time. A frequency domain tool called a spectrum analyzer is a more commonplace tool most often used during site surveys.

10. A, C, D. This is a tough question to answer because many of the same mediums can cause several different propagation behaviors. Metal will always bring about reflection. Water is a major source of absorption; however, large bodies of water can also cause reflection. Flat surfaces such as asphalt roads, ceilings, and walls will also result in reflection behavior.

11. A, B, C, D. Multipath is a propagation phenomenon that results in two or more paths of a signal arriving at a receiving antenna at the same time or within nanoseconds of each other. Because of the natural broadening of the waves, the propagation behaviors of reflection, scattering, diffraction, and refraction can all result in multiple paths of the same signal. The propagation behavior of reflection is usually considered to be the main cause of high-multipath environments.

12. B. Scattering, or scatter, is defined as an RF signal reflecting in multiple directions when encountering an uneven surface.

13. C. The noise floor is a signal strength measurement of all unwanted sources of noise.

14. A, B, C, D. Air stratification is a leading cause of refraction of an RF signal. Changes in air temperature, changes in air pressure, and water vapor are all causes of refraction. Smog can cause a density change in the air pressure as well as increased moisture.

15. A, D. Because of the natural broadening of the wave front, electromagnetic signals lose amplitude as they travel away from the transmitter. The rate of free space path loss is logarithmic and not linear. Attenuation of RF signals as they pass through different mediums does occur but is not a function of FSPL.

16. D. The time difference due to a reflected signal taking a longer path is known as the delay spread. The delay spread can cause intersymbol interference, which results in data corruption and layer 2 retransmissions.

17. C. A spectrum analyzer is a frequency domain tool that can be used to measure amplitude in a finite frequency spectrum. An oscilloscope is a time domain tool.

18. A, C. Firewall doors are made of metal and will cause reflections, which can lead to multipath issues. People are composed primarily of water, and a high user density could affect signal performance because of absorption. Drywall will attenuate a signal but not to the extent of water, cinder blocks, or other dense mediums. Air temperature has no significance during an indoor site survey.

19. A. There is an inverse relationship between frequency (f) and wavelength. A simplified explanation is that the higher the frequency of an RF signal, the shorter the wavelength will be of that signal. The longer the wavelength of an RF signal, the lower the frequency of that signal.

20. A. Refraction is the bending of an RF signal when it encounters a medium.

Radio Frequency Components, Measurements, and Mathematics

IN THIS CHAPTER, YOU WILL LEARN ABOUT THE FOLLOWING:

✓ **Components of RF communications**

- Transmitter
- Receiver
- Antenna
- Isotropic radiator
- Intentional radiator (IR)
- Equivalent isotropically radiated power (EIRP)

✓ **Units of power and comparison**

- Watt
- Milliwatt
- Decibel (dB)
- dBi
- dBd
- dBm
- Inverse square law

✓ **RF mathematics**

- Rule of 10s and 3s

✓ **Received signal strength indicator (RSSI)**

✓ **Link budget**

✓ **Fade margin/system operating margin**

To put it simply, data communication is the transferring of information between computers. No matter what form of communication is being used, many components are required to achieve a successful communication. Before we look at some of the individual components, let's initially keep things simple and look at the three basic requirements for successful communications:

- Two or more devices want to communicate.
- There must be a medium, a means, or a method for them to use to communicate.
- There must be a set of rules for them to use when they communicate. (This is covered in Chapter 8, "802.11 Medium Access.")

These three basic requirements are the same for all forms of communication, whether a group of people are having a conversation at a dinner party, two computers are transmitting data via a dial-up modem, or many computers are communicating via a wireless network.

The existence of a computer network essentially implies that the first requirement is met. If we did not have two or more devices that wanted to share data, we wouldn't need to create the network in the first place. The CWNA certification program also assumes this and is therefore rarely if ever concerned specifically with the data itself. It is assumed that we have data, and our concern is to transmit it.

This chapter focuses on the second requirement for successful communications: the medium, means, or method to communicate. We cover the components of radio frequency (RF), which make up what we refer to as the medium for wireless communications. Here we are concerned with the transmission of the RF signal and the role of each device and component along the transmission path. We also show how each device or component affects the transmission.

In Chapter 2, "Radio Frequency Fundamentals," you learned that there are many RF behaviors that affect the signal as it leaves the transmitter and travels toward the receiver. As the signal moves through the different components and through the air, the signal's amplitude changes. Some components increase the power of the signal (gain), while other components decrease the power (loss). In this chapter, you will learn how to quantify and measure the power of the waves and calculate how the waves are affected by both internal and external influences. Through these calculations, you will be able to accurately determine whether you will have the means to communicate between devices.

RF Components

Many components contribute to the successful transmission and reception of an RF signal. Figure 3.1 shows the key components that are covered in this section. In addition to understanding the function of the components, it is important to understand how the strength of the signal is specifically affected by each of the components.

FIGURE 3.1 RF components

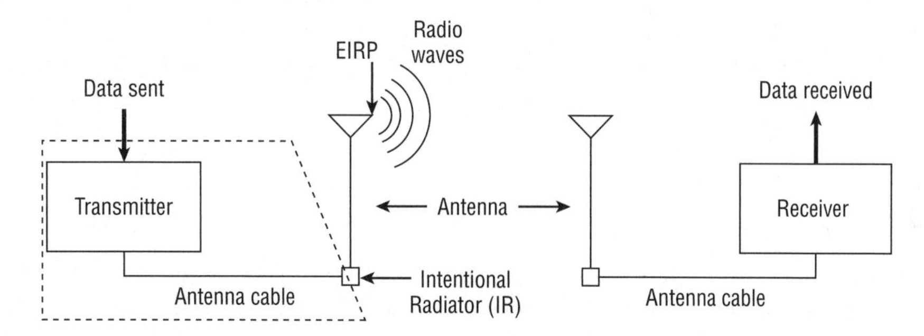

Later in this chapter, when we discuss RF mathematics, we will show you how to calculate the effect that each of the components has on the signal.

Transmitter

The *transmitter* is the initial component in the creation of the wireless medium. The computer hands the data off to the transmitter, and it is the transmitter's job to begin the RF communication.

In Chapter 1, "Overview of Wireless Standards, Organizations, and Fundamentals," you learned about carrier signals and modulation methods. When the transmitter receives the data, it begins generating an alternating current (AC) signal. This AC signal determines the frequency of the transmission. For an 802.11 (legacy DSSS or FHSS), 802.11b (HR-DSSS), or 802.11g (ERP) transmission, the AC signal oscillates around 2.4 billion times per second. For an 802.11a (OFDM) transmission, the AC signal oscillates around 5 billion times per second. This oscillation determines the frequency of the radio wave.

The exact frequencies used are covered in Chapter 6, "Wireless Networks and Spread Spectrum Technologies."

The transmitter takes the data provided and modifies the AC signal by using a modulation technique to encode the data into the signal. This modulated AC signal is now a carrier signal, containing the data to be transmitted. The carrier signal is then transported either directly to the antenna or through a cable to the antenna.

In addition to generating a signal at a specific frequency, the transmitter is responsible for determining the original transmission amplitude, or what is more commonly referred to as the *power level*, of the transmitter. The higher the amplitude of the wave, the more powerful the wave is and the farther it will travel. The power levels that the transmitter is allowed to generate are determined by the local regulatory domain authorities, such as the Federal Communications Commission (FCC) in the United States.

Although we are explaining the transmitter and receiver separately in this chapter, and although functionally they are different components, typically they are one device that is referred to as a *transceiver* (transmitter/receiver). Typical wireless devices that have transceivers built into them are access points, bridges, and client adapters.

Antenna

An *antenna* provides two functions in a communication system. When connected to the transmitter, it collects the AC signal that it receives from the transmitter and directs, or radiates, the RF waves away from the antenna in a pattern specific to the antenna type. When connected to the receiver, the antenna takes the RF waves that it receives through the air and directs the AC signal to the receiver. The receiver converts the AC signal to bits and bytes. As you will see later in this chapter, the signal that is received is much less than the signal that is generated. This signal loss is analogous to two people trying to talk to each other from opposite ends of a football field. Because of distance alone (free space), the yelling from one end of the field may be heard as barely louder than a whisper on the other end.

The signal of an antenna is usually compared or referenced to an isotropic radiator. An *isotropic radiator* is a *point source* that radiates signal equally in all directions. The sun is probably one of the best examples of an isotropic radiator. It generates equal amounts of energy in all directions. Unfortunately, it is not possible to manufacture an antenna that is a perfect isotropic radiator. The structure of the antenna itself influences the output of the antenna, similar to the way the structure of a lightbulb affects the bulb's ability to emit light equally in all directions.

There are two ways to increase the power output from an antenna. The first is to generate more power at the transmitter, as stated in the previous section. The other is to direct, or focus, the RF signal that is radiating from the antenna. This is similar to how you can focus light from a flashlight. If you remove the lens from the flashlight, the bulb is typically not very bright and radiates in almost all directions. To make the light brighter, you could use more-powerful batteries, or you could put the lens back on. The lens is not actually creating more light. It is focusing the light that was radiating in all different directions into a narrow area. Some antennas radiate waves as the bulb without the lens does, while some radiate focused waves as the flashlight with the lens does.

In Chapter 4, "Radio Frequency Signal and Antenna Concepts," you will learn about the types of antennas and how to properly and most effectively use them.

Receiver

The *receiver* is the final component in the wireless medium. The receiver takes the carrier signal that is received from the antenna and translates the modulated signals into 1s and 0s. It then takes this data and passes it to the computer to be processed. The job of the receiver is not always an easy one. The signal that is received is a much less powerful signal than what was transmitted because of the distance it has traveled and the effects of free space path loss (FSPL). The signal is also often altered due to interference from other RF sources and multipath.

Intentional Radiator (IR)

The FCC Code of Federal Regulations (CFR) Part 15 defines an *intentional radiator (IR)* as "a device that intentionally generates and emits radio frequency energy by radiation or induction." Basically, it's something that is specifically designed to generate RF as opposed to something that generates RF as a by-product of its main function, such as a motor that incidentally generates RF noise.

Regulatory bodies such as the FCC limit the amount of power that is allowed to be generated by an IR. The IR consists of all the components from the transmitter to the antenna but not including the antenna, as seen in Figure 3.1. The power output of the IR is thus the sum of all the components from the transmitter to the antenna, again not including the antenna. The components making up the IR include the transmitter, all cables and connectors, and any other equipment (grounding, lightning arrestors, amplifiers, attenuators, and so forth) between the transmitter and the antenna. The power of the IR is measured at the connecter that provides the input to the antenna. Because this is the point where the IR is measured and regulated, we often refer to this point alone as the IR. Using the flashlight analogy, the IR is all of the components up to the lightbulb socket but not the bulb and lens. This is the raw power, or signal, that is provided, and now the bulb and lens can focus the signal. This power level is typically measured in milliwatts (mW) or decibels relative to 1 milliwatt (dBm).

Equivalent Isotropically Radiated Power (EIRP)

Equivalent isotropically radiated power (EIRP) is the highest RF signal strength that is transmitted from a particular antenna. To understand this better, think of our flashlight example for a moment. Let's assume that the bulb without the lens generates 1 watt of power. When you put the lens on the flashlight, it focuses that 1 watt of light. If you were to look at the light now, it would appear much brighter. If you were to measure the brightest point of the light that was being generated by the flashlight, because of the effects of the lens, it may be equal to the brightness of an 8-watt bulb. So by focusing the light, you are able to make the equivalent isotropically radiated power of the focused bulb equal to 8 watts.

It is important for you to know that you can find other references to EIRP as *equivalent isotropic radiated power and effective isotropic radiated power.* The use of EIRP in this book is consistent with the FCC definition, "equivalent isotropically radiated power, the product of the power supplied to the antenna and the antenna gain in a given direction relative to an isotropic antenna." Even though the terms that the initials stand for differ, the definition of EIRP is consistent.

As you learned earlier in this chapter, antennas are capable of focusing, or directing, RF energy. This focusing capability can make the effective output of the antenna much greater than the signal entering the antenna. Because of this ability to amplify the output of the RF signal, regulatory bodies such as the FCC limit the amount of EIRP from an antenna.

In the next section of this chapter, you will learn how to calculate how much power is actually being provided to the antenna (IR) and how much power is coming out of the antenna (EIRP).

Real World Scenario

Why Are IR and EIRP Measurements Important?

As you learned in Chapter 1, the regulatory domain authority in an individual country or region is responsible for maximum transmit power regulations. The FCC and other domain authorities usually define maximum power output for the intentional radiator (IR) and a maximum equivalent isotropically radiated power (EIRP) that radiates from the antenna. In laymen's terms, the FCC regulates the maximum amount of power that goes into an antenna and the maximum amount of power that comes out of an antenna.

You will need to know the definitions of IR and EIRP measurements. However, the CWNA exam (PW0-104) will not test you on any power regulations because they vary from country to country. It is advisable to educate yourself about the maximum transmit power regulations of the country where you plan on deploying a WLAN so that no violations occur.

Units of Power and Comparison

When an 802.11 wireless network is designed, two key components are coverage and performance. A good understanding of RF power, comparison, and RF mathematics can be very helpful during the network design phase.

In this section, we will introduce you to an assortment of units of power and units of comparison. It is important to know and understand the different types of units of measurement and how they relate to each other. Some of the numbers that you will be working with will represent actual units of power, and others will represent relative units of comparison. Actual units are ones that represent a known or set value.

To say that a man is 6 feet tall is an example of an actual measurement. Since the man's height is a known value, in this case feet, you know exactly how tall he is. Relative units are comparative values comparing one item to a similar type of item. For example, if you wanted to tell someone how tall the man's wife is by using comparative units of measurement, you could say that she is 5/6 his height. You now have a comparative measurement: If you know the actual height of either one, you can then determine how tall the other is.

Comparative units of measurement are useful when working with units of power. As you will see later in this chapter, we can use these comparative units of power to compare the area that one access point can cover vs. another access point. Using simple mathematics, we can determine things such as how many watts are needed to double the distance of a signal from an access point.

Units of power are used to measure transmission amplitude and received amplitude. In other words, units of power measurements are *absolute* power measurements. Units of comparison are often used to measure how much gain or loss occurs because of the introduction of cabling or an antenna. Units of comparison are also used to represent a difference in power from point A to point B. In other words, units of comparison are measurements of *change in power.*

Table 3.1 categorizes the different units that are covered in the following sections.

TABLE 3.1 Units of Measure

Units of Power (Absolute)	Units of Comparison (Relative)
watt (W)	decibel (dB)
milliwatt (mW)	dBi
dBm	dBd

Watt

A *watt (W)* is the basic unit of power, named after James Watt, an 18th-century Scottish inventor. One watt is equal to 1 ampere (amp) of current flowing at 1 volt. To give a better explanation of a watt, we will use a modification of the classic water analogy.

Many of you are probably familiar with a piece of equipment known as a power washer. If you are not familiar with it, it is a machine that connects to a water source, such as a garden hose, and enables you to direct a stream of high-pressure water at an object, with the premise that the fast-moving water will clean the object. The success of a power washer is based on two components: the pressure applied to the water, and the volume of water used over a period of time, also known as flow. These two components provide the power of the water stream. If you increase the pressure, you will increase the power of the stream. If you increase the flow of the water, you will also increase the power of the stream. The power of the stream is equal to the pressure times the flow.

A watt is very similar to the output of the power washer. Instead of the pressure generated by the machine, electrical systems have voltage. Instead of water flow, electrical systems have current, which is measured in amps. So the amount of watts generated is equal to the volts times the amps.

Milliwatt (mW)

A *milliwatt (mW)* is also a unit of power. To put it simply, a milliwatt is 1/1,000 of a watt. The reason you need to be concerned with milliwatts is because most of the 802.11 equipment that you will be using transmits at power levels between 1 mW and 100 mW. Remember that the transmit power level of a radio will be attenuated by any cabling and will be amplified by the antenna. Although regulatory bodies such as the FCC may allow intentional radiator (IR) power output of as much as 1 watt, only rarely in point-to-point communications, such as in building-to-building bridge links, would you use 802.11 equipment with more than 250 mW of transmit power.

Decibel (dB)

The first thing you should know about the *decibel (dB)* is that it is a unit of comparison, not a unit of power. Therefore, it is used to represent a difference between two values. In other words, a dB is a relative expression and a measurement of change in power. In wireless networking, decibels are often used either to compare the power of two transmitters or, more often, to compare the difference or loss between the EIRP output of a transmitter's antenna and the amount of power received by the receiver's antenna.

Decibel is derived from the term *bel*. Employees at Bell Telephone Laboratories needed a way to represent power losses on telephone lines as power ratios. They defined a bel as the ratio of 10 to 1 between the power of two sounds. Let's look at an example: An access point transmits data at 100 mW. Laptop1 receives the signal at a power level of 10 mW, and laptop2 receives the signal at a power level of 1 mW. The difference between the signal from the access point (100 mW) to laptop1 (10 mW) is 100:10, or a 10:1 ratio, or 1 bel. The difference between the signal from laptop1 (10 mW) to laptop2 (1 mW) is 10:1, also a 10:1 ratio, or 1 bel. So the power difference between the access point and laptop2 is 2 bels.

Bels can be looked at mathematically by using logarithms. Not everyone understands or remembers logarithms, so we will review them. First, we need to look at raising a number to a power. If you take 10 and raise it to the third power ($10^3 = y$), what you are actually doing is multiplying three 10s ($10 \times 10 \times 10$). If you do the math, you will calculate that y is equal to 1,000. So the solution is $10^3 = 1{,}000$. When calculating logarithms, you change the formula to $10^y = 1{,}000$. Here you are trying to figure out what power 10 needs to be raised to in order to get to 1,000. You know in this example that the answer is 3. You can also write this equation as $y = \log_{10}(1{,}000)$ or $y = \log_{10}1{,}000$. So the complete equation is $3 = \log_{10}(1{,}000)$. Here are some examples of power and log formulas:

$10^1 = 10$	$\log_{10}(10) = 1$
$10^2 = 100$	$\log_{10}(100) = 2$
$10^3 = 1{,}000$	$\log_{10}(1{,}000) = 3$
$10^4 = 10{,}000$	$\log_{10}(10{,}000) = 4$

Now let's go back and calculate the bels from the access point to the laptop2 example by using logarithms. Remember that bels are used to calculate the ratio between two powers. So let's refer to the power of the access point as P_{AP} and the power of laptop1 as P_{L1}. So the formula for this example would be $y = \log_{10}(P_{AP}/P_{L1})$. If you plug in the power values, the formula becomes $y = \log_{10}(100/1)$, or $y = \log_{10}(100)$. So this equation is asking, 10 raised to what power equals 100? The answer is 2 bels ($10^2 = 100$).

OK, so this is supposed to be a section about decibels, but so far we have covered just bels. In certain environments, bels are not exact enough, which is why we use decibels instead. A decibel is equal to one-tenth of a bel. To calculate decibels, all you need to do is multiply bels by 10. So the formulas for bels and decibels are as follows:

$$\text{bels} = \log_{10}(P_1/P_2)$$

$$\text{decibels} = 10 \times \log_{10}(P_1/P_2)$$

Now let's go back and calculate the decibels for the example of the access point to laptop2. So the formula now is $y = 10 \times \log_{10}(P_{AP}/P_{L1})$. If you plug in the power values, the formula becomes $y = 10 \times \log_{10}(100/1)$, or $y = 10 \times \log_{10}(100)$. So the answer is +20 decibels.

You do not need to know how to calculate logarithms for the CWNA exam. These examples are here only to give you some basic understanding of what they are and how to calculate them. Later in this chapter, you will learn how to calculate decibels without using logarithms.

Now that you have learned about decibels, you are probably still wondering why you can't just work with milliwatts. You can if you want, but because power changes are logarithmic, the differences between values can become extremely large and more difficult to deal with. It is easier to say that a 100 mW signal decreased by 70 decibels than to say that it decreased to 0.00001 milliwatts. Table 3.2 compares milliwatts and decibel change, using 1 mW as the reference point. Because of the scale of the numbers, you can see why decibels can be easier to work with.

TABLE 3.2 Comparison of Milliwatts and Decibel Change (Relative to 1 mW)

Milliwatts	Decibel Change
0.0001	–40
0.001	–30
0.01	–20
0.1	–10
1	0
10	+10
100	+20
1,000	+30
10,000	+40
100,000	+50

Real World Scenario

Why Should You Use Decibels?

In Chapter 2, you learned that there are many behaviors of waves that can adversely affect a wave. One of the behaviors that you learned about was free space path loss.

If an access point is transmitting at 100 mW, and a laptop is 100 meters (0.1 kilometer) away from the access point, the laptop is receiving only about 0.000001 milliwatts of power. The difference between the numbers 100 and 0.000001 is so large that it doesn't have much relevance to someone looking at it. Additionally, it would be easy for someone to accidentally leave out a zero when writing or typing 0.00001 (as we just did).

If you use the free space path loss formula to calculate the decibel loss for this scenario, the formula would be

$$\text{decibels} = 32.4 + (20\log_{10}(2{,}400)) + (20\log_{10}(0.1))$$

The answer is a loss of 80.004 dB, which is approximately 80 decibels of loss. This number is easier to work with and less likely to be miswritten or mistyped.

dBi

Earlier in this chapter, we compared an antenna to an isotropic radiator. Theoretically, an isotropic radiator can radiate an equal signal in all directions. An antenna cannot do this because of construction limitations. In other instances, you do not want an antenna to radiate in all directions because you want to focus the signal of the antenna in a particular direction. Whichever the case may be, it is important to be able to calculate the radiating power of the antenna so that you can determine how strong a signal is at a certain distance from the antenna. You may also want to compare the output of one antenna to that of another.

The gain, or increase, of power from an antenna when compared to what an isotropic radiator would generate is known as *decibels isotropic (dBi)*. Another way of phrasing this is *decibel gain referenced to an isotropic radiator* or *change in power relative to an antenna*. Since antennas are measured in gain, not power, you can conclude that dBi is a relative measurement and not an absolute power measurement. dBi is simply a measurement of antenna gain. The dBi value is measured at the strongest point, or the focus point, of the antenna signal. Since antennas always focus their energy more in one direction than another, the dBi value of an antenna is always a positive gain and not a loss. There are, however, antennas with a dBi value of 0, which are often referred to as *no-gain*, or *unity-gain*, antennas.

A common antenna used on access points is the half-wave dipole antenna. The half-wave dipole antenna is a small, typically rubber-encased, general-purpose antenna. A 2.4 GHz half-wave dipole antenna has a dBi value of 2.14.

Any time you see *dBi*, think *antenna gain*.

dBd

The antenna industry uses two dB scales to describe the gain of antennas. The first scale, which you just learned about, is dBi, which is used to describe the gain of an antenna relative to a theoretical isotropic antenna. The other scale used to describe antenna gain is *decibels dipole (dBd)*, or *decibel gain relative to a dipole antenna.* So a dBd value is the increase in gain of an antenna when it is compared to the signal of a dipole antenna. As you will learn in Chapter 4, dipole antennas are also omnidirectional antennas. Therefore, a dBd value is a measurement of omnidirectional antenna gain and not unidirectional antenna gain. Because dipole antennas are measured in gain, not power, you can also conclude that dBd is a relative measurement and not a power measurement.

The definition of dBd seems simple enough, but what happens when you want to compare two antennas, and one is represented with dBi and the other with dBd? This is actually quite simple. A standard dipole antenna has a dBi value of 2.14. If an antenna has a value of 3 dBd, this means that it is 3 dB greater than a dipole antenna. Because the value of a dipole antenna is 2.14 dBi, all you need to do is add 3 to 2.14. So a 3 dBd antenna is equal to a 5.14 dBi.

Don't forget that dB, dBi, and dBd are comparative, or relative, measurements and not units of power.

Real World Scenario

The Real Scoop on dBd

When working with 802.11 equipment, it is not often that you will have an antenna with a dBd value. 802.11 antennas typically are measured using dBi. On the rare occasion that you do run into one, just add 2.14 to the dBd value and you will know the antenna's dBi value.

dBm

Earlier when you read about bels and decibels, you learned that they measured differences or ratios between two signals. Regardless of the type of power that was being transmitted, all you really knew was that the one signal was greater or less than the other by a particular number of bels or decibels. dBm also provides a comparison, but instead of comparing a signal to another signal, it is used to compare a signal to 1 milliwatt of power. *dBm* means *decibels relative to 1 milliwatt.* So what you are doing is setting dBm to 0 (zero) and equating that to 1 milliwatt of power. Because dBm is a measurement that is compared to a known value, 1 milliwatt, then dBm is actually a measure of power. Because decibels (relative) are referenced to 1 milliwatt (absolute), think of a dBm as an absolute assessment that measures change of power referenced to 1 milliwatt. You can now state that 0 dBm is equal to 1 milliwatt. Using the formula $\text{dBm} = 10 \times \log_{10}(P_{mW})$, you can determine that 100 mW of power is equal to +20 dBm.

If you happen to have the dBm value of a device and want to calculate the corresponding milliwatt value, you can do that too. The formula is $P_{mW} = \log^{-1}(P_{dBm} \div 10)$.

Remember that 1 milliwatt is the reference point and that 0 dBm is equal to 1 mW. Any absolute power measurement of +dBm indicates amplitude greater than 1 mW. Any absolute power measurement of –dBm indicates amplitude less than 1 mW. For example, we stated earlier that the transmission amplitude of most 802.11 radios usually ranges from 1 mW to 100 mW. A transmission amplitude of 100 mW is equal to +20 dBm. Because of free space path loss, received signals will always measure below 1 mW. A very strong received signal is –40 dBm, which is the equivalent of 0.0001 mW (1/10,000th of 1 milliwatt).

It might seem a little ridiculous to have to deal with both milliwatts and dBm. If milliwatts are a valid measurement of power, why not just use them? Why do you have to, or want to, also use dBm? These are good questions that are asked often by students. One reason is simply that dBm absolute measurements are often easier to grasp than measurements in the millionths and billionths of a single milliwatt. Most 802.11 radios can interpret received signals from –30 dBm (1/1,000th of 1 mW) to as low as –100 dBm (1/10 of a billionth of 1 mW). The human brain can grasp –100 dBm much easier than 0.0000000001 milliwatts. During a site survey, WLAN engineers will always determine coverage zones by recording the received signal strength in –dBm values.

Another very practical reason to use dBm can be shown using the free space path loss formula again. Following are two free space path loss equations. The first equation calculates the decibel loss of a 2.4 GHz signal at 100 meters (0.1 kilometer) from the RF source, and the second calculates the decibel loss of a 2.4 GHz signal at 200 meters (0.2 kilometer) from the RF source:

$$FSPL = 32.4 + (20\log_{10}(2{,}400)) + (20\log_{10}(0.1)) = 80.00422 \text{ dB}$$

$$FSPL = 32.4 + (20\log_{10}(2{,}400)) + (20\log_{10}(0.2)) = 86.02482 \text{ dB}$$

In this example, by doubling the distance from the RF source, the signal decreased by about 6 dB. If you double the distance between the transmitter and the receiver, the received signal will decrease by 6 dB. No matter what numbers are chosen, if the distance is doubled, the decibel loss will be 6 dB. This rule also implies that if you increase the amplitude by 6 dB, the usable distance will double. This *6 dB rule* is very useful for comparing cell sizes or estimating the coverage of a transmitter. The 6 dB rule is also useful for understanding antenna gain, because every 6 dBi of extra antenna gain will double the usable distance of an RF signal. Remember, if you were working with milliwatts, this rule would not be relevant. By converting milliwatts to dBm, you have a more practical way to compare signals.

Remember the *6 dB rule:* +6 dB doubles the distance of the usable signal; –6 dB halves the distance of the usable signal.

Using dBm also makes it easy to calculate the effects of antenna gain on a signal. If a transmitter generates a +20 dBm signal and the antenna adds 5 dBi of gain to the signal, then the power that is radiating from the antenna (EIRP) is equal to the sum of the two numbers, which is +25 dBm.

Inverse Square Law

You just learned about the 6 dB rule, which states that a +6 dB change in signal will double the usable distance of a signal, and a –6 dB change in signal will halve the usable distance of a signal. This rule and these numbers are based on the *inverse square law*, originally developed by Isaac Newton.

This law states that the change in power is equal to 1 divided by the square of the change in distance. In other words, as the distance from the source of a signal doubles, the energy is spread out over four times the area, resulting in one-fourth of the original intensity of the signal.

This means that if you are receiving a signal at a certain power level and a certain distance (D) and you double the distance (2 × D), the new power level will change by $1 \div (2 \times D)^2$. If at a distance of 1 foot (call this D) you were receiving a signal of 4 mW, then at a distance of 2 feet (2 × D) the power would change by $1 \div 2^2$, which is ¼. So the power at 2 feet is 4 mW × ¼, which is equal to 1 mW.

Let us also review the formula for free space path loss:

$$FSPL = 36.6 + (20\log_{10}(f)) + (20\log_{10}(D))$$

FSPL = path loss in dB

f = frequency in MHz

D = distance in miles between antennas

$$FSPL = 32.4 + (20\log_{10}(f)) + (20\log_{10}(D))$$

FSPL = path loss in dB

f = frequency in MHz

D = distance in kilometers between antennas

The concept of free space path loss is also based on Newton's inverse square law. The main variable for the inverse square law is simply distance. The FSPL formula is also based on distance but adds another variable, which is frequency.

RF Mathematics

When the topic of RF mathematics is discussed, most people cringe and panic because they expect formulas that have logarithms in them. Fear not. You are about to learn RF math, without having to use logarithms. If you want to refresh yourself on some of your math skills prior to going through this section, review the following:

- Addition and subtraction using the numbers 3 and 10
- Multiplication and division using the numbers 2 and 10

No, we are not kidding. If you know how to add and subtract using 3 and 10 and if you know how to multiply and divide using 2 and 10, you have all of the math skills you need to perform RF math. Read on, and we will teach you how.

Rule of 10s and 3s

Before you fully delve into the *rule of 10s and 3s*, it is important to know that this rule may not give you the exact same answers that you would get if you used the logarithmic formulas. The rule of 10s and 3s provides approximate values, not necessarily exact values. If you are an engineer creating a product that must conform to RF regulatory guidelines, you will need to use logarithms to calculate the exact values. However, if you are a network designer planning a network for your company, you will find that the rule of 10s and 3s will provide you with the numbers you need to properly plan your network.

This section will take you step-by-step through numerous calculations. All of the calculations will be based on the following four rules of the 10s and 3s:

- For every 3 dB of gain (relative), double the absolute power (mW).
- For every 3 dB of loss (relative), halve the absolute power (mW).
- For every 10 dB of gain (relative), multiply the absolute power (mW) by a factor of 10.
- For every 10 dB of loss (relative), divide the absolute power (mW) by a factor of 10.

For example, if your access point is configured to transmit at 100 mW and the antenna is rated for 3 dBi of passive gain, the amount of power that will radiate out of the antenna (EIRP) will be 200 mW. Following the rule that you just learned, you will see that the 3 dB of gain from the antenna caused the 100 mW signal from the access point to double. Conversely, if your access point is configured to transmit at 100 mW and is attached to a cable that introduces 3 dB of loss, the amount of absolute amplitude at the end of the cable will be 50 mW. Here you can see that the 3 dB of loss from the cable caused the 100 mW signal from the access point to be halved.

In another example, if your access point is configured to transmit at 40 mW and the antenna is rated for 10 dBi of passive gain, the amount of power that radiates out of the antenna (EIRP) will be 400 mW. Here you can see that the 10 dB of gain from the antenna caused the 40 mW signal from the access point to increase by a factor of 10. Conversely, if your access point is configured to transmit at 40 mW and is attached to a cable that introduces 10 dB of loss, the amount of absolute amplitude at the end of the cable will be 4 mW. Here you can see that the 10 dB of loss from the cable caused the 40 mW signal from the access point to be decreased by a factor of 10.

If you remember these rules, you will be able to quickly perform RF calculations. After reviewing these rules, continue reading this chapter for a step-by-step procedure for using the rule of 10s and 3s. As you work through the step-by-step procedures, remember that dBm is a unit of power and that dB is a unit of change. dB is a value of change that can be applied to dBm. So if you have +10 dBm and it increases by 3 dB, you can add these two numbers together to get a result of +13 dBm.

EXERCISE 3.1

Step-by-Step Procedure

1. On a sheet of paper, create two columns. The header of the first column should be **dBm**, and the header of the second column should be **mW**.

2. Next to the dBm header, place a + and – sign, and next to the mW header place a × and ÷ sign. These will help you to remember that all math performed on the dBm column is addition or subtraction, and all math performed on the mW column is multiplication or division.

	+ / –	dBm	mW	× / ÷	
	+			×	
	–	dBm	mW	÷	

3. To the left of the + and – signs, write the numbers **3** and **10**, and to the right of the × and ÷ signs, write the numbers **2** and **10**. Any addition or subtraction to the dBm column can be performed using only the numbers 3 and 10. Any multiplication or division to the mW column can be performed using only the numbers 2 and 10.

3	+			×	2
10	–	dBm	mW	÷	10

4. If there is a + on the left, there needs to be a × on the right. If there is a – on the left, there needs to be a ÷ on the right. If you are adding or subtracting a 3 on the left, you must be multiplying or dividing by a 2 on the right. If you are adding or subtracting a 10 on the left, you must be multiplying or dividing by a 10 on the right.

5. The last thing you need to do is to put a **0** under the dBm column and a **1** under the mW column. Remember that the definition of dBm is *decibels relative to 1 milliwatt.* So now the chart shows that 0 dBm is equal to 1 milliwatt.

3	+			×	2
10	–	dBm	mW	÷	10
		0	1		

Before we continue with other examples, it is important to emphasize that a change of ±3 dB equates to a doubling or halving of the power, no matter what power measurement is being used. In our usage of the rule of 10s and 3s, we are dealing with milliwatts because that is the typical transmission amplitude measurement used by 802.11 equipment. However, it is important to remember that a +3 dB increase means a doubling of the power regardless of the power scale used. So a +3 dB increase of 1.21 gigawatts of power would result in 2.42 gigawatts of power.

An animated explanation of the rule of 10s and 3s—as well as explanations of the following examples—has been created using Microsoft PowerPoint and can be run from the CD that you received with this book. If you do not have PowerPoint on your computer, you can download from Microsoft's website a PowerPoint Viewer that will allow you to view any PowerPoint file.

EXERCISE 3.2

Rule of 10s and 3s, Example 1

In this example, you will begin at 1 mW and double the power three times. In addition to calculating the new power level in milliwatts, you will calculate the power level in dBms.

1. The first thing to do is create the initial chart, as in Exercise 3.1.

3 + × 2
10 − ÷ 10

dBm	mW
0	1

2. Now you want to double the power for the first time. So to the right of the 1 mW and on the next line, write × 2. Then below the 1, perform the calculation.

3 + × 2
10 − ÷ 10

dBm	mW	
0	1 →	
	= 2 ←	× 2

3. You are not finished yet with this new line. Remember that for whatever is done to one side of the chart, there must be a correlative mathematical equation on the other side. Because you multiplied by 2 on the right side, you must add 3 to the left side. So you have just calculated that +3 dBm is equal to 2 mW.

3 + × 2
10 − ÷ 10

	dBm	mW	
	← 0	1 →	
+ 3 →	= 3	= 2 ←	× 2

EXERCISE 3.2 *(continued)*

4. You have just completed the first doubling of the power. Now you will double it two more times and perform the necessary mathematical commands. Since this is the first time using this process, all of the steps have been shown using arrows. Future examples will not contain these arrows.

3 +			× 2
10 −	dBm	mW	÷ 10
	0	1	
	3	2	
+3	= 6	= 4	× 2
+3	= 9	= 8	× 2

You have just calculated that 4 mW = +6 dBm, and 8 mW = +9 dBm. If you had used the formula for dBm instead of the rule of 10s and 3s, the actual answers would be 4 mW = +6.0206 dBm, and 8 mW = +9.0309 dBm. As you can see, this set of rules is accurate but not exact.

EXERCISE 3.3

Rule of 10s and 3s, Example 2

You have a wireless bridge that generates a 100 mW signal. The bridge is connected to an antenna via cable that creates –3 dB of signal loss. The antenna provides 10 dBi of signal gain. In this example, calculate the IR and EIRP values.

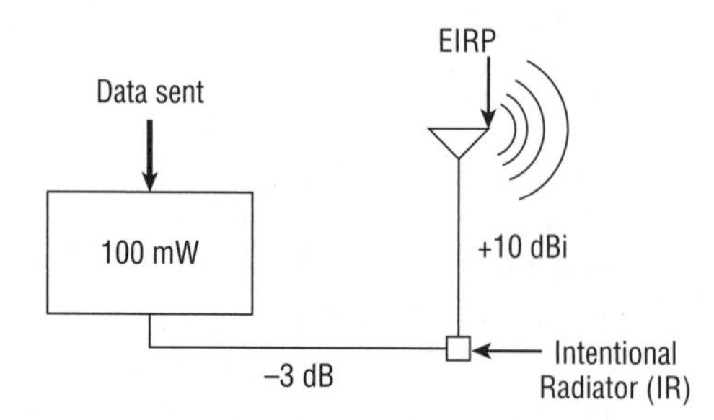

As a reminder, and as seen in the graphic, the IR is the signal up to but not including the antenna, and the EIRP is the signal radiating from the antenna.

EXERCISE 3.3 *(continued)*

1. The first step is to determine whether by using 10 or 2, and × or ÷, you can go from 1 mW to 100 mW. It is not too difficult to realize that multiplying 1 by 10 twice will give you 100. So the bridge is generating 100 mW, or +20 dBm, of power.

3 + / 10 -	dBm	mW	× 2 / ÷ 10
	0	1	
+ 10	10	10	× 10
+ 10	20	100	× 10

2. Next you have the antenna cable, which is introducing –3 dB of loss to the signal. After you calculate the effect of the –3 dB loss, you know the value of the IR. You can represent the IR as either +17 dBm or 50 mW.

3 + / 10 -	dBm	mW	× 2 / ÷ 10
	0	1	
+ 10	10	10	× 10
+ 10	20	100	× 10
- 3	17	50	÷ 2

3. Now all that is left is to calculate the increase on the signal due to the gain from the antenna. Because the gain is 10 dBi, you add 10 to the dBm column and multiply the mW column by 10. This gives you an EIRP of +27 dBm, or 500 mW.

3 + / 10 -	dBm	mW	× 2 / ÷ 10
	0	1	
+ 10	10	10	× 10
+ 10	20	100	× 10
- 3	17	50	÷ 2
+ 10	27	500	× 10

So far all of the numbers chosen in the examples have been very straightforward, using the values that are part of the template. However, in the real world, this will not be the case. Using a little creativity, you can calculate gain or loss for any integer. Unfortunately, the rule of 10s and 3s does not work for fractional or decimal numbers. For those numbers, you need to use the logarithmic formula.

dB gain or loss is cumulative. If, for example, you had three sections of cable connecting the transceiver to the antenna and each section of cable provided 2 dB of loss, all three cables would create 6 dB of loss. All you have to do to calculate the loss is to subtract 3 twice. Decibels are very flexible. As long as you come up with the total that you need, they don't care how you do it.

Table 3.3 shows how to calculate all integer dB loss and gain from –10 to +10 by using combinations of just 10s and 3s. Take a moment to look at these values and you will realize that with a little creativity, you can calculate the loss or gain of any integer.

TABLE 3.3 dB Loss and Gain (–10 through +10)

Loss or Gain (dB)	Combination of 10s and 3s
–10	–10
–9	–3 –3 –3
–8	–10 –10 +3 +3 +3 +3
–7	–10 +3
–6	–3 –3
–5	–10 –10 +3 +3 +3 +3 +3
–4	–10 +3 +3
–3	–3
–2	–3 –3 –3 –3 +10
–1	–10 +3 +3 +3
+1	+10 –3 –3 –3
+2	+3 +3 +3 +3 –10
+3	+3
+4	+10 –3 –3
+5	+10 +10 –3 –3 –3 –3 –3
+6	+3 +3
+7	+10 –3
+8	+10 +10 –3 –3 –3 –3
+9	+3 +3 +3
+10	+10

EXERCISE 3.4

Rule of 10s and 3s, Example 3

This example is a little more complicated. You have an access point that is transmitting at 50 mW. The signal loss between the access point and the antenna is –1 dB, and the access point is using a +5 dBi antenna. In this example, calculate the IR and the EIRP values.

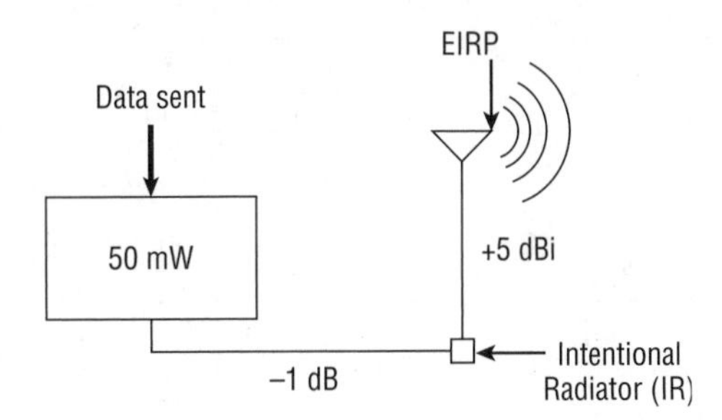

1. The first step after drawing up the template is to convert the 1 mW to 50 mW. This can be done by multiplying the 1 mW by 10 twice and then dividing by 2.

2. The dBm column then needs to be adjusted by adding 10 twice and subtracting 3. When the calculations are more complex, it's useful to separate and label the different sections.

3 + / 10 –	dBm	mW	× 2 / ÷ 10	
	0	1		
+10	10	10	× 10	
+10	20	100	× 10	Transmitter
– 3	17	50	÷ 2	

3. The signal loss between the access point and the antenna is –1 dB. Table 3.3 shows that –1 dB can be calculated by subtracting 10 and adding 3 three times.

4. The mW column will need to be adjusted by dividing by 10 and then multiplying by 2 three times. So the IR is either +16 dBm or 40 mW.

3 + / 10 –	dBm	mW	× 2 / ÷ 10	
	0	1		
+10	10	10	× 10	
+10	20	100	× 10	Transmitter
– 3	17	50	÷ 2	
– 10	7	5	÷ 10	
+3	10	10	× 2	Connector
+3	13	20	× 2	
+3	16	40	× 2	

EXERCISE 3.4 *(continued)*

5. The antenna adds a gain of 5 dBi. Table 3.3 shows that +5 dBi can be calculated by adding 10 twice and subtracting 3 five times.

6. The mW column will need to be adjusted by multiplying by 10 twice and dividing by 2 five times. The EIRP is therefore either +21 dBm or 125 mW.

3 + / 10 −	dBm	mW	× 2 / ÷ 10	
	0	1		
+ 10	10	10	× 10	
+ 10	20	100	× 10	Transmitter
− 3	17	50	÷ 2	
− 10	7	5	÷ 10	
+ 3	10	10	× 2	Connector
+ 3	13	20	× 2	
+ 3	16	40	× 2	
+ 10	26	400	× 10	
+ 10	36	4000	× 10	Antenna
− 3	33	2000	÷ 2	
− 3	30	1000	÷ 2	
− 3	27	500	÷ 2	
− 3	24	250	÷ 2	
− 3	21	125	÷ 2	

EXERCISE 3.5

Rule of 10s and 3s, Example 4

In this example, you have an access point that is providing coverage to a specific area of a warehouse via an external directional antenna. The access point is transmitting at 30 mW. The cable and connector between the access point and the antenna creates 3 dB of signal loss. The antenna provides 20 dBi of signal gain. In this example, you will calculate the IR and EIRP values.

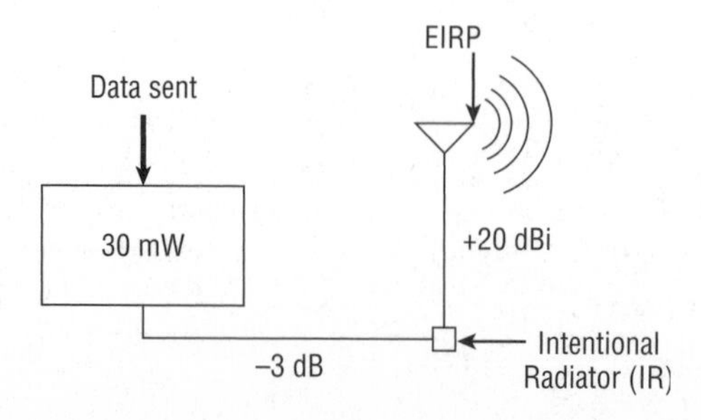

EXERCISE 3.5 *(continued)*

It's not always possible to calculate both sides of the chart by using the rule of 10s and 3s. In some cases, no matter what you do, you cannot calculate the mW value by using 10 or 2. This is one of those cases. You cannot set the mW and dBm values to be equal, but you can still calculate the mW values by using the information provided.

1. Instead of creating the template and setting 0 dBm equal to 1 mW, enter the value of the transmitter, in this case 30 mW. In the dBm column, just write **unknown**. Even though you will not know the dBm value, you can still perform all of the necessary mathematics.

3 + 10 -	dBm	mW	× 2 ÷ 10
	unknown	30	

2. The cable and connectors introduce 3 dB of loss, so subtract 3 from the dBm column and divide the mW column by 2. So the output of the IR is 15 mW.

3 + 10 -	dBm	mW	× ÷ 1
	unknown	30	
- 3	unknown - 3	15	÷ 2

3. The 20 dBi gain from the antenna increases the dBm by 20, so add 10 twice to the dBm column, and multiply the mW column by 10 twice. So the output of the EIRP is 1,500 mW.

3 + 10 -	dBm	mW	× 2 ÷ 10
	unknown	30	
- 3	unknown - 3	15	÷ 2
+ 10	unknown + 7	150	× 10
+ 10	unknown + 17	1,500	× 10

RF Math Summary

Many concepts, formulas, and examples were covered in the RF mathematics section, so we will bring things together and summarize what was covered. It is important to remember that the bottom line is that you are trying to calculate the power at different points in the RF system and the effects caused by gain or loss. If you want to perform the RF math calculations by using the logarithmic formulas, here they are:

$$\text{dBm} = 10 \times \log_{10}(\text{mW})$$

$$\text{mW} = \log^{-1}(\text{dBm} \div 10) = 10^{(\text{dBm} \div 10)}$$

If you want to use the rule of 10s and 3s, just remember these four simple tasks, and you will not have a problem.

- 3 dB gain = mW × 2
- 3 dB loss = mW ÷ 2
- 10 dB gain = mW × 10
- 10 dB loss = mW ÷ 10

Table 3.4 provides a quick reference guide comparing the absolute power measurements of milliwatts to the absolute power dBm values.

TABLE 3.4 dBm and Milliwatt Conversions

dBm	Milliwatts	
+ 36 dBm	4,000 mW	4 watts
+ 30 dBm	1,000 mW	1 watt
+ 20 dBm	100 mW	1/10th of 1 watt
+ 10 dBm	10 mW	1/100th of 1 watt
0 dBm	1 mW	1/1,000th of 1 watt
–10 dBm	0.1 mW	1/10th of 1 milliwatt
–20 dBm	0.01 mW	1/100th of 1 milliwatt
–30 dBm	0.001 mW	1/1,000th of 1 milliwatt
–40 dBm	0.0001 mW	1/10,000th of 1 milliwatt
–50 dBm	0.00001 mW	1/100,000th of 1 milliwatt
–60 dBm	0.000001 mW	1 millionth of 1 milliwatt
–70 dBm	0.0000001 mW	1 ten-millionth of 1 milliwatt
–80 dBm	0.00000001 mW	1 hundred-millionth of 1 milliwatt
–90 dBm	0.000000001 mW	1 billionth of 1 milliwatt

Received Signal Strength Indicator (RSSI)

Receive sensitivity refers to the power level of an RF signal required to be successfully received by the receiver radio. The lower the power level that the receiver can successfully process, the better the receive sensitivity. Think of this as being at a hockey game. There is an ambient level of noise that exists from everything around you. There is a certain volume that you have to speak for your neighbor to hear you. That level is the receiver sensitivity. It is the weakest signal that the transceiver can decode under normal circumstances. With that said, if the noise in a particular area is louder than normal, then the minimum level you have to yell gets louder.

In WLAN equipment, receive sensitivity is usually defined as a function of network speed. Wi-Fi vendors will usually specify their receive sensitivity thresholds at various data rates, as seen in Table 3.5. For any given receiver, more power is required by the receiver radio to support the higher data rates. Different speeds use different modulation techniques and encoding methods, and the higher data rates use encoding methods that are more susceptible to corruption. The lower data rates use modulation-encoding methods that are less susceptible to corruption.

TABLE 3.5 Receive Sensitivity Thresholds (Example)

Data Rate	Received Signal Amplitude
54 Mbps	–50 dBm
48 Mbps	–55 dBm
36 Mbps	–61 dBm
24 Mbps	–74 dBm
18 Mbps	–70 dBm
12 Mbps	–75 dBm
9 Mbps	–80 dBm
6 Mbps	–86 dBm

The 802.11-2007 standard defines the *received signal strength indicator (RSSI)* as a relative metric used by 802.11 radios to measure signal strength (amplitude). The 802.11 RSSI measurement parameter can have a value from 0 to 255. The RSSI value is designed to be used by the WLAN hardware manufacturer as a relative measurement of the RF signal strength that is received by an 802.11 radio. RSSI metrics are typically mapped to receive sensitivity thresholds expressed in absolute dBm values, as shown in Table 3.6. For example, an RSSI metric of 255 might represent –30 dBm of received signal amplitude. The RSSI metric of 0 might be mapped to –110 dBm of received signal amplitude.

TABLE 3.6 Received Signal Strength Indicator (RSSI) Metrics (Vendor Example)

RSSI	Receive Sensitivity Threshold	Signal Strength (%)	Signal-to-Noise Ratio	Signal Quality (%)
30	–30 dBm	100%	70 dB	100%
25	–41 dBm	90%	60 dB	100%
20	–52 dBm	80%	43 dB	90%
21	–52 dBm	80%	40 dB	80%
15	–63 dBm	60%	33 dB	50%
10	–75 dBm	40%	25 dB	35%
5	–89 dBm	10%	10 dB	5%
0	–110 dBm	0%	0 dB	0%

The 802.11-2007 standard also defines another metric called *signal quality (SQ)*, which is a measure of pseudonoise (PN) code correlation quality received by a radio. In simpler terms, the signal quality could be a measurement of what might affect coding techniques such as the Barker code or Complementary Code Keying (CCK). Anything that might increase the bit error rate (BER) such as a low SNR or multipath might trigger SQ metrics.

Information parameters from both RSSI and SQ metrics can be passed along from the PHY layer to the MAC sublayer. Some SQ parameters might also be used in conjunction with RSSI as part of a clear channel assessment (CCA) scheme. Although SQ metrics and RSSI metrics are technically separate measurements, most Wi-Fi vendors refer to both together as simply *RSSI metrics*. For the purposes of this book, whenever we refer to RSSI metrics, we are referring to both SQ and RSSI metrics.

Although not technically correct, many Wi-Fi vendors define signal quality as the *signal-to-noise ratio (SNR)*. The signal-to-noise ratio is simply the difference in decibels between the received signal and the background noise (noise floor, or most important, the receiver threshold). For example, if a radio receives a signal of –85 dBm and the noise floor is measured at –100 dBm, the difference between the received signal and the background noise is 15 dB. The SNR is 15 dB. Data transmissions can become corrupted with a very low SNR. If the amplitude of the noise floor is too close to the amplitude of the received signal, data corruption will occur and result in layer 2 retransmissions. An SNR of 25 dB or greater is considered good signal quality, and an SNR of 10 dB or lower is considered poor signal quality.

Although RSSI metrics are meant to represent only signal amplitude, WLAN vendors often use RSSI metrics to represent both received signal amplitude and SNR. For example, an RSSI metric of 125 might represent –65 dBm of received signal amplitude and an SNR value of 30 dB. The RSSI metric of 124 might also be mapped to –65 dBm of received signal amplitude but be mapped to an SNR value of 29 dB. According to the 802.11-2007 standard, "the RSSI is a measure of the RF energy received. Mapping of the RSSI values to actual received power is implementation dependent." In other words, WLAN vendors can define RSSI metrics in a proprietary manner. The actual range of the RSSI value is from 0 to a maximum value (less than or equal to 255) that each vendor can choose on its own (known as RSSI_Max). Many vendors publish their implementation of RSSI values in product documents and/or on the vendor's website. Some WLAN vendors do not publish their RSSI metrics.

Because the implementation of RSSI metrics is proprietary, two problems exist when trying to compare RSSI values between different manufacturers' wireless cards. The first problem is that the manufacturers may have chosen two different values as the RSSI_Max. So WLAN vendor A may have chosen a scale from 0 to 100, whereas WLAN vendor B may have chosen a scale from 0 to 30. Because of the difference in scale, WLAN vendor A may indicate a signal with an RSSI value of 25, whereas vendor B may indicate that same signal with a different RSSI value of 8. Also, the radio card manufactured by WLAN vendor A uses more RSSI metrics and is probably more sensitive when evaluating signal quality and SNR.

The second problem with RSSI is that the manufacturer could take their range of RSSI values and compare them to a different range of values. So WLAN vendor A may take its 100-number scale and relate it to dBm values of –110 dBm to –10 dBm, whereas WLAN vendor B may take its 60-number scale and relate it to dBm values of –95 dBm to –35 dBm. So not only do we have different numbering schemes, but we also have different ranges of values.

Although the way in which Wi-Fi vendors implement RSSI may be proprietary, most vendors are alike in that they use RSSI thresholds for very important mechanisms such as roaming and dynamic rate switching. During the *roaming* process, clients make the decision to move from one access point to the next. RSSI thresholds are key factors for clients when they initiate the roaming handoff. RSSI thresholds are also used by vendors to implement *dynamic rate switching (DRS)*, which is a process used by 802.11 radios to shift between data rates. Roaming is discussed in several chapters of this book, and dynamic rate switching is discussed in greater detail in Chapter 12 "WLAN Troubleshooting."

Link Budget

When deploying radio communications, a *link budget* is the sum of all gains and losses from the transmitting radio, through the RF medium, to the receiver radio. The purpose of link budget calculations is to guarantee that the final received signal amplitude is above the receiver sensitivity threshold of the receiver radio.

Link budget calculations include original transmit gain, passive antenna gain, and active gain from RF amplifiers. All gain must be accounted for, including RF amplifiers and antennas, and all losses must be accounted for, including attenuators, FSPL, and *insertion loss*. Any hardware device installed in a radio system adds a certain amount of signal attenuation called insertion loss. Cabling is rated for dB loss per 100 feet, and connectors typically add about 0.5 dB of insertion loss.

You have already learned that RF also attenuates as it travels through free space. Figure 3.2 depicts a point-to-point wireless bridge link and shows that loss occurs as the signal moves through various RF components as well as the signal loss caused by FSPL.

FIGURE 3.2 Link budget components

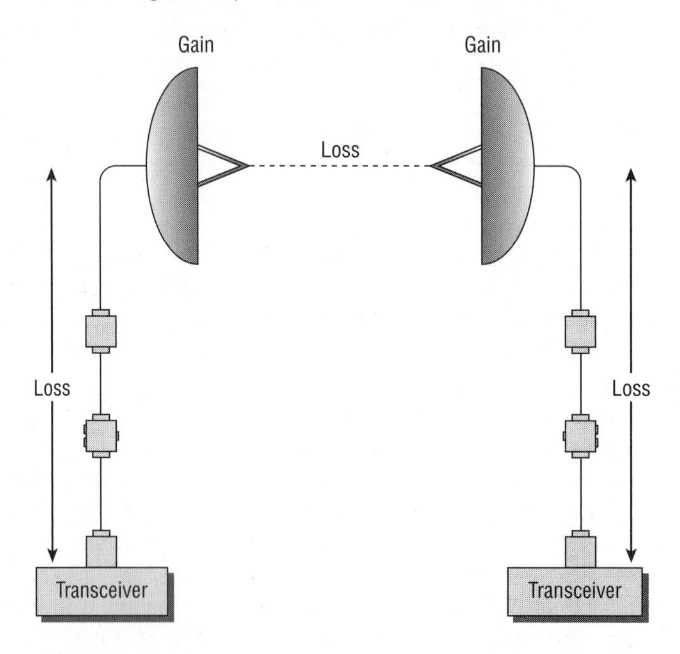

Let's look at the link budget calculations of a point-to-point wireless bridge link, as seen in Figure 3.3 and Table 3.7. In this case, the two antennas are 10 kilometers apart, and the original transmission is +10 dBm. Notice the amount of insertion loss caused by each RF component such as the cabling and the lightning arrestors. The antennas passively amplify the signal, and the signal attenuates as it travels through free space. The final received signal at the receiver end of the bridge link is –65.5 dBm. Now, let's assume that the receive sensitivity threshold of the receiver radio is –80 dBm. Any signal received with amplitude above –80 dBm can be understood by the receiver radio, while any amplitude below –80 dBm cannot be understood. The link budget calculations determined that the final received signal is –65.5 dBm, which is well above the receive sensitivity threshold of –80 dBm. There is almost a 15 dB buffer between the final received signal and the receive sensitivity threshold. The 15 dB buffer that was determined during link budget calculations is known as the *fade margin*, which is discussed in the next section of this chapter.

FIGURE 3.3 Point-to-point link budget gain and loss

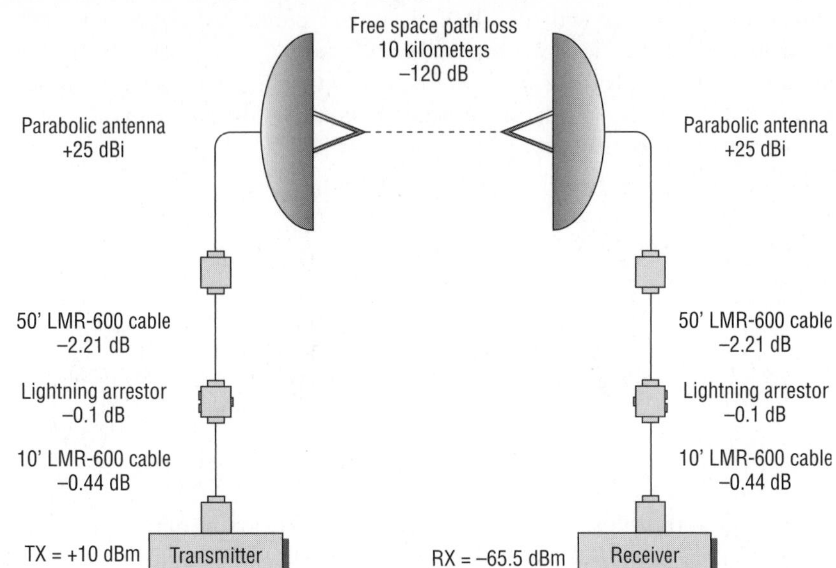

TABLE 3.7 Link Budget Calculations

Component	Gain or Loss	Signal Strength
Transceiver (original transmission signal)		+10 dBm
10' LMR-600 cable	–0.44 dB	+9.56 dBm
Lightning arrestor	–0.1 dB	+9.46 dBm
50' LMR-600 cable	–2.21 dB	+7.25 dBm
Parabolic antenna	+25 dBi	+32.25 dBm
Free space path loss	–120 dB	–87.75 dBm
Parabolic antenna	+25 dBi	–62.75 dBm
50' LMR-600 cable	–2.21 dB	–64.96 dBm
Lightning arrestor	–0.1 dB	–65.06 dBm
10' LMR-600 cable	–0.44 dB	–65.5 dBm
Receiver (final received signal)		–65.5 dBm

You may be wondering why these numbers are negative when up until now most of the dBm numbers you have worked with have been positive. Figure 3.4 shows a simple summary of the gains and losses in an office environment. Until now you have worked primarily with calculating the IR and EIRP. It is the effect of free space path loss that makes the values negative, as you will see in the calculations based on Figure 3.4. In this example, the received signal is the sum of all components, which is

+20 dBm + 5 dBi – 73.98 dB + 2.14 dBi = –46.84 dBm

FIGURE 3.4 Office link budget gain and loss

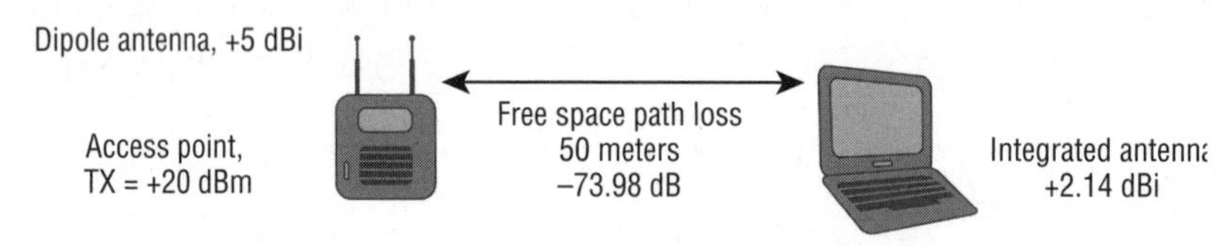

Although the initial transmission amplitude will almost always be above 0 dBm (1 mW), the final received signal amplitude will always be well below 0 dBm (1 mW) because of free space path loss.

Fade Margin/System Operating Margin

Fade margin is a level of desired signal above what is required. A good way to explain fade margin is to think of it as a comfort zone. If a receiver has a receive sensitivity of –80 dBm, a transmission will be successful as long as the signal received is greater than –80 dBm. The problem is that the signal being received fluctuates because of many outside influences such as interference and weather conditions. To accommodate for the fluctuation, it is a common practice to plan for a 10 dB to 25 dB buffer above the receive sensitivity threshold of a radio used in a bridge link. The 10 dB to 25 dB buffer above the receive sensitivity threshold is the fade margin.

A fade margin of 10 dB is an absolute minimum. This would only be acceptable for links less than 3 miles or so. Up to 5 miles should have at least a 15 dB fade margin, and links greater than that should be higher. A fade margin of 25 dB is recommended for links greater than 5 miles.

Let's say that a receiver has a sensitivity of –80 dBm, and a signal is typically received at –76 dBm. Then under normal circumstances, this communication is successful. However, because of outside influences, the signal may fluctuate by ± 10 dB. This means that most of the time, the communication is successful, but on those occasions that the signal has fluctuated to –86 dBm, the communication will be unsuccessful. By adding a fade margin of 20 dB in your link budget calculations, you are now stating that for your needs, the receive sensitivity is –60 dBm, and you will plan your network so that the received signal is greater than –60 dBm. If the received signal fluctuates, you have already built in some padding, in this case 20 dB.

If you look back at Figure 3.3, and you required a fade margin of 10 dB above the receive sensitivity of –80 dBm, the amount of signal required for the link would be –70 dBm. Since the signal is calculated to be received at –65.5 dBm, you will have a successful communication. However if you chose a fade margin of 20 dB, the amount of signal required would be –60 dBm, and based on the configuration in Figure 3.3, you would not have enough signal to satisfy the link budget plus the 20 dB fade margin.

Because RF communications can be affected by many outside influences, it is common to have a fade margin to provide a level of link reliability. By increasing the fade margin, you are essentially increasing the reliability of the link. Think of a fade margin as a buffer or a margin of error for received signals when designing an RF system. The fade margin buffer is also known as a *system operating margin (SOM)* when calculating a link budget.

Real World Scenario

When Are Fade Margin/SOM Calculations Needed?

Whenever an outdoor WLAN bridge link is designed, link budget and SOM calculations will be an absolute requirement. For example, an RF engineer may perform link budget calculations for a 25-mile point-to-point bridge link and determine that the final received signal is 5 dB above the receive sensitivity threshold of a radio at one end of a bridge link. It would seem that RF communications will be just fine; however, because of downfade caused by multipath and weather conditions, a fade margin buffer is needed. A torrential downpour can attenuate a signal as much as 0.08 dB per mile (0.05 dB per kilometer) in both the 2.4 GHz and 5 GHz frequency ranges. Over long-distance bridge links, an SOM of 25 dB is usually recommended to compensate for attenuation due to changes in RF behaviors such as multipath, and due to changes in weather conditions such as rain, fog, or snow.

When deploying a WLAN indoors where high-multipath or high-noise floor conditions exist, the best practice is to plan for an SOM of about 5 dB above the vendor's recommended receive sensitivity amplitude. For example, a –65 dBm or stronger signal falls above the RSSI threshold for the 11 Mbps data rate for most WLAN vendor radios. During the indoor site survey, RF measurements of –65 dBm will often be used to determine coverage areas with a data rate of 11 Mbps. In a high-multipath or noisy environment, RF measurements of –60 dBm utilizing a 5 dB fade margin is a recommended best practice.

EXERCISE 3.6

Link Budget and Fade Margin

In this exercise, you will use a Microsoft Excel file to calculate a link budget and fade margin. You will need Excel installed on your computer.

1. From the CD that is included with this book, copy the file `LinkBudget.xls` to your desktop. Open the Excel file from your desktop.

EXERCISE 3.6 *(continued)*

2. In row 10, enter a link distance of 25 kilometers. Note that the path loss due to a 25 kilometer link is now 128 dB in the 2.4 GHz frequency. In row 20, enter 128 for path loss in dB.
3. In row 23, change the radio receiver sensitivity to –80 dBm. Notice that the final received signal is now –69 dBm, and the fade margin is only 11 dB. Try to change the various components such as antenna gain and cable loss to ensure a fade margin of 20 dB.

Summary

This chapter covered six key areas of RF communications:

- RF components
- RF measurements
- RF mathematics
- RSSI thresholds
- Link budgets
- Fade margins

It is important to understand how each of the RF components affects the output of the transceiver. Whenever a component is added, removed, or modified, the output of the RF communications is changed. You need to understand these changes and make sure that the system conforms to regulatory standards. The following RF components were covered in this chapter:

- Transmitter
- Receiver
- Antenna
- Isotropic radiator
- Intentional radiator (IR)
- Equivalent isotropically radiated power (EIRP)

In addition to understanding the components and their effects on the transmitted signal, you must know the different units of power and comparison that are used to measure the output and the changes to the RF communications:

- Units of power
  - Watt
  - Milliwatt
  - dBm
- Units of comparison
  - dB
  - dBi
  - dBd

After you become familiar with the RF components and their effects on RF communications, and you know the different units of power and comparison, you need to understand how to perform the actual calculations and determine whether your RF communication will be successful. It is important to know how to perform the calculations and some of the terms and concepts involved with making sure that the RF link will work properly. These concepts and terms are as follows:

- Rule of 10s and 3s
- Receive sensitivity
- Received signal strength indicator (RSSI)
- Link budget
- System operating margin (SOM)/fade margin

Exam Essentials

Understand the RF components. Know the function of each of the components and which components add gain and which components add loss.

Understand the units of power and comparison. Make sure you are very comfortable with the difference between units of power (absolute) and units of comparison (relative). Know all of the units of power and comparison, what they measure, and how they are used.

Be able to perform RF mathematics. There will be no logarithms on the test; however, you must know how to use the rule of 10s and 3s. You will need to be able to calculate a result based on a scenario, power value, or comparative change.

Understand the practical uses of RF mathematics. When all is said and done, the ultimate question is, Will the RF communication work? This is where an understanding of RSSI, SOM, fade margin, and link budget is important.

Define RSSI. Understand that RSSI metrics are used by radios to interpret signal strength and quality. 802.11 radios use RSSI metrics for decisions such as roaming and dynamic rate switching.

Understand the necessity of a link budget and fade margin. A link budget is the sum of all gains and losses from the transmitting radio, through the RF medium, to the receiver radio. The purpose of link budget calculations is to guarantee that the final received signal amplitude is above the receiver sensitivity threshold of the receiver radio. Fade margin is a level of desired signal above what is required.

Key Terms

Before you take the exam, be certain you are familiar with the following terms:

6 dB rule

antenna

bel

dBm

decibel (dB)

decibels dipole (dBd)

decibels isotropic (dBi)

dynamic rate switching(DRS)

equivalent isotropically radiated power (EIRP)

fade margin

insertion loss

intentional radiator (IR)

inverse square law

isotropic radiator

link budget

milliwatt (mW)

point source

receive sensitivity

received signal strength indicator (RSSI)

receiver

roaming

rule of 10s and 3s

signal quality (SQ)

signal-to-noise ratio (SNR)

system operating margin (SOM)

transceiver

transmitter

watt (W)

Review Questions

1. What RF component is responsible for generating the AC signal?

    A. Antenna

    B. Receiver

    C. Transmitter

    D. Transponder

2. A point source that radiates signal equally in all directions is known as what?

    A. Omnidirectional signal generator

    B. Omnidirectional antenna

    C. Intentional radiator

    D. Nondirectional transmitter

    E. Isotropic radiator

3. When calculating the link budget and system operating margin of a point-to-point outdoor WLAN bridge link, what factors should be taken into account? (Choose all that apply.)

    A. Distance

    B. Receive sensitivity

    C. Transmit amplitude

    D. Antenna height

    E. Cable loss

    F. Frequency

4. The sum of all the components from the transmitter to the antenna, not including the antenna, is known as what? (Choose two.)

    A. IR

    B. Isotropic radiator

    C. EIRP

    D. Intentional radiator

5. The highest RF signal strength that is transmitted from an antenna is known as what?

    A. Equivalent isotropically radiated power

    B. Transmit sensitivity

    C. Total emitted power

    D. Antenna radiated power

6. Select the units of power. (Choose all that apply.)

   A. Watt

   B. Milliwatt

   C. Decibel

   D. dBm

   E. Bel

7. Select the units of comparison (relative). (Choose all that apply.)

   A. dBm

   B. dBi

   C. Decibel

   D. dBd

   E. Bel

8. 2 dBd is equal to how many dBi?

   A. 5 dBi

   B. 4.41 dBi

   C. 4.14 dBi

   D. The value cannot be calculated.

9. 23 dBm is equal to how many mW?

   A. 200 mW

   B. 14 mW

   C. 20 mW

   D. 23 mW

   E. 400 mW

10. A wireless bridge is configured to transmit at 100 mW. The antenna cable and connectors produce a 3 dB loss and are connected to a 16 dBi antenna. What is the EIRP?

    A. 20 mW

    B. 30 dBm

    C. 2,000 mW

    D. 36 dBm

    E. 8 W

11. A WLAN transmitter that emits a 400 mW signal is connected to a cable with a 9 dB loss. If the cable is connected to an antenna with 19 dBi of gain, what is the EIRP?

    A. 4 W

    B. 3,000 mW

    C. 3,500 mW

    D. 2 W

12. WLAN vendors use RSSI thresholds to trigger which radio card behaviors?
    - A. Receive sensitivity
    - B. Roaming
    - C. Retransmissions
    - D. Dynamic rate switching

13. Received signal strength indicator (RSSI) metrics are used by 802.11 radios to define which RF characteristics? (Choose all that apply.)
    - A. Signal strength
    - B. Phase
    - C. Frequency
    - D. Signal quality
    - E. Modulation

14. dBi is a measure of what?
    - A. The output of the transmitter
    - B. The signal increase caused by the antenna
    - C. The signal increase strength of the intentional transmitter
    - D. The comparison between an isotropic radiator and the transceiver
    - E. The strength of the intentional radiator

15. Which of the following are valid calculations when using the rule of 10s and 3s? (Choose all that apply.)
    - A. For every 3 dB of gain (relative), double the absolute power (mW).
    - B. For every 10 dB of loss (relative), divide the absolute power (mW) by a factor of 10.
    - C. For every 10 dB of loss (absolute), multiply the relative power (mW) by a factor of 10.
    - D. For every 10 mW of loss (relative), multiply the absolute power (dB) by a factor of 10.
    - E. For every 10 dB of loss (relative), halve the absolute power (mW).

16. A WLAN transmitter that emits a 100 mW signal is connected to a cable with a 3 dB loss. If the cable is connected to an antenna with 7 dBi of gain, what is the EIRP at the antenna element?
    - A. 200 mW
    - B. 250 mW
    - C. 300 mW
    - D. 400 mW

17. In a normal wireless bridged network, the greatest loss of signal is caused by what component?
    - A. Receive sensitivity
    - B. Antenna cable loss
    - C. Lightning arrestor
    - D. Free space path loss

18. To double the distance of a signal, the EIRP must be increased by how many dBs?

A. 3 dB

B. 6 dB

C. 10 dB

D. 20 dB

19. During a site survey in a manufacturing plant, the WLAN engineer determines that the noise floor is extremely high because of all the machinery that is operating in the plant. The engineer is worried about a low SNR and poor performance due to the high noise floor. What is a suggested best practice to deal with this scenario?

A. Increase the access points' transmission amplitude.

B. Mount the access points higher.

C. Double the distance of the signal with 6 dBi of antenna gain.

D. Plan for coverage cells with a 5 dB fade margin.

E. Increase the transmission amplitude of the client radios.

20. Which value should not be used to compare wireless network cards manufactured by different WLAN vendors?

A. Receive sensitivity

B. Transmit power range

C. Antenna dBi

D. RSSI

Answers to Review Questions

1. C. The transmitter generates the AC signal and modifies it by using a modulation technique to encode the data into the signal.
2. E. An isotropic radiator is also known as a point source.
3. A, B, C, E, F. When deploying radio communications, a link budget is the sum of all gains and losses from the transmitting radio, through the RF medium, to the receiver radio. Link budget calculations include original transmit gain and passive antenna gain. All losses must be accounted for, including free space path loss. Frequency and distance are needed to calculate free space path loss. The height of an antenna has no significance when calculating a link budget.
4. A, D. *IR* is the abbreviation for *intentional radiator.* The components making up the IR include the transmitter, all cables and connectors, and any other equipment (grounding, lightning arrestors, amplifiers, attenuators, and so forth) between the transmitter and the antenna. The power of the IR is measured at the connecter that provides the input to the antenna.
5. A. Equivalent isotropically radiated power, also known as EIRP, is a measure of the strongest signal that is radiated from an antenna.
6. A, B, D. Watts, milliwatts, and dBms are all absolute power measurements. One watt is equal to 1 ampere (amp) of current flowing at 1 volt. A milliwatt is one-thousandth of 1 watt. dBm is decibels relative to 1 milliwatt.
7. B, C, D, E. The unit of measurement known as a bel is a relative expression and a measurement of change in power. A decibel (dB) is equal to one-tenth of a bel. Antenna gain measurements of dBi and dBd are relative measurements. dBi is defined as decibels referenced to an isotropic radiator. dBd is defined as decibels referenced to a dipole.
8. C. To convert any dBd value to dBi, simply add 2.14 to the dBd value.
9. A. To convert to mW, first calculate how many 10s and 3s are needed to add up to 23, which is 0 + 10 + 10 + 3. To calculate the mW, you must multiply 1 × 10 × 10 × 2, which calculates to 200 mW.

   Note: The CD has a PowerPoint presentation that also explains this answer.
10. C. To reach 100 mW, you can use 10s and 2s and multiplication and division. Multiplying by two 10s will accomplish this. This means that on the dBm side, you must add two 10s, which equals 20 dBm. Then subtract the 3 dB of cable loss for a dBm of 17. Because you subtracted 3 from the dBm side, you must divide the 100 mW by 2, giving you a value of 50 mW. Now add in the 16 dBi by adding a 10 and two 3s to the dBm column, giving a total dBm of 33. Because you added a 10 and two 3s, you must multiply the mW column by 10 and two 2s, giving a total of 2,000 mW, or 2 W. Since the cable and connector loss is 3 dB and the antenna gain is 16 dBi, you can add the two together for a cumulative gain of 13 dB; then apply that gain to the 100 mW transmit signal to calculate an EIRP of 2,000 mW, or 2W.

    Note: The CD has a PowerPoint presentation that also explains this answer.

11. A. If the original transmit power is 400 mW and cabling induces a 9 dB loss, the power at the opposite end of the cable will be 50 mW. The first 3 dB of cable loss halved the absolute power to 200 mW. The second 3 dB of cable loss halved the absolute power to 100 mW. The final 3 dB of cable loss halved the power to 50 mW. The antenna with 19 dBi of gain passively amplified the 50 mW signal to 4,000 mW. The first 10 dBi of antenna boosts the signal to 500 mW. The next 9 dBi of antenna gain doubles the signal three times to a total of 4 watts. Since the cable loss is 9 dB and the antenna gain is 19 dBi, you could add the two together for a cumulative gain of 10 dB, and then apply that gain to the 400 mW transmit signal to calculate an EIRP of 4,000 mW, or 4W.

12. B, D. RSSI thresholds are a key factor for clients when they initiate the roaming handoff. RSSI thresholds are also used by vendors to implement dynamic rate switching, which is a process used by 802.11 radios to shift between data rates.

13. A. The received signal strength indicator (RSSI) is a metric used by 802.11 radio cards to measure signal strength (amplitude). Some vendors use a proprietary scale to also correlate to signal quality. Most vendors erroneously define signal quality as the signal-to-noise ratio (SNR). The signal-to-noise ratio is simply the difference in decibels between the received signal and the background noise (noise floor).

14. B. dBi is defined as "decibel gain referenced to an isotropic radiator" or "change in power relative to an antenna." dBi is the most common measurement of antenna gain.

15. A, B. The four rules of the 10s and 3s are as follows: For every 3 dB of gain (relative), double the absolute power (mW). For every 3 dB of loss (relative), halve the absolute power (mW). For every 10 dB of gain (relative), multiply the absolute power (mW) by a factor of 10. For every 10 dB of loss (relative), divide the absolute power (mW) by a factor of 10.

16. B. If the original transmit power is 100 mW and cabling induces a 3 dB loss, the power at the opposite end of the cable will be 50 mW. The 3 dB of cable loss halved the absolute power to 50 mW. An antenna with 10 dBi of gain would boost the signal to 500 mW. We also know that 3 dB of loss halves the absolute power. Therefore, an antenna with 7 dBi of gain would amplify the signal to half that of a 10 dBi antenna. The antenna with 7 dBi of gain passively amplified the 50 mW signal to 250 mW.

17. D. A distance of as little as 100 meters will cause free space path loss of 80 dB, far greater than any other component. RF components such as connectors, lightning arrestors, and cabling all introduce insertion loss. However, FSPL will always be the reason for the greatest amount of loss.

18. B. The 6 dB rule states that increasing the amplitude by 6 decibels, will double the usable distance of an RF signal. The 6 dB rule is very useful for understanding antenna gain because every 6 dBi of extra antenna gain will double the usable distance of an RF signal.

19. D. In a high-multipath or noisy environment, a common best practice is to add a 5 dB fade margin when designing for coverage based on a vendor's recommend received signal strength or the noise floor, whichever is louder.

20. D. WLAN vendors execute RSSI metrics in a proprietary manner. The actual range of the RSSI value is from 0 to a maximum value (less than or equal to 255) that each vendor can choose on its own (known as RSSI_Max). Therefore, RSSI metrics should not be used to compare different WLAN vendor radios because there is no standard for the range of values or a consistent scale.

Radio Frequency Signal and Antenna Concepts

IN THIS CHAPTER, YOU WILL LEARN ABOUT THE FOLLOWING:

✓ **Active and passive gain**

✓ **Azimuth and elevation charts (antenna radiation envelopes)**

✓ **Interpreting polar charts**

✓ **Beamwidth**

✓ **Antenna types**

- Omnidirectional antennas
- Semidirectional antennas
- Highly directional antennas
- Phased array antennas
- Sector antennas

✓ **Visual line of sight**

✓ **RF line of sight**

✓ **Fresnel zone**

✓ **Earth bulge**

✓ **Antenna polarization**

✓ **Antenna diversity**

✓ **Multiple-input multiple-output (MIMO)**

✓ **Antenna connection and installation**

- Voltage standing wave ratio (VSWR)
- Signal loss
- Antenna mounting

✓ **Antenna accessories**

- Cables
- Connectors
- Splitters
- Amplifiers
- Attenuators
- Lightning arrestors
- Grounding rods and wires

In order to be able to communicate between two or more transceivers, the radio frequency (RF) signal must be radiated from the antenna of the transmitter with enough power so that it is received and understood by the receiver. The installation of antennas has the greatest ability to affect whether the communication is successful or not. Antenna installation can be as simple as placing an access point in the middle of a small office, providing full coverage for your company, or it can be as complex as installing an assortment of directional antennas, kind of like piecing together a jigsaw puzzle. Do not fear this process; with proper understanding of antennas and how they function, you may find successfully planning for and installing antennas in a wireless network to be a skillful and rewarding task.

This chapter focuses on the categories and types of antennas and the different ways that they can direct an RF signal. Choosing and installing antennas is like choosing and installing lighting in a home. When installing home lighting, you have many choices: table lamps, ceiling lighting, narrow- or wide-beam directional spotlights. In Chapter 3, "Radio Frequency Components, Measurements, and Mathematics," you were introduced to the concept of antennas focusing RF signal. In this chapter, you will learn about the different types of antennas, their radiation patterns, and how to use the different antennas in different environments.

You will also learn that even though we often use light to explain RF radiation, there are differences between the way the two behave. You will learn about aiming and aligning antennas, and you will learn that what you see is not necessarily what you will get.

In addition to learning about antennas, you will learn about the accessories that may be needed for proper antenna installation. In office environments, you may simply need to connect the antenna to the access point. In outdoor installations, you will need special cable and connectors, lightning arrestors, and special mounting brackets. In this chapter, we will introduce you to the components necessary for successfully installing an antenna.

To summarize, in this chapter you will gain the knowledge that will enable you to properly select, install, and align antennas. These skills will help you successfully implement a wireless network, whether it is a point-to-point network between two buildings or a network providing wireless coverage throughout an office building.

Active and Passive Gain

In the preceding chapter, you learned that you can increase the signal that is radiated out of the antenna (EIRP) by increasing the output of the transmitter, which in turn increases the amount of power provided to the antenna (intentional radiator) and thus the amount of power from the antenna (EIRP). When the power is increased by some type of electrical device, such as the transmitter or—as you will learn later in this chapter—an amplifier, the increase is referred to as *active gain*.

Another method of increasing power, also discussed in the preceding chapter, is to direct or focus the power. When power is focused, the amount provided to the antenna does not change. Instead, the antenna acts like a lens on a flashlight that increases the power output by concentrating the RF signal in a specific direction. Because the gain from the antenna was created by shaping or concentrating the signal, and not by increasing the overall power, this increase is referred to as *passive gain*.

Passive gain is caused by focusing the existing power, whereas active gain is caused by adding more power.

When trying to decide whether gain is active or passive, determine whether the gain is due to a total increase in power from an electronic device (active gain) or whether it is due to the power being focused or directed (passive gain).

Azimuth and Elevation Charts (Antenna Radiation Envelopes)

There are many types of antennas designed for many different purposes, just as there are many types of lights designed for many different purposes. When purchasing lighting for your home, it is easy to compare two lamps by turning them on and looking at the way each disperses the light.

Unfortunately, it is not possible to compare antennas in the same way. Actual side-by-side comparison requires you to walk around the antenna with an RF meter, take numerous signal measurements, and then plot the measurements either on the ground or on a piece of paper that represents the environment. Besides the fact that this is a time-consuming task, the results could be skewed by outside influences on the RF signal, such as furniture or other RF signals in the area. To assist potential buyers with their purchasing decision, antenna manufacturers create *azimuth charts* and *elevation charts*, commonly known as radiation patterns, for their antennas. These radiation patterns are created in controlled environments where the results cannot be skewed by outside influences and represent the signal pattern that is radiated by a particular model of antenna. These charts are commonly known as *polar charts* or *antenna radiation envelopes*.

Figure 4.1 shows the azimuth and elevation charts of an omnidirectional antenna. The azimuth chart, labeled H-plane, shows the top-down view of the radiation pattern of the antenna. The elevation chart, labeled E-plane, shows the side view of the radiation pattern of the antenna. There is no standard that requires the antenna manufacturers to align the degree marks of the chart with the direction that the antenna is facing, so unfortunately it is up to the reader of the chart to understand and interpret it.

Here are a few statements that will help you interpret the radiation charts:

- In either chart, the antenna is placed at the middle of the chart.
- Azimuth chart = H-plane = top-down view
- Elevation chart = E-plane = side view

The outer ring of the chart usually represents the strongest signal of the antenna. The chart does not represent distance or any level of power or strength. It represents only the relationship of power between different points on the chart.

One way to think of the chart is to consider the way a shadow behaves. If you were to move a flashlight closer or farther from your hand, the shadow of your hand would grow larger or smaller. The size of the shadow does not represent the size of the hand. The shadow shows only the relationship between the hand and the fingers. With an antenna, the radiation pattern will grow larger or smaller depending on how much power the antenna receives, but the shape and the relationships represented by the patterns will always stay the same.

FIGURE 4.1 Azimuth and elevation charts

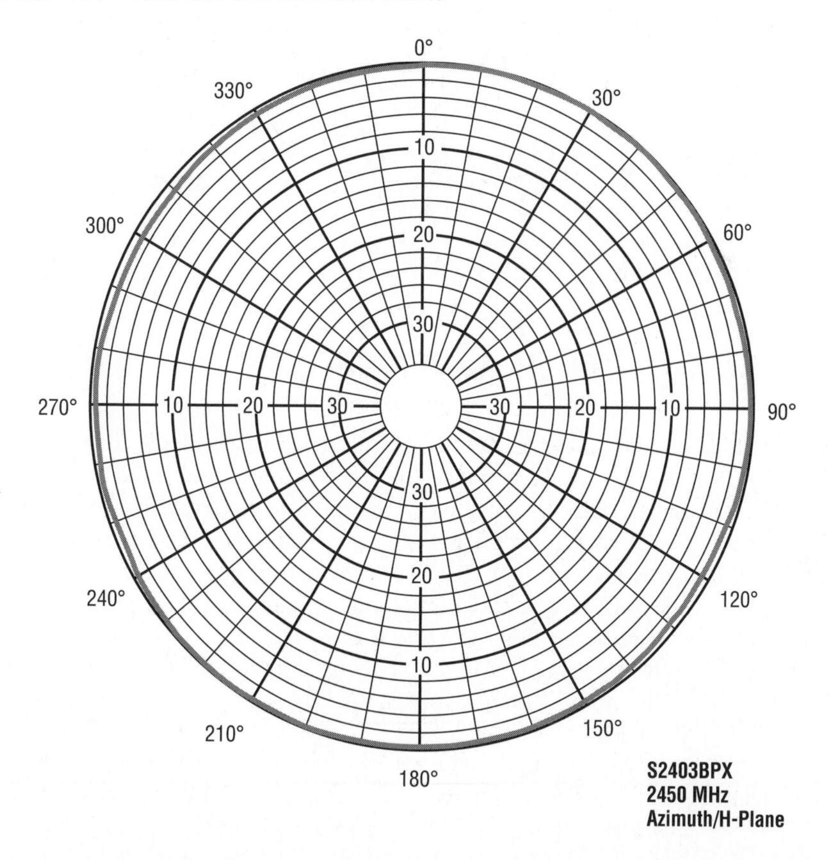

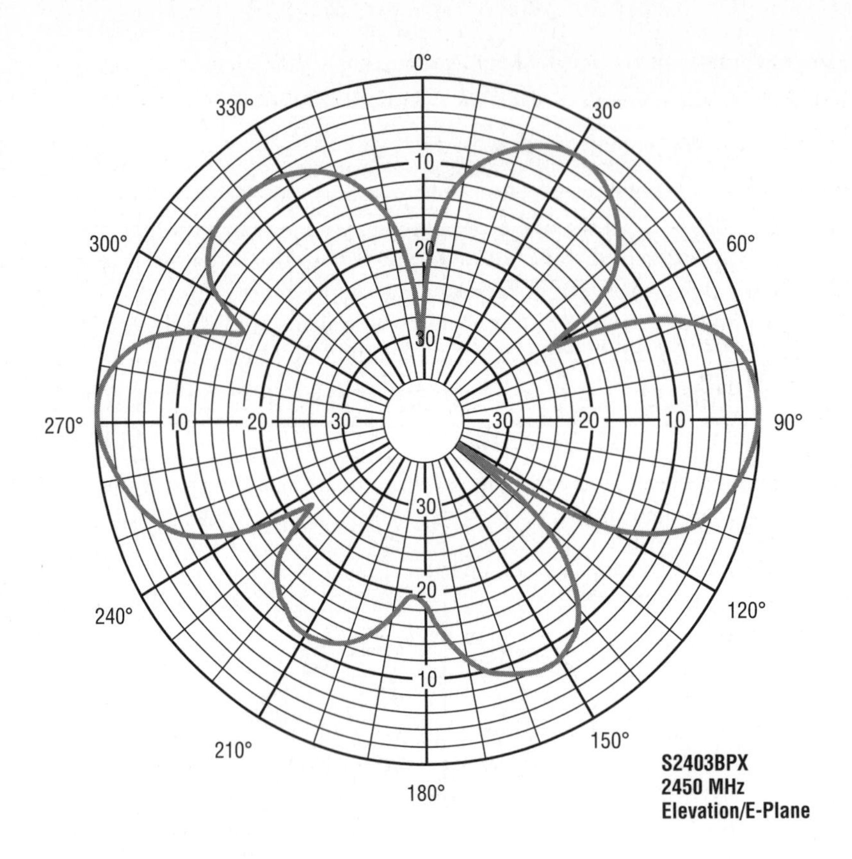

Interpreting Polar Charts

As we have stated, the antenna azimuth (H-plane) and elevation (E-plane) charts are commonly referred to as polar charts. These charts are often misinterpreted and misread. One of the biggest reasons these charts are misinterpreted is that the chart represents the decibel mapping of the antenna coverage. This dB mapping represents the radiation pattern of the antenna; however, it does this using a logarithmic scale instead of a linear scale. Remember that the logarithmic scale is a variable scale, based on exponential values; so the polar chart is actually a visual representation using a variable scale.

Let us try to explain this by using Figure 4.2. The top drawing shows four boxes. The number inside the box tells you how long and wide the box is. So, even though visually in our drawing we are representing these boxes as the same size, in reality each one is twice as long and wide as the previous one. It is easier for us to draw the four boxes as the same physical size and just put the number in the box to represent the actual size of the box. In the middle drawing, we draw the boxes showing the actual size of the four boxes.

FIGURE 4.2 Logarithmic/linear comparison

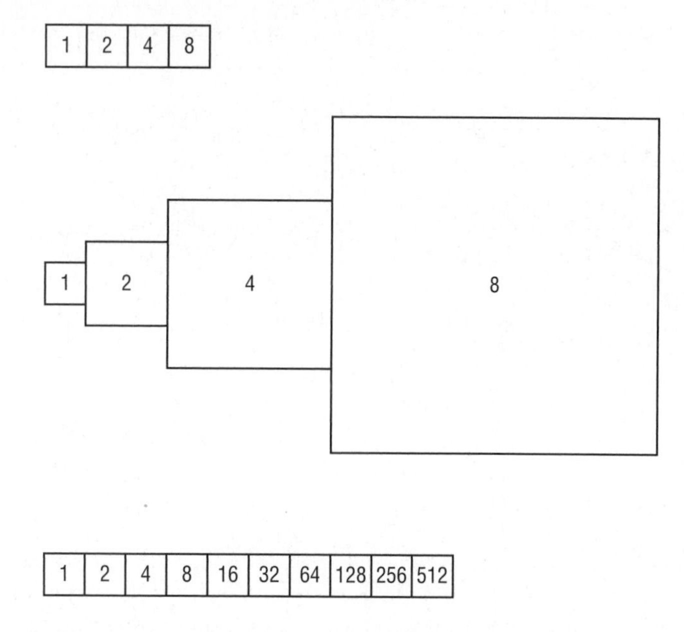

What if we had more boxes, say 10? By representing each box by using the same-sized drawing, it is easier to illustrate the boxes, as seen in the lower drawing. In this example, if we tried to show the actual differences in size, as we did in the middle drawing, we could not fit this drawing on the page in the book. In fact, the room that you are in may not have enough space for you to even draw this. Because the scale changes so drastically, it is necessary to not draw the boxes to scale so that we can still represent the information.

In Chapter 3, you learned about RF math. In that chapter, one of the rules that you learned was the rule of 6 dB, which indicates that a 6 dB decrease of power decreases the distance the signal travels by half. A 10 dB decrease of power decreases the distance the signal travels by approximately 70 percent. In Figure 4.3, the left polar chart displays the logarithmic representation of the elevation chart of an omnidirectional antenna. This is what you are typically looking at on an antenna brochure or specification sheet. Someone who is untrained in reading these charts would look at the chart and be impressed with how much vertical coverage the antenna provides but would likely be disappointed with the actual coverage. When reading the logarithmic chart, you must remember that for every 10 dB decrease from the peak signal, the actual distance decreases by 70 percent. Each concentric circle on the logarithmic chart represents a change of 5 dB. So if you look at Figure 4.3, the first *side lobe* is 10 dB weaker than the main lobe. Remember to compare where the lobes are relative to the concentric circles. This 10 dB decrease on the logarithmic chart is equal to a 70 percent decrease in range on the linear chart. Comparing both charts, you see that the side lobes on the logarithmic chart are essentially insignificant when adjusted to the linear chart. As you can see, this omnidirectional antenna has very little vertical coverage.

FIGURE 4.3 Omnidirectional polar chart (E-plane)

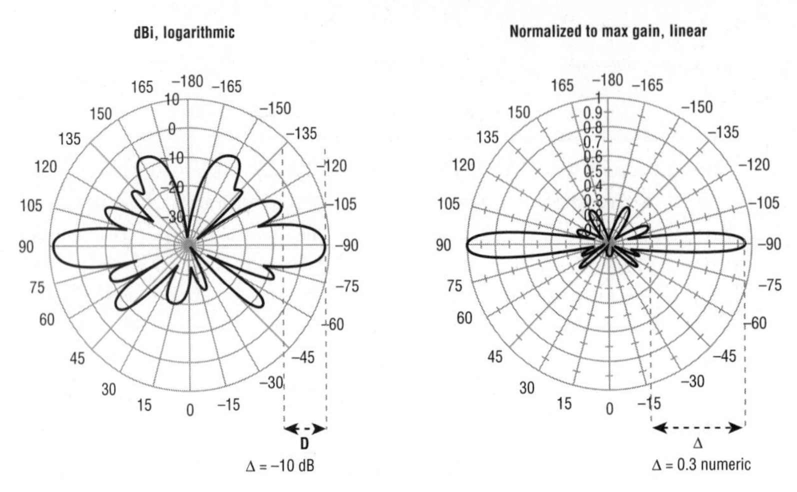

To give you another comparison, Figure 4.4 shows the logarithmic pattern of the elevation chart of a directional antenna along with a linear representation of the vertical coverage area of this antenna. We rotated the polar chart on its side so that you could better visualize the antenna mounted on the side of a building and aiming at another building.

FIGURE 4.4 Directional polar chart (E-plane)

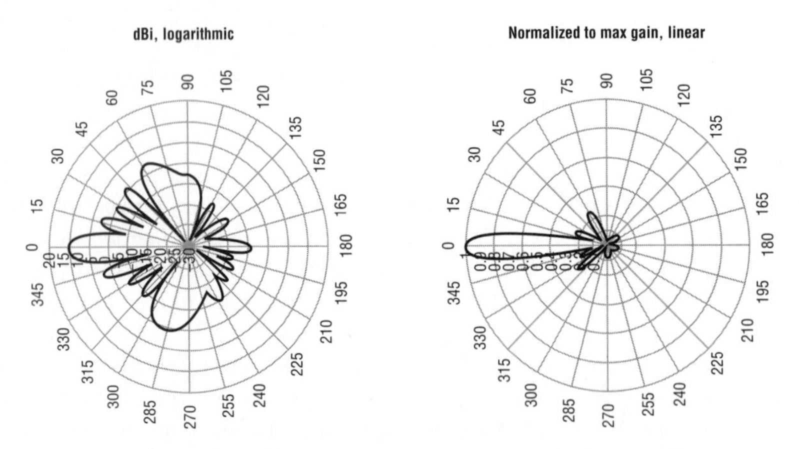

Beamwidth

Many flashlights have adjustable lenses, enabling the user to widen or tighten the concentration of light that is radiating from them. RF antennas are capable of focusing the power that is radiating from them, but unlike flashlights, antennas are not adjustable. The user must decide how much focus is desired prior to the purchase of the antenna.

Beamwidth is the measurement of how broad or narrow the focus of an antenna is—and is measured both horizontally and vertically. It is the measurement from the center, or strongest point, of the antenna signal to each of the points along the horizontal and vertical axes where the signal decreases by half power (–3 dB), as seen in Figure 4.5. These –3 dB points are often referred to as *half-power points*. The distance between the two half-power points on the horizontal axis is measured in degrees, giving the horizontal beamwidth measurement. The distance between the two half-power points on the vertical axis is also measured in degrees, giving the vertical beamwidth measurement.

FIGURE 4.5 Antenna beamwidth

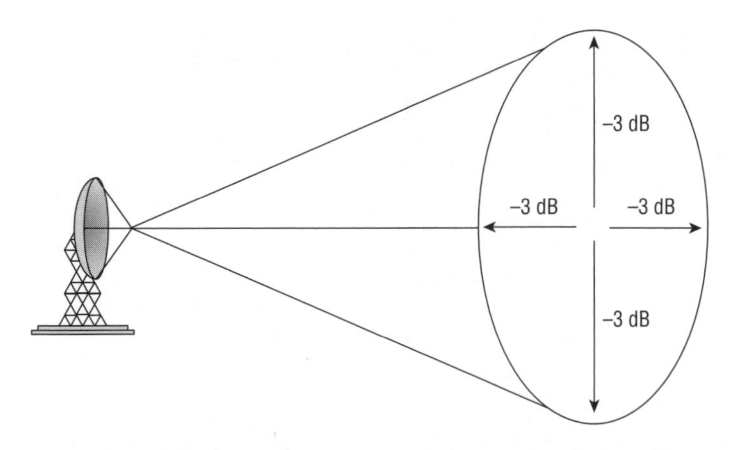

Most of the time when you are deciding which antenna will address your communications needs, you will look at the manufacturer's brochure to determine the technical specifications of the antenna. In these brochures, the manufacturer typically includes the numerical values for the horizontal and vertical beamwidths of the antenna. It is important for you to understand how these numbers are calculated.

Figure 4.6 illustrates the process. First determine the scale of the polar chart. On this chart, you can see that the solid circles represent the –10, –20, and –30 dB lines, and the dotted circles therefore represent the –5, –15, and –25 dB lines. These represent the dB decrease from the peak signal. Now to determine the beamwidth of this antenna, first locate the point on the chart where the antenna signal is the strongest. In this example, the signal is strongest where the number 1 arrow is pointing. Move along the antenna pattern away from the peak signal (as shown by the two number 2 arrows) until you reach the point where the antenna pattern is 3 dB closer to the center of the diagram (as shown by the two number 3 arrows). This is why you needed to know the scale of the chart first. Draw a line from each of these points to the middle of the polar chart (as shown by the dark dotted lines) and measure the distance in degrees between these lines to calculate the beamwidth of the antenna. In this example, the beamwidth of this antenna is about 30 degrees.

FIGURE 4.6 Beamwidth calculation

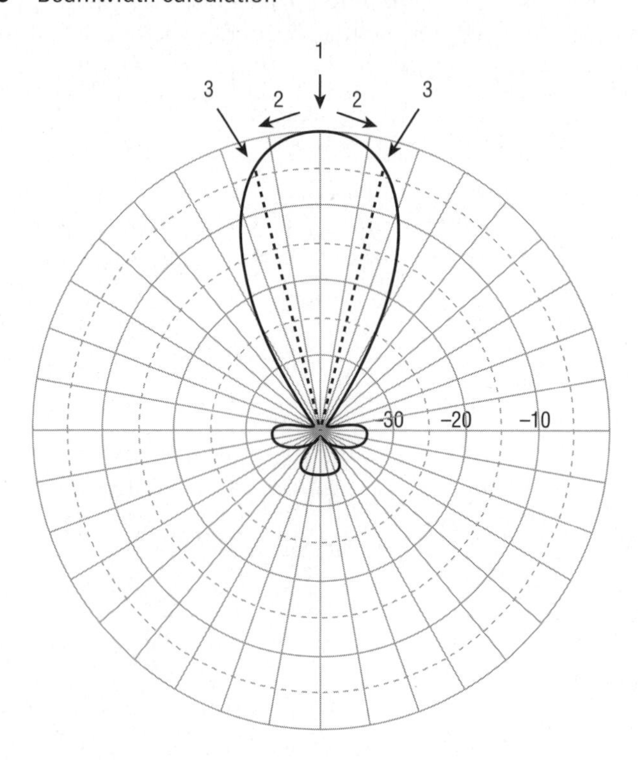

It is important to realize that even though the majority of the RF signal that is generated is focused within the beamwidth of the antenna, a significant amount of signal can still radiate from outside the beamwidth, from what is known as the antenna's side or rear lobes. As you look at the azimuth charts of different antennas, you will notice that some of these side and rear lobes are fairly significant. Although the signal of these lobes is drastically less than the signal of the main beamwidth, they are dependable, and in certain implementations very functional. It is important when aligning point-to-point antennas that you make sure they are actually aligned to the main lobe and not a side lobe.

Table 4.1 shows the different types of antennas that are used in 802.11 communications.

Table 4.1 provides reference information that will be useful as you learn about different types of antennas in this chapter.

TABLE 4.1 Antenna Beamwidth

Antenna Types	Horizontal Beamwidth (in Degrees)	Vertical Beamwidth (in Degrees)
Omnidirectional	360	7 to 80
Patch/panel	30 to 180	6 to 90
Yagi	30 to 78	14 to 64
Sector	60 to 180	7 to 17
Parabolic dish	4 to 25	4 to 21

Antenna Types

There are three main categories of antennas:

Omnidirectional *Omnidirectional antennas* radiate RF in a fashion similar to the way a table or floor lamp radiates light. They are designed to provide general coverage in all directions.

Semidirectional *Semidirectional antennas* radiate RF in a fashion similar to the way a wall sconce radiates light away from the wall or the way a street lamp shines light down on a street or a parking lot, providing a directional light across a large area.

Highly directional *Highly directional antennas* radiate RF in a fashion similar to the way a spotlight focuses light on a flag or a sign.

Each type of antenna is designed with a different objective in mind.

It is important to keep in mind that this section is discussing types of antennas and not lighting. Although it is useful to refer to lighting to provide analogies to antennas, it is critical to remember that unlike lighting, RF signals can travel through solid objects such as walls and floors.

In addition to antennas acting as radiators and focusing signals that are being transmitted, they also focus signals that are received. If you were to walk outside and look up at a star, it would appear fairly dim. If you were to look at that same star through binoculars, it would appear brighter. If you were to use a telescope, it would appear even brighter. Antennas function in a similar way. Not only do they amplify signal that is being transmitted, they also amplify signal that is being received. High-gain microphones operate in the same way, enabling us to not only watch the action of our favorite sport on television, but to also hear the action.

Antennas or Antennae?

Although it is not a matter of critical importance, many are often curious whether the plural of *antenna* is *antennas* or *antennae*. The simple answer is both, but the complete answer is it depends. When *antenna* is used as a biological term, the plural is *antennae*, such as the antennae of a bug. When it is used as an electronics term, the plural is *antennas*, such as the antennas on an access point.

Omnidirectional Antennas

Omnidirectional antennas radiate RF signal in all directions. The small, rubber *dipole antenna*, often referred to as a *rubber duck* antenna, is the classic example of an omnidirectional antenna and is the default antenna of most access points. A perfect omnidirectional antenna would radiate RF signal like the theoretical isotropic radiator from Chapter 2, "Radio Frequency Fundamentals." The closest thing to an isotropic radiator is the omnidirectional dipole antenna.

An easy way to explain the radiation pattern of a typical omnidirectional antenna is to hold your index finger straight up (this represents the antenna) and place a bagel on it as if it were a ring (this represents the RF signal). If you were to slice the bagel in half horizontally, as if you were planning to spread butter on it, the cut surface of the bagel would represent the azimuth chart, or H-plane, of the omnidirectional antenna. If you took another bagel and sliced it vertically instead, essentially cutting the hole that you are looking through in half, the cut surface of the bagel would now represent the elevation, or E-plane, of the omnidirectional antenna.

In Chapter 3, you learned that antennas can focus or direct the signal that they are transmitting. It is important to know that the higher the dBi or dBd value of an antenna, the more focused the signal. When discussing omnidirectional antennas, it is not uncommon to initially question how it is possible to focus a signal that is radiated in all directions. With higher-gain omnidirectional antennas, the vertical signal is decreased and the horizontal power is increased.

Figure 4.7 shows the elevation view of three theoretical antennas. Notice that the signal of the higher-gain antennas is elongated, or more focused horizontally. The horizontal beamwidth of omnidirectional antennas is always 360 degrees, and the vertical beamwidth ranges from 7 to 80 degrees, depending on the particular antenna.

Because of the narrower vertical coverage of the higher-gain omnidirectional antennas, it is important to carefully plan how they are used. Placing one of these higher-gain antennas on the first floor of a building may provide good coverage to the first floor, but because of the narrow vertical coverage, the second and third floors may receive minimal signal. In some installations, you may want this; in others, you may not. Indoor installations typically use low-gain omnidirectional antennas with gain of about 2.14 dBi.

FIGURE 4.7 Vertical radiation patterns of omnidirectional antennas

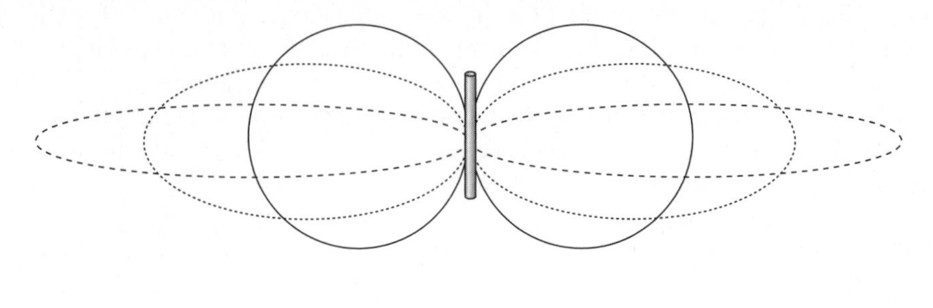

Antennas are most effective when the length of the element is an even fraction (such as ¼ or ½) or a multiple of the wavelength (λ). A 2.4 GHz half-wave dipole antenna (see Figure 4.8) consists of two elements, each ¼ λ in length (about 1 inch), running in the opposite direction from each other. Although this drawing of a dipole is placed horizontally, the antenna is always placed in a vertical orientation. Higher-gain omnidirectional antennas are typically constructed by stacking multiple dipole antennas on top of each other and are known as *collinear antennas*.

FIGURE 4.8 Half-wave dipole antenna

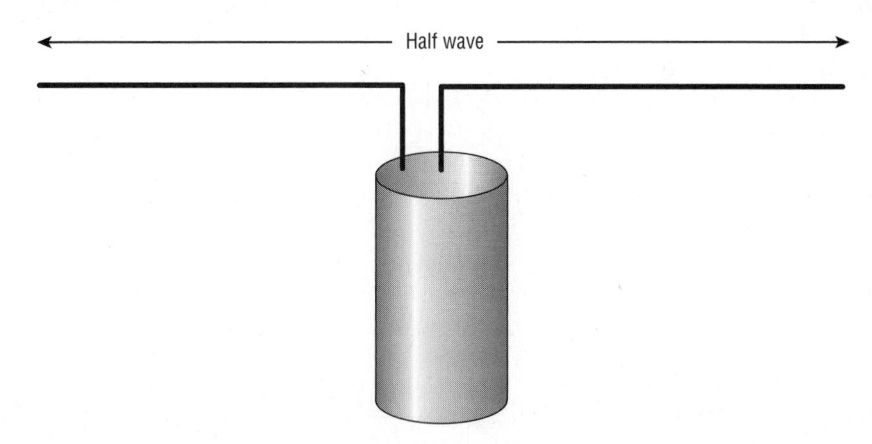

Omnidirectional antennas are typically used in point-to-multipoint environments. The omnidirectional antenna is connected to a device (such as an access point) that is placed at the center of a group of client devices, providing central communications capabilities to the

surrounding clients. High-gain omnidirectional antennas can also be used outdoors to connect multiple buildings together in a point-to-multipoint configuration. A central building would have an omnidirectional antenna on its roof, and the surrounding buildings would have directional antennas aimed at the central building. In this configuration, it is important to make sure that the gain of the omnidirectional antenna is high enough to provide the coverage necessary but not so high that the vertical beamwidth is too narrow to provide an adequate signal to the surrounding buildings.

Figure 4.9 shows an installation where the gain is too high. The building to the left will be able to communicate, but the building on the right is likely to have problems. To solve the problem that is pictured in Figure 4.9, sector arrays using a downtilt configuration are used instead of high-gain omnidirectional antennas. Sector antennas are discussed later in this chapter.

FIGURE 4.9 Improperly installed omnidirectional antenna

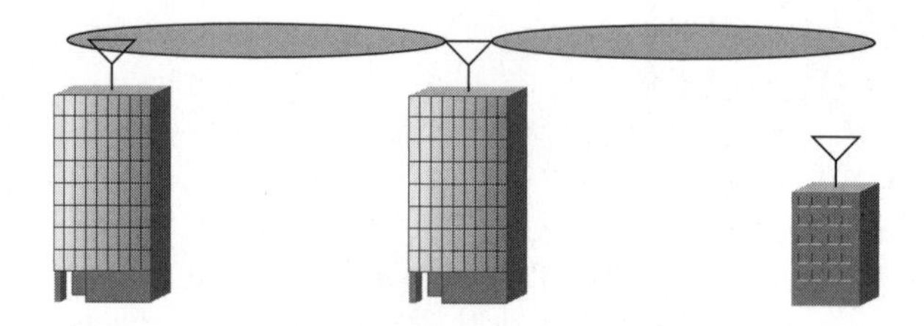

Semidirectional Antennas

Unlike omnidirectional antennas that radiate RF signals in all directions, semidirectional antennas are designed to direct a signal in a specific direction. Semidirectional antennas are used for short- to medium-distance communications, with long-distance communications being served by highly directional antennas.

It is common to use semidirectional antennas to provide a network bridge between two buildings in a campus environment or down the street from each other. Longer distances would be served by highly directional antennas.

There are three types of antennas that fit into the semidirectional category:

- Patch
- Panel
- Yagi (pronounced *YAH-gee*)

Patch and panel antennas, as shown in Figure 4.10, are more accurately classified or referred to as planar antennas. *Patch* refers to a particular way of designing the radiating elements inside the antenna. Unfortunately, it has become common practice to use the terms *patch antenna* and *panel antenna* interchangeably. If you are unsure of the antenna's specific design, it is better to refer to it as a planar antenna.

These antennas can be used for outdoor point-to-point communications up to about a mile but are more commonly used as a central device to provide unidirectional coverage from the access point to the clients in an indoor environment. It is common for patch or panel antennas to be connected to access points to provide directional coverage within a building. Planar antennas can be used effectively in libraries, warehouses, and retail stores with long aisles of shelves. Because of the tall, long shelves, omnidirectional antennas often have difficulty providing RF coverage effectively. In contrast, planar antennas can be placed high on the side walls of the building, aiming through the rows of shelves. The antennas can be alternated between rows, with every other antenna being placed on the opposite wall. Since planar antennas have a horizontal beamwidth of 180 degrees or less, a minimal amount of signal will radiate outside of the building. With the antenna placement alternated and aimed from opposite sides of the building, the RF signal is more likely to radiate down the rows, providing the necessary coverage.

FIGURE 4.10 The exterior of a patch antenna and the internal antenna element

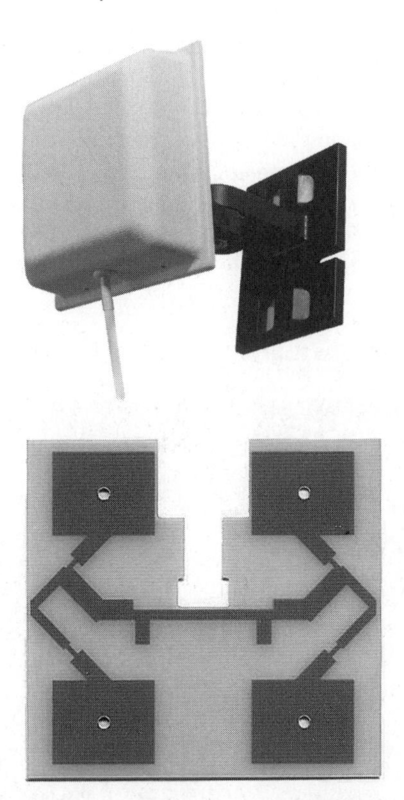

Planar antennas are also often used to provide coverage for long hallways with offices on each side or hospital corridors with patient rooms on each side. A planar antenna can be placed at the end of the hall and aimed down the corridor. A single planar antenna can provide RF signal to some or all of the corridor and the rooms on each side and some coverage to the floors above and below. How much coverage will depend on the power of the transmitter, the gain and beamwidth (both horizontal and vertical) of the antenna, and the attenuation properties of the building.

Real World Scenario

Using Antennas to Deal with Multipath

The use of indoor planar antennas is also highly recommended in high-multipath environments. For example, many warehouses, manufacturing plants, and retail outlets have long aisles with metal racks and inventory that cause many reflections. As you learned in Chapter 2, reflections cause multipath, which can lead to data corruption and layer 2 retransmissions. Using a unidirectional planar antenna will cut down on reflections and thereby decrease the negative effects of multipath.

Yagi antennas, as seen in Figure 4.11, are not as unusual as they sound. The traditional television antenna that is attached to the roof of a house or apartment is a Yagi antenna. The television antenna looks quite different because it is designed to receive signals of many frequencies (different channels) and the length of the elements vary according to the wavelength of the different frequencies. A *Yagi antenna* that is used for 802.11 communications is designed to support a very narrow range of frequencies, so the elements are all about the same length. Yagi antennas are commonly used for short- to medium-distance point-to-point communications of up to about 2 miles, although high-gain Yagi antennas can be used for longer distances.

FIGURE 4.11 The exterior of a Yagi antenna and the internal antenna element

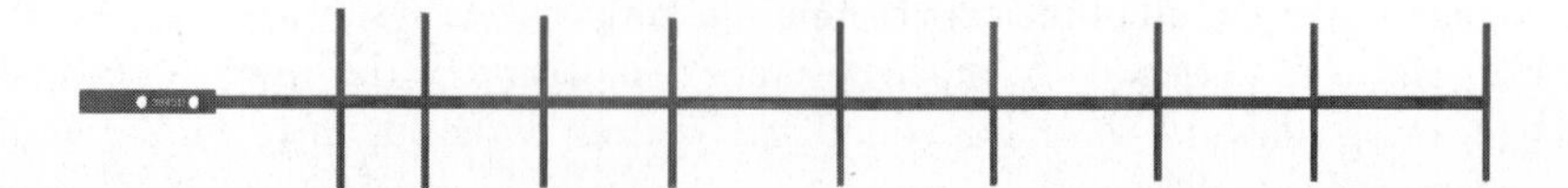

Another benefit of semidirectional antennas is that they can be installed high on a wall and tilted downward toward the area to be covered. This cannot be done with an omnidirectional antenna without causing the signal on the other side of the antenna to be tilted upward. Since the only RF signal that radiates from the back of a semidirectional antenna is incidental, the ability to aim it vertically is an additional benefit.

Figure 4.12 shows the radiation patterns of a typical semidirectional panel antenna that was discussed in this section. Remember that these are actual azimuth and elevation charts from a specific antenna and that every manufacturer and model of antenna will have a slightly different radiation pattern.

FIGURE 4.12 Radiation pattern of a typical semidirectional panel antenna

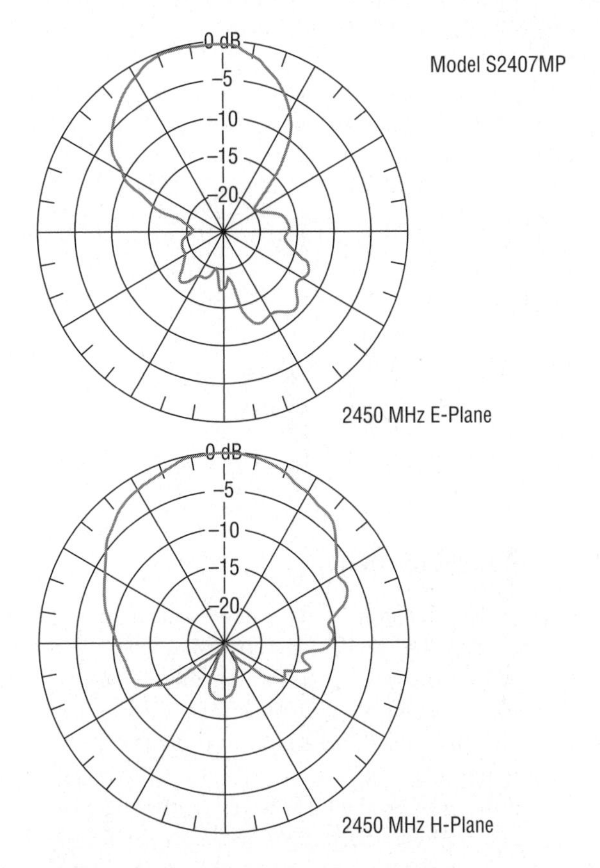

Highly Directional Antennas

Highly directional antennas are strictly used for point-to-point communications, typically to provide network bridging between two buildings. They provide the most focused, narrow beamwidth of any of the antenna types.

There are two types of highly-directional antennas: parabolic dish antennas and grid antennas. The *parabolic dish antenna* is similar in appearance to the small digital satellite TV antennas that can be seen on the roofs of many houses. As pictured in Figure 4.13, the *grid antenna* resembles the grill of a barbecue, with the edges slightly curved inward. The spacing of the wires on a grid antenna is determined by the wavelength of the frequencies that the antenna is designed for. Because of the high gain of highly directional antennas, they are ideal for long-distance point-to-point communications as far as 35 miles (58 km).

Because of the long distances and narrow beamwidth, highly directional antennas are affected more by antenna wind loading, which is antenna movement or shifting caused by wind. Even slight movement of a highly directional antenna can cause the RF beam to be aimed away from the receiving antenna, interrupting RF communications. In high-wind environments, grid antennas, because of the spacing between the wires, are less susceptible to wind load and may be a better choice.

Another option in high-wind environments is to choose an antenna with a wider beamwidth. In this situation, if the antenna were to shift slightly, the signal would still be received because of its wider coverage area. Keep in mind that a wider beam means less gain. If a solid dish is used, it is highly recommended that a radome be used to help offset some of the effects of the wind. No matter which type of antenna is installed, the quality of the mount and antenna will have a huge effect in reducing wind load.

FIGURE 4.13 Grid antenna

Phased Array Antennas

A *phased array antenna* is actually an antenna system made up of multiple antennas that are connected to a signal processor. The processor feeds the individual antennas with signals of different relative phases, creating a directed beam of RF signal aimed at the client device. Because this antenna is capable of creating narrow beams, it is also able to transmit multiple beams to multiple users simultaneously. Phased array antennas do not behave like other antennas, because they can transmit multiple signals at the same time. Because of this unique capability, they are often regulated differently by the local RF regulatory agency.

Phased array antennas are extremely specialized, expensive, and have not been commonly used in the 802.11 market. It is an interesting and capable technology; however, time will tell whether it has a future in the 802.11 market. The 802.11n draft amendment proposes an optional PHY capability called *transmit beamforming (TxBF)*. The technology uses phased-array antenna technology and is often referred to as *smart antenna technology.*

Sector Antennas

Sector antennas are a special type of high-gain, semidirectional antenna that provides a pie-shaped coverage pattern. These antennas are typically installed in the middle of the area where RF coverage is desired and placed back to back with other sector antennas. Individually, each antenna services its own piece of the pie, but as a group, all of the pie pieces fit together and provide omnidirectional coverage for the entire area. As pictured in Figure 4.14, combining multiple sector antennas to provide 360 degrees of horizontal coverage is known as a *sectorized array*.

FIGURE 4.14 Sectorized array

Unlike other semidirectional antennas, a sector antenna generates very little RF signal behind the antenna (*back lobe*) and therefore does not interfere with the other sector antennas that it is working with. The horizontal beamwidth of a sector antenna is from 60 to 180 degrees, with a narrow vertical beamwidth of 7 to 17 degrees. Sector antennas typically have a gain of at least 10 dBi.

Installing a group of sector antennas to provide omnidirectional coverage for an area provides many benefits over installing a single omnidirectional antenna. To begin with, sector antennas can be mounted high over the terrain and tilted slightly downward, with the tilt of each antenna at an angle appropriate for the terrain it is covering. Omnidirectional antennas can also be mounted high over the terrain; however, if an omnidirectional antenna is tilted downward on one side, the other side will be tilted upward.

Since each antenna covers a separate area, each antenna can be connected to a separate transceiver and can transmit and receive independently of the other antennas. This provides the capability for all of the antennas to be transmitting at the same time, providing much greater throughput. A single omnidirectional antenna is capable of transmitting to only one device at a time. The last benefit of the sector antennas over a single omnidirectional antenna is that the gain of the sector antennas is much greater than the gain of the omnidirectional antenna, providing a much larger coverage area.

Sector antennas are used extensively for cellular telephone communications and are starting to be used for 802.11 networking.

Real World Scenario

Cellular Sector Antennas Are Everywhere

As you walk or drive around your town or city, look for radio communications towers. Many of these towers have what appear to be rings of antennas around them. These rings of antennas are sector antennas. If a tower has more than one grouping or ring around it, then multiple cellular carriers are using the same tower.

Visual Line of Sight

When light travels from one point to another, it travels across what is perceived to be an unobstructed straight line, known as the visual *line of sight (LOS)*. For all intents and purposes, it is a straight line, but because of the possibility of light refraction, diffraction, and reflection, there is a slight chance that it is not. If you have been outside on a summer day and looked across a hot parking lot at a stationary object, you may have noticed that, because of the heat rising from the pavement, the object that you were looking at seemed to be moving. This is an example of how visual LOS is sometimes altered slightly. When it comes to RF communications, visual LOS has no bearing on whether the RF transmission is successful.

RF Line of Sight

Point-to-point RF communication also needs to have an unobstructed line of sight between the two antennas. So the first step for installing a point-to-point system is to make sure that from the installation point of one of the antennas, you have a clear direct path to the other antenna. Unfortunately, for RF communications to work properly, this is not sufficient. An additional area around the visual LOS needs to remain clear of obstacles and obstructions. This area around the visual LOS is known as the Fresnel zone and is often referred to as RF line of sight.

Fresnel Zone

The *Fresnel zone* (pronounced *FRUH-nel*—the *s* is silent) is an imaginary football-shaped area (American football) that surrounds the path of the visual LOS between two point-to-point antennas. Figure 4.15 shows an illustration of the Fresnel zone's football-like shape.

FIGURE 4.15 Fresnel zone

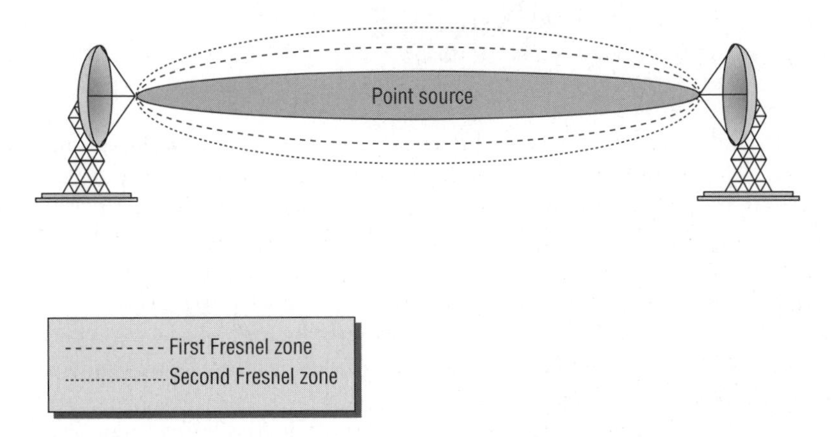

Theoretically, there are an infinite number of Fresnel zones, or concentric ellipsoids (the football shape), that surround the visual LOS. The closest ellipsoid is known as the first Fresnel zone, the next one is the second Fresnel zone, and so on, as seen in Figure 4.15. For simplicity's sake, and because they are the most relevant for this section, only the first two Fresnel zones are displayed in the figure. The subsequent Fresnel zones have very little effect on communications.

If the first Fresnel zone becomes even partly obstructed, the obstruction will negatively influence the integrity of the RF communication. In addition to the obvious reflection and scattering that can occur if there are obstructions between the two antennas, the RF signal can be diffracted or bent as it passes an obstruction of the Fresnel zone. This diffraction of the signal decreases the amount of RF energy that is received by the antenna and may even cause the communications link to fail.

Figure 4.16 illustrates a link that is 1 mile long. The top solid line is a straight line from the center of one antenna to the other. The dotted line shows 60 percent of the bottom half of the first Fresnel zone. The bottom solid line shows the bottom half of the first Fresnel zone. The trees are potential obstructions along the path.

Under no circumstances should you allow any object or objects to encroach more than 40 percent into the first Fresnel zone of an outdoor point-to-point bridge link. Anything more than 40 percent is likely to make the communications link unreliable. Even less than 40 percent obstruction is likely to impair the performance of the link. Therefore, it is recommended that you try not to allow more than 20 percent obstruction of the first Fresnel zone, particularly in wooded areas where the growth of trees may obstruct the Fresnel zone further in the future. A solid design will leave the first Fresnel zone completely free.

FIGURE 4.16 60 percent and 100 percent Fresnel zone clearances

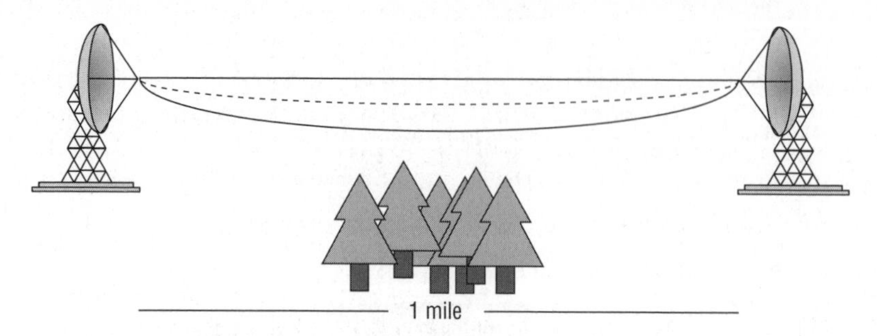

The typical obstacles that you are likely to encounter are trees and buildings. It is important to periodically visually check your link to make sure that trees have not grown into the Fresnel zone or that buildings have not been constructed that encroach into the Fresnel zone. Do not forget that the Fresnel zone exists below, to the sides, and above the visual LOS. If the Fresnel zone does become obstructed, you will need to either move the antenna (usually raise it) or remove the obstacle (usually with a chain saw—just kidding).

To determine whether an obstacle is encroaching into the Fresnel zone, you need to be familiar with a few formulas that enable you to calculate its radius. Don't fret; you will not be tested on these formulas.

The first formula enables you to calculate the radius of the first Fresnel zone at the midpoint between the two antennas. This is the point where the Fresnel zone is the largest. This formula is as follows:

radius = $72.2 \times \sqrt{[D \div (4 \times F)]}$

D = distance of the link in miles

F = transmitting frequency in GHz

This is the optimal clearance that you want along the signal path. Although this is the ideal radius, it is not always feasible or practical. Therefore, the next formula will be very useful. It can be used to calculate the radius of the Fresnel zone that will enable you to have 60 percent of the Fresnel zone unobstructed. This is the minimum amount of clearance you need at the midpoint between the antennas. Here is this formula:

radius (60%) = $43.3 \times \sqrt{[D \div (4 \times F)]}$

D = distance of the link in miles

F = transmitting frequency in GHz

Both of these formulas are very useful, but in addition to their benefits, they have major shortcomings. These formulas calculate the radius of the Fresnel zone at the midpoint between the antennas. Since this is the point where the Fresnel zone is the largest, these numbers can be used to determine the minimum height the antennas need to be above the ground. You need to know this number, because if you place the antennas too low, the ground would encroach on the Fresnel zone and cause degradation to the communications. The problem is that if there is a known object somewhere other than the midpoint between the antennas, it is not possible to calculate the radius of the Fresnel zone at that point by using these equations. The following formula can be used to calculate the radius of any Fresnel zone at any point between the two antennas:

radius = $72.2 \times \sqrt{[(N \times d1 \times d2) \div (F \times D)]}$

N = which Fresnel zone you are calculating (usually 1 or 2)

d1 = distance from one antenna to the location of the obstacle in miles

d2 = distance from the obstacle to the other antenna in miles

D = total distance between the antennas in miles (D = d1 + d2)

F = frequency in GHz

To look at an example, Figure 4.17 shows a point-to-point communications link that is 10 miles long. There is an obstacle (tree) that is 3 miles away from one antenna and 40 feet tall. So the values and the formula to calculate the radius of the Fresnel zone at a point 3 miles from the antenna are as follows:

N = 1 (for first Fresnel zone)

d1 = 3 miles

d2 = 7 miles

D = 10 miles

F = 2.4 GHz

radius at 3 miles = $72.2 \times \sqrt{[(1 \times 3 \times 7) \div (2.4 \times 10)]}$

radius at 3 miles = $72.2 \times \sqrt{(21 \div 24)}$

radius at 3 miles = 67.53 feet

So if the obstacle is 40 feet tall and the Fresnel zone at that point is 67.53 feet tall, the antennas need to be mounted at least 108 feet (40' + 67.53' = 107.53', we rounded up) above the ground to have complete clearance. If we are willing to allow the obstruction to encroach up to 40 percent into the Fresnel zone, we need to keep 60 percent of the Fresnel zone clear. So 60 percent of 67.53 feet is 40.52 feet. The absolute minimum height of the antennas will need to be 81 feet (40' + 40.52' = 80.52', again we rounded up). In the next section, you will learn that because of the curvature of the earth, you will need to raise the antennas even higher to compensate for the earth's bulge.

FIGURE 4.17 Point-to-point communication with potential obstacle

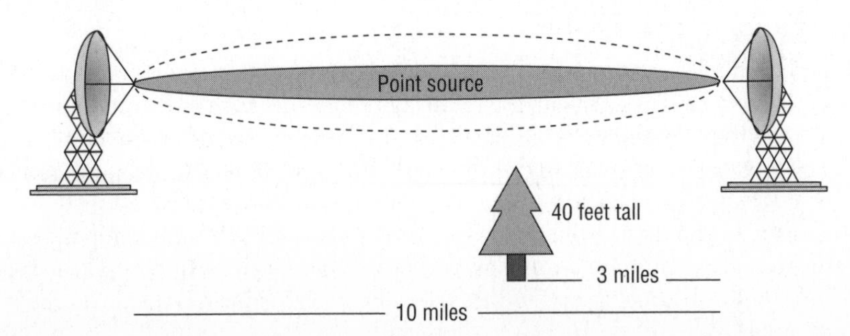

When highly directional antennas are used, the beamwidth of the signal is smaller, causing a more focused signal to be transmitted. Many people think that a smaller beamwidth would decrease the size of the Fresnel zone. This is not the case. The size of the Fresnel zone is a function of the frequency being used and the distance of the link. Since the only variables in the formula are frequency and distance, the size of the Fresnel zone will be the same regardless of the antenna type or beamwidth. The first Fresnel zone is technically the area around the point source, where the waves are in phase with the point source signal. The second Fresnel zone is then the area beyond the first Fresnel zone, where the waves are out of phase with the point source signal. All of the odd-numbered Fresnel zones are in phase with the point source signal, and all of the even-numbered Fresnel zones are out of phase.

If an RF signal of the same frequency but out of phase with the primary signal intersects the primary signal, the out-of-phase signal will cause degradation or even cancellation of the primary signal (this is covered in Chapter 2 and demonstrated using the EMANIM software). One of the ways that an out-of-phase signal can intercept the primary signal is by reflection. It is therefore important to consider the second Fresnel zone when evaluating point-to-point communications. If the height of the antennas and the layout of the geography are such that the RF signal from the second Fresnel zone is reflected toward the receiving antenna, it can cause degradation of the link. Although this is not a common occurrence, the second Fresnel zone should be considered when planning or troubleshooting the connection, especially in flat, arid terrain like a desert. You should also be cautious of metal roofing along the Fresnel zone.

Please understand that the Fresnel zone is three-dimensional. Can something impede on the Fresnel zone from above? Although trees do not grow from the sky, a point-to-point bridge link could be shot under a railroad trestle or a freeway. In these rare situations, consideration would have to be given to proper clearance of the upper radius of the first Fresnel zone. A more common scenario would be the deployment of point-to-point links in an urban city environment. Very often building-to-building links must be shot between other buildings. In these situations, other buildings have the potential of impeding the side radiuses of the Fresnel zone.

Until now, all of the discussion about the Fresnel zone has related to point-to-point communications. The Fresnel zone exists in all RF communications; however, it is in outdoor point-to-point communications where it can cause the most problems. Indoor environments have so many walls and other obstacles where there is already so much reflection, refraction, diffraction, and scattering that the Fresnel zone does not play a very big part in the success or failure of the link.

Earth Bulge

When you are installing long-distance point-to-point RF communications, another variable that must be considered is the curvature of the earth, also known as the *earth bulge*. Because the landscape varies throughout the world, it is impossible to specify an exact distance for when the curvature of the earth will affect a communications link. The recommendation is that if the antennas are more than 7 miles away from each other, you should take into consideration the earth bulge, because after 7 miles, the earth itself begins to impede on the Fresnel zone. The following formula can be used to calculate the additional height that the antennas will need to be raised to compensate for the earth bulge:

$H = D^2 \div 8$

H = height of the earth bulge in feet

D = distance between the antennas in miles

You now have all of the pieces to estimate how high the antennas need to be installed. Remember, this is an estimate that is being calculated, because it is assumed that the terrain between the two antennas does not vary. You need to know or calculate the following three things:

- The 60 percent radius of the first Fresnel zone
- The height of the earth bulge
- The height of any obstacles that may encroach into the Fresnel zone, and the distance of those obstacles from the antenna

Taking these three pieces and adding them together gives you the following formula, which can be used to calculate the antenna height:

H = obstacle height + earth bulge + Fresnel zone

$H = OB + (D^2 \div 8) + (43.3 \times \sqrt{[D \div (4 \times F)]})$

OB = obstacle height

D = distance of the link in miles

F = transmitting frequency in GHz

As an example, Figure 4.18 shows a point-to-point link that spans a distance of 12 miles. In the middle of this link is an office building that is 30 feet tall. A 2.4 GHz signal is being used to communicate between the two towers. Using the formula, we calculate that each of the antennas needs to be installed at least 96.4 feet above the ground:

$H = 30 + (12^2 \div 8) + (43.3 \times \sqrt{[12 \div (4 \times 2.4)]})$

$H = 30 + 18 + 48.4$

$H = 96.4$

Although these formulas are useful, the good news is that you do not need to know them for the test.

FIGURE 4.18 Calculating antenna height

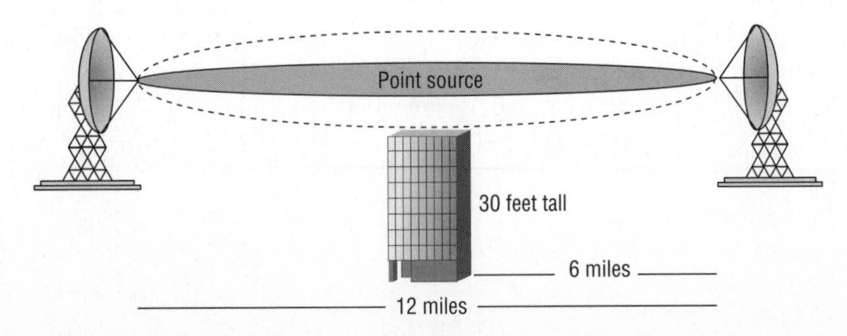

Antenna Polarization

Another consideration when installing antennas is *antenna polarization*. Although it is a lesser-known concern, it is extremely important for successful communications. Proper polarization alignment is vital when installing any type of antenna. Whether the antennas are installed with horizontal or vertical polarization is irrelevant, as long as both antennas are aligned with the same polarization.

Polarization is not as important for indoor communications because the polarization of the RF signal often changes when it is reflected, which is a common occurrence indoors. Most access points use low-gain omnidirectional antennas, and they should be polarized vertically when mounted from the ceiling. Laptop manufacturers build diversity antennas into the sides of the monitor. When the laptop monitor is in the upright position, the internal antennas are vertically polarized as well.

When aligning a point-to-point or point-to-multipoint bridge, proper polarization is extremely important. If the best received signal level (RSL) you receive when aligning the antennas is 15 to 20 dB less than your estimated RSL, then there is a good chance you have cross-polarization. If this difference exists on only one side and the other has higher signal, you are aligned to a side lobe.

On the CD that is included with this book is an excellent video, *Beam Patterns and Polarization of Directional Antennas*. This 3-minute video explains and demonstrates the effects of antenna side lobes and polarization. The filename of the video is `Antenna Properties.wmv`.

Antenna Diversity

Wireless networks, especially indoor networks, are prone to multipath signals. To help compensate for the effects of multipath, antenna diversity, also called spatial diversity, is commonly implemented in wireless networking equipment such as access points (APs). *Antenna diversity* exists when an access point has two antennas and receivers functioning together to minimize the negative effects of multipath. Figure 4.19 shows a picture of an access point that uses antenna diversity.

Because the wavelengths of 802.11 wireless networks are less than 5 inches long, the antennas can be placed very near each other and still allow antenna diversity to be effective. When the access point senses an RF signal, it compares the signal that it is receiving on both antennas and uses whichever antenna has the higher signal strength to receive the frame of data. This sampling is performed on a frame-by-frame basis, choosing whichever antenna has the higher signal strength.

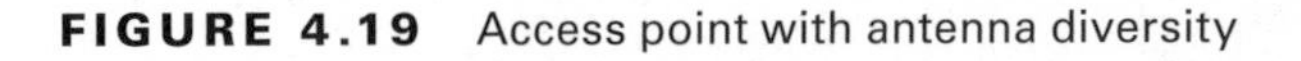

FIGURE 4.19 Access point with antenna diversity

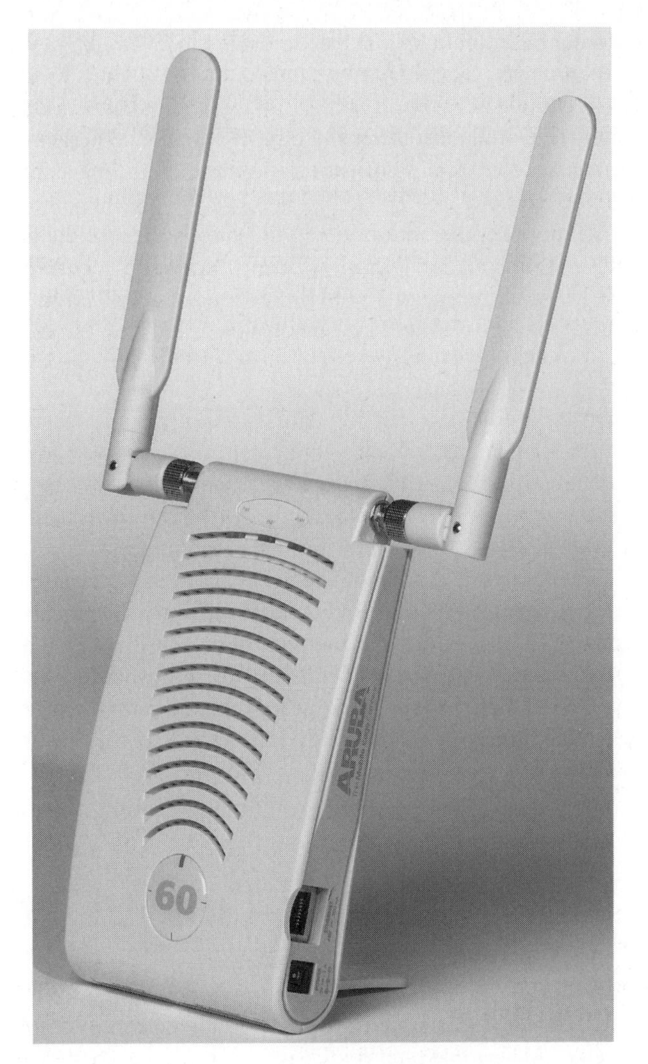

Most pre-802.11n radios use *switched diversity*. When receiving incoming transmissions, switched diversity listens with multiple antennas. Multiple copies of the same signal arrive at the receiver antennas with different amplitudes. The signal with the best amplitude is chosen, and the other signals are ignored.

The method of listening for the best received signal is known as *receive diversity*. Switched diversity is also used when transmitting, but only one antenna is used. The transmitter will transmit out of the diversity antenna where the best amplitude signal was last heard. The method of transmitting out of the antenna where the last best received signal was heard is known as *transmit diversity*.

Note that when an access point has two antenna ports for antenna diversity, the antennas should be installed in the same location and with the same orientation. You should not be running antenna cables to antennas in opposite directions to try to provide better coverage. The distance between the antennas should be a factor of the wavelength (¼, ½, 1, 2).

Because the antennas are so close to each other, it is not uncommon to doubt that antenna diversity is actually beneficial. As you may recall from Chapter 3, the amount of RF signal that is received is often less than 0.00000001 milliwatts. At this level of signal, the slightest difference between the signals that each antenna receives can be significant. Other factors to remember are that the access point is often communicating with multiple client devices at different locations. These clients are not always stationary, thus further affecting the path of the RF signal.

The access point has to handle transmitting data differently than receiving data. When the access point needs to transmit data back to the client, it has no way of determining which antenna the client would receive from the best. An access point can handle transmitting data by using the antenna that it used most recently to receive data. This is often referred to as transmit diversity. Not all access points are equipped with this capability.

There are many kinds of antenna diversity. The most common implementation of antenna diversity utilizes one radio card, two connectors, and two antennas. The question often gets asked why client cards seem to have only one antenna. In reality, PCMCIA client cards typically have two diversity antennas encased inside the card. Laptops with internal cards have diversity antennas mounted inside the laptop monitor. Remember that because of the half-duplex nature of the RF medium, when antenna diversity is used, only one antenna is operational at any given time. In other words, a radio card transmitting a frame with one antenna cannot be receiving a frame with the other antenna at the same time.

Multiple-Input Multiple-Output (MIMO)

Multiple-input multiple-output (MIMO) is another, more sophisticated form of antenna diversity. Unlike conventional antenna systems, where multipath propagation is an impairment, MIMO (pronounced *MY-moh*) systems take advantage of multipath. There is much research and development currently happening with this technology. MIMO can safely be described as any RF communication system that has multiple antennas at both ends of the communications link, with the antennas being used concurrently. Complex signal-processing techniques known as Space Time Coding (STC) are often associated with MIMO. These techniques send data by using multiple simultaneous RF signals, and the receiver then reconstructs the data from those signals. The proposed 802.11n amendment includes MIMO technology. MIMO is discussed in much further detail in Chapter 18, “HT (High Throughput) and 802.11n,” as it is a key component of 802.11n.

Antenna Connection and Installation

In addition to the physical antenna being a critical component in the wireless network, the installation and connection of the antenna to the wireless transceiver is critical. If the antenna is not properly connected and installed, any benefit that the antenna introduces to the network can be instantly wiped out. Three key components associated with the proper installation of the antenna are voltage standing wave ratio (VSWR), signal loss, and the actual mounting of the antenna.

Voltage Standing Wave Ratio (VSWR)

Voltage standing wave ratio (VSWR) is a measurement of the change in impedances to an AC signal. Voltage standing waves exist because of impedance mismatches or variations between devices in an RF communications system. Impedance is a value of ohms of electrical resistance to an AC signal. A standard unit of measurement of electrical resistance is the ohm, named after German physicist Georg Ohm. When the transmitter generates the AC radio signal, the signal travels along the cable to the antenna. Some of this incident (or forward) energy is reflected back toward the transmitter because of impedance mismatch.

Mismatches may occur anywhere but are usually due to abrupt impedance changes between the radio transmitter and cable and between the cable and the antenna. The amount of energy reflected depends on the level of mismatch between the transmitter, cable, and antenna. The ratio between the voltage of the reflected wave and the voltage of the incident wave, *at the same point along the cable,* is called the *voltage reflection coefficient*, usually designated by the Greek letter rho (ρ).

When this quantity is expressed in dB, it is called *return loss*. So in an ideal system, where there are no mismatches (the impedance is the same everywhere), all of the incident energy will be delivered to the antenna (except for resistive losses in the cable itself) and there will be no reflected energy. The cable is said to be *matched*, and the voltage reflection coefficient is exactly zero and the return loss, in dB, is infinite. The combination of incident and reflected waves traveling back and forth along the cable creates a resulting *standing wave* pattern along the length of the line. The standing wave pattern is periodic (it repeats) and exhibits multiple peaks and troughs of voltage, current, and power.

VSWR is a numerical relationship between the measurement of the maximum voltage along the line (what is generated by the transmitter) and the measurement of the minimum voltage along the line (what is received by the antenna). VSWR is therefore a ratio of impedance mismatch, with 1:1 (no impedance) being optimal but unobtainable, and typical values from 1.1:1 to as much as 1.5:1. VSWR military specs are 1.1:1.

$$VSWR = V_{max} \div V_{min}$$

When the transmitter, cable, and antenna impedances are matched (that is, there are no standing waves), the voltage along the cable will be constant. This matched cable is also referred to as a *flat line* because there are no peaks and troughs of voltage along the length of the cable. In this case, VSWR is 1:1. As the degree of mismatch increases, the VSWR increases with a corresponding decrease in the power delivered to the antenna. Table 4.2 shows this effect.

TABLE 4.2 Signal Loss Caused by VSWR

VSWR	Radiated Power	Lost Power	dB Power Loss
1:1	100%	0%	0 dB
1.5:1	96%	4%	Nearly 0 dB
2:1	89%	11%	< 1 dB
6:1	50%	50%	3 dB

If VSWR is large, this means that a large amount of voltage is being reflected back toward the transmitter. This of course means a decrease in power or amplitude (loss) of the signal that is supposed to be transmitted. This loss of forward amplitude is known as *return loss* and can be measured in dB. Additionally, the power that is being reflected back is then directed back into the transmitter. If the transmitter is not protected from excessive reflected power or large voltage peaks, it can overheat and fail. Understand that VSWR may cause decreased signal strength, erratic signal strength, or even transmitter failure.

The first thing that can be done to minimize VSWR is to make sure that the impedance of all of the wireless networking equipment is matched. Most wireless networking equipment has an impedance of 50 ohms; however, you should check the manuals to confirm this. When attaching the different components, make sure that all connectors are installed and crimped properly and that they are snugly tightened.

Signal Loss

When connecting an antenna to a transmitter, the main objective is to make sure that as much of the signal that is generated by the transmitter is received by the antenna to be transmitted. To achieve this, it is important to pay particular attention to the cables and connectors that connect the transmitter to the antenna. In the "Antenna Accessories" section later in this chapter, we review the cables, connectors, and many other components that are used when installing antennas. If inferior components are used, or if the components are not installed properly, the access point will most likely function below its optimal capability.

Antenna Mounting

As stated earlier in this chapter, proper installation of the antenna is one of the most important tasks to ensure an optimally functioning network. The following are key areas to be concerned with when installing antennas:

- Placement
- Mounting

- Appropriate use
- Orientation and alignment
- Safety
- Maintenance

Placement

The proper placement of an antenna is dependent on the type of antenna. When installing omnidirectional antennas, it is important to place the antenna at the center of the area where you want coverage. Remember that lower-gain omnidirectional antennas provide broader vertical coverage, while higher-gain omnidirectional antennas provide wider but much flatter coverage. Be careful not to place high-gain omnidirectional antennas too high above the ground, because the narrow vertical coverage may cause the antenna to provide insufficient signal to clients located on the ground.

When installing directional antennas, make sure that you know both the horizontal and vertical beamwidths so that you can properly aim the antennas. Also make sure that you are aware of the amount of gain that the antenna is adding to the transmission. If the signal is too strong, it will overshoot the area that you are looking to provide coverage to. This can be a security risk, and you should decrease the amount of power that the transceiver is generating to reduce the coverage area. Not only can it be a security risk, overshooting your coverage area is considered rude.

If you are installing an outdoor directional antenna, in addition to concerns regarding the horizontal and vertical beamwidths, make sure that you have correctly calculated the Fresnel zone and mounted the antenna accordingly.

Mounting

After deciding *where* to place the antenna, the next step is to decide *how* to mount it. Many antennas, especially outdoor antennas, are mounted on masts or towers. It is common to use mounting clamps and U-bolts to attach the antennas to the masts. For mounting directional antennas, specially designed tilt-and-swivel mounting kits are available to make it easier to aim and secure the antenna. If the antenna is being installed in a windy location (and what rooftop or tower isn't windy?), make sure that you take into consideration wind load and properly secure the antenna.

There are numerous ways of mounting antennas indoors. Two common concerns are aesthetics and security. Many organizations, particularly ones that provide hospitality-oriented services such as hotels and hospitals, are concerned about the aesthetics of the installation of the antennas. Specialty enclosures and ceiling tiles can help to hide the installation of the access points and antennas. Other organizations, particularly schools and public environments, are concerned with securing the access points and antennas from theft or vandalism. An access point can be locked in a secure enclosure, with a short cable connecting it to the antenna. There are even ceiling tiles with antennas built into them, invisible to anyone walking by. If security is a concern, mounting the antenna high on the wall or ceiling can also minimize unauthorized access.

Appropriate Use

Make sure that indoor antennas are not used for outdoor communications. Outdoor antennas are specifically built to withstand the wide range of temperatures that they may be exposed to. Outdoor antennas are also built to stand up to other elements, such as rain, snow, and fog. In addition to installing the proper antenna, make sure that the mounts you use are designed for the environment in which you are installing them.

Orientation and Alignment

Before installing an antenna, make sure you read the manufacturer's recommendations for mounting it. This suggestion is particularly important when installing directional antennas. Since directional antennas may have different horizontal and vertical beamwidths, and because directional antennas can be installed with different polarization, proper orientation can make the difference between being able to communicate or not.

First, make sure you have decided on a polarization. Next, decide on the mounting technique and ensure that it is compatible with the mounting location. Then align the antennas. After the antennas are aligned, the cables and connectors can be weatherproofed and secured from movement.

Safety

We can't emphasize enough the importance of being careful when installing antennas. Most of the time, the installation of an antenna requires climbing ladders, towers, or rooftops. Gravity and wind have a way of making an installation difficult for both the climber and the people below helping.

Plan the installation before you begin, making sure you have all of the tools and equipment that you will need to install the antenna. Unplanned stoppages of the installation and relaying forgotten equipment up and down the ladder add to the risk of injury.

Be careful when working with your antenna or near other antennas. Highly directional antennas are focusing high concentrations of RF energy. This large amount of energy can be dangerous to your health. Do not power on your antenna while you are working on it, and do not stand in front of other antennas that are near where you are installing your antenna.

When installing antennas (or any device) on ceilings, rafters, or masts, make sure they are properly secured. Even a 1-pound antenna can be deadly if it falls from the rafters of a warehouse.

If you will be installing antennas as part of your job, we recommend that you take an RF health and safety course. These courses will teach you the FCC and the U.S. Department of Labor Occupational Safety and Health Administration (OSHA) regulations and how to be safe and compliant with the standards.

If you need an antenna installed on any elevated structure, such as a pole, tower, or even a roof, consider hiring a professional installer. Professional climbers and installers are trained and in some places certified to perform these types of installations. In addition to the training, they have the necessary safety equipment and proper insurance for the job.

If you are planning to install wireless equipment as a profession, you should develop a safety policy that is blessed by your local occupational safety representative. You should also receive certified training on climbing safety in addition to RF safety training. First aid and CPR training are also highly recommended.

Maintenance

There are two types of maintenance: preventative and diagnostic. When installing an antenna, it is important to prevent problems from occurring in the future. This seems like simple advice, but since antennas are often difficult to get to after they have been installed, it is especially prudent advice. Two key problems that can be minimized with proper preventative measures are wind damage and water damage. When installing the antenna, make sure all of the nuts, bolts, screws, and so on are tightened. Also make sure all of the cables are properly secured so that they are not thrashed about by the wind.

To help prevent water damage, cold-shrink tubing or coaxial sealant can be used to minimize the risk of water getting into the cable or connectors. Another common method is a combination of electrical tape and mastic, installed in layers to provide a completely watertight installation.

Heat-shrink tubing should not be used because the cable can be damaged by the heat that is necessary to shrink the wrapping. Silicone also should not be used, because air bubbles can form under the silicone and moisture can collect.

Another cabling technique is the drip loop. A drip loop prevents water from flowing down the cable and onto a connector or into the hole where a cable exits the building. Any water that is flowing down the cable will continue to the bottom of the loop and then drip off.

Antennas are typically installed and forgotten about until they break. It is advisable to periodically perform a visual inspection of the antenna. If the antenna is not easily accessible, a pair of binoculars or a camera with a very high zoom lens can make this a simple task.

Antenna Accessories

In Chapter 3, we introduced the components of RF communications. In that chapter, the main components were reviewed; however, there are other components that are either not as significant or not always installed as part of the communications link. Important specifications for all antenna accessories include frequency response, impedance, VSWR, maximum input power, and insertion loss. This section will discuss some of these components and accessories.

Cables

Improper installation or selection of cables can detrimentally affect the RF communications more than just about any other component or outside influence. It is important to remember this when installing antenna cables. The following list addresses some concerns when selecting and installing cables:

- Make sure you select the correct cable. The impedance of the cable needs to match the impedance of the antenna and transceiver. If there is an impedance mismatch, the return loss from VSWR will affect the link.
- Make sure the cable you select will support the frequencies that you will be using. Typically, cable manufacturers list cutoff frequencies, which are the lowest and highest frequencies that the cable supports. This is often referred to as frequency response. For instance, LMR cable is a popular brand of coaxial cable used in RF communications. LMR-1200 will not work with 5 GHz transmissions. LMR-900 is the highest you can use. However, you can use LMR-1200 for 2.4 GHz operations.
- Cables introduce signal loss into the communications link. To determine how much loss, cable vendors provide charts or calculators to assist you. Figure 4.20 is an attenuation chart for LMR cable produced by Times Microwave Systems. The left side of the chart lists different types of LMR cable. The farther you move down the list, the better the cable is. The better cable is typically thicker, stiffer, more difficult to work with, and of course, more expensive. The chart shows how much decibel loss the cable will add to the communications link. The column headers list the frequencies that may be used with the cable. For example, 100 feet of LMR-400 cable used on a 2.5 GHz network (2,500 MHz) would decrease the signal by 6 dB.
- Attenuation increases with frequency. If you convert from 802.11b to 802.11a, the loss caused by the cable will be greater.
- Either purchase the cables precut and preinstalled with the connectors or hire a professional cabler to install the connections. Improperly installed connectors will add more loss to the communications link, which can nullify the extra money you spend for the better-quality cable. It can also introduce return loss in the cable due to reflections.

Connectors

Many types of connectors are used to connect antennas to 802.11 equipment. Part of the reason for this is that the FCC Report & Order 04-165 requires that amplifiers have either unique connectors or electronic identification systems to prevent the use of noncertified antennas. This requirement was created to prevent people from connecting higher-gain antennas, either intentionally or unintentionally, to a transceiver. An unauthorized high-gain antenna could exceed the maximum EIRP that is allowed by the FCC or other regulatory body.

In response to this regulation, cable manufacturers sell *pigtail* adapter cables. These pigtail cables are usually short segments of cable (typically about 2 feet long) with different connectors on each end. They act as adapters, changing the connector, and allowing a different antenna to be used. The use of these adapter cables often violates the rules of the local regulatory body. They are typically used by Wi-Fi hobbyists or network installers for testing purposes. Remember that these pigtails usually violate RF regulations and are not recommended or condoned.

FIGURE 4.20 Coaxial cable attenuation chart

Times Microwave Systems

LMR Cable	30	50	150	220	450	900	1,500	1,800	2,000	2,500	5,800
100A	3.94	5.10	8.95	10.90	15.83	22.84	30.08	33.22	35.19	39.81	64.10
195	1.97	2.55	4.44	5.40	7.78	11.13	14.53	15.99	16.90	19.02	29.90
195UF	2.34	3.03	5.28	6.42	9.25	13.23	17.28	19.01	20.10	22.62	35.57
200	1.77	2.29	3.98	4.83	6.96	9.92	12.92	14.21	15.01	16.87	26.35
200UF	2.12	2.74	4.78	5.80	8.35	11.91	15.51	17.05	18.01	20.24	31.62
240	1.34	1.73	3.01	3.66	5.28	7.56	9.87	10.87	11.49	12.93	20.35
240UF	1.60	2.07	3.62	4.40	6.34	9.07	11.85	13.04	13.78	15.52	24.42
300	1.06	1.37	2.40	2.92	4.22	6.06	7.93	8.74	9.24	10.42	16.53
300UF	1.27	1.65	2.88	3.50	5.06	7.26	9.51	10.48	11.08	12.50	19.81
400	0.68	0.88	1.54	1.87	2.71	3.90	5.13	5.66	5.99	6.76	10.82
400UF	0.81	1.05	1.84	2.25	3.25	4.68	6.15	6.79	7.19	8.12	12.99
500	0.54	0.70	1.22	1.49	2.17	3.13	4.13	4.57	4.84	5.48	8.86
500UF	0.64	0.84	1.47	1.79	2.60	3.76	4.96	5.48	5.81	6.58	10.64
600	0.42	0.55	0.96	1.18	1.72	2.50	3.32	3.67	3.90	4.43	7.26
600UF	0.51	0.66	1.16	1.41	2.06	3.00	3.98	4.41	4.68	5.31	8.71
900	0.29	0.37	0.66	0.80	1.17	1.70	2.25	2.48	2.64	2.99	4.87
900UF	0.35	0.45	0.79	0.96	1.41	2.04	2.70	2.98	3.16	3.59	5.85
1200	0.21	0.27	0.48	0.59	0.87	1.27	1.69	1.87	1.99	2.27	not supported
1700	0.15	0.20	0.35	0.43	0.63	0.94	1.27	1.41	1.50	1.72	not supported

UF = Ultraflex (more flexible cable)

Many of the same principles of cables apply to the connectors and to many of the other accessories. RF connectors need to be of the correct impedance to match the other RF equipment. They also support specific ranges of frequencies. The connectors add signal loss to the RF link, and lower-quality connectors are more likely to cause connection or VSWR problems. RF connectors on average add about ½ dB of insertion loss.

Splitters

Splitters are also known as signal splitters, RF splitters, power splitters, and power dividers. A splitter takes an RF signal and divides it into two or more separate signals. Only in an unusually special or unique situation would you need to use an RF splitter. One such situation would be if you were connecting sector antennas to one transceiver. If you had three 120-degree antennas aimed away from a central point to provide 360-degree coverage, you could connect each antenna to its own transceiver or you could use a three-way splitter and equal-length cables to connect the antennas to a single transceiver. When you install a splitter in this type of configuration, not only will the signal be degraded because it is being split three times, known as through loss, but also each connector will add its own insertion loss to the signal. There are so many variables and potential problems with this configuration that we recommend that this type of installation be attempted only by a very RF knowledgeable person and only for temporary installations.

A more practical, but again rare, use of a splitter is to monitor the power that is being transmitted. The splitter can be connected to the transceiver and then split to the antenna and a power meter. This would enable you to actively monitor the power that is being sent to the antenna.

Amplifiers

An RF *amplifier* takes the signal that is generated by the transceiver, increases it, and sends it to the antenna. Unlike the antenna providing an increase in gain by focusing the signal, an amplifier provides an overall increase in power by adding electrical energy to the signal, which is referred to as active gain.

Amplifiers can be purchased as either unidirectional or bidirectional devices. Unidirectional amplifiers perform the amplification in only one direction, either when transmitting or when receiving. Bidirectional amplifiers perform the amplification in both directions.

The amplifier's increase in power is created using one of two methods: fixed gain or fixed output. With the fixed-gain method, the output of the transceiver is increased by the amount of the amplifier. A fixed-output amplifier does not add to the output of the transceiver. The fixed-output amplifier simply generates a signal equal to the output of the amplifier regardless of the power generated by the transceiver. Adjustable variable-gain amplifiers also exist, but using adjustable-gain amplifiers is not a recommended practice. Unauthorized adjustment of a variable-rate amplifier may result in either violation of power regulations or insufficient transmission amplitude.

Since most regulatory bodies have a maximum power regulation of 1 watt or less at the intentional radiator (IR), the main purpose of using amplifiers is to compensate for cable loss as opposed to boosting the signal for range. Therefore, when installing an amplifier, install it as close to the antenna as possible. Because the antenna cable adds loss to the signal, the shorter antenna cable will produce less loss and allow more signal to the antenna.

Additionally, it is important to note that an amplifier increases noise as well as signal strength. It is not uncommon for an amplifier to raise the noise floor by 10 dB or more.

Amplifiers must be certified with the system in use according to regulatory bodies such as the FCC. It is far better to further engineer the system than to use an amplifier.

Attenuators

In some situations, it may be necessary to decrease the amount of signal that is radiating from the antenna. You could be installing a short point-to-point link and want to reduce the output to minimize interference to other RF equipment in the area. In some instances, even the lowest power setting of the transceiver may generate more signal than you want. In this situation, you can add a fixed-loss or a variable-loss *attenuator.* Attenuators are typically small devices about the size of a C-cell battery, with cable connectors on both sides. Attenuators absorb energy, decreasing the signal as it travels through. Fixed-loss attenuators provide a set amount of loss. A variable-loss attenuator has a dial or switch configuration on it that enables you to adjust the amount of energy that is absorbed.

Variable-loss attenuators are often used during outdoor site surveys to simulate loss caused by various grades of cabling and different cable lengths. Another interesting use of a variable attenuator is to test the actual fade margin on a point-to-point link. By gradually increasing the attenuation until there is no more link, you can use that number to determine the actual fade margin of the link.

Lightning Arrestors

The purpose of a *lightning arrestor* is to redirect (shunt) transient currents caused by nearby lightning strikes or ambient static away from your electronic equipment and into the ground. Lightning arrestors are used to protect electronic equipment from the sudden surge of power that a nearby lightning strike or static buildup can cause. You may have noticed the use of the phrase *nearby lightning strike.* This wording is used because lightning arrestors are not capable of protecting against a direct lightning strike. Lightning arrestors can typically protect against surges of up to 5,000 amperes at up to 50 volts. The IEEE specifies that lightning arrestors should be capable of redirecting the transient current in less than 8 microseconds. Most lightning arrestors are capable of doing it in less than 2 microseconds.

The lightning arrestor is installed between the transceiver and the antenna. Any devices that are installed between the lightning arrestor and the antenna will not be protected by the lightning arrestor. Therefore, the lightning arrestor is typically placed closer to the antenna, with all other communications devices (amplifiers, attenuators, etc.) installed between the lightning arrestor and the transceiver. Figure 4.21 shows a properly grounded radio, cabling, and antenna. After a lightning arrestor has performed its job by protecting the equipment from an electrical surge, it will have to be replaced, or it may have a replaceable gas discharge tube (like a fuse). Most installations place the lighting arrestor at the egress to the building. Cable grounding kits can be installed near the antenna and at every 100 feet.

Fiber-optic cable can also be used to provide additional lightning protection. A short piece of fiber-optic cable can be inserted into the Ethernet cable that connects the wireless bridge to the rest of the network. Ethernet-to-fiber adapters, known as transceivers, convert the electrical Ethernet signal to a light-based fiber signal and then back to Ethernet. Because fiber-optic cable is constructed of glass and it uses light and not electricity to transmit data, it does not conduct electricity. It is important to make sure that the power supply for the adapters is protected as well.

The fiber-optic cable acts as a kind of safety net should the lightning arrestor fail due to a much higher transient current or even a direct lightning strike. Realize that if there is a direct lightning strike to the antenna, you can plan on replacing all of the components from the fiber-optic cable to the antenna. Furthermore, a direct lightning strike may also arc over the fiber link and still cause damage to equipment on the opposite side of the fiber link. Grounding the RF cables as well can help prevent this from happening.

FIGURE 4.21 Installation of lightning protection equipment

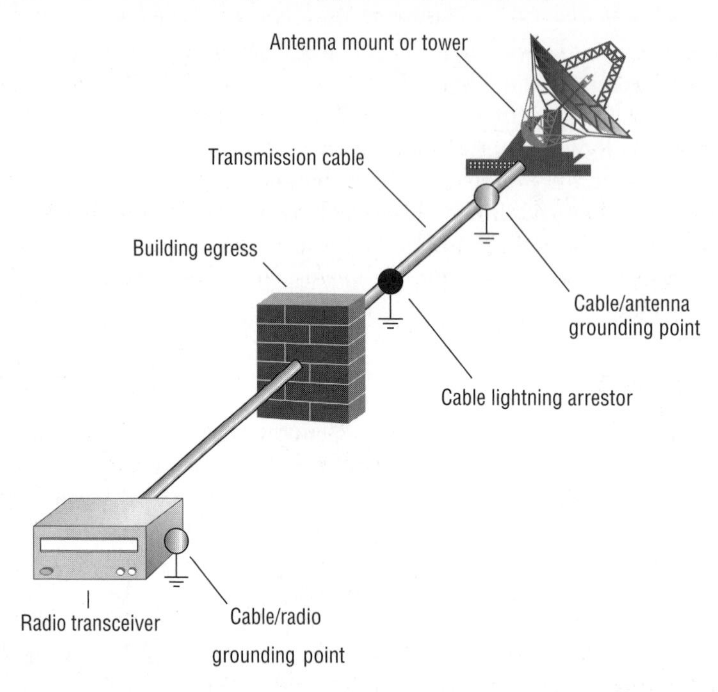

Real World Scenario

Not Only Is Lightning Unpredictable, the Results Are Too!

A business in a five-story, 200-year-old brick brownstone in the North End neighborhood of Boston had a lightning strike, or a nearby lightning strike. This building was not even one of the tallest buildings in the area, and it was at the bottom of a small hill and surrounded by other similar buildings. An electrical current traveled down the water vent pipe, past a bundle of Ethernet cables. A transient current on the Ethernet cables damaged the transceiver circuits on the Ethernet cards in the PCs and on the individual ports on the Ethernet hub. About half of the Ethernet devices in the company failed, and about half of the ports on the hub were no longer functioning. Yet all of the software recognized the cards, and all of the power and port lights worked flawlessly. The problem appeared to be cabling related.

You often will not know that the problem is lightning related, and the symptoms may be misleading. Testing the lightning arrestors can help with your diagnosis.

Grounding Rods and Wires

When lightning strikes an object, it is looking for the path of least resistance, or more specifically, the path of least impedance. This is where lightning protection and grounding equipment come into play. A grounding system, which is made up of a grounding rod and wires, provides a low-impedance path to the ground. This low-impedance path is installed to encourage the lightning to travel through it instead of through your expensive electronic equipment.

Grounding rods and wires are also used to create what is referred to as a *common ground*. One way of creating a common ground is to drive a copper rod into the ground and connect your electrical and electronic equipment to this rod by using wires or straps (grounding wires). The grounding rod should be at least 6 feet long and should be fully driven into the ground, leaving enough of the rod accessible to attach the ground wires to it. By creating a common ground, you have created a path of least impedance for all of your equipment should lightning cause an electrical surge.

On tower structures, a grounding rod should be placed off of each leg with a No. 2 tinned copper wire. These connections should be exothermically welded to the tower legs. A No. 2 tinned copper wire should also form a ring around the grounding rods, as illustrated in Figure 4.22. The dashed lines are No. 2 tinned copper wire and the circles are grounding rods. Ice bridges and building grounds should also be bonded to this ring to provide equal grounding potentials.

FIGURE 4.22 Grounding ring

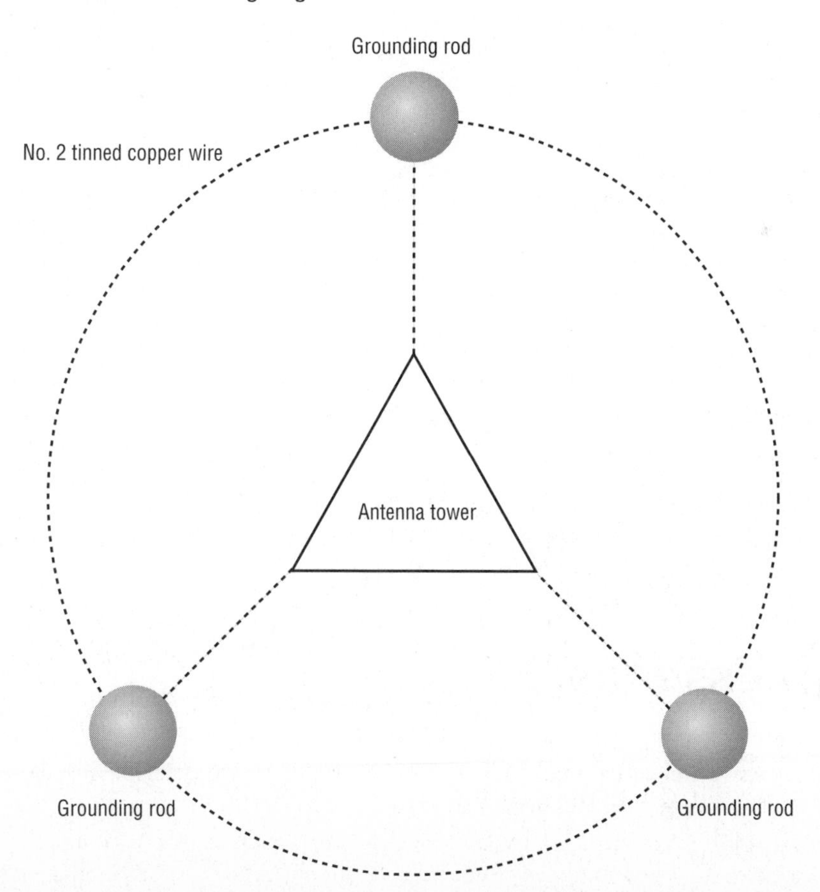

Section 12.2.2 of the EIA/TIA 222F standard states that "a minimum ground shall consist of two ⅝" diameter galvanized steel rods driven not less than 8 feet into the ground....The ground rods shall be bonded with a lead of not smaller than No. 6 tinned bare copper to the nearest leg or to the metal base of the structure." Current standards dictate a No. 2 solid wire from bus bars and other items to the ground ring. The ground ring is often 2/0 stranded bare wire.

Summary

This chapter focused on RF signal and antenna concepts. The antenna is a key component of successful RF communications. There are five types of antennas that are used with 802.11 networks:

- Omnidirectional (dipole, collinear)
- Semidirectional (patch, panel, Yagi)
- Highly directional (parabolic dish, grid)
- Phased array
- Sector

The antenna types produce different signal patterns, which can be viewed on azimuth and elevation charts.

This chapter also reviewed some of the key concerns when installing point-to-point communications:

- Visual LOS
- RF LOS
- Fresnel zone
- Earth bulge
- Antenna polarization

The final section of this chapter covered VSWR and antenna mounting issues, along with antenna accessories and their roles.

Exam Essentials

Understand passive and active gain. Understand how antennas provide passive gain and how transceivers and amplifiers provide active gain.

Know the different categories and types of antennas, how they radiate signals, and what type of environment they are used in. Make sure you know the three main categories of antennas and the different types of antennas. Know the similarities and differences between them, and understand when and why you would use one antenna over another. Make sure that you understand azimuth and elevation charts, beamwidth, antenna polarization, and antenna diversity.

Fully understand the Fresnel zone. Make sure you understand all of the issues and variables involved with installing point-to-point communications. You are not required to memorize the Fresnel zone or earth bulge formulas; however, you will need to know the principles regarding these topics and when and why you would use the formulas.

Understand the concerns associated with connecting and installing antennas and the antenna accessories. Every cable, connector, and device between the transceiver and the antenna affects the signal that gets radiated from the antenna. Understand which devices provide gain and which devices provide loss. Understand what VSWR is and what values are good or bad. Know the different antenna accessories, what they do, and why and when you would use them.

Key Terms

Before you take the exam, be certain you are familiar with the following terms:

active gain
amplifier
antenna diversity
antenna polarization
antenna radiation envelopes
attenuator
azimuth charts
back lobe
beamwidth
dipole antenna
earth bulge
elevation charts
Fresnel zone
grid antenna
highly directional antennas
lightning arrestor
line of sight (LOS)
multiple-input multiple-output (MIMO)
omnidirectional antennas
panel antenna
parabolic dish antenna
passive gain
patch antenna
phased array antenna
polar charts
receive diversity
return loss
sector antennas
sectorized array
semidirectional antennas
side lobe
splitters
switched diversity
transmit beamforming (TxBF)
transmit diversity
voltage standing wave ratio (VSWR)
Yagi antenna

Review Questions

1. Which of the following devices produce active gain? (Choose all that apply.)
   - **A.** RF transceiver
   - **B.** Parabolic dish
   - **C.** RF amplifier
   - **D.** Sector antenna
2. The azimuth chart represents a view of an antenna's radiation pattern from which direction?
   - **A.** Top
   - **B.** Side
   - **C.** Front
   - **D.** Both top and side
3. What is the definition of the horizontal beamwidth of an antenna?
   - **A.** The measurement of the angle of the main lobe as represented on the azimuth chart.
   - **B.** The distance between the two points on the horizontal axis where the signal decreases by a third. This distance is measured in degrees.
   - **C.** The distance between the two –3 dB power points on the horizontal axis, measured in degrees.
   - **D.** The distance between the peak power and the point where the signal decrease by half. This distance is measured in degrees.
4. Which antennas are highly directional? (Choose all that apply.)
   - **A.** Yagi
   - **B.** Patch
   - **C.** Panel
   - **D.** Parabolic dish
   - **E.** Grid
   - **F.** Sector
5. Semidirectional antennas are often used for which of the following purposes? (Choose all that apply.)
   - **A.** Providing short-distance point-to-point communications
   - **B.** Providing long-distance point-to-point communications
   - **C.** Providing unidirectional coverage from an access point to clients in an indoor environment
   - **D.** Reducing reflections and the negative effects of multipath

6. The Fresnel zone should not be blocked by more than what percentage to maintain a reliable communications link?

A. 20 percent

B. 40 percent

C. 50 percent

D. 60 percent

7. The size of the Fresnel zone is controlled by what factors? (Choose all that apply.)

A. Antenna beamwidth

B. RF line of sight

C. Distance

D. Frequency

8. When a long-distance point-to-point link is installed, earth bulge should be considered beyond what distance?

A. 5 miles

B. 7 miles

C. 10 miles

D. 30 miles

9. A network administrator replaced some coaxial cabling used in an outdoor bridge deployment after water damaged the cabling. After replacing the cabling, the network administrator noticed that the EIRP increased drastically and is possibly violating the maximum EIRP power regulation mandate. What are the possible causes of the increased amplitude? (Choose all that apply.)

A. The administrator installed a shorter cable.

B. The administrator installed a lower-grade cable.

C. The administrator installed a higher-grade cable.

D. The administrator installed a longer cable.

E. The administrator used a different-color cable.

10. Which of the following are true for antenna diversity? (Choose all that apply.)

A. The transceiver combines the signal from both antennas to provide better coverage.

B. Transceivers can transmit from both antennas at the same time.

C. The transceiver samples both antennas and chooses the best received signal from one antenna.

D. Transceivers can transmit from only one of the antennas at a time.

11. In order to establish a 4-mile point-to-point bridge link in the 2.4 GHz ISM band, what factors should be taken under consideration? (Choose all that apply.)
    A. Fresnel zone with 40 percent or less blockage
    B. Earth bulge calculations
    C. Minimum of 16 dBi of passive gain
    D. Proper choice of semidirectional antennas
    E. Proper choice of highly directional antennas

12. The ratio between the maximum peak voltage and minimum voltage on a line is known as what?
    A. Signal flux
    B. Return loss
    C. VSWR
    D. Signal incidents

13. What are some of the possible negative effects of an impendence mismatch? (Choose all that apply.)
    A. Signal reflection
    B. Blockage of the Fresnel zone
    C. Erratic signal strength
    D. Decreased signal amplitude
    E. Amplifier/transmitter failure

14. When determining the mounting height of a long-distance point-to-point antenna, which of the following needs to be considered? (Choose all that apply.)
    A. Frequency
    B. Distance
    C. Visual line of sight
    D. Earth bulge
    E. Antenna beamwidth
    F. RF line of sight

15. Which of the following are true about cables? (Choose all that apply.)
    A. They cause impedance on the signal.
    B. They work regardless of the frequency.
    C. Attenuation decreases as frequency increases.
    D. They add loss to the signal.

16. Amplifiers can be purchased with which of the following features? (Choose all that apply.)

A. Bidirectional amplification

B. Unidirectional amplification

C. Fixed gain

D. Fixed output

17. The signal between the transceiver and the antenna will be reduced by which of the following methods? (Choose all that apply.)

A. Adding an attenuator

B. Increasing the length of the cable

C. Shortening the length of the cable

D. Using cheaper-quality cable

18. Lightning arrestors will defend against which of the following?

A. Direct lighting strikes

B. Power surges

C. Transient currents

D. Improper common grounding

19. The radius of the second Fresnel zone is ________________. (Choose all that apply.)

A. Out of phase with the point source

B. In phase with the point source

C. Smaller than the first Fresnel zone

D. Larger than the first Fresnel zone

20. While aligning a directional antenna, you notice that the signal drops as you turn the antenna away from the other antenna, but then it increases a little. This increase in signal is cause by what?

A. Signal reflection

B. Frequency harmonic

C. Side band

D. Side lobe

Answers to Review Questions

1. A, C. A parabolic dish and a sector antenna are both antennas that are capable of producing only passive gain.
2. A. The azimuth is the top-down view of an antenna's radiation pattern, also known as the H-plane.
3. C. The beamwidth is the distance in degrees between the –3 dB (half-power) point on one side of the main signal and the –3 dB point on the other side of the main signal, measured along the horizontal axis. These are sometimes known as half-power points.
4. D, E. A parabolic dish and a grid are highly directional. The rest of the antennas are semi-directional, and the sector antenna is a special type of semidirectional antenna.
5. A, C, D. Semidirectional antennas provide too wide of a beamwidth to support long-distance communications but will work for short distances. They are also useful for providing unidirectional coverage from the access point to clients in an indoor environment.
6. B. Any more than 40 percent encroachment into the Fresnel zone is likely to make a link unreliable. The clearer the Fresnel zone, the better.
7. C, D. The distance and frequency determine the size of the Fresnel zone; these are the only variables in the Fresnel zone formula.
8. B. The distance when the curvature of the earth should be considered is 7 miles.
9. A, C. Installing a shorter cable of the same grade will result in less loss and thus more amplitude being transmitted out the antenna. A higher-grade cable rated for less dB loss will have the same result.
10. C, D. A transceiver using antenna diversity can transmit from only one antenna at a time. If it transmitted from both antennas, the two signals would interfere with each other. A transceiver can also interpret only one signal at a time, so it samples the signals received by both antennas and chooses the better signal to be received.
11. A, D. Point-to-point bridge links require a minimum Fresnel zone clearance of 60 percent. Semidirectional antennas such as patch antennas or Yagi antennas are used for short-to-medium-distance bridge links. Highly directional antennas are used for long-distance bridge links. Compensating for earth bulge is not a factor until 7 miles.
12. C. Voltage standing wave ratio (VSWR) is the difference between these voltages and is represented as a ratio, for example, 1.5:1.
13. C, D, E. The reflected voltage caused by an impendence mismatch can result in a decrease in power or amplitude (loss) of the signal that is supposed to be transmitted. If the transmitter is not protected from excessive reflected power or large voltage peaks, it can overheat and fail. Understand that VSWR may cause decreased signal strength, erratic signal strength, or even transmitter failure.

14. A, B, D, F. Frequency and distance are needed to determine the Fresnel zone. Visual line of sight is not needed as long as you have RF line of sight. You may not be able to see the antenna because of fog, but the fog will not prevent RF line of sight. Earth bulge will need to be considered. The beamwidth is not needed to determine the height, although it is useful when aiming the antenna.

15. A, D. Cables must be selected that support the frequency you are using. Attenuation actually increases with frequency.

16. A, B, C, D. These are all possible capabilities of RF amplifiers.

17. A, B, D. Adding an attenuator is an intentional act to add loss to the signal. Since cable adds loss, increasing the length will add more loss, whereas shortening the length will reduce the loss. Better-quality cables produce less signal loss.

18. C. Lightning arrestors will not stand up to a direct lightning strike, only transient currents caused by nearby lightning strikes.

19. A, D. The first Fresnel zone is in phase with the point source. The second Fresnel zone begins at the point where the signals transition from being in phase to being out of phase. Because the second Fresnel zone begins where the first Fresnel zone ends, the radius of the second Fresnel zone is larger than the radius of the first Fresnel zone.

20. D. Side lobes are areas of coverage (other than the coverage provided by the main signal) that have a stronger signal than would be expected when compared with the areas around them. Side lobes are best seen on an azimuth chart. Side bands and frequency harmonics have nothing to do with antenna coverage.

IEEE 802.11 Standards

IN THIS CHAPTER, YOU WILL LEARN ABOUT THE FOLLOWING:

✓ **IEEE 802.11 standard**

✓ **IEEE 802.11 ratified amendments**

- 802.11b
- 802.11a
- 802.11g
- 802.11d
- 802.11F
- 802.11h
- 802.11i
- 802.11j
- 802.11e
- 802.11k
- 802.11r

✓ **IEEE 802.11 draft amendments**

- 802.11m
- 802.11n
- 802.11p
- 802.11s
- 802.11T
- 802.11u
- 802.11v
- 802.11w
- 802.11y
- 802.11z
- 802.11aa

As discussed in Chapter 1, "Overview of Wireless Standards, Organizations, and Fundamentals," the Institute of Electrical and Electronics Engineers (IEEE) is the professional society that creates and maintains standards that we use for communications, such as the 802.3 Ethernet standard for wired networking. The IEEE has assigned working groups for several wireless communication standards. For example, the 802.15 Working Group is responsible for personal area network (PAN) communications using radio frequencies. Some of the technologies defined within the 802.15 standard include Bluetooth and ZigBee. Another example is the 802.16 standard, which is overseen by the Broadband Wireless Access Working Group; this technology is often referred to as WiMAX. The focus of this book is the technology as defined by the IEEE 802.11 standard, which provides for local area network (LAN) communications using radio frequencies (RF).

The 802.11 Working Group comprises more than 250 wireless companies and has over 650 active members. It consists of standing committees, study groups, and numerous *task groups*. For example, the Standing Committee—Publicity (PSC) is in charge of finding means to better publicize the 802.11 standard. The 802.11 Study Group (SG) is in charge of investigating the possibility of putting new features and capabilities into the 802.11 standard.

IEEE 802.11: More about the Working Group and 2007 Standard

A quick guide to the IEEE 802.11 Working Group can be found at

`http://grouper.ieee.org/groups/802/11/QuickGuide_IEEE_802_WG_and_Activities.htm.`

The 802.11-2007 standard can be downloaded from

`http://standards.ieee.org/getieee802/802.11.html.`

Various 802.11 task groups are in charge of revising and amending the original standard that was developed by the MAC Task Group (MAC) and the PHY Task Group (PHY). Each group is assigned a letter from the alphabet, and it is common to hear the term *802.11 alphabet soup* when referring to all the amendments created by the multiple 802.11 task groups. Quite a few of the 802.11 task group projects have been completed, and amendments to the original standard have been ratified. Other 802.11 task group projects still remain active and exist as draft amendments.

In this chapter, we discuss the original 802.11 standard, the ratified amendments (most of which now have been incorporated into the 802.11-2007 standard), and the draft amendments of various 802.11 task groups.

Overview of the IEEE 802.11 Standard

The original 802.11 standard was published in June 1997 as IEEE Std. 802.11-1997, and it is often referred to as 802.11 Prime because it was the first WLAN standard. The standard was revised in 1999, reaffirmed in 2003, and published as IEEE Std. 802.11-1999 (R2003). On March 8, 2007, a new iteration of the standard was approved, IEEE Std. 802.11-2007. This new standard is an update of the IEEE Std. 802.11-1999 revision. The following documents have been rolled into this latest revision, providing users a single document with all of the amendments that had been published to date. This new standard includes the following:

- IEEE Std 802.11-1999 (R2003)
- IEEE Std 802.11a-1999
- IEEE Std 802.11b-1999
- IEEE Std 802.11d-2001
- IEEE Std 802.11g-2003
- IEEE Std 802.11h-2003
- IEEE Std 802.11i-2004
- IEEE Std 802.11j-2004
- IEEE Std 802.11e-2005

This revision also includes corrections, clarifications, and enhancements.

The IEEE specifically defines 802.11 technologies at the Physical layer and the MAC sublayer of the Data-Link layer. By design, the 802.11 standard does not address the upper layers of the OSI model, although there are interactions between the 802.11 MAC layer and the upper layers for parameters such as quality of service (QoS). The PHY Task Group worked in conjunction with the MAC Task Group to define the original 802.11 standard. The PHY Task Group defined three original Physical layer specifications:

Infrared (IR) *Infrared (IR)* technology uses a light-based medium. Although an infrared medium was indeed defined in the original 802.11 standard, the implementation is obsolete. More information about modern implementations of infrared technology can be found at the Infrared Data Association's website, at `www.irda.org`. The scope of this book focuses on the 802.11 RF mediums. Infrared devices are known as *clause 16 devices*.

Frequency hopping spread spectrum (FHSS) Radio frequency signals can be defined as narrowband signals or as spread spectrum signals. An RF signal is considered *spread spectrum* when the bandwidth is wider than what is required to carry the data. *Frequency hopping spread spectrum (FHSS)* is a spread spectrum technology that was first patented during

World War II. Frequency hopping 802.11 radio cards are often called *clause 14 devices* because of the clause that referenced them in the original 802.11 standard.

Direct sequence spread spectrum (DSSS) *Direct sequence spread spectrum (DSSS)* is another spread spectrum technology that is frequently used and easiest to implement. DSSS 802.11 radio cards are often known as *clause 15 devices.*

What Is an IEEE Clause?

The IEEE standards are very organized, structured documents. A standards document is hierarchically structured, with each section numbered. The highest level (such as 7) is referred to as a *clause,* with the lower-level sections such as 7.3.2.4 referred to as *subclauses.* As amendments are created, the sections in the amendment are numbered relative to the original standard, even though the amendment is a separate document. When a standard and its amendments are rolled into a new version of the standard, as was recently done with IEEE Std. 802.11-2007, the clauses and subclauses of all of the individual documents are unique, enabling the documents to be combined without having to change any of the section (clause/subclause) numbers.

As defined by 802.11 Prime, the frequency space in which either FHSS or DSSS radio cards can transmit is the license-free 2.4 GHz *Industrial, Scientific, and Medical (ISM) band.* DSSS 802.11 radio cards can transmit in channels subdivided from the entire 2.4 GHz to 2.4835 GHz ISM band. The IEEE is more restrictive for FHSS radio cards, which are permitted to transmit on 1 MHz subcarriers in the 2.402 GHz to 2.480 GHz range of the 2.4 GHz ISM band.

Chances are that you will not be working with older legacy 802.11 equipment because most WLAN deployments use technologies as defined by newer 802.11 amendments. WLAN companies had the choice of manufacturing either clause 14 FHSS radio cards or clause 15 DSSS radio cards. Because these spread spectrum technologies differ, they cannot communicate with each other and often have a hard time coexisting. These spread spectrum signals are analogous to oil and water in that they do not mix well. Therefore, it is important to understand that an 802.11 DSSS (clause 15) radio cannot communicate with an 802.11 (clause 14) FHSS radio. The majority of legacy WLAN deployments used frequency hopping, but some DSSS solutions were available as well.

What about the speeds? Data rates defined by the original 802.11 standard were 1 Mbps and 2 Mbps regardless of which spread spectrum technology was used. A *data rate* is the number of bits per second the Physical layer carries during a single-frame transmission, normally stated as a number of millions of bits per second (Mbps). Keep in mind that a data rate is the *speed* and not actual *throughput.* Because of medium access methods, aggregate throughput is typically one-half or less of the available data rate speed.

FHSS and DSSS are discussed in more detail in Chapter 6, "Wireless Networks and Spread Spectrum Technologies."

IEEE 802.11 Ratified Amendments

In the years that followed the publishing of the original 802.11 standard, new task groups were assembled to address potential enhancements to the standard. At the time this book was written, 11 amendments to the standard had been ratified and published by the distinctive task groups. These ratified supplements will now be discussed in a somewhat chronological order.

In 2007, the IEEE consolidated the majority of the ratified amendments with the original standard, creating a single document that is now published as the *802.11-2007 standard*. The Wi-Fi Alliance and most WLAN professionals still refer to the ratified amendments by name.

However, the CWNA exam (PW0-104) never refers to any of the amendments by name. The exam will quiz you on the technologies used by each amendment. For example, 802.11b is a ratified amendment that has now been integrated into the 802.11-2007 standard. The technology that was originally defined by the 802.11b amendment is called High-Rate DSSS (HR-DSSS). Although the name *802.11b* effectively remains the more commonly used marketing term, any exam questions will use the technical term *HR-DSSS* instead of 802.11b.

For the CWNA exam (PW0-104), you should understand the differences between technologies and how each one works. An understanding of which technologies are defined by each of the amendments will also be helpful for your career. Remember, the CWNP program maintains an updated current list of the exam terms that can be downloaded from the following URL: `www.cwnp.com/exams/exam_terms.html`.

802.11b

Although the Wi-Fi consumer market continues to grow at a tremendous rate, 802.11b-compatible WLAN equipment gave the industry the first needed huge shot in the arm. In 1999, the IEEE Task Group b (TGb) published the IEEE Std. 802.11b-1999, which was later amended and corrected as IEEE Std. 802.11b-1999/Cor1-2001. All aspects of the 802.11b ratified amendment can now be found in clause 18 of the 802.11-2007 standard.

The Physical layer medium that is defined by 802.11b is *High-Rate DSSS (HR-DSSS)*. The frequency space in which 802.11b radio cards can operate is the unlicensed 2.4 GHz to 2.4835 GHz ISM band.

Real World Scenario

Will 802.11b Devices Work with Legacy 802.11 Devices?

802.11b radio cards are known as clause 18 devices. The 802.11b amendment specifies the use of only a DSSS-type physical medium and does not specify FHSS. Because a good portion of the legacy 802.11 deployments used FHSS, 802.11b radio cards are not backward compatible with those systems and cannot be used. However, 802.11b clause 18 radio cards are backward compatible with the legacy 802.11 DSSS clause 15 devices. 802.11b HR-DSSS WLAN equipment should be able to communicate with legacy 802.11 DSSS WLAN equipment. The caveat to this is that, depending on the manufacturer, the devices might not use the same interpretation of the IEEE standards. Many of the legacy devices did not undergo any compatibility testing such as that provided by the Wi-Fi Alliance.

The TGb's main goal was to achieve higher data rates within the 2.4 GHz ISM band. 802.11b radio devices accomplish this feat by using a different spreading/coding technique called *Complementary Code Keying (CCK)* and modulation methods using the phase properties of the RF signal. 802.11 devices used a spreading technique called the *Barker code*. The end result is that 802.11b radio devices support data rates of 1, 2, 5.5, and 11 Mbps. 802.11b systems are backward compatible with the 802.11 DSSS data rates of 1 Mbps and 2 Mbps. The transmission data rates of 5.5 Mbps and 11 Mbps are known as HR-DSSS. Once again, understand that the supported data rates refer to available bandwidth and not aggregate throughput. An optional technology called *Packet Binary Convolutional Code (PBCC)* is also defined under clause 18.

The Barker code and CCK spreading techniques, as well as applicable modulation methods, are discussed further in Chapter 6. A brief examination of PBCC can also be found in Chapter 6.

802.11a

During the same year that the 802.11b amendment was approved, another important amendment was also ratified and published as IEEE Std. 802.11a-1999. The engineers in the Task Group a (TGa) set out to define how 802.11 technologies would operate in the newly allocated *Unlicensed National Information Infrastructure (UNII)* frequency bands. 802.11a radio cards can transmit in three different 100 MHz unlicensed frequency bands in the 5 GHz range. A total of 12 channels are available in the three UNII bands. All aspects of the 802.11a ratified amendment can now be found in clause 17 of the 802.11-2007 standard.

The 2.4 GHz ISM band is a much more crowded frequency space than the 5 GHz UNII bands. Microwave ovens, Bluetooth devices, cordless phones, and numerous other devices all operate in the 2.4 GHz ISM band and are potential sources of interference. In addition, the sheer number of 2.4 GHz WLAN deployments has often been a problem in environments such as multitenant office buildings.

One big advantage of using 802.11a WLAN equipment is that it operates in the less-crowded 5 GHz UNII bands. Eventually, the three UNII bands will also become crowded. Regulatory bodies such as the FCC are opening up more frequency space in the 5 GHz range, and the IEEE addressed this in the 802.11h amendment.

Further discussion about both the ISM and UNII bands occurs in Chapter 6.

802.11a radio cards operating in the 5 GHz UNII bands are classified as clause 17 devices. As defined by the 802.11a amendment, these devices are required to support data rates of 6, 12, and 24 Mbps with a maximum of 54 Mbps. With the use of a spread spectrum technology called *Orthogonal Frequency Division Multiplexing (OFDM)*, data rates of 6, 9, 12, 18, 24, 36, 48, and 54 Mbps are supported in most manufacturers' radio cards. It should be noted that an 802.11a radio does not have to support all of these rates, and one vendor may have an implementation of data rates that is not compatible with another vendor.

OFDM is discussed further in Chapter 6.

It should also be noted that 802.11a radio cards cannot communicate with 802.11 legacy, 802.11b, or 802.11g radio cards for two reasons. First, 802.11a radio cards use a different spread spectrum technology than 802.11 legacy or 802.11b devices. Second, 802.11a devices transmit in the 5 GHz UNII bands, while the 802.11/802.11b/802.11g devices operate in the 2.4 GHz ISM band. The good news is that 802.11a can coexist in the same physical space with 802.11, 802.11b, or 802.11g devices because these devices transmit in separate frequency ranges. In Figure 5.1, you see an access point (AP) with both a 2.4 GHz 802.11b/g radio and a 5 GHz 802.11a radio. Many enterprise wireless deployments run both 802.11a and 802.11b/g networks simultaneously.

802.11g

Another amendment that generated a lot of excitement in the Wi-Fi marketplace was published as IEEE Std. 802.11g-2003. The IEEE defines 802.11g cards as clause 19 devices, which transmit in the 2.4 GHz to 2.4835 GHz ISM frequency band. Clause 19 defines a technology called *Extended Rate Physical (ERP)*. All aspects of the 802.11g ratified amendment can now be found in clause 19 of the 802.11-2007 standard.

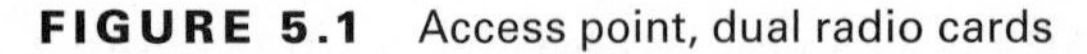

FIGURE 5.1 Access point, dual radio cards

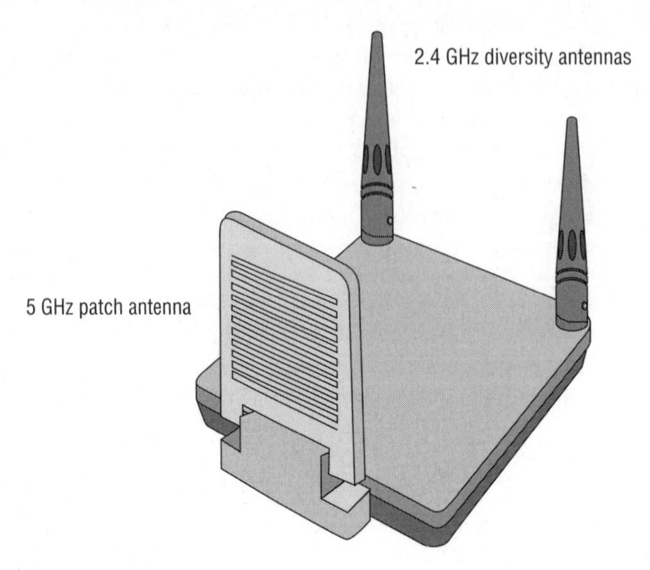

The main goal of the Task Group g (TGg) was to enhance the 802.11b Physical layer to achieve greater bandwidth yet remain compatible with the 802.11 MAC. Two mandatory and two optional ERP physical layers (PHYs) are defined by the 802.11g amendment.

The mandatory PHYs are ERP-OFDM and ERP-DSSS/CCK. To achieve the higher data rates, a PHY technology called *Extended Rate Physical OFDM (ERP-OFDM)* is mandated. Data rates of 6, 9, 12, 18, 24, 36, 48, and 54 Mbps are possible using this technology, although once again the IEEE requires only the data rates of 6, 12, and 24 Mbps. To maintain backward compatibility with 802.11 (DSSS only) and 802.11b networks, a PHY technology called *Extended Rate Physical DSSS (ERP-DSSS/CCK)* is used with support for the data rates of 1, 2, 5.5, and 11 Mbps.

What Is the Difference between ERP-DSSS/CCK, DSSS, and HR-DSSS?

From a technical viewpoint, there is no difference between ERP-DSSS/CCK and DSSS and HR-DSSS. A key point of the 802.11g amendment was to maintain backward compatibility with older 802.11 (DSSS only) and 802.11b radios while at the same time achieving higher data rates. 802.11g devices (clause 19 radios) use ERP-OFDM for the higher data rates. ERP-DSSS/CCK is effectively the same technology as the DSSS that is used by legacy 802.11 devices (clause 15 radios), and HR-DSSS that is used by 802.11b devices (clause 18 radios). Mandated support for ERP-DSSS/CCK allows for backward compatibility with older 802.11 (DSSS only) and 802.11b radios. The technology is explained further in Chapter 6.

The 802.11g ratified amendment also defined two optional PHYs called *ERP-PBCC* and *DSSS-OFDM*. These optional technologies are beyond the scope of this book and rarely used by WLAN vendors.

What Is the Difference between OFDM and ERP-OFDM?

From a technical viewpoint, there is no difference between OFDM and ERP-OFDM. The only difference is the transmit frequency. OFDM refers to 802.11a devices (clause 17 radios) that transmit in the 5 GHz UNII-1, UNII-2, and UNII-3 frequency bands. ERP-OFDM refers to 802.11g devices (clause 19 radios) that transmit in the 2.4 GHz ISM frequency band. The technology is explained further in Chapter 6.

Real World Scenario

What Are the Vendor Operational Modes of an 802.11g Access Point and What Is the Effective Throughput?

While the 802.11g amendment mandates support for both ERP-DSSS/CCK and ERP-OFDM, Wi-Fi vendors typically allow an 802.11g access point to be configured in three very distinct modes:

B-only mode When an 802.11g AP is running in this operational mode, support for DSSS, HR-DSSS and ERP-DSSS/CCK technology is solely enabled. Effectively, the access point has been configured to be an 802.11b access point, and only 802.11b clients will be able to communicate with the AP using data rates of 11, 5.5, 2 and 1 Mbps. Aggregate throughput will be the same as that achieved in an 802.11b network.

G-only mode APs configured as G Only will communicate with 802.11g client stations using only ERP-OFDM technology. Support for ERP-DSSS/CCK, HR-DSSS and DSSS is disabled, and therefore 802.11b clients will not be able to associate with the access point. Only ERP (802.11g) radios will be able to communicate with the access point using data rates of 6–54 Mbps. The aggregate throughput of an AP with a data rate of 54 Mbps might be about 19 Mbps to 20 Mbps. A *G Only* WLAN is sometimes referred to as a *Pure G* network.

B/G mode This is the default operational mode of most 802.11g access points and is often called *mixed mode*. Support for both ERP-DSSS/CCK and ERP-OFDM is enabled. Therefore, 802.11 DSSS, 802.11b, and 802.11g clients can communicate with the access point. However, a price must be paid for the coexistence of these two very different technologies. As soon as the first 802.11 DSSS or 802.11b HR-DSSS station attempts to associate, the access point signals to all the 802.11g stations to enable "protection." Although the protection mechanism does allow for 802.11 (DSSS only), 802.11b, and 802.11g clients to coexist, the result is an immediate and significant degradation in throughput. An 802.11b/g access point with a data rate of 54 Mbps might see a decrease in aggregate throughput from 20 Mbps down to as little as 8 Mbps the instant the protection mechanism is enabled. A thorough discussion of the protection mechanism can be found in Chapter 9, "802.11 MAC Architecture."

As you have learned, the 802.11g amendment requires support for both ERP-DSSS/CCK and ERP-OFDM. The good news is that an 802.11g AP can communicate with 802.11g client stations as well as 802.11 (DSSS only) or 802.11b stations. The ratification of the 802.11g amendment triggered monumental sales of Wi-Fi gear in both the small office, home office (SOHO) and enterprise markets because of both the higher data rates and the backward compatibility with older equipment. As mentioned earlier in this chapter, different spread spectrum technologies cannot communicate with each other, yet the 802.11g amendment mandates support for both ERP-DSSS/CCK and ERP-OFDM. In other words, ERP-OFDM and ERP-DSSS/CCK technologies can coexist, yet they cannot speak to each other. Therefore, the 802.11g amendment calls for a *protection mechanism* that allows the two technologies to coexist. The goal of the protection mechanism is to prevent older 802.11b HR-DSSS or 802.11 DSSS radio cards from transmitting at the same time as 802.11g (ERP) radio cards. Table 5.1 shows a brief overview and comparison of 802.11, 802.11b, 802.11g, and 802.11a.

TABLE 5.1 802.11 Amendment Comparison

	802.11 Legacy	802.11b	802.11g	802.11a
Frequency	2.4 GHz ISM band	2.4 GHz ISM band	2.4 GHz ISM band	5 GHz UNII-1, UNII-2, and UNII-3 bands
Spread Spectrum Technology	FHSS or DSSS	HR-DSSS	ERP: ERP-OFDM and ERP-DSSS/CCK are mandatory	OFDM
		PBCC is optional.	ERP-PBCC and DSSS-OFDM are optional.	
Data Rates	1, 2 Mbps	DSSS: 1, 2 Mbps HR-DSSS: 5.5 and 11 Mbps	ERP-DSSS/CCK: 1, 2, 5.5, and 11 Mbps ERP-OFDM: 6, 12 and 24 Mbps are mandatory. Also supported are 9, 18, 36, 48, and 54 Mbps. ERP-PBCC: 22 and 33 Mbps	6, 12, and 24 Mbps are mandatory. Also supported are 9, 18, 36, 48, and 54 Mbps.
Backward Compatibility	N/A	802.11 DSSS only	802.11b HR-DSSS and 802.11 DSSS	None
Ratified	1997	1999	2003	1999

802.11d

The original 802.11 standard was written for compliance with the regulatory domains of the United States, Japan, Canada, and Europe. Regulations in other countries might define different limits on allowed frequencies and transmit power. The 802.11d amendment, which was published as IEEE Std. 802.11d-2001, added requirements and definitions necessary to allow 802.11 WLAN equipment to operate in areas not served by the original standard.

Country code information is delivered in fields inside two wireless frames called *beacons* and *probe responses*. This information is then used by 802.11d-compliant devices to ensure that they are abiding by a particular country's frequency and power rules. Figure 5.2 shows an AP configured for use in Mongolia and a capture of a beacon frame containing the country code, frequency, and power information.

FIGURE 5.2 802.11d settings

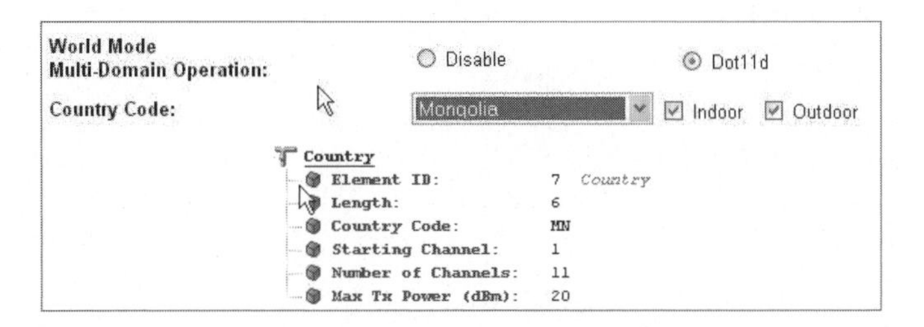

The 802.11d amendment also defines other information specific to configuration parameters of a FHSS access point. FHSS parameters such as hopping patterns might vary from country to country, and the information needs to be once again delivered via the beacon or probe response frames. This information would be useful only in legacy deployments using FHSS technology. All aspects of the 802.11d ratified amendment can now be found in clause 9.8 of the 802.11-2007 standard.

A detailed discussion of beacons, probes, and other wireless frames can be found in Chapter 9.

802.11F

The IEEE Task Group F (TGF) published IEEE Std. 802.11F-2003 as a recommended practice in 2003. The amendment was never ratified and was withdrawn in February 2006.

The use of an uppercase letter designation for an IEEE task group, such as that in IEEE Task Group F, indicates that this amendment (F) is considered a recommended practice and not part of the 802.11-2007 standard.

The original published 802.11 standard mandated that vendor access points support *roaming*. A mechanism is needed to allow client stations that are already communicating through one AP to be able to jump from the coverage area of the original AP and continue communications through a new AP. A perfect analogy is the roaming that occurs when using a cellular telephone. When you are talking on a cell phone to your best friend while driving in your car, your phone will roam between cellular towers to allow for seamless communications and hopefully an uninterrupted conversation. Seamless roaming allows for mobility, which is the heart and soul of true wireless networking and connectivity.

In Figure 5.3, you see a station downloading a file through AP-1 from an FTP server residing on a wired network backbone. Please note that the access points have overlapping areas of coverage. As the station moves closer to AP-2, which has a stronger signal, the station may roam to AP-2 and continue the FTP transfer through the portal supplied by the new access point.

FIGURE 5.3 Roaming

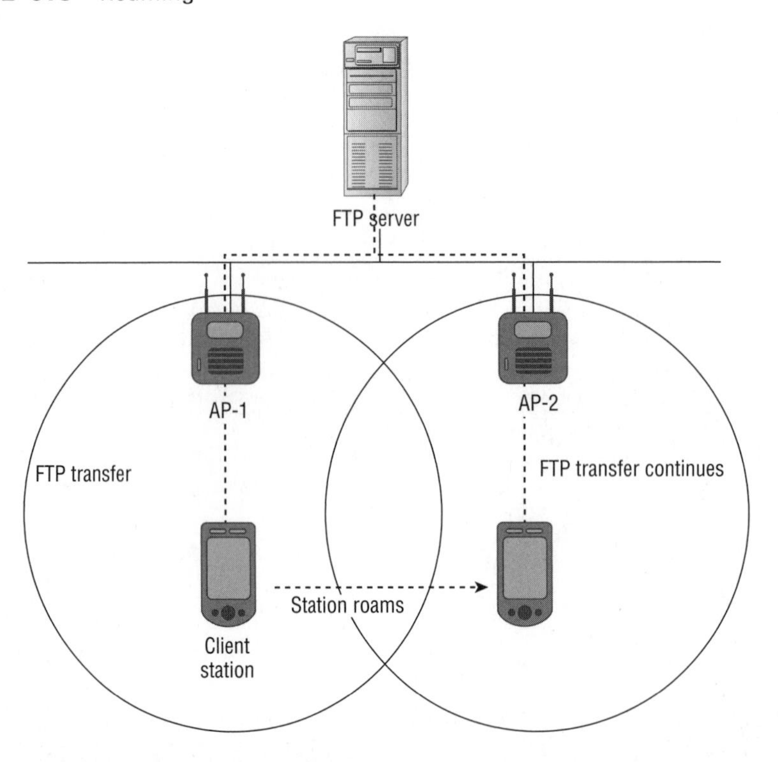

Although the handover that occurs during roaming can be measured in milliseconds, data packets intended for delivery to the station that has roamed to a new access point might still be buffered at the original access point. In order for the buffered data packets to find their way to the station, two things must happen:

1. The new access point must inform the original access point about the station that has roamed and request any buffered packets.

2. The original access point must forward the buffered packets to the new access point via the distribution system for delivery to the client who has roamed.

Figure 5.4 illustrates these two needed tasks.

FIGURE 5.4 Roaming-distribution system medium

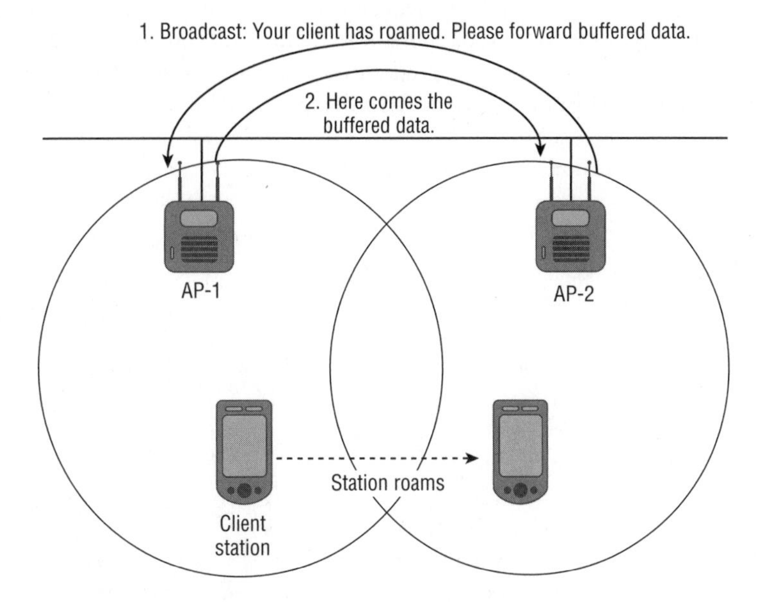

Will Roaming Work If I Mix and Match Different Vendors' Access Points?

The answer is maybe. If the access points of different vendors both support IAPP, then the roaming handover should indeed work. However, one of the reasons that the 802.11F amendment became only a recommended practice is that vendors want customers to purchase only the brand of AP that the vendor sells and not the competition's brand of AP. As a matter of fact, the use of IAPP is not required for certification with the Wi-Fi Alliance. It is therefore the "recommended practice" of this book not to mix different vendors' autonomous access points on the same wired network segment. 802.11F was intended to address roaming interoperability between *autonomous access points* from different vendors. Because most WLAN vendors now use controller-based systems with *lightweight access points,* IAPP is really no longer needed. It should be noted that mixing controller-based solutions from different WLAN vendors is also not advisable. A further discussion of autonomous and lightweight access points can be found in Chapter 10, "Wireless Devices." Roaming is discussed in further detail in Chapter 7, "Wireless LAN Topologies," Chapter 9, "802.11 MAC Architecture" and Chapter 12, "WLAN troubleshooting."

Although the original 802.11 standard calls for the support of roaming, it fails to dictate how roaming should actually transpire. The IEEE initially intended for vendors to have flexibility in implementing proprietary AP-to-AP roaming mechanisms. The 802.11F amendment was an attempt to standardize how roaming mechanisms work behind the scenes on the distribution system medium, which is typically an 802.3 Ethernet network using TCP/IP networking protocols. 802.11F addressed "vendor interoperability" for AP-to-AP roaming. The final result was a recommended practice to use the *Inter-Access Point Protocol (IAPP)*. The IAPP protocol uses announcement and handover processes that results in APs informing other APs about roamed clients as well as delivery for buffered packets.

802.11h

Published as IEEE Std. 802.11h-2003, this amendment defines mechanisms for *dynamic frequency selection (DFS)* and *transmit power control (TPC)*. It was originally proposed to satisfy regulatory requirements for operation in the 5 GHz band in Europe and to detect and avoid interference with 5 GHz satellite and radar systems. Many of these same regulatory requirements have now also been adopted by the FCC in the United States. The main purpose of DFS and TPC is to provide services where 5 GHz 802.11 radio transmissions will not cause interference with 5 GHz satellite and radar transmissions.

The 802.11h amendment also introduced the capability for 802.11 radios to transmit in a new frequency band called UNII-2 Extended with 11 more channels, as seen in Table 5.2. The 802.11h amendment effectively is an extension of the 802.11a amendment. OFDM transmission technology is used in all of the UNII bands. The radar detection and avoidance technologies of DFS and TPC are defined by the IEEE. However, the RF regulatory organizations in each country still define the RF regulations. In the United States and Europe, radar detection and avoidance is required in both the UNII-2 and UNII-2 Extended bands.

TABLE 5.2 Unlicensed National Information Infrastructure

Band Frequency Range	Amendment	Channels
UNII-1 (lower) 5.150 GHz–5.250 GHz	802.11a	4
UNII-2 (middle) 5.250 GHz–5.350 GHz	802.11a	4
UNII-2 Extended 5.47 GHz–5.725 GHz	802.11h	11
UNII-3 (upper) 5.725 GHz–5.825 GHz	802.11a	4

DFS is used for spectrum management of 5 GHz channels by OFDM radio devices. The European Radiocommunications Committee (ERC) and now the FCC mandate that radio cards operating in the 5 GHz band implement a mechanism to avoid interference with radar systems as well as provide equable use of the channels. The DFS service is used to meet these regulatory requirements.

The dynamic frequency selection (DFS) service provides for the following:

- An AP will allow client stations to associate based on the supported channel of the access point. The term *associate* means that a station has become a member of the AP's wireless network.
- An AP can quiet a channel to test for the presence of radar.
- An AP may test a channel for the presence of radar before using the channel.
- An AP can detect radar on the current channel and other channels.
- An AP can cease operations after radar detection to avoid interference.
- When interference is detected, the AP may choose a different channel to transmit on and inform all the associated stations.

TPC is used to regulate the power levels used by OFDM radio cards in the of 5 GHz frequency bands. The ERC mandates that radio cards operating in the 5 GHz band use TPC to abide by a maximum regulatory transmit power, and are able to alleviate transmission power to avoid interference. The TPC service is used to meet the regulatory transmission power requirements.

The transmit power control (TPC) service provides for the following:

- Stations can associate with an AP based on their transmit power.
- Designation of the maximum transmit power levels permitted on a channel, as permitted by regulations.
- An AP can specify the transmit power of any or all stations that are associated with the access point.
- An AP can change transmission power on stations based on factors of the physical RF environment such as path loss.

The information used by both DFS and TPC is exchanged between stations and access points inside of management frames. The 802.11h amendment effectively introduced two major enhancements: more frequency space with the introduction of the UNII-2 Extended band, and radar avoidance and detection technologies. All aspects of the 802.11h ratified amendment can now be found in clauses 11.8 and 11.9 of the 802.11-2007 standard.

802.11i

From 1997 to 2004, not much was defined in terms of security in the original 802.11 standard. Two key components of any wireless security solution are *data privacy* (encryption) and *authentication* (identity verification). For seven years, the only defined method of encryption in an 802.11 network was the use of 64-bit static encryption called *Wired Equivalent Privacy (WEP)*.

WEP encryption has long been cracked and is not considered an acceptable means of providing data privacy. The original 802.11 standard defined two methods of authentication. The default method is *Open System authentication*, which verifies the identity of everyone regardless. Another defined method is called *Shared Key authentication*, which opens up a whole new can of worms and potential security risks.

The 802.11i amendment, which was ratified and published as IEEE Std. 802.11i-2004, has finally defined stronger encryption and better authentication methods. The 802.11i amendment defined a *robust security network (RSN)*.The intended goal of an RSN was to better hide the data flying through the air while at the same time placing a bigger guard at the front door. The 802.11i security amendment is without a doubt one of the most important enhancements to the original 802.11 standard because of the seriousness of properly protecting a wireless network. The major security enhancements addressed in 802.11i are as follows:

Data privacy Confidentiality needs have been addressed in 802.11i with the use of a stronger encryption method called *Counter Mode with Cipher Block Chaining Message Authentication Code Protocol (CCMP)*, which uses the *Advanced Encryption Standard (AES)* algorithm. The encryption method is often abbreviated as CCMP/AES, AES CCMP, or often just CCMP. The 802.11i supplement also defines an optional encryption method known as *Temporal Key Integrity Protocol (TKIP)*, which uses the RC-4 stream cipher algorithm and is basically an enhancement of WEP encryption.

Authentication 802.11i defines two methods of authentication using either an IEEE *802.1X* authorization framework or *preshared keys (PSKs)*. An 802.1X solution requires the use of an *Extensible Authentication Protocol (EAP)*, although the 802.11i amendment does not specify what EAP method to use.

Robust security network (RSN) This defines the entire method of establishing authentication, negotiating security associations, and dynamically generating encryption keys for clients and access points.

The Wi-Fi Alliance also has a certification known as *Wi-Fi Protected Access 2 (WPA2)*, which is a mirror of the IEEE 802.11i security amendment. WPA version 1 was considered a preview of 802.11i, whereas WPA version 2 is fully compliant with 802.11i. All aspects of the 802.11i ratified security amendment can now be found in clause 8 of the 802.11-2007 standard.

Wi-Fi security is the top priority when deploying any WLAN, and that is why there is another valued certification called Certified Wireless Security Professional (CWSP). At least 10 percent of the CWNA test will involve questions regarding Wi-Fi security. Therefore, wireless security topics such as 802.1X, EAP, AES CCMP, TKIP, WPA, and more are described in more detail in Chapter 13, "802.11 Network Security Architecture," and Chapter 14, "Wireless Attacks, Intrusion Monitoring, and Policy."

802.11j

The main goal set out by the IEEE Task Group j (TGj) was to obtain Japanese regulatory approval by enhancing the 802.11 MAC and 802.11a PHY to additionally operate in Japanese 4.9 GHz and 5 GHz bands. The 802.11j amendment was approved and published as IEEE Std. 802.11j-2004.

In Japan, 802.11a radio cards can transmit in the lower UNII band at 5.15 GHz to 5.25 GHz as well as a Japanese licensed/unlicensed frequency space of 4.9 GHz to 5.091 GHz.

802.11a radio cards use OFDM technology with required channel spacing of 20 MHz. When 20 MHz channel spacing is used, data rates of 6, 9, 12, 18, 24, 36, 48, and 54 Mbps are possible using OFDM technology. Japan also has the option of using OFDM channel spacing of 10 MHz, which results in available bandwidth data rates of 3, 4.5, 6, 9, 12, 18, 24, and 27 Mbps. The data rates of 3, 6, and 12 Mbps are mandatory when using 10 MHz channel spacing.

802.11e

Since the adoption of the original 802.11 standard, there have not been any adequate *quality of service (QoS)* procedures defined for the use of time-sensitive applications such as *Voice over IP (VoIP)*. Voice over Wireless IP (VoWIP) is also known as Voice over Wireless LAN (VoWLAN) and as *Voice over Wi-Fi (VoWiFi)*. The terminology used by most vendors and the CWNP program is Voice over Wi-Fi (VoWiFi). Application traffic such as voice, audio, and video have a lower tolerance for latency and jitter and requires priority before standard application data traffic. The 802.11e amendment defines the layer 2 MAC methods needed to meet the QoS requirements for time-sensitive applications over IEEE 802.11 WLANs.

The original 802.11 standard defined two methods in which an 802.11 radio card may gain control of the half-duplex medium. The default method, *Distributed Coordination Function (DCF)*, is a completely random method determining who gets to transmit on the wireless medium next. The original standard also defines another medium access control method called *Point Coordination Function (PCF)*, where the access point briefly takes control of the medium and polls the clients.

Chapter 8, "802.11 Medium Access," describes the DCF and PCF methods of medium access in greater detail.

The 802.11e amendment defines enhanced medium access methods to support QoS requirements. *Hybrid Coordination Function (HCF)* is an additional coordination function that is applied in an 802.11e QoS wireless network. HCF has two access mechanisms to provide QoS. *Enhanced Distributed Channel Access (EDCA)* is an extension to DCF. The EDCA medium access method will provide for the "prioritization of frames" based on upper-layer protocols. Application traffic such as voice or video will be transmitted in a timely fashion on the 802.11 wireless medium, meeting the necessary latency requirements.

Hybrid Coordination Function Controlled Channel Access (HCCA) is an extension of PCF. HCCA gives the access point the ability to provide for "prioritization of stations." In other words, certain client stations will be given a chance to transmit before others.

The Wi-Fi Alliance also has a certification known as *Wi-Fi Multimedia (WMM)*. The WMM standard is a "mirror" of 802.11e and defines traffic prioritization in four access categories with varying degrees of importance. All aspects of the 802.11e ratified QoS amendment can now be found in clause 9.9 of the 802.11-2007 standard.

802.11e and WMM are covered in more detail in Chapter 8.

802.11k

The goal of the 802.11 Task Group k (TGk) is to provide a means of radio resource measurement (RRM). This amendment calls for measurable client statistical information in the form of requests and reports for the Physical layer 1 and the MAC sublayer of the Data-Link layer 2. 802.11k defines mechanisms in which client station resource data is gathered and processed by an access point or *WLAN controller* (WLAN controllers are covered in Chapter 10. For now, think of a WLAN controller as a core device that manages many access points.) In some instances, the client may also request information from an access point or WLAN controller. The following are some of the key radio resource measurements defined under 802.11k:

Transmit power control (TPC) The 802.11h amendment defined the use of TPC for the 5 GHz band to reduce interference. Under 802.11k, TPC will also be used in other frequency bands and in areas governed by other regulatory agencies.

Client statistics Physical layer information such as signal-to-noise ratio, signal strength, and data rates can all be reported back to the access point or WLAN controller. MAC information such as frame transmissions, retries, and errors may all be reported back to the access point or WLAN controller as well.

Channel statistics Clients may gather noise-floor information based on any RF energy in the background of the channel and report this information back to the access point. Channel-load information may also be collected and sent to the AP. The access point or WLAN controller may use this information for channel management decisions.

Neighbor reports Mobile Assisted Handover (MAHO) is a technique used by digital phones and cellular systems working together to provide better handover between cells. 802.11k gives access points or WLAN controllers the ability to direct stations to perform the sort of tasks that a cellular network requires its handhelds to do when using MAHO.

Using proprietary methods, a client station keeps a table of known access points and makes decisions on when to roam to another access point. As defined by 802.11k, the access point or WLAN controller will request a station to listen for neighbor access points on other channels and gather information. The current AP or WLAN controller will then process that information and generate a *neighbor report* detailing available access points from best to worst. Before a station roams, it will request the neighbor report from the current AP or controller and then decide whether to roam to one of the access points on the neighbor report. The recently ratified 802.11k amendment in conjunction with the recently ratified 802.11r "fast roaming" amendment have the potential to greatly improve roaming performance in 802.11 wireless networks.

The 802.11k amendment is not part of the 802.11-2007 standard. However, it was ratified in June of 2008 and is published as IEEE 802.11k-2008.

802.11r

The 802.11r amendment is known as the *fast basic service set transition (FT)* amendment. The technology is more often referred to as *fast secure roaming* because it defines faster handoffs when roaming occurs between cells in a WLAN using the strong security defined by a robust secure network (RSN). 802.11r was proposed primarily because of the time constraints of applications such as VoIP. Average time delays of hundreds of milliseconds occur when a client station roams from one access point to another access point.

Roaming can be especially troublesome when using a WPA-Enterprise or WPA2-Enterprise security solution, which requires the use of a RADIUS server for 802.1X/EAP authentication and often takes 700 milliseconds or greater for the client to authenticate. VoWiFi requires a handoff of 150 milliseconds or less to avoid a degradation of the quality of the call, or, even worse, a loss of connection. Currently 802.1X/EAP security solutions are rare in time-critical environments because of the latency problems caused by the long roaming handoff times.

Under 802.11r, a station will be able to establish a QoS stream and set up a security association with a new access point before actually roaming to the new access point. The station will be able to achieve these tasks either over the wire via the original access point or through the air. Eventually, the station will complete the roaming process and move to the new access point. The time saved from prearranging security associations and QoS services will drastically speed up the handoffs between WLAN cells.

The 802.11r amendment is not part of the 802.11-2007 standard. However, it was ratified in July of 2008 and is published as IEEE 802.11r-2008. Tactical enterprise deployments of this technology will be extremely important for providing more-secure communications for VoWiFi. The details of this technology will soon be a heavily tested topic on the CWSP exam.

IEEE 802.11 Draft Amendments

What does the future hold in store for us with 802.11 wireless networking? The draft amendments are a looking glass into the enhancements and capabilities that might be available in the near future for 802.11 wireless networking devices. Greater throughput, client control, improved roaming, mesh networking, and more await us on the wireless horizon.

It is important to remember that draft amendments are proposals that have yet to be ratified. Although some vendors are already selling products that have some of the capabilities described in the following sections, these features are still considered proprietary. Even though a vendor might be marketing these pre-ratified capabilities, there is no guarantee that their current product will work with future products that are certified as compliant with the forthcoming ratified amendment. One exception to this is the 802.11n draft amendment. The Wi-Fi Alliance currently has a vendor certification program called Wi-Fi CERTIFIED 802.11n draft 2.0. This certification program currently tests most of the high throughput (HT) capabilities that will eventually be seen in the 802.11n amendment when it is ratified.

The CWNA exam (PW0-104) currently covers all of the technologies defined in the ratified amendments that are now part of the 802.11-2007 standard. You will not be tested on the draft amendments except for 802.11n, because most enterprise vendors are already implementing 802.11n technology, and the Wi-Fi Alliance is certifying the technology. Detailed discussion about 802.11n and MIMO technology can be found in Chapter 18, "High Throughput (HT) and 802.11n."

The recent ratification of the 802.11k and 802.11r amendments combined with the future approval of the 802.11n and 802.11v drafts is expected to spark a major convergence of data, voice, and video over the wireless medium. The remaining pages of this chapter provide a glimpse into the future of more-advanced and sophisticated Wi-Fi products.

Once again, please remember that because these IEEE amendments are still draft documents, they will likely be different from the final ratified amendments.

802.11m

The IEEE Task Group m (TGm) started an initiative in 1999 for internal maintenance of the 802.11 standard's technical documentation. 802.11m is often referred to as *802.11 housekeeping* because of its mission of clarifying and correcting the 802.11 standard. Unless you are a member of TGm, this amendment is of little significance.

802.11n

An event that is sure to have a major impact on the Wi-Fi marketplace will be the passage of the 802.11n amendment. Since 2004, the 802.11 Task Group n (TGn) has been working on improvements to the 802.11 standard to provide for greater throughput. Many of the IEEE 802.11 amendments in the past have addressed bandwidth data rates. However, the objective of the 802.11n amendment is to increase the throughput in both the 2.4 GHz and 5 GHz frequency bands. The 802.11n amendment defines a new operation known as *High*

Throughput (HT), which provides PHY and MAC enhancements to support throughput of 100 Mbps or greater.

HT clause 20 radios use *multiple-input multiple-output (MIMO)* technology in unison with OFDM technology. MIMO uses multiple receiving and transmitting antennas and actually capitalizes on the effects of multipath as opposed to compensating for or eliminating them. The beneficial consequences of using MIMO are increased throughput and even greater range. As stated earlier in this chapter, the Wi-Fi Alliance currently has a vendor certification program called Wi-Fi CERTIFIED 802.11n draft 2.0. This certification program currently tests most of the HT capabilities that will eventually be seen in the ratified 802.11n amendment.

Further discussion about 802.11n and MIMO technology can be found in Chapter 18.

802.11p

The mission of the 802.11 Task Group p (TGp) is to define enhancements to the 802.11 standard to support Intelligent Transportation Systems (ITS) applications. Data exchanges between high-speed vehicles will be possible in the licensed ITS band of 5.9 GHz. Additionally, communications between vehicles and roadside infrastructure will be supported in the 5 GHz bands, specifically the 5.850 GHz to 5.925 GHz band within North America.

Communications may be possible at speeds of up to 200 kilometers per hour (124 mph) and within a range of 1,000 meters (3281 feet). Very short latencies will also be needed as some applications must guarantee data delivery within 4 to 50 milliseconds.

802.11p is also known as Wireless Access in Vehicular Environments (WAVE) and is the possible foundation for a U.S. Department of Transportation project called Dedicated Short Range Communications (DSRC). The DSRC project envisions a nationwide vehicle and roadside communication network utilizing applications such as vehicle safety services, traffic jam alerts, toll collections, vehicle collision avoidance, and adaptive traffic light control. 802.11p will also be applicable to marine and rail communications.

802.11s

802.11 access points typically act as portal devices to a *distribution system (DS)* that is usually a wired 802.3 Ethernet medium. The 802.11-2007 standard, however, does not mandate that the distribution system use a wired medium. Access points can therefore act as portal devices to a *wireless distribution system (WDS)*. The 802.11s amendment proposes the use of a protocol for adaptive, autoconfiguring systems that support broadcast, multicast, and unicast traffic over a multihop mesh WDS.

The 802.11 Task Group s (TGs) has set forth the pursuit of standardizing *mesh networking* using the IEEE 802.11 MAC/PHY layers. The 802.11s amendment defines the use of mesh points which are 802.11 QoS stations that support mesh services. A *mesh point (MP)* is capable of using a mandatory mesh routing protocol called *Hybrid Wireless Mesh Protocol (HWMP)* that uses a default path selection metric. Vendors may also use proprietary mesh routing protocols and metrics. As depicted in Figure, 5.5, a *mesh access point (MAP)* is a device that provides both mesh functionalities and AP functionalities simultaneously. A *mesh point portal (MPP)* is a device that acts as a gateway to one or more external networks such as an 802.3 wired backbone.

FIGURE 5.5 Mesh points, mesh APs and mesh portal

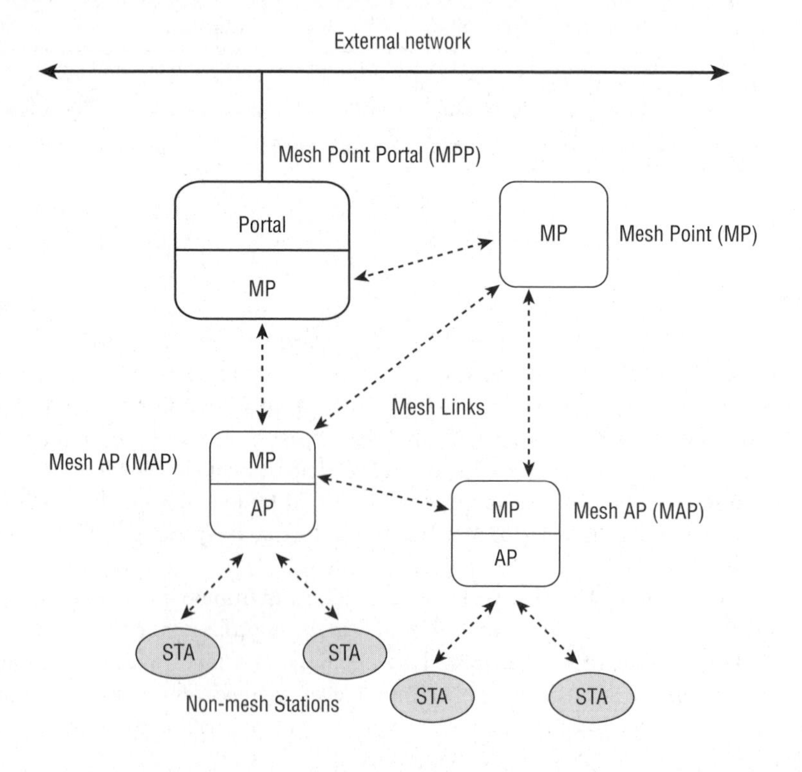

Further discussion on distribution systems (DS) and wireless distribution systems (WDS) can be found in Chapter 7. Further discussion on 802.11 mesh networking can be found in Chapter 10.

802.11T

The goal of the IEEE 802.11 Task Group T (TGT) is to develop performance metrics, measurement methods, and test conditions to measure the performance of 802.11 wireless networking equipment.

The uppercase *T* in the name *IEEE 802.11T* indicates that this amendment will be considered a recommended practice and not a standard.

The 802.11T draft is also called Wireless Performance Prediction (WPP). Its final objective is consistent and universally accepted WLAN measurement practices. These 802.11 performance benchmarks and methods could be used by independent test labs, manufacturers, and even end users.

Real World Scenario

Are Throughput Results the Same among Vendors?

Multiple factors can affect throughput in a wireless network, including the physical environment, range, and type of encryption. Another factor that can affect throughput is simply the vendor radio device that is being used for transmissions. Even though the 802.11-2007 standard clearly defines frequency bandwidths, data rate speeds, and medium access methods, throughput results vary widely from vendor to vendor. A throughput performance test using two radio cards from one vendor will most often yield very different results than the same throughput performance test using two radio cards from another vendor. Typically, you will see better throughput results when sticking with one vendor as opposed to mixing vendor equipment. However, sometimes mixing vendor equipment will produce the unexplained consequence of increased throughput. Ratification of the 802.11T amendment will at the very least provide a more accurate means of performance testing.

802.11u

The primary objective of the 802.11 Task Group u (TGu) is to address interworking issues between an IEEE 802.11 access network and any external network to which it is connected. A common approach is needed to integrate IEEE 802.11 access networks with external networks in a generic and standardized manner. 802.11u is also often referred to as Wireless InterWorking with External Networks (WIEN). This amendment may well address seamless handoff and session persistence with other external networks such as the Internet, cellular phone networks, and WiMAX.

Proprietary equipment (such as hybrid telephones) that allows for roaming between an enterprise WLAN and a wide area cellular network is currently being developed. The 802.11u draft may one day standardize the procedures needed for interworking between two very different networks.

802.11v

While 802.11k defines methods of retrieving information from client stations, 802.11v will give us the ability to configure client stations wirelessly from a central point of management. The main goal of the IEEE Task Group v (TGv) is for WLAN infrastructure (access points and wireless switches) to take improved control of wireless client stations. The following list includes some of the 802.11v proposals currently being discussed:

Wireless client control SNMP Management Information Bases (MIBs) for client station attributes are currently being defined under 802.11v. This will give the ability to configure and manage clients wirelessly from a WLAN infrastructure device.

Load balancing Enterprise WLAN deployments often encounter disproportionate associations of client stations between access points. This can cause an uneven distribution of available bandwidth and result in throughput problems. Currently, some vendors implement proprietary load-balancing procedures to help alleviate these problems. 802.11v may standardize and simplify load balancing.

Network selection In order to join a wireless network, a client station radio card must be preconfigured with a profile that exactly matches the security credentials on the infrastructure side. 802.11v may provide mechanisms to implement client-side security settings from the WLAN infrastructure.

Virtual APs Most WLAN vendors currently allow a physical AP to advertise multiple SSIDs, where each SSID is mapped to a unique BSSID. 802.11v may standardize this practice in the future.

802.11w

A common type of attack on an 802.11 WLAN is a denial-of-service attack (DoS attack). There are a multitude of DoS attacks that can be launched against a wireless network; however, a very common DoS attack occurs at layer 2 using 802.11 management frames. Currently, it is simple for an attacker to edit deauthentication or disassociation frames and then retransmit the frames into the air, effectively shutting down the wireless network.

The IEEE Task Group w (TGw) is working on a "protected" management frame amendment with a goal of delivering management frames in a secure manner. The end result will hopefully prevent some of the layer 2 denial-of-service attacks that currently exist.

A discussion about both layer 1 and layer 2 DoS attacks can be found in Chapter 14.

802.11y

The objective of the IEEE Task Group y (TGy) is to standardize the mechanisms required to allow high-powered, shared 802.11 operation with other users in the 3650 MHz–3700 MHz band in the United States and possibly other frequencies in other countries.

802.11z

The purpose of IEEE Task Group z (TGz) is to establish and standardize a Direct Link Setup (DLS) mechanism to allow operation with non-DLS-capable access points. In most WLAN environments, all frame exchanges between client stations that are associated to the same access point must pass through the access point. DLS allows client stations to bypass the access point and communicate with direct frame exchanges. Some of the earlier amendments have defined DLS communications. However, they are rarely used in enterprise WLANs. The 802.11z draft proposes enhancements to DLS communications.

802.11aa

The 802.11aa amendment specifies enhancements to the 802.11 Media Access Control (MAC) for robust audio video streaming, while maintaining coexistence with other types of traffic.

Neither 802.11l nor 802.11o amendments exist because they are considered typologically problematic. Also, it should be noted that there is no amendment with the name of 802.11x. The term *802.11x* sometimes is used to refer to all the 802.11 standards. The IEEE 802.1X standard, which is a port-based access-control standard, is often incorrectly called 802.11x.

Summary

This chapter covered the original 802.11 standard as well as its many ratified enhancements, which have been incorporated in the current 802.11-2007 standard. This chapter also discussed possible future enhancements. We covered the following:

- All the defined PHY and MAC layer requirements of the original 802.11 Prime standard
- All the approved enhancements to the 802.11 standard in the form of ratified amendments, including higher data rates, different spread spectrum technologies, quality of service, and security
- Future capabilities and improvements as proposed in the 802.11 draft documents, including increased throughput, mesh networking, client management, and more

Although many proprietary Wi-Fi solutions exist and will continue to exist in the foreseeable future, standardization brings stability to the marketplace. The 802.11-2007 standard and all the future enhanced supplements provide a much needed foundation for vendors, network administrators, and end users.

The CWNA exam will test your knowledge of the original 802.11-2007 standard and all the related technologies. Technologies discussed in the 802.11n draft amendment are also covered in the CWNA exam. Your primary focus should be on the 802.11-2007 standard. However, please understand that when 802.11 draft amendments become approved, they will be weighted heavier in future versions of the CWNA exam.

Exam Essentials

Know the defined spread spectrum technologies of the original 802.11 standard and the subsequent 802.11-2007 standard. Although the original 802.11 standard defined infrared, FHSS, and DSSS, later amendments that are now incorporated in the 802.11-2007 standard also define HR-DSSS, OFDM, and ERP.

Remember both the required data rates and supported data rates of each PHY. DSSS and FHSS require and support data rates of 1 and 2 Mbps. Other PHYs offer a wider support for data rates. For example, OFDM and ERP-OFDM support data rates of 6, 9, 12, 18, 24, 36, 48, and 54 Mbps, but only the rates of 6, 12 and 24 Mbps are mandatory. Please understand that data rates are speeds and not aggregate throughput.

Know the frequency bands used by each PHY as defined by the 802.11-2007 standard. OFDM equipment operates in the 5 GHz UNII bands. DSSS, FHSS, HR-DSSS, and ERP devices transmit and receive in the 2.4 GHz ISM band.

Explain the three vendor operational modes of ERP (802.11g) and the consequences of each mode. An 802.11g access point may be configured as B-only mode, G-only mode, or B/G mixed mode. The three modes support different spread spectrum technologies and have different aggregate throughput results.

Know the mandatory and optional technologies used in an ERP WLAN. ERP (802.11g) defines two mandatory PHYs, ERP-OFDM and ERP-DSSS/CCK. The two optional PHYs are ERP-PBCC and DSSS-OFDM.

Define *transmit power control* and *dynamic frequency selection*. TPC and DFS are often mandated for use in the 5 GHz band. Both technologies are used as a means to avoid interference with radar transmissions.

Explain the defined wireless security standards both pre-802.11i and post-802.11i. Before the passage of 802.11i, WEP encryption and either Open System or Shared Key authentication were defined. The 802.11i amendment calls for the use of CCMP/AES for encryption. For authentication, 802.11i defines either an 802.1X/EAP solution or the use of preshared keys.

Define the Inter-Access Point Protocol and why it was originally proposed. IAPP is a "vendor interoperability" roaming protocol that is outlined in the 802.11F recommended practice.

Explain the purpose of the 802.11e amendment and the medium access methods it requires. The 802.11e amendment addresses quality of service (QoS) issues by mandating the use of Enhanced Distributed Channel Access (EDCA) and Hybrid Coordination Function (HCF).

Understand the purpose of each 802.11 draft proposal. Each draft has a specific intended goal. The 802.11s draft, for example, outlines mesh networking. 802.11n proposes throughput enhancements using MIMO technology.

Key Terms

Before you take the exam, be certain you are familiar with the following terms:

802.11-2007 standard
802.1X
Advanced Encryption Standard (AES)
authentication
autonomous access points
Barker code
Complementary Code Keying (CCK)
Counter Mode with Cipher Block Chaining Message Authentication Code Protocol (CCMP)
data privacy
data rate
Direct sequence spread spectrum (DSSS)
Distributed Coordination Function (DCF)
distribution system
distribution system
DSSS-OFDM
dynamic frequency selection (DFS)
Enhanced Distributed Channel Access (EDCA)
ERP-PBCC
Extended Rate Physical (ERP)
Extended Rate Physical DSSS (ERP-DSSS/CCK)
Extended Rate Physical OFDM (ERP-OFDM)
Extensible Authentication Protocol (EAP)
fast basic service set transition (FT)
fast secure roaming
frequency hopping spread spectrum (FHSS)
High Throughput (HT)
High-Rate DSSS (HR-DSSS)
Hybrid Coordination Function (HCF)
Hybrid Coordination Function Controlled Channel Access (HCCA)
Hybrid Wireless Mesh Protocol (HWMP)
Industrial, Scientific, and Medical (ISM) band
Infrared (IR)
Inter-Access Point Protocol (IAPP)
lightweight access points
mesh access point (MAP)
mesh networking
mesh point (MP)
mesh point portal (MPP)
mixed mode
multiple-input multiple-output (MIMO)
Open System authentication
Orthogonal Frequency Division Multiplexing (OFDM)
Packet Binary Convolutional Code (PBCC)
Point Coordination Function (PCF)
preshared keys (PSKs)
protection mechanism
quality of service (QoS)
roaming
robust security network (RSN)
Shared Key authentication
task groups
Temporal Key Integrity Protocol (TKIP)
transmit power control (TPC)
Unlicensed National Information Infrastructure (UNII)
Voice over IP (VoIP)

Voice over Wi-Fi (VoWiFi)
Wi-Fi Multimedia (WMM)
Wi-Fi Protected Access 2 (WPA2)
Wired Equivalent Privacy (WEP)
wireless distribution system (WDS)
WLAN controller

Review Questions

1. An ERP network mandates support for which two spread spectrum technologies? (Choose all that apply.)

   A. ERP-OFDM

   B. FHSS

   C. ERP-PBCC

   D. ERP-DSSS/CCK

   E. CSMA/CA

2. The 802.11-2007 standard using an ERP-DSSS/CCK radio supports which data rates?

   A. 3, 6, and 12 Mbps

   B. 6, 9, 12, 18, 24, 36, 48, and 54 Mbps

   C. 6, 12, 24, and 54 Mbps

   D. 6, 12, and 24 Mbps

   E. 1, 2, 5.5, and 11 Mbps

3. Which types of devices were defined in a legacy 802.11 WLAN network? (Choose all that apply.)

   A. Clause 17 OFDM

   B. Clause 15 DSSS

   C. Clause 18 HR-DSSS

   D. Clause 16 IR

   E. Clause 14 FHSS

   F. Clause 19 ERP

4. The 802.11F recommended practice requires the use of which protocol?

   A. TPC

   B. EDCAF

   C. IAPP

   D. CSMA/CA

   E. DFS

5. A robust security network (RSN) requires the use of which security mechanisms? (Choose all that apply.)
   A. 802.11x
   B. WEP
   C. IPsec
   D. CCMP/AES
   E. CKIP
   F. 802.1X

6. An 802.11a radio card can transmit on ________ frequency and uses ________ spread spectrum technology.
   A. 5 MHz, OFDM
   B. 2.4 GHz, HR-DSSS
   C. 2.4 GHz, ERP-OFDM
   D. 5 GHz, OFDM
   E. 5 GHz, DSSS

7. What are the required data rates of an OFDM station?
   A. 3, 6, and 12 Mbps
   B. 6, 9, 12, 18, 24, 36, 48, and 54 Mbps
   C. 6, 12, 24, and 54 Mbps
   D. 6, 12, and 24 Mbps
   E. 1, 2, 5.5, and 11 Mbps

8. When implementing an 802.1X/EAP RSN network with a VoWiFi solution, what is needed to avoid latency issues?
   A. Inter-Access Point Protocol
   B. Fast BSS Transition
   C. Distributed Coordination Function
   D. Roaming Coordination Function
   E. Lightweight access points

9. The original 802.11 standard requires which data rates?
   A. 1, 2, 5.5, and 11 Mbps
   B. 6, 12, and 24 Mbps
   C. 1 and 2 Mbps
   D. 6, 9, 12, 18, 24, 36, 48, and 54 Mbps
   E. 3, 6, and 12 Mbps

10. What is the primary reason that OFDM (802.11a) radios cannot communicate with ERP (802.11g) radios?
    - **A.** 802.11a uses OFDM, and 802.11g uses DSSS.
    - **B.** 802.11a uses DSSS, and 802.11g uses OFDM.
    - **C.** 802.11a uses OFDM, and 802.11g uses CCK.
    - **D.** 802.11a operates at 5 GHz, and 802.11g operates at 2.4 GHz.
    - **E.** 802.11a requires Dynamic Frequency Selection, and 802.11g does not.

11. What two technologies are used to prevent 802.11 radios from interfering with radar transmissions at 5 GHz? (Choose all that apply.)
    - **A.** Dynamic frequency selection
    - **B.** Enhanced Distributed Channel Access
    - **C.** Direct sequence spread spectrum
    - **D.** Temporal Key Integrity Protocol
    - **E.** Transmit power control

12. Which 802.11 draft amendment gives network administrators the ability to configure client stations wirelessly from a central point of management?
    - **A.** 802.11aa
    - **B.** 802.11v
    - **C.** 802.11u
    - **D.** 802.11w
    - **E.** 802.11s
    - **F.** None of the above

13. As defined by the 802.11-2007 standard, which equipment is compatible? (Choose all that apply.)
    - **A.** ERP and HR-DSSS
    - **B.** HR-DSSS and FHSS
    - **C.** OFDM and ERP
    - **D.** 802.11a and 802.11h
    - **E.** DSSS and HR-DSSS

14. Maximum data rates of ________ are permitted using OFDM radios.
    - **A.** 108 Mbps
    - **B.** 22 Mbps
    - **C.** 24 Mbps
    - **D.** 54 Mbps
    - **E.** 11 Mbps

15. What are the security options available as defined in the original IEEE Std. 802.11-1999 (R2003)? (Choose all that apply.)
    A. CCMP/AES
    B. Open System authentication
    C. Preshared keys
    D. Shared Key authentication
    E. WEP
    F. TKIP

16. The 802.11u draft amendment is also known as _________.
    A. Wireless InterWorking with External Networks (WIEN)
    B. Wireless Local Area Networking (WLAN)
    C. Wireless Performance Prediction (WPP)
    D. Wireless Access in Vehicular Environments (WAVE)
    E. Wireless Access Protocol (WAP)

17. The 802.11-2007 standard defines which two technologies for quality of service (QoS) in a WLAN? (Choose all that apply.)
    A. EDCA
    B. PCF
    C. Hybrid Coordination Function Controlled Channel Access
    D. VoIP
    E. Distributed Coordination Function
    F. VoWiFi

18. The 802.11h amendment (now part of the 802.11-2007 standard) introduced what two major changes for 5 GHz radios?
    A. UNII-2 Extended
    B. IAPP
    C. Radar detection
    D. Transmit Frequency Avoidance
    E. Frequency hopping spread spectrum

19. The 802.11b amendment defines the _________ PHY.
    A. HR-DSSS
    B. FHSS
    C. OFDM
    D. PBCC
    E. EIRP

20. Which layers of the OSI model are referenced in the 802.11 standard? (Choose all that apply.)

A. Application

B. Data-Link

C. Presentation

D. Physical

E. Transport

F. Network

Answers to Review Questions

1. A, D. Support for both Extended Rate Physical DSSS (ERP-DSSS/CCK) and Extended Rate Physical Orthogonal Frequency Division Multiplexing (ERP-OFDM) are required in an ERP WLAN, also known as an 802.11g WLAN. Support for ERP-PBCC and DSSS-OFDM PHYs are optional in an ERP WLAN.

2. E. ERP (802.11g) radios mandate the support for both ERP-DSSS/CCK and ERP-OFDM spread spectrum technologies. ERP-DSSS/CCK supports data rates of 1, 2, 5.5, and 11 Mbps and is backward compatible with HR-DSSS (802.11b) and DSSS (802.11 legacy).

3. B, D, E. The original 802.11 standard defines three Physical layer specifications. An 802.11 legacy network could use FHSS, DSSS, or infrared. FHSS 802.11 radio cards are often known as clause 14 devices. DSSS 802.11 radio cards are often known as clause 15 devices. Infrared devices are known as clause 16 devices.

4. C. The 802.11F recommended practice addresses "vendor interoperability" for AP-to-AP roaming. The Inter-Access Point Protocol (IAPP) uses announcement and handover processes that result in how autonomous APs inform other autonomous APs about roamed clients and defines a method of delivery for buffered packets.

5. D, F. The required encryption method defined by an RSN wireless network (802.11i) is Counter Mode with Cipher Block Chaining Message Authentication Code Protocol (CCMP), which uses the Advanced Encryption Standard (AES) algorithm. An optional choice of encryption is the Temporal Key Integrity Protocol (TKIP). The 802.11i amendment also requires the use of an 802.1X/EAP authentication solution or the use of preshared keys.

6. D. 802.11a radio cards operate in the 5 GHz Unlicensed National Information Infrastructure (UNII) 1–3 frequency bands using Orthogonal Frequency Division Multiplexing (OFDM).

7. D. The IEEE 802.11-2007 standard requires data rates of 6, 12, and 24 Mbps for both OFDM and ERP-OFDM radios. Data rates of 6, 9, 12, 18, 24, 36, 48, and 54 Mbps are typically supported. 54 Mbps is the maximum defined rate.

8. B. Fast basic service set transition (FT), also known as fast secure roaming, defines fast handoffs when roaming occurs between cells in a WLAN using the strong security defined in a robust security network (RSN). Applications such as VoIP that require timely delivery of packets, require the roaming handoff to occur in 150ms or less.

9. C. The legacy 802.11 standard, also known as 802.11 Prime, specified data rates of 1 and 2 Mbps using either DSSS or FHSS radios.

10. D. Both 802.11a and 802.11g use OFDM technology but because they operate at different frequencies, they cannot communicate with each other. 802.11a equipment operates in the 5 GHz UNII bands, while 802.11g equipment operates in the 2.4 GHz ISM band.

11. A, E. The 802.11-2007 standard defines mechanisms for dynamic frequency selection (DFS) and transmit power control (TPC) that may be used to satisfy regulatory requirements for operation in the 5 GHz band. This technology was originally defined in the 802.11h amendment, which is now part of the 802.11-2007 standard.

12. B. The 802.11v draft amendment defines SNMP-like Management Information Bases (MIBs), which could give administrators the ability to configure client stations wirelessly from a central point of management.

13. A, D, E. ERP (802.11g) requires the use of ERP-OFDM and ERP-DSSS/CCK in the 2.4 GHz ISM band, and is backward compatible with 802.11b HR-DSSS and DSSS equipment. 802.11b uses HR-DSSS in the 2.4 GHz ISM band and is backward compatible with only legacy DSSS equipment and not legacy FHSS equipment. The 802.11h amendment defines use of TPC and DFS in the 5 GHz UNII bands and is an enhancement of the 802.11a amendment. OFDM technology is used with all 802.11a- and 802.11h-compliant radios.

14. D. The 802.11-2007 standard using OFDM or ERP-OFDM radios requires data rates of 6, 12, and 24 Mbps. Data rates of 6, 9, 12, 18, 24, 36, 48, and 54 Mbps are typically supported. 54 Mbps is the maximum defined rate.

15. B, D, E. The original 802.11 standard defined the use of WEP for encryption. The original 802.11 standard also defined two methods of authentication: Open System authentication and Shared Key authentication.

16. A. The 802.11u draft amendment defines integration of IEEE 802.11 access networks with external networks in a generic and standardized manner. 802.11u is often referred to as Wireless InterWorking with External Networks (WIEN).

17. A, C. The 802.11e amendment (now part of the 802.11-2007 standard) defined two enhanced medium access methods to support quality of service (QoS) requirements. Enhanced Distributed Channel Access (EDCA) is an extension to DCF. Hybrid Coordination Function Controlled Channel Access (HCCA) is an extension to PCF.

18. A, C. The 802.11h amendment effectively introduced two major enhancements: more frequency space in the UNII-2 extended band, and radar avoidance and detection technologies. All aspects of the 802.11h ratified amendment can now be found in clauses 11.8 and 11.9 of the 802.11-2007 standard.

19. A. The 802.11b amendment defines systems that can transmit at data rates of 5.5 Mbps and 11 Mbps using High-Rate DSSS (HR-DSSS). 802.11b devices are also compatible with 802.11 DSSS devices and can transmit at data rates of 1 and 2 Mbps.

20. B, D. The IEEE specifically defines 802.11 technologies at the Physical layer and the MAC sublayer of the Data-Link layer. By design, anything that occurs at the upper layers of the OSI model is insignificant to 802.11 communications.

Wireless Networks and Spread Spectrum Technologies

IN THIS CHAPTER, YOU WILL LEARN ABOUT THE FOLLOWING:

✓ **Industrial, Scientific, and Medical bands (ISM)**

- 900 MHz ISM band
- 2.4 GHz ISM band
- 5.8 GHz ISM band

✓ **Unlicensed National Information Infrastructure bands (UNII)**

- UNII-1 (lower band)
- UNII-2 (middle band)
- UNII-2 Extended
- UNII-3 (upper band)

✓ **Narrowband and spread spectrum**

✓ **Frequency hopping spread spectrum (FHSS)**

- Hopping sequence
- Dwell time
- Hop time
- Modulation

✓ **Direct sequence spread spectrum (DSSS)**

- DSSS data encoding
- Modulation

✓ **Packet Binary Convolutional Code (PBCC)**

- ✓ **Orthogonal Frequency Division Multiplexing (OFDM)**
  - Convolutional coding
  - Modulation
- ✓ **2.4 GHz channels**
- ✓ **5 GHz channels**
- ✓ **Adjacent, nonadjacent, and overlapping channels**
- ✓ **Throughput vs. bandwidth**
- ✓ **Communication resilience**

In this chapter, you will learn about the different spread spectrum transmission technologies and frequency ranges that are supported by the 802.11 standard and amendments. You will learn how these frequencies are divided into different channels, and some of the proper and improper ways of using the channels. Additionally, you will learn about the different types of spread spectrum technologies. You will also learn about Orthogonal Frequency Division Multiplexing (OFDM) and the similarities and differences between OFDM and spread spectrum.

Industrial, Scientific, and Medical (ISM) Bands

The IEEE 802.11 standard and the subsequent 802.11b and 802.11g amendments all define communications in the frequency range between 2.4 GHz and 2.4835 GHz. This frequency range is one of three frequency ranges known as the *Industrial, Scientific, and Medical (ISM) bands*. The ISM bands are as follows:

- 902–928 MHz (26 MHz wide)
- 2.4000–2.4835 GHz (83.5 MHz wide)
- 5.725–5.875 GHz (150 MHz wide)

The ISM bands are defined by the ITU Telecommunication Standardization Sector (ITU-T) in S5.138 and S5.150 of the Radio Regulations. Although the FCC governs the use of the ISM bands defined by the ITU-T in the United States, their usage in other countries may be different because of local regulations. The 900 MHz band is known as the Industrial band, the 2.4 GHz band is known as the Scientific band, and the 5.8 GHz band is known as the Medical band.

It should be noted that all three of these bands are license-free bands, and there are no restrictions on what types of equipment can be used in any of them. For example, a radio card used in medical equipment can be used in the 900 MHz Industrial band.

900 MHz ISM Band

The 900 MHz ISM band is 26 MHz wide and spans from 902 MHz to 928 MHz. In the past, this band was used for wireless networking. However, most wireless networking now uses higher frequencies, which are capable of higher throughput.

Another factor limiting the use of the 900 MHz ISM band is that in many parts of the world, part of the 900 MHz frequency range has already been allocated to the Global System for Mobile Communications (GSM) for use by cellular phones. Although the 900 MHz ISM band is rarely used for networking, many products such as baby monitors, wireless home telephones, and wireless headphones use this frequency range.

802.11 radio cards do not operate in the 900 MHz ISM band, but many older legacy deployments of wireless networking did operate in this band. Some vendors still manufacture non-802.11 wireless networking devices that operate in the 900 MHz ISM band. This is a particularly popular frequency that is used for wireless ISPs because of its superior foliage penetration over the 2.4 GHz and 5 GHz frequency ranges.

2.4 GHz ISM Band

The 2.4 GHz ISM band is the most common band used for wireless networking communications. The 2.4 GHz ISM band is 83.5 MHz wide and spans from 2.4000 GHz to 2.4835 GHz. Use of the 2.4 GHz ISM for wireless LANs is defined by the IEEE in the 802.11-2007 standard. The bulk of Wi-Fi radios currently transmit in the 2.4 GHz ISM band, including radios that use the following technologies:

- 802.11 (FHSS clause 14 radios or DSSS clause 15 radios)
- 802.11b (HR-DSSS clause 18 radios)
- 802.11g (ERP clause 19 radios)
- 802.11n draft (HT clause 20 radios)

In addition to being used by wireless networking equipment, the 2.4 GHz ISM band is also used by microwave ovens, cordless home telephones, baby monitors, and wireless video cameras. The 2.4 GHz ISM band is heavily used, and one of the big disadvantages of using 802.11b/g radios is the potential for interference.

Please keep in mind that not every country's RF regulatory body will allow for transmissions across the entire 2.4–2.4835 GHz ISM band. The IEEE 802.11-2007 standard allows for WLAN transmissions in this band across 14 channels. However, each country can determine which channels can be used. A discussion of all the 2.4 GHz channels occurs later in this chapter.

5.8 GHz ISM Band

The 5.8 GHz ISM band is 150 MHz wide and spans from 5.725 GHz to 5.875 GHz. As with the other ISM bands, the 5.8 GHz ISM band is used by many of the same types of consumer products: baby monitors, cordless telephones, and cameras. It is not uncommon for novices to confuse the 5.8 GHz ISM band with the UNII-3 band, which spans from 5.725 GHz to 5.825 GHz. Both unlicensed bands span the same frequency space. However, the 5.8 ISM band is 50 MHz larger.

The IEEE 802.11a amendment (now part of 802.11-2007) states, "the OFDM PHY shall operate in the 5 GHz band, as allocated by a regulatory body in its operational region." Most countries allow for OFDM transmissions in channels of the various UNII bands, which are discussed in this chapter. However, the United States also allows for OFDM transmissions on ISM channel 165, whose center frequency is 5.825 GHz, which is at the upper edge of the UNII-3 band. Channel 165 resides squarely within the FCC's 5.725–5.875 GHz ISM band. It should be noted that channel 165 is rarely used in WLAN deployments.

Because of less-restrictive FCC power regulations, the 5.8 GHz ISM band is a preferred spectrum for long-distance wireless bridging. The UNII-3 band is also used for outdoor bridging. However, more-stringent power regulations limit distances in the UNII-3 band.

Unlicensed National Information Infrastructure (UNII) Bands

The IEEE 802.11a amendment designated WLAN transmissions within the frequency space of the three 5 GHz bands, each with four channels. These frequency ranges are known as the *Unlicensed National Information Infrastructure (UNII) bands*. The 802.11a amendment defined three groupings, or bands, of UNII frequencies, often known as the lower, middle, and upper UNII bands. These three bands are typically designated as UNII-1 (lower), UNII-2 (middle), and UNII-3 (upper). All three of these bands are 100 MHz wide, which is a useful fact when trying to remember their frequency ranges.

When the 802.11h amendment was ratified, the IEEE designated more frequency space for WLAN transmissions. This frequency space, which consists of 11 additional channels, is often referred to as UNII-2 Extended. Unlike the other three UNII bands that are 100 MHz wide, this new band is 255 MHz wide.

Although we use UNII as the abbreviation, many documents will show U-NII as the abbreviation. Both abbreviations are common and acceptable.

Wi-Fi radios that currently transmit in the 5 GHz UNII bands include radios that use the following technologies:

- 802.11a (OFDM clause 17 radios)
- 802.11h (TPC and DFS)
- 802.11n draft (HT clause 20 radios)

Please keep in mind that not every country's RF regulatory body will allow for transmissions in all these bands. The IEEE 802.11-2007 standard allows for WLAN transmissions in all four of the bands across 23 channels. However, each country may be different. A more detailed discussion of all the 5 GHz channels occurs later in this chapter.

UNII-1 (Lower Band)

UNII-1, the lower UNII band, is 100 MHz wide and spans from 5.150 GHz to 5.250 GHz. This band is typically used indoors with a maximum allowed output power of 50 mW at the intentional radiator (IR) as defined by the FCC. The IEEE has implemented a transmit power cap of 40 mW, which complies with the FCC maximum. Prior to 2004, the FCC required that all UNII-1-capable devices have permanently attached antennas. This meant that any 802.11a device that supported UNII-1 could not have a detachable antenna, even if the device supported other frequencies or standards.

In 2004, the FCC changed the regulations to allow detachable antennas, providing that the antenna connector is unique. This requirement is similar to the antenna requirements for the other UNII bands and the 2.4 GHz ISM band. Some access point manufacturers allow the ability to configure the device as a bridge and to work in the lower UNII band. Care must be taken to make sure that you do not exceed the limitations of your local regulatory body.

UNII-2 (Middle Band)

UNII-2, the middle UNII band, is 100 MHz wide and spans from 5.250 GHz to 5.350 GHz. The FCC allows this band to be used for indoor or outdoor communications, with a maximum allowed output power of 250 mW. The IEEE has implemented a restriction on 802.11 devices of only 200 mW at the intentional radiator, which complies with the FCC maximum. Local regulatory agencies may impose other restrictions that you will need to comply with.

UNII-2 Extended

The UNII-2 Extended band is 255 MHz wide and spans from 5.470 GHz to 5.725 GHz. This band can be used for indoor or outdoor communications, with a maximum allowed output power of 250 mW as defined by the FCC. The IEEE has restricted that to 200 mW at the intentional radiator, which complies with the FCC maximum. Local regulatory agencies may impose other restrictions that you will need to comply with. Operations for WLAN communications were first allowed in this band with the ratification of the 802.11h amendment. Prior to the ratification of this amendment, 5 GHz WLAN communications were allowed in only UNII-1, UNII-2, and UNII-3.

In Chapter 5, "IEEE 802.11 Standards," you learned that the 802.11h amendment defined the use of transmit power control (TPC) and dynamic frequency selection (DFS) to avoid interference with radar transmissions. Any 5 GHz WLAN products that ship in the United States or Canada on or after July 20, 2007, are required to support dynamic frequency selection. FCC Rule # 15.407(h)(2) requires that WLAN products operating in the UNII-2 and UNII-2 Extended bands must support DFS, to protect WLAN communications from interfering with military or weather radar systems. Europe also requires DFS safeguards. Once again, the local regulatory agencies determine how TPC and DFS restrictions are imposed in any of the UNII bands.

UNII-3 (Upper Band)

UNII-3, the upper UNII band, is 100 MHz wide and spans from 5.725 GHz to 5.825 GHz. This band is typically used for outdoor point-to-point communications but can also be used indoors in some countries, including the United States. Europe does not use the UNII-3 band for WLAN communications. The maximum allowed output power by the FCC is 1000 mW.

The IEEE has implemented a power restriction of 800 mW at the intentional radiator, which complies with the FCC maximum.

In Table 6.1, notice that the starting frequency of UNII-3 is the same as the 5.8 GHz ISM band. Remember that the UNII-3 band is 100 MHz wide, and the 5.8 GHz ISM band is 150 MHz wide.

TABLE 6.1 The 5 GHz UNII Bands

UNII-1	Lower	5.15–5.25 GHz	4 channels
UNII-2	Middle	5.25–5.35 GHz	4 channels
UNII-2 Extended	Extended	5.47–5.725 GHz	11 channels
UNII-3	Upper	5.725–5.825 GHz	4 channels

The CWNA Exam (PW0-104) will not test you on any power regulations because they vary from country to country. It is advisable to educate yourself about the maximum transmit power regulations of the country where you plan on deploying a WLAN so that no violations occur. The CWNA exam will test you on your knowledge of the frequency ranges of all the ISM and UNII bands.

Narrowband and Spread Spectrum

There are two primary radio frequency (RF) transmission methods: *narrowband* and *spread spectrum*. A narrowband transmission uses very little bandwidth to transmit the data that it is carrying, whereas a spread spectrum transmission uses more bandwidth than is necessary to carry its data. Spread spectrum technology takes the data that is to be transmitted and spreads it across the frequencies that it is using. For example, a narrowband radio might transmit data on 2 MHz of frequency space at 80 watts, while a spread spectrum radio might transmit data over a 22 MHz frequency space at 100 milliwatts.

Figure 6.1 shows a rudimentary comparison of how a narrowband and spread spectrum signal relate to each other. Because narrowband signals take up a single or very narrow band of frequencies, intentional jamming or unintentional interference of this frequency range is likely to cause disruption in the signal. Because spread spectrum uses a wider range of frequency space, it is typically less susceptible to intentional jamming or unintentional interference from outside sources, unless the interfering signal was also spread across the range of frequencies used by the spread spectrum communications.

FIGURE 6.1 Overlay of narrowband and spread spectrum frequency use

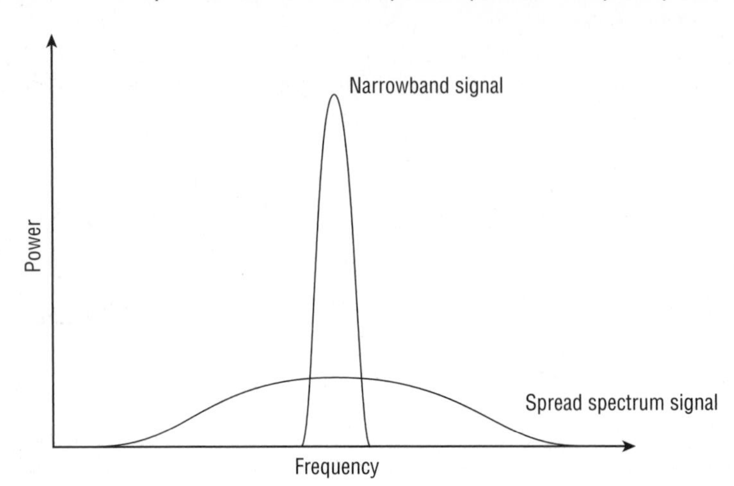

Narrowband signals are transmitted using much higher power than spread spectrum signals. Typically, the FCC or other local regulatory bodies require that narrowband transmitters are licensed to minimize the risk of two narrowband transmitters interfering with each other. AM and FM radio stations are examples of narrowband transmitters that are licensed to make sure that two stations in the same or nearby market are not transmitting on the same frequency.

Spread spectrum signals are transmitted using very low power levels. Because the power levels are so low, they are less likely to cause interference with other systems and therefore are typically not required to be licensed by the local regulatory body.

Real World Scenario

Who Invented Spread Spectrum?

Spread spectrum was originally patented on August 11, 1942, by actress Hedy Kiesler Markey (Hedy Lamarr) and composer George Antheil and was originally designed to be a radio guidance system for torpedoes, a purpose for which it was never used. The idea of spread spectrum was ahead of its time. It was not until 1957 that further development on spread spectrum occurred, and in 1962 frequency hopping spread spectrum was used for the first time between the U.S. ships at the blockade of Cuba during the Cuban Missile Crisis.

If you would like to learn more about the interesting history of spread spectrum, search the Internet for *Lamarr and Antheil*. There are many websites with articles about these two inventors and even copies of the original patent. Neither inventor made any money from their patent because it expired before the technology was developed.

One of the problems that can occur with RF communications is multipath interference. Multipath occurs when a reflected signal arrives at the receiving antenna after the primary signal. This is similar to the way an echo is heard after the original sound.

Let's use an example of yelling to a friend across a canyon. Let's assume you are going to yell, "Hello, how are you?" to your friend. To make sure that your friend understands your message, you might pace your message and yell each word 1 second after the previous word. If your friend heard the echo (multipath reflection of your voice) a half-second after the main sound arrived, your friend would hear "HELLO hello HOW how ARE are YOU you" (echoes are represented by lowercase). Your friend would be able to interpret the message because the echo arrived between the main signals, or the sound of your voice. However, if the echo arrived 1 second after the main sound, the echo for the word *hello* would arrive at the same time the word *HOW* arrives. With both sounds arriving at the same time, it may not be possible to understand the message.

RF data communications behave the same way as the sound example. The delay between the main signal and the reflected signal is known as the *delay spread*. If the delay spread is long enough that the reflected signal interferes with the next piece of data from the main signal, this is referred to as *intersymbol interference (ISI)*. Spread spectrum systems are not as susceptible to ISI because they spread their signals across a range of frequencies. These different frequencies produce different delays in multipath, such that some wavelengths may be affected by ISI whereas others may not. Because of this behavior, spread spectrum signals are typically more tolerant of multipath interference than narrowband signals.

802.11 (DSSS), 802.11b (HR-DSSS), and 802.11g (ERP) are tolerant of delay spread only to a certain extent. 802.11 (DSSS) and 802.11b (HR-DSSS) can tolerate delay spread of up to 500 nanoseconds. Even though the delay spread can be tolerated, performance is much better when the delay spread is lower. The 802.11b transmitter will drop to a lower data rate when the delay spread increases. Longer symbols are used when transmitting at the lower data rates. When longer symbols are used, longer delays can occur before ISI occurs. According to some of the 802.11b vendors, 65 nanoseconds or lower delay spread is required for 802.11b at 11 Mbps.

Because of OFDM's greater tolerance of delay spread, an 802.11g transmitter can maintain 54 Mbps with a delay spread of up to about 150 nanoseconds. This depends on the 802.11g chipset that is being used in the transmitter and receiver. Some chipsets are not as tolerant and switch to a lower data rate at a lower delay spread value.

Frequency Hopping Spread Spectrum (FHSS)

Frequency hopping spread spectrum (FHSS) was used in the original 802.11 standard and provided 1 and 2 Mbps RF communications using the 2.4 GHz ISM band for legacy clause 14 radios. The majority of legacy FHSS radios were manufactured between 1997 and 1999. The IEEE specified that in North America, 802.11 FHSS would use 79 MHz of frequencies, from 2.402 GHz to 2.480 GHz.

Generally, the way FHSS works is that it transmits data by using a small frequency carrier space, then hops to another small frequency carrier space and transmits data, then to another frequency, and so on, as illustrated in Figure 6.2. More specifically, frequency hopping spread spectrum transmits data by using a specific frequency for a set period of time, known as the *dwell time*. When the dwell time expires, the system changes to another frequency and begins to transmit on that frequency for the duration of the dwell time. Each time the dwell time is reached, the system changes to another frequency and continues to transmit.

FIGURE 6.2 FHSS components

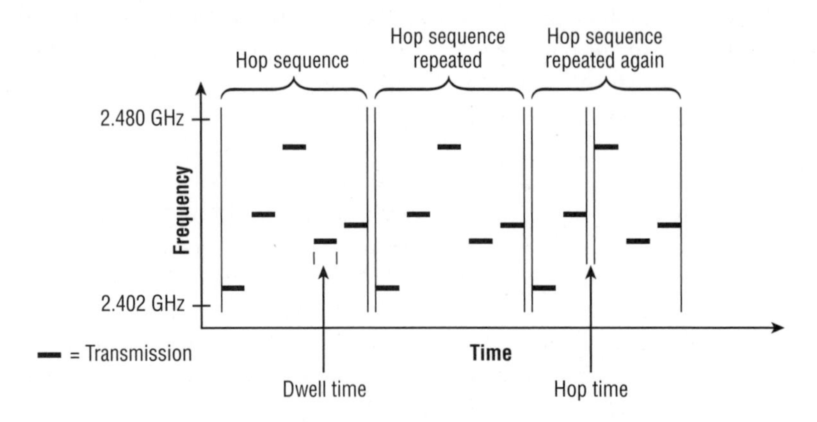

Hopping Sequence

FHSS radios use a predefined *hopping sequence* (also called a hopping pattern or hopping set) comprising a series of small carrier frequencies, or *hops*. Instead of transmitting on one set channel or finite frequency space, an FHSS radio card transmits on a sequence of subchannels called hops. Each time the hop sequence is completed, it is repeated. Figure 6.2 shows a make-believe hopping sequence that consists of five hops.

The original IEEE 802.11 standard mandates that each hop is 1 MHz in size. These individual hops are then arranged in predefined sequences. In North America and most of Europe, the hopping sequences contain at least 75 hops, but no greater than 79 hops. Other countries have different requirements; for example, France uses 35 hops, while Spain and Japan use 23 hops in a sequence. For successful transmissions to occur, all FHSS transmitters and receivers must be synchronized on the same carrier hop at the same time. The 802.11 standard defines hopping sequences that can be configured on an FHSS access point, and the hopping sequence information is delivered to client stations via the beacon management frame.

Dwell Time

Dwell time is a defined amount of time that the FHSS system transmits on a specific frequency before it switches to the next frequency in the hop set. The local regulatory body typically limits the amount of dwell time. For example, the FCC specifies a maximum dwell time of 400 milliseconds (ms) per carrier frequency during any 30-second period of time. Typical dwell times are around 100ms to 200ms. The IEEE 802.11 standard specifies that a hopping sequence must consist of at least 75 frequencies, 1 MHz wide. Because the standard specifies a maximum bandwidth of 79 MHz, the maximum number of hops possible for a hop set would be 79. With an FHSS hop sequence consisting of 75 hops and a dwell time of 400ms, it would take about 30 seconds to complete the hop sequence. After the hop sequence is complete, it is repeated.

Hop Time

Hop time is not a specified period of time but rather a measurement of the amount of time it takes for the transmitter to change from one frequency to another. Hop time is typically a fairly small number, often about 200 to 300 microseconds (µs). With typical dwell times of 100 to 200ms, hop times of 200 to 300µs are insignificant. Insignificant or not, the hop time is essentially wasted time, or overhead, and takes up the same amount of time regardless of the dwell time. The longer the dwell time, the less often the transmitter has to waste time hopping to another frequency, thus greater throughput. If the dwell time is shorter, the transmitter has to hop more frequently, thus decreasing throughput.

Modulation

FHSS uses Gaussian Frequency Shift Keying (GFSK) to encode the data. Two-level GFSK (2GFSK) uses two frequencies to represent a 0 or a 1 bit. Four-level GFSK (4GFSK) uses four frequencies, with each frequency representing 2 bits (00, 01, 10, or 11). Because it takes cycles before the frequency can be determined, the symbol rate (rate that the data is sent) is only about 1 or 2 million symbols per second, a fraction of the 2.4 GHz carrier frequency.

What Is the Significance of the Dwell Time?

Because FHSS transmissions jump inside a frequency range of 79 MHz, a narrowband signal or noise would disrupt only a small range of frequencies and would produce only a minimal amount of throughput loss. Decreasing the dwell time can further reduce the effect of interference. Conversely, because the radio card is transmitting data during the dwell time, the longer the dwell time, the greater the throughput.

Direct Sequence Spread Spectrum (DSSS)

Direct sequence spread spectrum (DSSS) was originally specified in the primary, or root, 802.11 standard and provides 1 and 2 Mbps RF communications using the 2.4 GHz ISM band. DSSS was also specified in the 802.11b addendum and provides 5.5 and 11Mbps RF communications using the same 2.4 GHz ISM band. The 802.11b 5.5 and 11 Mbps speeds are known as *High-Rate DSSS (HR-DSSS)*.

802.11b clause 18 devices are backward compatible with the legacy 802.11 DSSS clause 15 devices. This means that an 802.11b device can transmit using DSSS at 1 and 2 Mbps and using HR-DSSS at 5.5 and 11Mbps. However, 802.11b devices are not capable of transmitting using FHSS; therefore, they are not backward compatible with 802.11 FHSS clause 14 devices.

DSSS 1 and 2 Mbps are specified in clause 15 of the 802.11-2007 standard. HR-DSSS 5.5 and 11 Mbps are specified in clause 18 of the 802.11-2007 standard.

Unlike FHSS, where the transmitter jumped between frequencies, DSSS is set to one channel. The data that is being transmitted is spread across the range of frequencies that make up the channel. The process of spreading the data across the channel is known as *data encoding*.

DSSS Data Encoding

In Chapter 2, "Radio Frequency Fundamentals," you learned about many ways that RF signals can get altered or corrupted. Because 802.11 is an unbounded medium with a huge potential for RF interference, it had to be designed to be resilient enough that data corruption could be minimized. To achieve this, each bit of data is encoded and transmitted as multiple bits of data.

The task of adding additional, redundant information to the data is known as *processing gain*. In this day and age of data compression, it seems strange that we would use a technology that adds data to our transmission, but by doing so, the communication is more resistant to data corruption. The system converts the 1 bit of data into a series of bits that are referred to as *chips*. To create the chips, a Boolean XOR is performed on the data bit, and a fixed-length bit sequence pseudo-random number (PN) code. Using a PN code known as the Barker code, the binary data 1 and 0 are represented by the following chip sequences:

Binary data 1 = 1 0 1 1 0 1 1 1 0 0 0

Binary data 0 = 0 1 0 0 1 0 0 0 1 1 1

This sequence of chips is then spread across a wider frequency space. While 1 bit of data might need only 2 MHz of frequency space, the 11 chips will require 22 MHz of frequency carrier. This process of converting a single data bit into a sequence is often called *spreading* or *chipping*. The receiving radio card converts, or *de-spreads*, the chip sequence back into a single data bit. When the data is converted to multiple chips and some of the chips are not received properly, the radio will still be able to interpret the data by looking at the chips that were received properly. When the Barker code is used, as many as 9 of the 11 chips can be corrupted, yet the receiving radio card will still be able to interpret the sequence and convert them back into a single data bit. This chipping process also makes the communication less likely to be affected by inter-symbol interference because it uses more bandwidth.

After the Barker code is applied to data, a series of 11 bits, referred to as chips, represent the original single bit of data. This series of encoded bits makes up 1 bit of data. To help prevent confusion, it is best to think of and refer to the encoded bits as *chips*.

The Barker code uses an 11-chip PN; however, the length of the code is irrelevant. To help provide the faster speeds of HR-DSSS, another more-complex code, *Complementary Code Keying (CCK)*, is utilized. CCK uses an 8-chip PN, along with using different PNs for different bit sequences. CCK can encode 4 bits of data with 8 chips (5.5 Mbps) and can encode 8 bits of data with 8 chips (11 Mbps). Although it is interesting to learn about, a thorough understanding of CCK is not required for the CWNA exam.

Modulation

After the data has been encoded using a chipping method, the transmitter needs to modulate the signal to create a carrier signal containing the chips. *Differential Binary Phase Shift Keying (DBPSK)* utilizes two phase shifts, one that represents a 0 chip and another that represents a 1 chip. To provide faster throughput, *Differential Quadrature Phase Shift Keying (DQPSK)* utilizes four phase shifts, allowing each of the four phase shifts to modulate 2 chips (00, 01, 10, 11) instead of just 1 chip, doubling the speed.

Table 6.2 shows a summary of the data encoding and modulation techniques used by 802.11 and 802.11b.

TABLE 6.2 DSSS and HR-DSSS Encoding and Modulation Overview

	Data Rate (Mbps)	Encoding	Chip Length	Bits Encoded	Modulation
DSSS	1	Barker coding	11	1	DBPSK
DSSS	2	Barker coding	11	1	DQPSK
HR-DSSS	5.5	CCK coding	8	4	DQPSK
HR-DSSS	11	CCK coding	8	8	DQPSK

Packet Binary Convolutional Code (PBCC)

Packet Binary Convolutional Code (PBCC) is a modulation technique that supports data rates of 5.5, 11, 22, and 33 Mbps; however, both the transmitter and receiver must support the technology to achieve the higher speeds. PBCC was developed by Alantro Communications, which was purchased by Texas Instruments. PBCC modulation was originally defined as optional under the 802.1b amendment. The introduction of the 802.11g amendment allowed for two additional optional ERP-PBCC modulation modes with payload data rates of 22 and 33 Mbps.

PBCC and ERP-PBCC technology was seen for a short time in the SOHO marketplace. However, the technology was rarely deployed in an enterprise environment.

Orthogonal Frequency Division Multiplexing (OFDM)

Orthogonal Frequency Division Multiplexing (OFDM) is one of the most popular communications technologies, used in both wired and wireless communications. The 802.11-2007 standard specifies the use of OFDM at 5 GHz and also specifies the use of ERP-OFDM at 2.4 GHZ. As mentioned in Chapter 5, OFDM and ERP-OFDM are the same technology. OFDM is not a spread spectrum technology, even though it has similar properties to spread spectrum, such as low transmit power and using more bandwidth than is required to transmit data. Because of these similarities, OFDM is often referred to as a spread spectrum technology even though technically that reference is incorrect. OFDM actually transmits across 52 separate, closely and precisely spaced frequencies, often referred to as *subcarriers*, as illustrated in Figure 6.3.

The frequency width of each subcarrier is 312.5 KHz. The subcarriers are also transmitted at lower data rates, but because there are so many subcarriers, overall data rates are higher. Also, because of the lower subcarrier data rates, delay spread is a smaller percentage of the symbol period, which means that ISI is less likely to occur. In other words, OFDM technology is more resistant to the negative effects of multipath than DSSS and FHSS spread spectrum technologies. Figure 6.4 represents four of the 52 subcarriers. One of the subcarriers is highlighted so that you can more easily understand the drawing. Notice that the frequency spacing of the subcarriers has been chosen so that the harmonics overlap and provide cancellation of most of the unwanted signals.

The 52 subcarriers are numbered from –26 to +26. Forty-eight of the subcarriers are used to transmit data. The other four, numbers –21, –7, +7, and +21, are known as *pilot carriers*. These four are used as references for phase and amplitude by the demodulator, allowing the receiver to compensate for distortion of the OFDM signal.

FIGURE 6.3 802.11 Channels and OFDM subcarriers

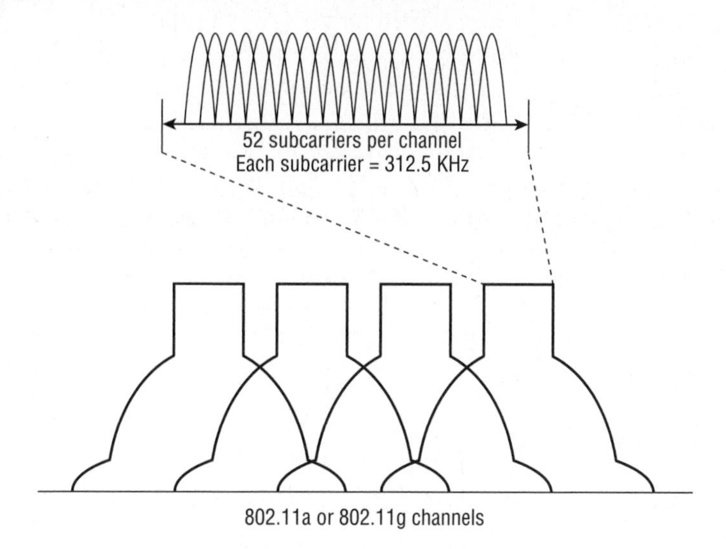

FIGURE 6.4 Subcarrier signal overlay

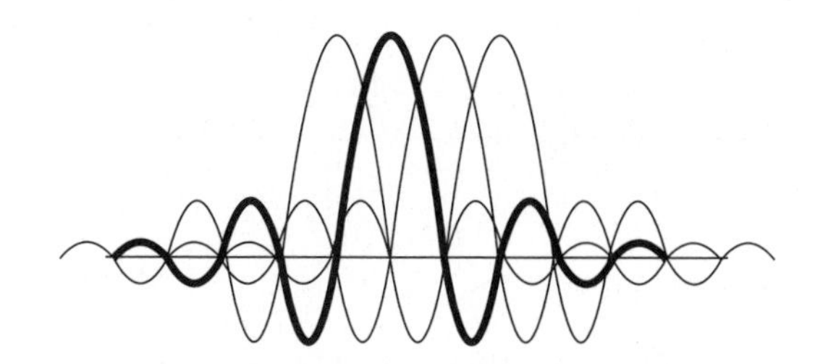

Convolutional Coding

In order to make OFDM more resistant to narrowband interference, a form of error correction known as *convolutional coding* is performed. The 802.11-2007 standard defines the use of convolutional coding as the error-correction method to be used with OFDM technology. It is a *forward error correction (FEC)* that allows the receiving system to detect and repair corrupted bits.

There are many levels of convolutional coding. Convolutional coding uses a ratio between the bits transmitted vs. the bits encoded to provide these different levels. The lower the ratio, the less resistant the signal is to interference and the greater the data rate will be. Table 6.3 displays a comparison between the technologies used to create the different data rates of both 802.11a and 802.11g. Notice that the data rates are grouped by pairs based on modulation technique and that the difference between the two speeds is caused by the different levels of convolutional coding. A detailed explanation of convolutional coding is extremely complex and far beyond the knowledge needed for the CWNA exam.

TABLE 6.3 802.11a and 802.11g Data Rate & Modulation Comparison Chart

Data Rates (Mbps)	Modulation Method	Coded Bits per Subcarrier	Data Bits per OFDM Symbol	Coded bits per OFDM Symbol	Coding Rate (Data Bits/ Coded Bits)
6	BPSK	1	24	48	1/2
9	BPSK	1	36	48	3/4
12	QPSK	2	48	96	1/2
18	QPSK	2	72	96	3/4
24	16-QAM	4	96	192	1/2
36	16-QAM	4	144	192	3/4
48	64-QAM	6	192	288	2/3
54	64-QAM	6	216	288	3/4

Modulation

OFDM uses Binary Phase Shift Keying (BPSK) and Quadrature Phase Shift Keying (QPSK) phase modulation for the lower ODFM data rates. The higher OFDM data rates use 16-QAM and 64-QAM modulation. *Quadrature amplitude modulation (QAM)* is a hybrid of phase and amplitude modulation.

2.4 GHz Channels

To better understand how legacy 802.11 (DSSS), 802.11b (HR-DSSS), and 802.11g (ERP) radios are used, it is important to understand how the IEEE 802.11-2007 standard divides the 2.4 GHz ISM band into 14 separate channels, as listed in Table 6.4. Although the 2.4 GHz ISM band is divided into 14 channels, the FCC or local regulatory body designates which channels are allowed to be used. Table 6.4 also shows what channels are supported in a sample of a few countries. As you can see, the regulations can vary greatly between countries.

TABLE 6.4 2.4 GHz Frequency Channel Plan

Channel ID	Center Frequency (GHz)	U.S. (FCC)	Canada (IC)	Europe (ETSI)
1	2.412	X	X	X
2	2.417	X	X	X
3	2.422	X	X	X
4	2.427	X	X	X
5	2.432	X	X	X
6	2.437	X	X	X
7	2.442	X	X	X
8	2.447	X	X	X
9	2.452	X	X	X
10	2.457	X	X	X
11	2.462	X	X	X
12	2.467			X
13	2.472			X
14	2.484			

X = supported channel

Channels are designated by their center frequency. Each channel is 22 MHz wide and is often referenced by the center frequency ± 11 MHz. For example, channel 1 is 2.412 GHz ± 11 MHz, which means that channel 1 spans from 2.401 GHz to 2.423 GHz. It should also be noted that within the 2.4 GHz ISM band, the distance between channel center frequencies is only 5 MHz. Because each channel is 22 MHz wide, and because the separation between center frequencies of each channel is only 5 MHz, the channels will have overlapping frequency space.

Figure 6.5 shows an overlay of all the channels and how they overlap. Channels 1, 6, and 11 have been highlighted because, as you can see, they are separated from each other by enough frequencies that they do not overlap. In order for two channels to not overlap, they must be separated by at least five channels or 25 MHz. Channels, such as 2 and 9, do not overlap, but by selecting 2 and 9, there is no additional legal channel that can be chosen that does not overlap either 2 or 9. In the United States and Canada, the only three simultaneously nonoverlapping channels are 1, 6, and 11. In regions where channels 1 through 13 are allowed to be used, there are different combinations of three nonoverlapping channels, although channels 1, 6, and 11 are usually chosen.

FIGURE 6.5 2.4 GHz channel overlay diagram

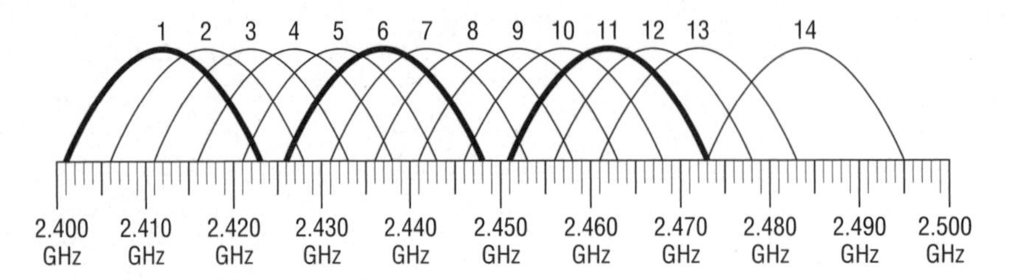

The IEEE 802.11-2007 definitions of nonoverlapping channels in the 2.4 GHz ISM band can be somewhat confusing if not properly explained. Legacy 802.11 (DSSS), 802.11b (HR-DSSS), and 802.11g (ERP) channels all use the same numbering schemes and have the same center frequencies. However, the individual channels' frequency space may overlap. Figure 6.6 shows channels 1, 6, and 11 with 25 MHz of spacing between the center frequencies. These are the most commonly used *nonoverlapping channels* in North America and most of the world for 802.11b/g networks.

FIGURE 6.6 HR-DSSS center frequencies

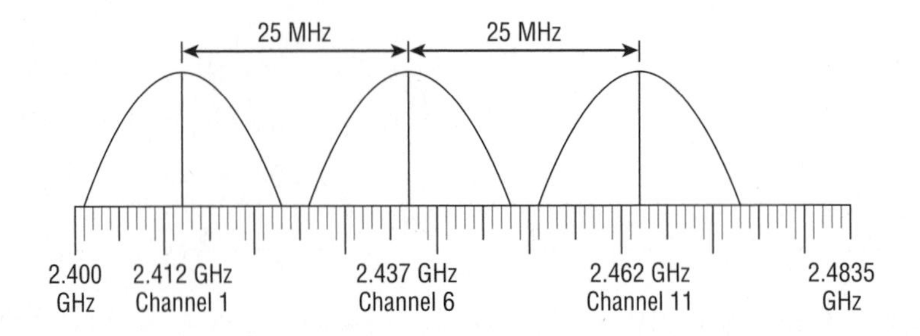

What exactly classifies DSSS or HR-DSSS channels as nonoverlapping? According to the original 802.11 standard, legacy DSSS channels had to have at least 30 MHz of spacing between the center frequencies to be considered nonoverlapping. In a deployment of legacy DSSS equipment using a channel pattern of 1, 6, and 11, the channels were considered overlapping because the center frequencies were only 25 MHz apart. Although DSSS channels 1, 6, and 11 were defined as overlapping, these were still the only three channels used in channel reuse patterns when legacy networks were deployed. This really is of little significance anymore because most 2.4 GHz deployments now use 802.b/g technology.

HR-DSSS was introduced under the 802.11b amendment, which states that channels need a minimum of 25 MHz of separation between the center frequencies to be considered nonoverlapping. Therefore, when 802.11b was introduced, channels 1, 6, and 11 were considered nonoverlapping.

The 802.11g amendment, which allows for backward compatibility with 802.11b HR-DSSS, also requires 25 MHz of separation between the center frequencies to be considered nonoverlapping. 802.11g Extended Rate Physical (ERP) channels also require 25 MHz of separation between the center frequencies to be considered nonoverlapping. Under the 802.11g amendment, channels 1, 6, and 11 are also considered nonoverlapping for both ERP-DSSS/CCK and ERP-OFDM.

Although it is very common to represent the RF signal of a particular channel with an arch-type line, this is not a true representation of the signal. To explain it simply, in addition to the main *carrier frequency*, or main frequency, sideband carrier frequencies are also generated, as shown in Figure 6.7. The IEEE defines a *transmit spectrum mask,* specifying that the first sideband frequency (–11 MHz to –22 MHz from the center frequency, and +11 MHz to +22 MHz from the center frequency) must be at least 30 dB less than the main frequency. The mask also specifies that any additional sideband carrier frequencies (–22 MHz from the center frequency and beyond, and +22 MHz from the center frequency and beyond) must be at least 50 dB less than the main frequency.

FIGURE 6.7 IEEE 802.11b transmit spectrum mask

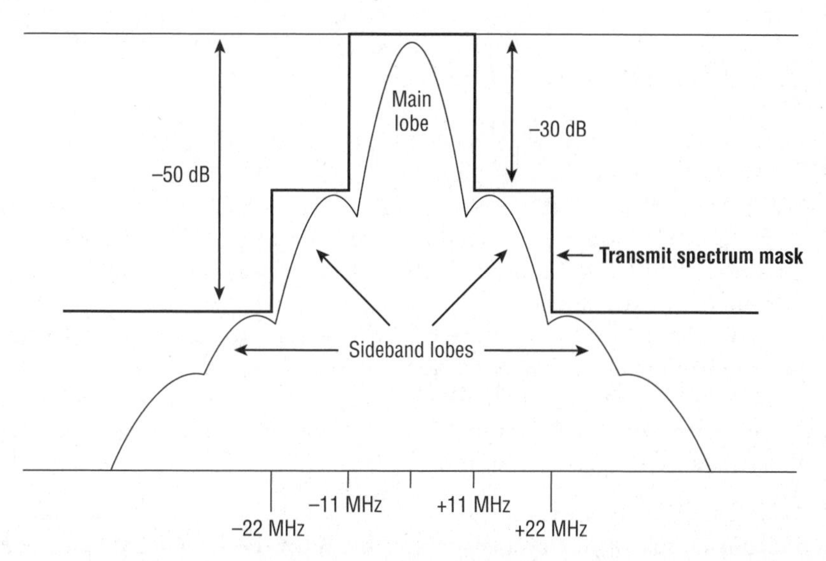

Figure 6.7 illustrates the transmit spectrum mask of an HR-DSSS channel at 2.4 GHz. The transmit spectrum mask is defined to minimize interference between devices on different frequencies. Even though the sideband carrier frequencies are mere whispers of signals compared to the main carrier frequency, even a whisper is noticeable when the person whispering is close to you. This is true for RF devices too.

Figure 6.8 represents RF signals on channels 1, 6, and 11. A signal-level line indicates an arbitrary level of reception by the access point on channel 6. At level 1, meaning the AP on channel 6 receives only the signals above the level 1 line, the signals from channel 1 and channel 11 do not intersect (interfere) with the signals on channel 6. However, at the level 2 line, the signals from channel 1 and channel 11 do intersect (interfere) slightly with the signals on channel 6. At the level 3 line, there is significant interference from the signals from channel 1 and channel 11. Because of the potential for this situation, it is important to separate access points (usually 5 to 10 feet is sufficient) so that interference from sideband frequencies does not occur. This is important both horizontally and vertically.

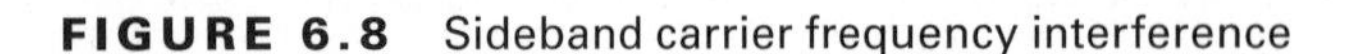

FIGURE 6.8 Sideband carrier frequency interference

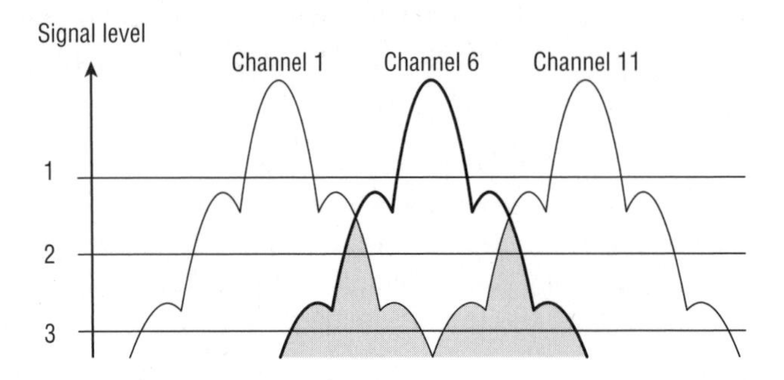

5 GHz Channels

The 802.11-2007 standard specifies the 5 GHz UNII bands: UNII-1, UNII-2, UNII-2 Extended, and UNII-3. The centers of the outermost channels must be 30 MHz from the band's edge in the UNII-1 and UNII-2 bands, and 20 MHz in the UNII-3 band. The original three UNII bands each have four nonoverlapping channels with 20 MHz separation between the center frequencies. The UNII-2 Extended band has eleven nonoverlapping channels with 20 MHz of separation between the center frequencies.

Figure 6.9 shows the eight UNII-1 and UNII-2 channels in the top graphic, the eleven UNII-2 Extended channels in the center graphic, and the four UNII-3 channels in the bottom graphic. Channel 36 is highlighted so that it is easier to distinguish a single carrier and its sideband frequencies. The IEEE defines the center frequency of each channel as follows, where n_{ch} is all values from 0 through 200:

$$5{,}000 + 5 \times n_{ch} \text{ (MHz)}$$

The IEEE does not specifically define a channel width, however the spectral mask of an OFDM channel is approximately 20 MHz.

As seen in Figure 6.10 of the OFDM spectrum mask, the sideband carrier frequencies do not drop off very quickly, and therefore the sideband frequencies of two adjacent valid channels overlap and are more likely to cause interference. The 802.11a amendment, which originally defined the use of OFDM (clause 17), required only 20 MHz of separation between the center frequencies for channels to be considered nonoverlapping. All 23 channels in the 5 GHz UNII bands use OFDM and have 20 MHz of separation between the center frequencies. Therefore, all 5 GHz OFDM channels are considered nonoverlapping by the IEEE. In reality, there is some sideband carrier frequency overlap between any two adjacent 5 GHz channels.

Luckily, due to the number of channels and the channel spacing of 802.11a, it is easier to separate adjacent channels and prevent interference.

FIGURE 6.9 UNII channel overview

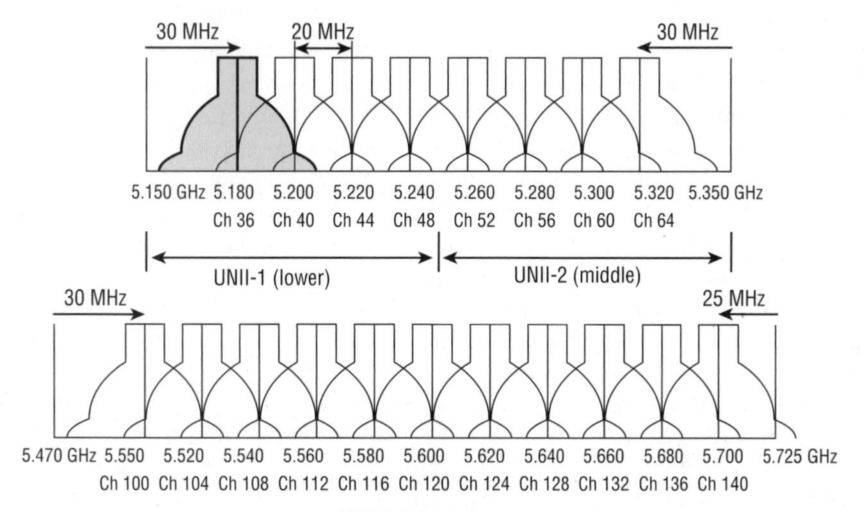

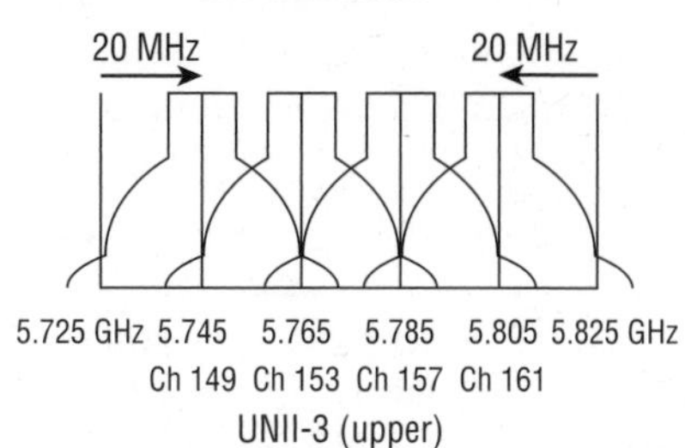

FIGURE 6.10 OFDM spectrum mask

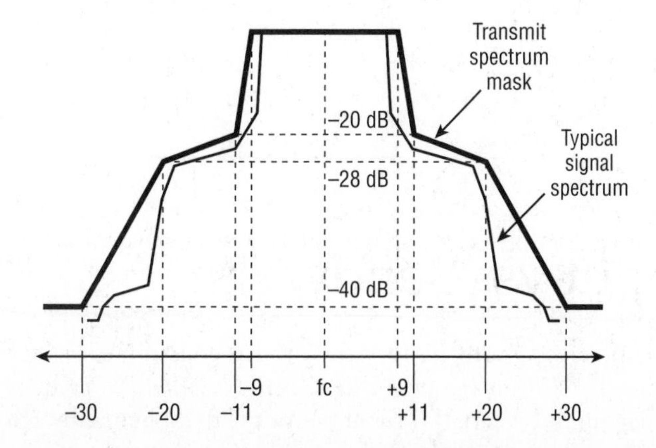

Adjacent, Nonadjacent, and Overlapping Channels

In the preceding paragraphs, you learned how the IEEE-2007 standard defines nonoverlapping channels. DSSS (legacy) channels require 30 MHz of separation between the center frequencies to be considered nonoverlapping. HR-DSSS (802.11b) and ERP (802.11g) channels require 25 MHz of separation between the center frequencies to be considered nonoverlapping. And finally, 5 GHz OFDM channels require 20 MHz of separation between the center frequencies to be considered nonoverlapping. Why are these definitions important? When deploying a WLAN, it is important to have overlapping cell coverage for roaming to occur. However, it is just as important for these coverage cells not to have overlapping frequency space. A channel reuse pattern is needed because overlapping frequency space causes degradation in performance. The design aspects of channel reuse patterns are discussed in great detail in Chapter 12, "WLAN Troubleshooting."

An often debated topic is what defines an *adjacent channel* and *nonadjacent channel*. The 802.11-2007 standard loosely defines an adjacent channel as any channel with nonoverlapping frequencies for the DSSS and HR-DSSS PHYs. With ERP and OFDM PHYs, the standard loosely defines an adjacent channel as the first channel with a nonoverlapping frequency space. In other words, the IEEE's definition of adjacent channels is almost exactly the same as the definition of nonoverlapping channels that has been discussed earlier. Confused? Table 6.5 illustrates the CWNP program's interpretation of these concepts.

TABLE 6.5 IEEE Adjacent vs. Nonadjacent

	DSSS clause 15	**HR-DSSS clause 18**	**ERP clause 19**	**OFDM clause 17**
Frequency Band	2.4 GHz ISM	2.4 GHz ISM	2.4 GHz ISM	UNII bands
Adjacent	≥ 30 MHz	≥ 25 MHz	= 25 MHz	= 20 MHz
Nonadjacent	N/A	N/A	> 25 MHz	> 20 MHz
Overlapping	< 30 MHz	< 25 MHz	< 25 MHz	N/A

Throughput vs. Bandwidth

Wireless communication is typically performed within a constrained set of frequencies known as a frequency band. This frequency band is the *bandwidth*. Frequency bandwidth does play a part in the eventual throughput of the data, but many other factors also determine throughput. In addition to frequency bandwidth, data encoding, modulation, medium contention, encryption, and many other factors also play a large part in data throughput.

What Is the Significance of Adjacent Channels?

The IEEE's loose definition of adjacent channels contradicts how the term *adjacent channel interference* is used in the WLAN marketplace. Most Wi-Fi vendors use the term adjacent channel interference to refer to the degradation of performance resulting from overlapping frequency space that occurs because of an improper channel reuse design. In the WLAN industry, an adjacent channel is considered to be the next or previous numbered channel. For example, channel 3 is adjacent to channel 2. The authors of this book recommend that you do not get caught up in the IEEE's definition of adjacent channels that was covered in the previous paragraph. The definition of adjacent channels that the Wi-Fi industry has adopted is much more commonplace. The concept of adjacent channel interference is discussed in great detail in Chapter 12.

Care should be taken not to confuse frequency bandwidth with data bandwidth. Data encoding and modulation determine data rates, which are sometimes also referred to as data bandwidth. Simply look at the 5 GHz channels and OFDM as an example. OFDM clause 17 radios can transmit at 6, 9, 12, 18, 24, 36, 48, or 54 Mbps, yet the frequency bandwidth for all the UNII band channels is the same for all of these speeds. What changes between all of these speeds (data rates) is the modulation and coding technique. The proper term for the changes in speed due to modulation and coding is *data rates*; however they are also often referred to as *data bandwidth*.

One of the surprising facts when explaining wireless networking to a layperson is the actual throughput that an 802.11 wireless network provides. When novices walk through a computer store and see the packages of 802.11 devices, they usually assume that a device that is labeled as 54 Mbps is going to provide throughput of 54 Mbps. A medium access method known as Carrier Sense Multiple Access with Collision Avoidance (CSMA/CA) helps to ensure that only one radio device can be transmitting on the medium at any given time. Because of the half-duplex nature of the medium and the overhead generated by CSMA/CA, the actual aggregate throughput is typically 50 percent or less of the data rate. In addition to the throughput being affected by the half-duplex nature of 802.11 communications, the throughput is affected differently based on the frequency used. 802.11a uses OFDM, and 802.11g uses ERP-OFDM, which are effectively the same technology. However, 802.11g does not perform as well because of the higher level of RF noise that is typical in the 2.4 GHz ISM band. It is also very important to understand that the 802.11 RF medium is a *shared* medium, meaning that in any discussion of throughput, it should be thought of as *aggregate throughput*. For example, if a data rate is 54 Mbps, because of CSMA/CA the aggregate throughput might be about 20 Mbps. If five client stations were all downloading the same file from an FTP server at the same time, the perceived throughput for each client station would be about 4 Mbps under ideal circumstances.

Many other things can add overhead and affect throughput. Security and encryption can both add additional processing requirements to encrypt and decrypt the data, along with increasing the frame size, thus increasing the communication overhead. Fragmentation of frames creates additional overhead by forcing the system to transmit smaller frames, each with a complete set of 802.11 headers. RTS/CTS (which you will learn about in Chapter 9, "802.11 MAC Architecture") can also affect throughput by adding communication overhead. In some environments, fragmentation and RTS/CTS can actually increase throughput if the initial throughput was low because of communication problems.

Variables at almost all layers of the OSI model can affect the throughput of 802.11 communications. It is important to understand the different causes, their effects, and what, if anything, can be done to minimize their effect on overall data throughput.

Communication Resilience

Many technologies that have been covered in this chapter either directly or indirectly provide resilience to 802.11 communications. Spread spectrum spreads the data across a range of frequencies, making it less likely for a narrowband RF signal to cause interference. FHSS is inherently more resilient to narrowband interference than OFDM, and OFDM is more resilient to narrowband interference than DSSS. Spread spectrum technology uses a range of frequencies, which inherently adds resilience because delay spread and ISI will vary between the different frequencies. Additionally, data encoding provides error recovery methods, helping to reduce the need for retransmission of the data.

Summary

This chapter focused on the technologies that make up wireless networking and spread spectrum. 802.11, 802.11b, and 802.11g radios use the 2.4 GHz ISM band, while 802.11a/h radios use 5 GHz UNII bands. 802.11n draft radios can use both the 2.4 GHz ISM band and the 5 GHz UNII bands. The ISM and UNII bands discussed in this chapter are as follows:

- ISM 902–928 MHz—Industrial
- ISM 2.4000–2.4835 GHz—Scientific
- ISM 5.725–5.875 GHz—Medical
- UNII-1 5.150–5.250 GHz—lower UNII
- UNII-2 5.250–5.350 GHz—middle UNII
- UNII-2 extended 5.470–5.725 GHz—Extended UNII
- UNII-3 5.725–5.825 GHz—upper UNII

Spread spectrum technology was introduced and described in detail along with OFDM and convolutional coding. The following are key spread spectrum technologies and terms that were discussed:

- FHSS
- Dwell time
- Hop time
- DSSS

This chapter ended with a comparison of throughput and bandwidth and a review of the communication resilience of the technologies used in 802.11.

Exam Essentials

Know the technical specifications of all the ISM and UNII bands. Make sure that you know all of the frequencies, bandwidth uses, channels, and center channel separation rules.

Know spread spectrum. Spread spectrum can be complicated and has different flavors. Understand FHSS, DSSS, and OFDM (although OFDM is not a spread spectrum technology, it has similar properties and you have to know it). Understand how coding and modulation work with spread spectrum and OFDM.

Understand the similarities and differences between the transmission methods discussed in this chapter. There are differences and similarities between many of the topics in this chapter. Carefully compare and understand them. Minor subtleties can be difficult to recognize when taking the test.

Key Terms

Before you take the exam, be certain you are familiar with the following terms:

adjacent channel
adjacent channel interference
bandwidth
carrier frequency
chips
Complementary Code Keying (CCK)
convolutional coding
delay spread
Differential Binary Phase Shift Keying (DB DBPSK)
Differential Quadrature Phase Shift Keying (DQPSK)
direct sequence spread spectrum (DSSS)
dwell time
forward error correction (FEC)
frequency hopping spread spectrum (FHSS)
High-Rate DSSS (HR-DSSS)
hop time
hopping sequence
Industrial, Scientific, and Medical (ISM) bands
intersymbol interference (ISI)
narrowband
nonadjacent channel
nonoverlapping channels
Orthogonal Frequency Division Multiplexing (OFDM)
Packet Binary Convolutional Code (PBCC)
processing gain
quadrature amplitude modulation (QAM)
spread spectrum
subcarriers
transmit spectrum mask
Unlicensed National Information Infrastructure (UNII) bands

Review Questions

1. Which of the following are valid ISM bands? (Choose all that apply.)
   - A. 902–928 MHz
   - B. 2.4–2.4835 GHz
   - C. 5.725–5.825 GHz
   - D. 5.725–5.875 GHz
2. Which of the following are valid UNII bands? (Choose all that apply.)
   - A. 5.150–5.250 GHz
   - B. 5.47–5.725 GHz
   - C. 5.725–5.825 GHz
   - D. 5.725–5.850 GHz
3. Which technologies are used in the 2.4 GHz ISM band? (Choose all that apply.)
   - A. FHSS
   - B. ERP
   - C. DSSS
   - D. HR-DSSS
4. 802.11n draft (HT clause 20 radios) can transmit in which frequency bands? (Choose all that apply.)
   - A. 2.4–2.4835 GHz
   - B. 5.47–5.725 GHz
   - C. 902–928 GHz
   - D. 5.15–5.25 GHz
5. In North America, the 802.11 standard designates what frequency space for use of FHSS transmissions?
   - A. 2.4 GHz–2.4835 GHz
   - B. 2.402 GHz–2.480 GHz
   - C. 5.725 GHz–5.825 GHz
   - D. 5.725 GHz–5.875 GHz
   - E. None of the above

6. The original 802.11 standard requires how much separation between center frequencies for DSSS channels to be considered nonoverlapping?

   A. 22 MHz

   B. 25 MHz

   C. 30 MHz

   D. 35 MHz

   E. 40 MHz

7. The 802.11-2007 standard requires how much separation between center frequencies for HR-DSSS (clause 18) channels to be considered nonoverlapping?

   A. 22 MHz

   B. 25 MHz

   C. 30 MHz

   D. 35 MHz

   E. 40 MHz

8. What best describes *hop time*?

   A. The period of time that the transmitter waits before hopping to the next frequency

   B. The period of time that the standard requires when hopping between frequencies

   C. The period of time that the transmitter takes to hop to the next frequency

   D. The period of time the transmitter takes to hop through all of the FHSS frequencies

9. As defined by the IEEE-2007 standard, how much separation is needed between center frequencies of channels in the UNII-2 Extended band?

   A. 10 MHz

   B. 20 MHz

   C. 22 MHz

   D. 25 MHz

   E. 30 MHz

10. When deploying an ERP-OFDM wireless network with only two access points, which of these channel groupings would be considered nonoverlapping? (Choose all that apply.)

    A. Channels 1 and 3

    B. Channels 7 and 10

    C. Channels 3 and 8

    D. Channels 5 and 11

    E. Channels 6 and 10

11. Which spread spectrum technology specifies data rates of 22 Mbps and 33 Mbps?

A. DSSS

B. ERP-PBCC

C. OFDM

D. PPtP

12. If data is corrupted by previous data from a reflected signal, this is known as what?

A. Delay spread

B. ISI

C. Forward error creation

D. Bit crossover

13. Assuming all channels are supported by a 5 GHz access point, how many possible channels can be configured on the access point?

A. 4

B. 11

C. 12

D. 24

14. Which of these technologies is the most resilient against the negative effects of multipath?

A. FHSS

B. DSSS

C. HR-DSSS

D. OFDM

15. HR-DSSS (clause 18) calls for data rates of 5.5Mbps, and 11Mbps. What is the average amount of aggregate throughput percentage at any data rate?

A. 80 percent

B. 75 percent

C. 50 percent

D. 100 percent

16. In an FHSS system, throughput can be increased by which of the following?

A. Shortening the dwell time and lengthening the hop time

B. Shortening the dwell time and shortening the hop time

C. Lengthening the dwell time and shortening the hop time

D. Lengthening the dwell time and lengthening the hop time

17. With the center frequency of channel 1 at 2.412 GHz, what is the center frequency of channel 2?

    A. 2.444 GHz

    B. 2.417 GHz

    C. 2.424 GHz

    D. 2.422 GHz

18. What are the modulation types used by OFDM technology? (Choose all that apply.)

    A. QAM

    B. Phase

    C. Frequency

    D. Amplitude

19. The Barker code converts a bit of data into a series of bits that are referred to as what?

    A. Chipset

    B. Chips

    C. Convolutional code

    D. Complementary code

20. OFDM uses how many 312.5 KHz subcarriers for transmitting data?

    A. 54

    B. 52

    C. 48

    D. 36

Answers to Review Questions

1. A, B, D. The ISM bands are 902–928 MHz, 2.4–2.4835 GHz, and 5.725–5.875 GHz. 5.725–5.825 GHz is the UNII-3 band.

2. A, B, C. The four UNII bands are 5.15–5.25 GHz, 5.25–5.35 GHz, 5.47–5.725 GHz and 5.725–5.825 GHz.

3. A, B, C, D. The 802.11-2007 standard allows for the use of FHSS clause 14 radios (802.11), DSSS clause 15 radios (802.11), HR-DSSS clause 18 radios (802.11b), and ERP clause 19 radios (802.11g).

4. A, B, D. The 802.11n draft amendment specifies that HT clause 20 radios can transmit in 2.4 GHz ISM band and all four of the 5 GHz UNII bands.

5. B. The IEEE specified that in North America, 802.11 FHSS would use 79 MHz of frequencies, from 2.402 GHz to 2.480 GHz.

6. C. According to the 802.11-2007 standard, legacy DSSS channels require at least 30 MHz of spacing between the center frequencies to be considered nonoverlapping. Current 2.4 GHz radios using either HR-DSSS or ERP technology require 25 MHz of separation between the center frequencies to be considered nonoverlapping.

7. B. HR-DSSS (clause 18) was introduced under the 802.11b amendment, which states that channels need a minimum of 25 MHz of separation between the center frequencies to be considered nonoverlapping.

8. C. The time that the transmitter waits before hopping to the next frequency is known as the dwell time. The hop time is not a required time but rather a measurement of how long the hop takes.

9. B. The 802.11a amendment, which originally defined the use of OFDM (clause 17), required only 20 MHz of separation between the center frequencies for channels to be considered nonoverlapping. All 23 channels in the 5 GHz UNII bands use OFDM and have 20 MHz of separation. Therefore, all 5 GHz OFDM channels are considered nonoverlapping by the IEEE. However, it should be noted that adjacent 5 GHz channels do have some sideband carrier frequency overlap.

10. C, D. In order for two ERP or HR-DSSS channels to be considered nonoverlapping, they require 25 MHz of separation between the center frequencies. Therefore, any two channels must have at least a five-channel separation. The simplest way to determine what channels are valid is to add five to it or subtract five from it. Any channel below or above the values you calculated are valid. Deployments of three or more access points in the 2.4 GHz ISM band normally use channels 1, 6, and 11, which are all considered non-overlapping.

11. B. Extended Rate Physical Packet Binary Convolutional Code (ERP-PBCC) is the optional modulation technique that specifies data rates of 22 and 33 Mbps.

12. B. The cause of the problem is delay spread resulting in intersymbol interference (ISI), which causes data corruption.

13. D. The 802.11-2007 standard states, "the OFDM PHY shall operate in the 5 GHz band, as allocated by a regulatory body in its operational region." The standard defines the use of four 5 GHz UNII bands. The three original UNII bands each have four channels and are 100 MHz wide. The UNII-2 extended band has 11 channels and is 255 MHz wide. A total of 23 channels are available in the UNII bands. The 24th channel resides in a different unlicensed band. In the United States, ODFM transmissions can also occur in channel 165, which is part of the 5.8 GHz ISM band.

14. D. Because of the lower subcarrier data rates, delay spread is a smaller percentage of the symbol period, which means that ISI is less likely to occur. In other words, OFDM technology is more resistant to the negative effects of multipath than DSSS and FHSS spread spectrum technologies.

15. C. A medium access method known as Carrier Sense Multiple Access with Collision Avoidance (CSMA/CA) helps to ensure that only one radio card can be transmitting on the medium at any given time. Because of the half-duplex nature of the medium and the overhead generated by CSMA/CA, the actual aggregate throughput is typically 50 percent or less of the data rate.

16. C. Lengthening the dwell time decreases the amount of hopping, and shortening the hop time decreases overhead. Increasing the dwell time increases throughput.

17. B. Each 2.4 GHz channel center frequency is 5 MHz above the previous channel's center frequency.

18. A, B. OFDM uses BPSK and QPSK modulation for the lower ODFM data rates. The higher OFDM data rates use 16-QAM and 64-QAM modulation. QAM modulation is a hybrid of phase and amplitude modulation.

19. B. When a data bit is converted to a series of bits, these bits that represent the data are known as chips.

20. C. OFDM uses 52 subcarriers, but only 48 of them are used to transport data. The other 4 subcarriers are used as pilot carriers.

Wireless LAN Topologies

IN THIS CHAPTER, YOU WILL LEARN ABOUT THE FOLLOWING:

✓ **Wireless networking topologies**

- Wireless wide area network (WWAN)
- Wireless metropolitan area network (WMAN)
- Wireless personal area network (WPAN)
- Wireless local area network (WLAN)

✓ **802.11 topologies**

- Access point
- Client station
- Integration service (IS)
- Distribution system (DS)
- Wireless distribution system (WDS)
- Service set identifier (SSID)
- Basic service set (BSS)
- Basic service set identifier (BSSID)
- Basic service area (BSA)
- Extended service set (ESS)
- Independent basic service set (IBSS)
- Nonstandard 802.11 topologies

✓ **802.11 configuration modes**

- Access point modes
- Client station modes

A computer network is a system that provides communications between computers. Computer networks can be configured as peer to peer, as client/server, or as centralized central processing units (CPUs) with distributed dumb terminals. A networking *topology* is defined simply as the physical and/or logical layout of nodes in a computer network. Any individual who has taken a networking basics class is already familiar with bus, ring, star, mesh, and hybrid topologies that are often used in wired networks.

All topologies have advantages and disadvantages. A topology may cover very small areas or can exist as a worldwide architecture. Wireless topologies also exist as defined by the physical and logical layout of wireless hardware. Many wireless technologies exist and can be arranged into four major wireless networking topologies. The 802.11-2007 standard defines one specific type of wireless communication. Within the 802.11 standard exist three types of topologies, known as *service sets*. Over the years, vendors have also used 802.11 hardware using nonstandard topologies to meet specific wireless networking needs. This chapter covers the topologies used by a variety of wireless technologies and covers 802.11-specific topologies, both standard and nonstandard.

Wireless Networking Topologies

Although the main focus of this Study Guide is 802.11 wireless networking, which is a local area technology, other wireless technologies and standards exist in which wireless communications span either smaller or larger areas of coverage. Examples of other wireless technologies are cellular telephone, Bluetooth, and ZigBee. All of these different wireless technologies may or may not be arranged into four major wireless topologies:

- Wireless wide area network (WWAN)
- Wireless metropolitan area network (WMAN)
- Wireless personal area network (WPAN)
- Wireless local area network (WLAN)

Additionally, although the 802.11-2007 standard is a WLAN standard, the same technology can sometimes be deployed in different wireless network architectures discussed in this section.

Wireless Wide Area Network (WWAN)

A wide area network (WAN) covers a vast geographical area. A WAN might traverse an entire state, region, or country, or even span worldwide. The best example of a WAN is the Internet. Many private and public corporate WANs consist of hardware infrastructure such as T1 lines, fiber optics, and routers. Protocols used for wired WAN communications include Frame Relay, ATM, MPLS, and others.

A *wireless wide area network (WWAN)* also covers broad geographical boundaries but obviously uses a wireless medium instead of a wired medium. Wireless wide area networks typically use cellular telephone technologies or proprietary licensed wireless bridging technologies. Cellular providers such as T-Mobile, Verizon, and Vodafone use a variety of competing technologies to carry data. Some examples of these cellular technologies are GPRS, CDMA, TDMA, and GSM. Data can be carried to a variety of devices such as cell phones, personal digital assistants (PDAs), and cellular networking cards (pictured in Figure 7.1).

FIGURE 7.1 A cellular networking card

Data rates and bandwidth using these technologies are relatively slow when compared to other wireless technologies, such as 802.11. However, as cellular technologies have improved, so have cellular data-transfer rates. It is important to understand that 802.11 wireless networking infrastructure cannot be deployed as a WWAN.

Wireless Metropolitan Area Network (WMAN)

A *wireless metropolitan area network (WMAN)* provides coverage to a metropolitan area such as a city and the surrounding suburbs. WMANs have been created for some time by matching different wireless technologies, and recent advancements have made this more practical. The wireless technology that is newly associated with a WMAN is defined by the 802.16 standard. This standard defines broadband wireless access and is sometimes referred to as Worldwide Interoperability for Microwave Access (WiMAX). The WiMAX Forum is responsible for compatibility and interoperability testing of wireless broadband equipment such as 802.16 hardware.

802.16 technologies are viewed as a direct competition to other broadband services such as DSL and cable. Although 802.16 wireless networking is typically thought of as a last-mile data-delivery solution, the technology might also be used to provide access to users over city-wide areas. Currently, several major WiMAX installations are being deployed throughout the country by companies such as Sprint and Cricket.

More information about the 802.16 standard can be found at `http://ieee802.org/16`. Information about WiMAX can be located at `www.wimaxforum.org`.

In the past, a lot of press was generated about the possibility of citywide deployments of Wi-Fi networks, giving city residents access to the Internet throughout a metropolitan area. Although 802.11 technology was never intended to be used to provide access over such a wide area, many cities such as Philadelphia and San Francisco had initiatives to achieve this very feat. The equipment that was being used for these large-scale 802.11 deployments was proprietary wireless mesh routers or mesh access points. At the time this book was written, most of the large-scale citywide Wi-Fi deployment plans have been scrapped. Currently some 802.11 WMAN deployments do exist in small towns or in sections of major cities. However, Wi-Fi technology by itself will probably not be used to provide citywide wireless access in most major metropolitan areas because the financial business models have failed and the technology was never meant to be scaled as a WMAN.

Wireless Personal Area Network (WPAN)

A *wireless personal area network (WPAN)* is a wireless computer network used for communication between computer devices within close proximity of a user. Devices such as laptops, PDAs, and telephones can communicate with each other by using a variety of wireless technologies. WPANs can be used for communication between devices or as portals to higher-level networks such as local area networks (LANs) and/or the Internet. The most common technologies in WPANs are Bluetooth and infrared. Infrared is a light-based medium, whereas Bluetooth is a radio-frequency medium that uses frequency hopping spread spectrum (FHSS) technology. Figure 7.2 shows a headset and a cellular telephone that use Bluetooth radios to provide wireless connectivity between the two devices.

FIGURE 7.2 Bluetooth communications

The IEEE 802.15 Working Group focuses on technologies used for WPANs such as Bluetooth and ZigBee. ZigBee is another RF medium that has the potential of low-cost wireless networking between devices in a WPAN architecture.

Further information about the 802.15 WPAN standards can be found at www.ieee802.org/15. Information about Bluetooth can be located at www.bluetooth.com. The ZigBee Alliance provides information about ZigBee technology at www.zigbee.org. The Infrared Data Association offers data about infrared communications at www.irda.org.

The best example of 802.11 radios being used in a wireless personal area networking scenario would be as peer-to-peer connections. We provide more information about 802.11 peer-to-peer networking later in this chapter, in the section "Independent Basic Service Set (IBSS)."

Wireless Local Area Network (WLAN)

As you learned in earlier chapters, the 802.11-2007 standard is defined as a *wireless local area network (WLAN)* technology. Local area networks provide networking for a building or campus environment. The 802.11 wireless medium is a perfect fit for local area networking simply because of the range and speeds that are defined by the 802.11-2007 standard and future amendments. The majority of 802.11 wireless network deployments are indeed LANs that provide access at businesses and homes.

WLANs typically use multiple 802.11 access points connected by a wired network backbone. In enterprise deployments, WLANs are typically used to provide end users access to network resources and network services and a gateway to the Internet. Although 802.11 hardware can be used in other wireless topologies, the majority of Wi-Fi deployments are indeed WLANs, which is how the technology is defined by the IEEE 802.11 Working Group. The discussion of WLANs usually refers to 802.11 solutions; however, other proprietary and competing WLAN technologies do exist.

802.11 Topologies

The main component of an 802.11 wireless network is the radio card, which is referred to by the 802.11 standard as a *station (STA)*. The radio card can reside inside an access point or be used as a client station. The 802.11 standard defines three separate 802.11 topologies, known as *service sets*, which describe how these radio cards may be used to communicate with each other. These three 802.11 topologies are known as a basic service set (BSS), extended service set (ESS), and independent basic service set (IBSS). 802.11 radio cards can also be used in topologies not defined under the 802.11 standard. Some examples of these nonstandard topologies are bridging, repeating, workgroup bridging, and mesh networking.

Before we discuss the different 802.11 topologies, we need to review a few basic networking terms that are often misunderstood: *simplex*, *half-duplex*, and *full-duplex*. These are three dialog methods that are used for communications between people and also between computer equipment.

In *simplex communications*, one device is capable of only transmitting, and the other device is capable of only receiving. FM radio is an example of simplex communications. Simplex communications are rarely used on computer networks.

In *half-duplex communications*, both devices are capable of transmitting and receiving; however, only one device can transmit at a time. Walkie-talkies, or two-way radios, are examples of half-duplex devices. IEEE 802.11 wireless networks use half-duplex communications.

In *full-duplex communications*, both devices are capable of transmitting and receiving at the same time. A telephone conversation is an example of a full-duplex communication. Most IEEE 802.3 equipment is capable of full-duplex communications. The only way to accomplish full-duplex communications in a wireless environment is to have a two-channel setup where all transmissions in one direction are receiving while all transmissions in the other direction are transmitting. Current 802.11 technologies do not employ this technology, contrary to some marketing literature.

In this section, we cover all the components that make up the three 802.11 service sets as well as components in nonstandard 802.11 topologies.

Access Point

A wired infrastructure device typically associated with half-duplex communications is an Ethernet hub. A wired hub is effectively a shared medium in which only one host device can transmit data at a time. Access points are half-duplex devices because the RF medium uses half-duplex communications that allow for only one radio card to be transmitting at any given time. In reality, an access point is simply a hub with a radio card and an antenna. The radio card inside an access point must contend for the half-duplex medium in the same fashion that the client station radio cards must contend for the medium.

The original CWNP definition of an *access point (AP)* was a half-duplex device with switchlike intelligence. That definition can still be used to characterize *autonomous access points*. In Chapter 10, "Wireless Devices," we discuss in detail the differences between autonomous access points that do indeed have switchlike intelligence versus *lightweight access points* that do not. With lightweight access points, the intelligence resides inside a WLAN controller instead of inside the lightweight access points that are managed by the controller.

The best example of switchlike intelligence used by access points or WLAN controllers is the ability to address and direct wireless traffic at layer 2. Managed wired switches maintain dynamic MAC address tables known as content addressable memory (CAM) tables that can direct frames to ports based on the destination MAC address of a frame. Similarly, an access point or WLAN controller is a portal device that directs traffic either to the network backbone or back into the wireless medium. The 802.11 header of a wireless frame typically has three MAC addresses, but it can have as many as four in certain situations. The access point uses the complicated layer 2 addressing scheme of the wireless frames to eventually forward the layer 3–7 information either to the integration service or to another wireless client station. The upper-layer information that is contained in the body of an 802.11 wireless data frame is called a *MAC Service Data Unit (MSDU)*. The forwarding of the MSDU is the switchlike intelligence that exists in either autonomous APs or WLAN controllers. The intelligence that is often compared to a CAM table is known as the distribution system services (DSS), which are described in more detail later in this chapter.

Because an autonomous access point operates in a half-duplex shared medium and possesses some switchlike intelligence, an autonomous AP is a hybrid device that might be humorously characterized as a wireless SWUB (half switch/half hub).

Many access points also support the use of virtual local area networks (VLANs). For example, although not defined by the 802.11 standard, an access point can support VLANs that can be created on a managed wired switch or a WLAN controller. VLANs are used to reduce the size of broadcast domains and to segregate the network for security purposes.

Client Station

A radio card that is not used in an access point is typically referred to as a *client station*. Client station radio cards can be used in laptops, PDAs, scanners, phones, and many other mobile devices. Client stations must contend for the half-duplex medium in the same manner that an access point radio card contends for the RF medium. When client stations have a layer 2 connection with an access point, they are known as *associated*.

Integration Service (IS)

The 802.11-2007 standard defines an *integration service (IS)* that enables delivery of MSDUs between the distribution system (DS) and a non-IEEE-802.11 LAN, via a portal. A simpler way of defining the integration service is to characterize it as a frame format transfer method. The portal is usually either an access point or a WLAN controller. As mentioned earlier, the payload of a wireless 802.11 data frame is the layer 3–7 information known as the MSDU. The eventual destination of this payload is usually to a wired network infrastructure. Because the wired infrastructure is a different physical medium, an 802.11 data frame payload must be effectively transferred into an 802.3 Ethernet frame. For example, a VoWiFi phone sends an 802.11 data frame to an autonomous access point. The MSDU payload of the frame is a VoIP packet with a final destination of a VoIP server that resides at the 802.3 network core. The job of the integration service is to remove the 802.11 header and trailer and then encase the MSDU VoIP payload inside an 802.3 frame. The 802.3 frame is then sent on to the Ethernet network. The integration service performs the same actions in reverse when an 802.3 frame payload must be transferred into an 802.11 frame that is eventually transmitted by the access point radio.

It is beyond the scope of the 802.11-2007 standard to define how the integration service operates. Normally, the integration service transfers data frame payloads between an 802.11 and 802.3 medium. However, the integration service could transfer an MSDU between the 802.11 medium and some sort of other medium such as an 802.5 token ring. The integration service mechanism happens at the edge when autonomous APs are deployed. The integration service mechanism takes place inside a WLAN controller when lightweight APs are deployed.

Distribution System (DS)

The 802.11-2007 standard also defines a *distribution system (DS)* that is used to interconnect a set of basic service sets (BSSs) via integrated LANs to create an extended service set (ESS). Service sets are described in detail later in this chapter. Access points by their very nature are portal devices. Wireless traffic can be destined back onto the wireless medium or forwarded to the integration service. The DS consists of two main components:

Distribution system medium (DSM) A logical physical medium used to connect access points is known as a *distribution system medium (DSM)*. The most common example is an 802.3 medium.

Distribution system services (DSS) System services built inside an access point usually in the form of software. The *distribution system services (DSS)* provide the switchlike intelligence mentioned earlier in this chapter. These software services are used to manage client station associations, reassociations, and disassociations. Distribution system services also use the layer 2 addressing of the 802.11 MAC header to eventually forward the layer 3–7 information (MSDU) either to the integration service or to another wireless client station. A full understanding of DSS is beyond the scope of the CWNA exam but is necessary at the Certified Wireless Network Expert (CWNE) certification level.

A single access point or multiple access points may be connected to the same distribution system medium. The majority of 802.11 deployments use an AP as a portal into an 802.3 Ethernet backbone, which serves as the distribution system medium. Access points are usually connected to a switched Ethernet network, which often also offers the advantage of supplying power to the access points via Power over Ethernet (PoE).

An access point may also act as a portal device into other wired and wireless mediums. The 802.11-2007 standard by design does not care, nor does it define onto which medium an access point translates and forwards data. Therefore, an access point can be characterized as a "translational bridge" between two mediums. The AP translates and forwards data between the 802.11 medium and whatever medium is used by the distribution system medium. Once again, the distribution system medium will almost always be an 802.3 Ethernet network, as pictured in Figure 7.3.

FIGURE 7.3 Distribution system medium

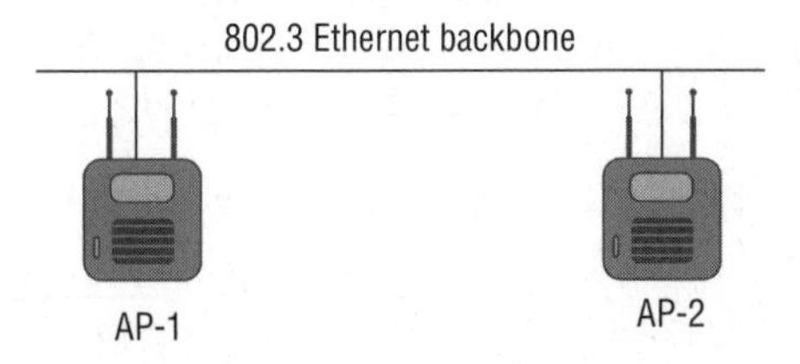

Although rare, 802.5 token ring access points do exist, and the distribution system medium would be the 802.5 token ring infrastructure. In the case of a wireless mesh network, the handoff is through a series of wireless devices, with the final destination typically being an 802.3 network.

Wireless Distribution System (WDS)

The 802.11-2007 standard defines a mechanism for wireless communication using a four-MAC-address frame format. The standard describes such a frame format but does not describe how such a mechanism or frame format would be used. This mechanism is known as a *wireless distribution system (WDS)*. Real-world examples of WDS include bridging, repeaters, and mesh networks. Another example of a WDS is when access points are deployed to provide both coverage and backhaul. Although the DS normally uses a wired Ethernet backbone, it is possible to use a wireless connection instead. A WDS can connect access points together using what is referred to as a *wireless backhaul*.

A WDS may operate by using access points with a single 802.11 radio or dual 802.11 radios. Figure 7.4 depicts two 802.11b/g access points, each with a single radio. The radios in the APs not only provide access to the client stations but also communicate with each other directly as a WDS. A disadvantage to this solution is that throughput can be adversely affected because of the half-duplex nature of the medium, particularly in a single-radio scenario, where an access point cannot be communicating with a client station and another access point at the same time. The end result is a degradation of throughput.

Which Distribution System Is Most Desirable?

Whenever possible, an 802.3 network will always be the best option for the distribution system. Because most enterprise deployments already have a wired 802.3 infrastructure in place, integrating a wireless network into a switched Ethernet network is the most logical solution. A wired distribution system medium does not encounter many of the problems that may affect a WDS, such as multipath and radio frequency interference. If the occasion does arise when a wired network cannot connect access points together, a WDS might be a viable alternative. The more-desirable WDS solution utilizes different frequencies and radios for client access and distribution.

In Figure 7.5, two dual-radio access points are shown, each with radios operating at different frequencies. The 2.4 GHz radios provide access for the client stations, and the 5 GHz radios serve as the WDS link between the two access points. Throughput is not adversely affected because the 2.4 GHz radio can communicate at the same time as the 5 GHz backhaul radios.

Wireless repeaters are another example of an 802.11 WDS. Repeaters are used to extend WLAN cell coverage to areas where it is not possible to provide an 802.3 Ethernet cable drop. As illustrated in Figure 7.6, a client station is associated and communicating via a repeater AP. The repeater provides coverage but is not connected to the wired backbone. When a client station sends a frame to the repeater, it is then forwarded to an access point that is connected to the wired backbone. The frame payload is converted into an 802.3 Ethernet frame and sent to a server on the backbone. The 802.11 communications between the repeater and the access point is a WDS. As pictured in Figure 7.7, a frame sent within a WDS requires four MAC addresses, a source address, a destination address, a transmitter address, and a receiver address.

FIGURE 7.4 Wireless distribution system, single radio

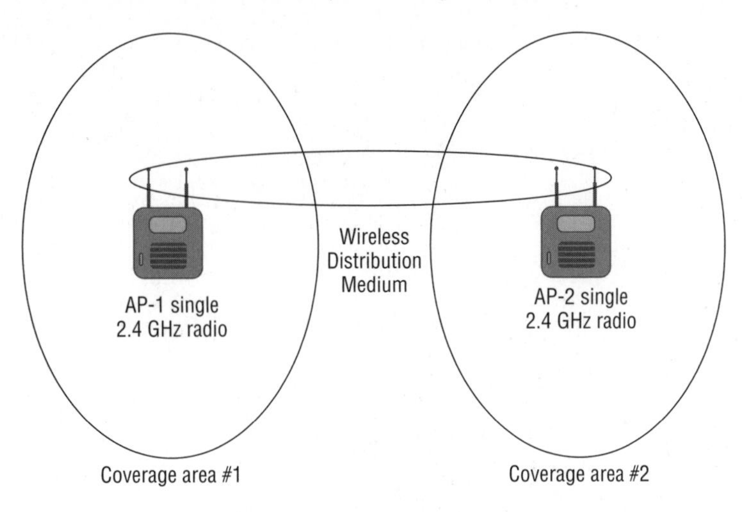

FIGURE 7.5 Wireless distribution system, dual radios

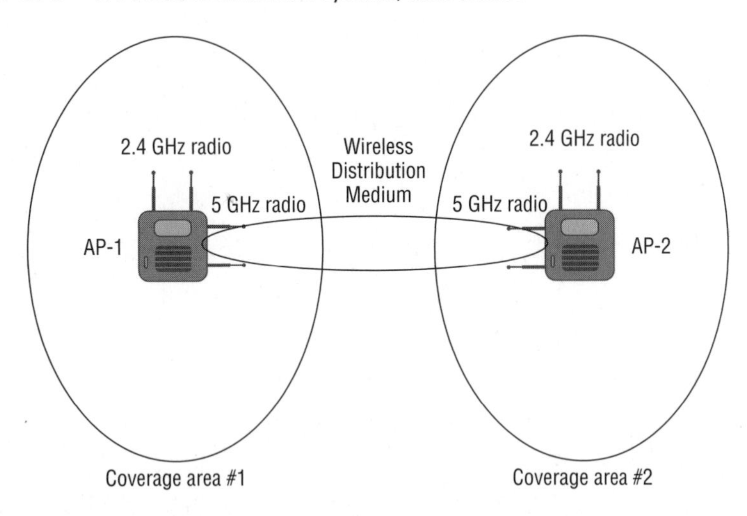

Repeaters effectively extend the cell coverage of the original access point; therefore, both the repeater and AP must be on the same channel. There must also be at least a 50 percent cell overlap between the coverage cells so that the repeater and AP can communicate with each other. Repeaters do provide coverage into areas where a cable drop is not possible. However, all frame transmissions must be sent twice, which decreases throughput and increases latency. Because the AP cell and the repeater cell are on the same channel and exist in the same layer 1 domain, all radios must contend for the medium. Repeater environments add extra medium contention overhead, which also affects performance. Because of the extra hops, deployment of VoWiFi phones is not usually supported in most WDS environments.

FIGURE 7.6 Repeater cell

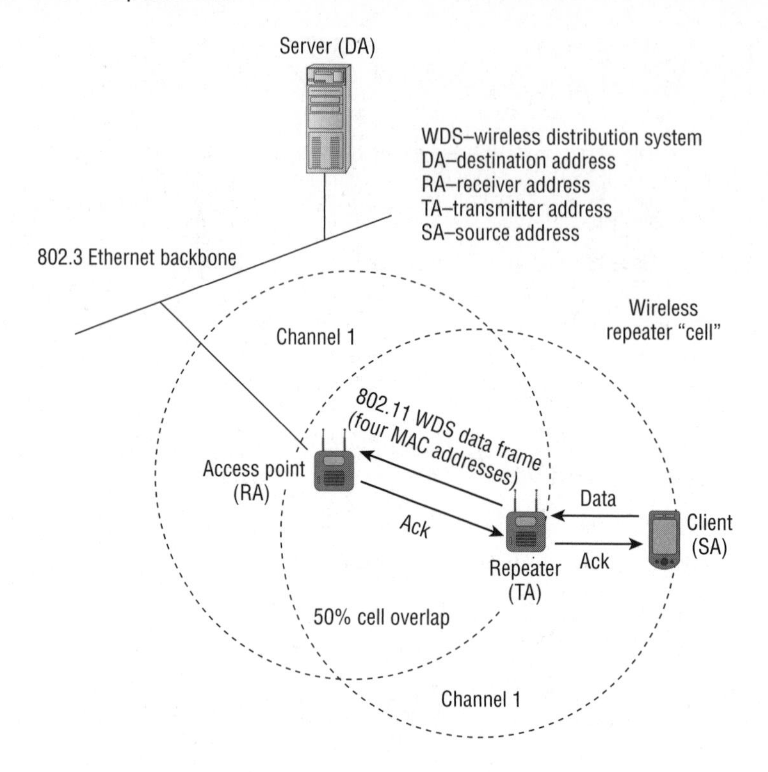

FIGURE 7.7 WDS frame header

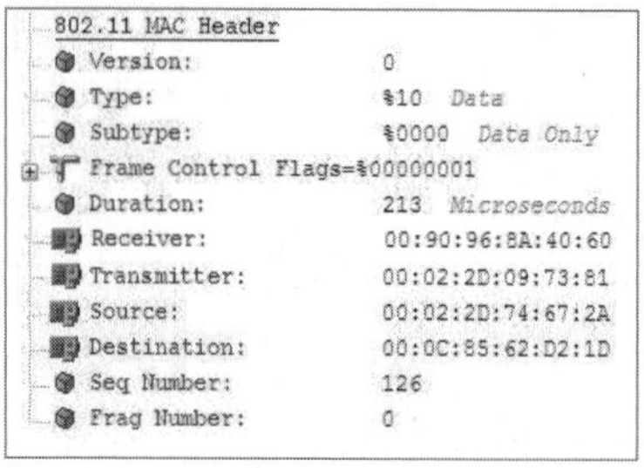

802.11 MAC Header	
Version:	0
Type:	%10 *Data*
Subtype:	%0000 *Data Only*
Frame Control Flags=%00000001	
Duration:	213 *Microseconds*
Receiver:	00:90:96:8A:40:60
Transmitter:	00:02:2D:09:73:81
Source:	00:02:2D:74:67:2A
Destination:	00:0C:85:62:D2:1D
Seq Number:	126
Frag Number:	0

Service Set Identifier (SSID)

The *service set identifier (SSID)* is a logical name used to identify an 802.11 wireless network. The SSID wireless network name is comparable to a Windows workgroup name. The three 802.11 topologies utilize the SSID so that radio cards may identify each other in a process known as *active scanning* or *passive scanning*. The SSID is a configurable setting on all radio cards, including access points and client stations. The SSID can be made up of as many as 32 characters and is case sensitive. Figure 7.8 shows an SSID configuration of an access point.

FIGURE 7.8 Service set identifier

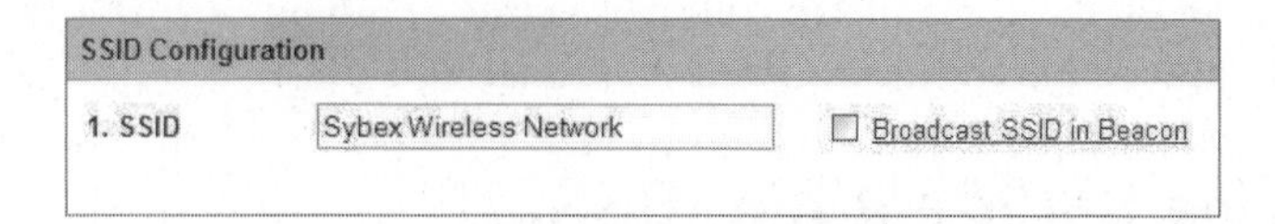

Most access points have the ability to cloak an SSID and keep the network name hidden from nonlegitimate end users. Hiding the SSID is a very weak attempt at security that is not defined by the 802.11-2007 standard. However, it is an option many administrators still choose to implement.

Active and passive scanning are discussed in detail in Chapter 9, "802.11 MAC Architecture." SSID cloaking is discussed in Chapter 13, "802.11 Network Security Architecture."

Basic Service Set (BSS)

The *basic service set (BSS)* is the cornerstone topology of an 802.11 network. The communicating devices that make up a BSS are solely one AP with one or more client stations. Client stations join the AP's wireless domain and begin communicating through the AP. Stations that are members of a BSS have a layer 2 connection and are called *associated*. Figure 7.9 depicts a standard basic service set.

FIGURE 7.9 Basic service set

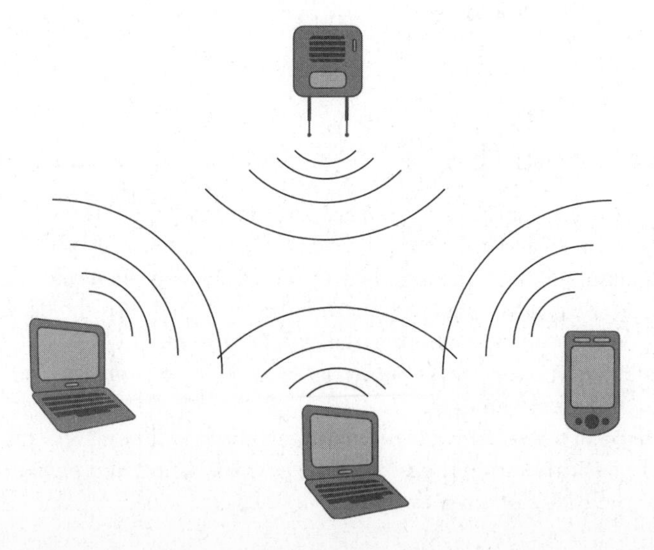

Typically the AP is connected to a distribution system medium, but that is not a requirement of a basic service set. If an AP is serving as a portal to the distribution system, client stations may communicate, via the AP, with network resources that reside on the DSM. It should also be noted that if client stations wish to communicate with each other, they must relay their data through the access point. Stations cannot communicate directly with each other unless they go through the access point.

Basic Service Set Identifier (BSSID)

The 48-bit (6-octet) MAC address of an access point's radio card is known as the *basic service set identifier (BSSID)*. The BSSID address is the layer 2 identifier of each individual BSS. Most often the BSSID is the MAC address of the access point.

Do not confuse the BSSID address with the SSID. The service set identifier (SSID) is the logical WLAN name that is user configurable, while the BSSID is the layer 2 MAC address of an AP provided by the hardware manufacturer. It should be noted that many vendors offer virtual BSSID configurations; these are explained in Chapter 10, "Wireless Devices."

As shown in Figure 7.10, the BSSID address is found in the header of most 802.11 wireless frames and is used for identification purposes. The BSSID address plays a role in directing 802.11 traffic within the basic service set. This address is also used as a unique layer 2 identifier of the basic service set. Furthermore, the BSSID address is needed during the roaming process.

FIGURE 7.10 Basic service set identifier

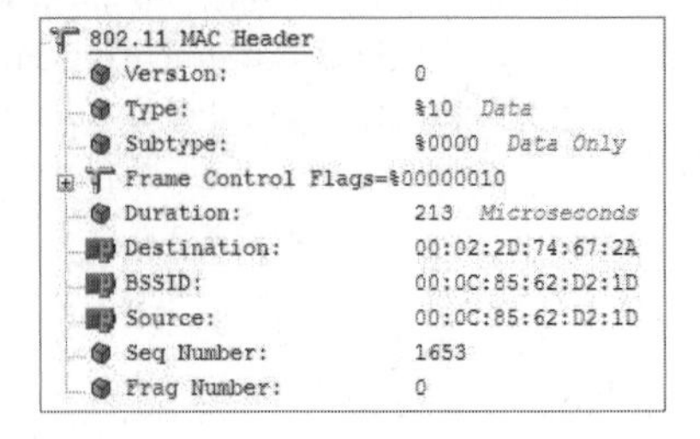

Basic Service Area (BSA)

The physical area of coverage provided by an access point in a BSS is known as the *basic service area (BSA)*. Figure 7.11 shows a typical BSA. Client stations can move throughout the coverage area and maintain communications with the AP as long the received signal between the radios remains above received signal strength indicator (RSSI) thresholds. Client stations can also shift between concentric zones of variable data rates that exist within the BSA. The process of moving between data rates is known as *dynamic rate switching* and is discussed in Chapter 12, "WLAN Troubleshooting."

The size and shape of a BSA depends on many variables, including AP transmit power, antenna gain, and physical surroundings. Because environmental and physical surroundings often change, the BSA can often be fluid.

FIGURE 7.11 Basic service area

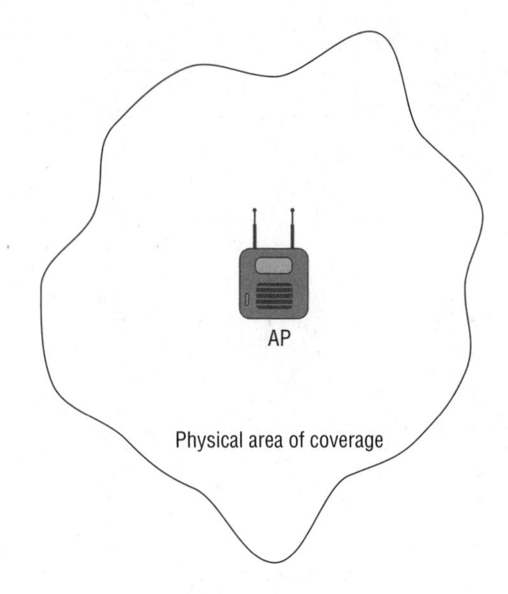

Extended Service Set (ESS)

While a BSS might be considered the cornerstone 802.11 topology, an *extended service set (ESS)* 802.11 topology is analogous to an entire stone building. An extended service set is one or more basic service sets connected by a distribution system medium. Usually an extended service set is a collection of multiple access points and their associated client stations, all united by a single DSM.

The most common example of an ESS has access points with partially overlapping coverage cells, as shown in Figure 7.12. The purpose behind an ESS with partially overlapping coverage cells is to provide seamless roaming to the client stations. Most vendors recommend cell overlap of at least 15 to 25 percent to achieve successful seamless roaming.

Although seamless roaming is usually a key aspect of WLAN design, there is no requirement for an ESS to guarantee uninterrupted communications. For example, an ESS can utilize multiple access points with nonoverlapping coverage cells, as pictured in Figure 7.13. In this scenario, a client station that leaves the basic service area of the first access point will lose connectivity. The client station will later reestablish connectivity as it moves into the coverage cell of the second access point. This method of station mobility between disjointed cells is sometimes referred to as *nomadic roaming*.

One final example of an ESS deploys multiple access points with totally overlapping coverage areas, as pictured in Figure 7.14. This 802.11 ESS topology is called *colocation*, and the intended goal is increased client capacity. Colocation is discussed in more detail in Chapter 12.

FIGURE 7.12 Extended service set, seamless roaming

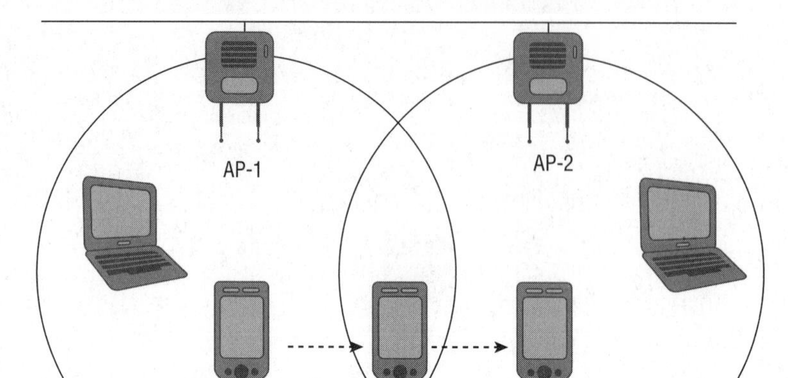

FIGURE 7.13 Extended service set, nomadic roaming

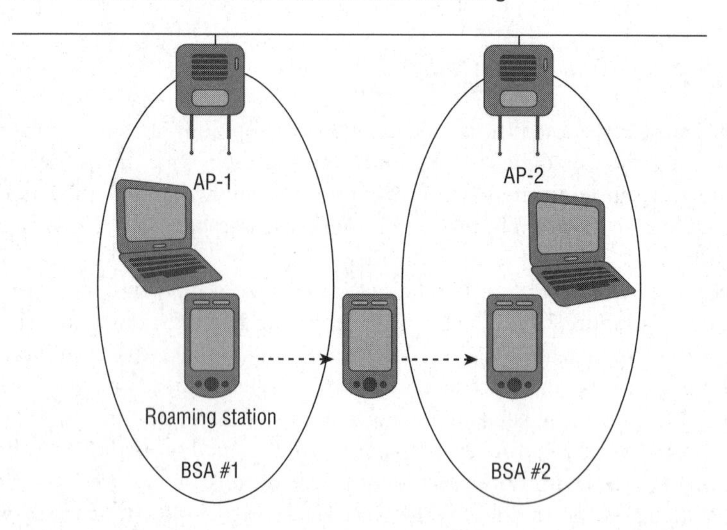

It should be noted that all three of the previously mentioned extended service sets share a distribution system. As stated earlier in this chapter, the distribution system medium is usually an 802.3 Ethernet network; however, the DS may use another type of medium. In the majority of extended service sets, the access points all share the same SSID WLAN name. The network name of an ESS is often called an *extended service set identifier (ESSID).* Although an ESSID is essentially synonymous with an SSID, there is no requirement for all the access points in an ESS to share the exact same WLAN name. Access points that share a DSM may have different SSIDs and still be classified as an extended service set. However, as pictured in Figure 7.15, access points in an ESS where roaming is required must all share the same logical name (SSID), but have unique layer 2 identifiers (BSSIDs) for each unique BSS coverage cell.

FIGURE 7.14 Extended service set, colocation

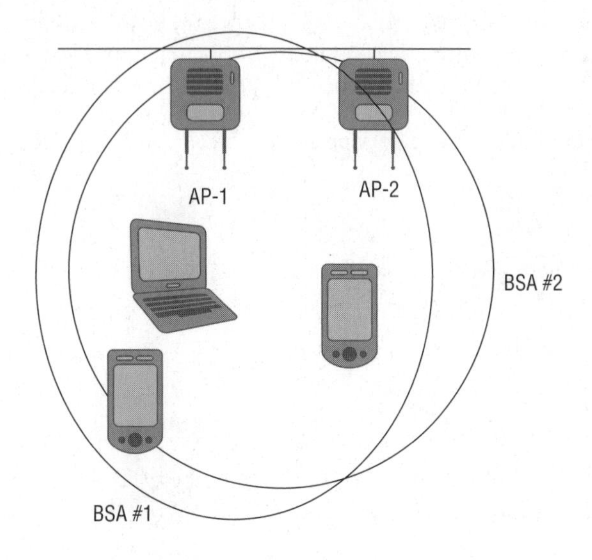

FIGURE 7.15 SSID and BSSIDs within an ESS

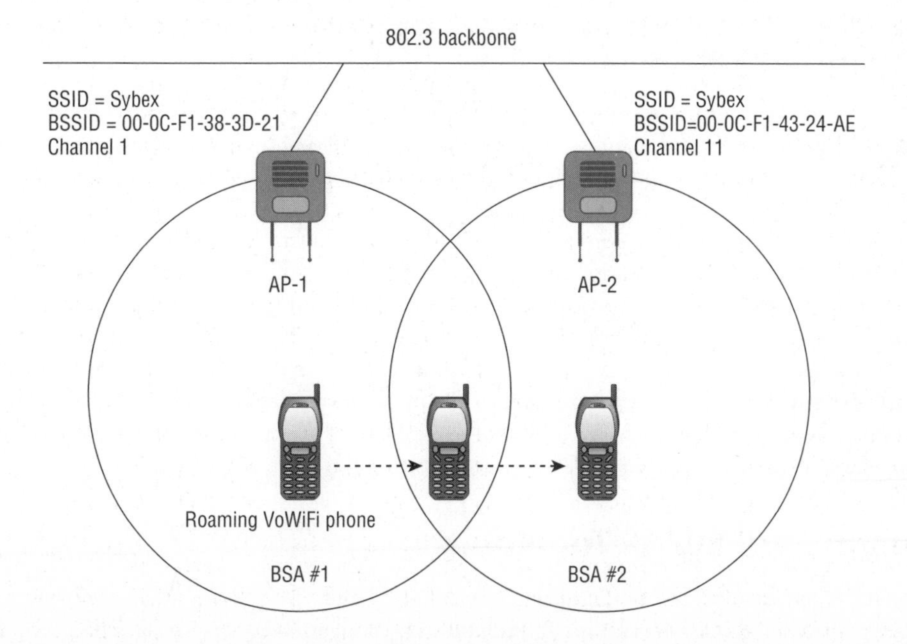

Independent Basic Service Set (IBSS)

The final service set topology defined by the 802.11 standard is an *independent basic service set (IBSS)*. The radio cards that make up an IBSS network consist solely of client stations (STAs), and no access point is deployed. An IBSS network that consists of just two STAs is analogous to a wired crossover cable. An IBSS can, however, have multiple client stations in one physical area communicating in an ad hoc fashion. Figure 7.16 depicts four client stations communicating with each other in a peer-to-peer fashion.

FIGURE 7.16 Independent basic service set

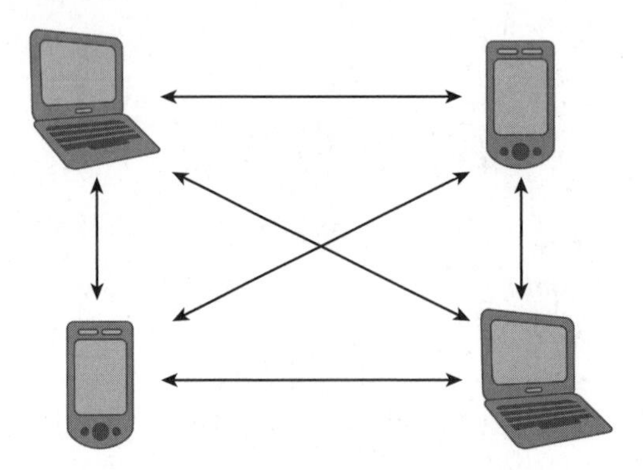

All of the stations transmit frames to each other directly and do not route their frames from one client to another. All client station frame exchanges in an IBSS are peer to peer. All stations in an IBSS must contend for the half-duplex medium, and at any given time only one STA can be transmitting.

The independent basic service set has two other names. Wi-Fi vendors often refer to an IBSS as either a *peer-to-peer network* or an *ad hoc network*.

In order for IBSS communications to succeed, all stations must be transmitting on the same frequency channel. Furthermore, this entire set of stand-alone wireless stations connected together as a group must share the same SSID WLAN name. Another caveat of an IBSS is that a BSSID address is created. Earlier in this chapter, we defined a BSSID as the MAC address of the radio card in an access point. So how can an independent basic service set have a BSSID if no access point is used in the IBSS topology? The first station that starts up in an IBSS randomly generates a BSSID in the MAC address format. This randomly generated BSSID is a virtual MAC address and is used for layer 2 identification purposes within the IBSS.

Nonstandard 802.11 Topologies

As you have just learned, the three service sets defined by the 802.11 standard are basic service set (BSS), extended service set (ESS), and independent basic service set (IBSS). Wi-Fi vendors also utilize 802.11 radio cards in nonstandard topologies while still remaining compliant with the 802.11-2007 standard.

The most common example is wireless bridging. 802.11 radios can be used to connect two wired networks together using a wireless bridged link. Most WDS deployments such as mesh networks are considered nonstandard 802.11 topologies. Although the 802.11-2007 standard does define a WDS frame format with four MAC addresses, the standard does not describe how such a mechanism or frame format would be used.

Another nonstandard 802.11 topology is the workgroup bridge (WGB). A workgroup bridge acts as a gateway for a small wired workgroup, yet the workgroup bridge is a client station associated with an access point. As mentioned earlier, a repeater is a special access point that forwards the data of client stations to a root access point. The net effect of a repeater is that the root access point's coverage cell is extended. Wireless mesh routers are essentially a combination of multiple repeaters using proprietary layer 2 routing protocols. Bridges, workgroup bridges, repeaters and mesh WLAN deployments are all examples of a WDS and are all considered to be nonstandard 802.11 topologies.

More-detailed information about all the devices discussed in the preceding paragraph can be found in Chapter 10.

Real World Scenario

Vendor Considerations When Using 802.11 Equipment in Nonstandard Topologies

Nowhere in the 802.11 standard are there any guidelines or characterizations for bridges, workgroup bridges, repeaters, or mesh routers. The 802.11s draft amendment details mesh networking, but it has not yet been ratified. Whenever equipment that uses 802.11 radios is deployed in nonstandard topologies, the recommended practice is to purchase the equipment from one vendor. A bridge from vendor A is not likely to work with a bridge from vendor B. Because none of these topologies are standardized, the likelihood of vendor interoperability is low.

802.11 Configuration Modes

While the 802.11-2007 standard defines all radios as stations (STAs), an access point (AP) radio and a client station radio can each be configured in a number of ways. The default configuration of an AP is to allow it to operate inside a basic service set (BSS). However, an AP can be configured to function in a nonstandard topology. Client stations can be configured to participate in either a BSS or an IBSS 802.11 service set. We these two methods in this section.

Access Point Modes

The only configuration mode of an access point that is compliant with the 802.11 standard is known as *root mode*. The main purpose of an AP is to serve as a portal to a distribution system. The normal default setting of an access point is root mode, which allows the AP to transfer data back and forth between the DS and the 802.11 wireless medium.

The default root configuration of an AP allows it to operate as part of a BSS. There are, however, other nonstandard modes in which an AP may be configured:

Bridge mode The AP is converted into a wireless bridge.

Workgroup Bridge mode The AP is transformed into a workgroup bridge.

Repeater mode The AP performs as a repeater access point.

Scanner mode The access point radio is converted into a sensor radio, allowing the access point to integrate into a wireless intrusion detection system (WIDS) architecture.

Because these configurations are all considered nonstandard, not all vendors support these modes. Figure 7.17 shows a screen capture of an access point's various configurable modes.

FIGURE 7.17 Access point configuration modes

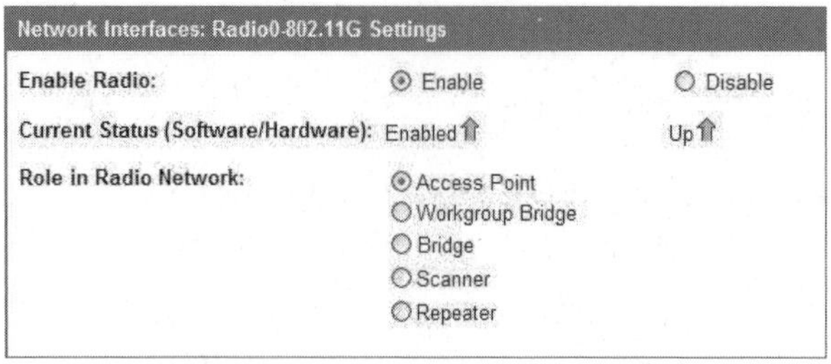

Client Station Modes

A client station may operate in one of two settings, as shown in the screen capture in Figure 7.18. The default mode for a client radio card is typically *Infrastructure mode*. When running in Infrastructure mode, the client station will allow communication via an access point. Infrastructure mode allows for a client station to participate in a basic service set or an extended service set. Clients that are configured in this mode may communicate, via the AP, with other wireless client stations within a BSS.

Clients may also communicate through the AP with other networking devices that exist on the distribution system, such as servers or wired desktops.

The second client station mode is called *Ad Hoc mode*. Other vendors may refer to this mode as Peer-to-Peer mode. Client cards set to Ad Hoc mode participate in an IBSS topology and do not communicate via an access point. All station transmissions and frame exchanges are peer to peer.

FIGURE 7.18 Client station configuration modes

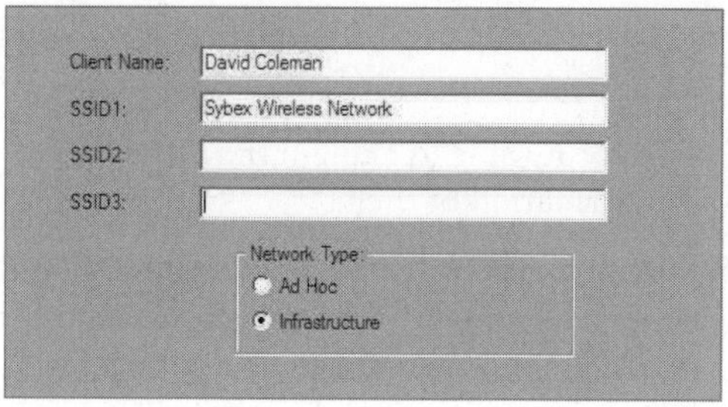

Summary

This chapter covered the major types of generic wireless topologies as well as the topologies specific to 802.11 wireless networking:

- The four wireless architectures that can be used by many different wireless technologies
- The three service sets as defined by the 802.11 standard, and the various aspects and purposes defined for each service set
- Standard and nonstandard configuration modes of both access points and client stations

As a wireless network administrator, it is important to have a full understanding of the defined 802.11 service sets and how they operate. Administrators typically oversee the design and management of an 802.11 ESS, but there is a good chance that they will also deploy 802.11 radios using a nonstandard topology.

Exam Essentials

Know the four major types of wireless topologies. Understand the differences between a WWAN, WLAN, WPAN, and WMAN.

Explain the three 802.11 service sets. Be able to fully expound on all the components, purposes, and differences of a basic service set, an extended service set, and an independent basic service set. Understand how the radio cards interact with each other in each service set.

Identify the various ways in which an 802.11 radio can be used. Understand that the 802.11 standard expects a radio card to be used either as a client station or inside an access point. Also understand that an 802.11 radio card can be used for other purposes, such as bridging, repeaters, and so on.

Explain the purpose of the distribution system. Know that the DS consists of two pieces: distribution system services (DSS) and the distribution system medium (DSM). Understand that the medium used by the DS can be any type of medium. Explain the functions of a wireless distribution system (WDS).

Define SSID, BSSID, and ESSID. Be able to explain the differences or similarities of all three of these addresses and the function of each.

Describe the various ways in which an ESS can be implemented and the purpose behind each design. Explain the three ways in which the coverage cells of the ESS access points can be designed and the purpose behind each design.

Demonstrate an understanding of the various nonstandard 802.11 topologies. Understand that alternative 802.11 topologies such as bridging and mesh networks exist. Further discussion of these nonstandard topologies can be found throughout this book.

Explain access point and client station configuration modes. Remember all the standard and nonstandard configuration modes of both an AP and a client station.

Key Terms

Before you take the exam, be certain you are familiar with the following terms:

access point (AP)

Ad Hoc mode

Ad Hoc network

autonomous access points

basic service area (BSA)

basic service set (BSS)

basic service set identifier (BSSID)

client station

distribution system (DS)

distribution system medium (DSM)

distribution system services (DSS)

extended service set (ESS)

extended service set identifier (ESSID)

full-duplex

infrastructure mode

integration service (IS)

lightweight access points

MAC Service Data Unit (MSDU)

peer-to-peer network

service set identifier (SSID)

service sets

simplex

station (STA)

topology

wireless distribution system (WDS)

wireless local area network (WLAN)

wireless metropolitan area network (WMAN)

wireless personal area network (WPAN)

Review Questions

1. An 802.11 wireless network name is known as which type of address? (Choose all that apply.)
   A. BSSID
   B. MAC address
   C. IP address
   D. SSID
   E. Extended service set identifier

2. Which two 802.11 topologies require the use of an access point? (Choose two.)
   A. WPAN
   B. IBSS
   C. Basic service set
   D. Ad hoc
   E. ESS

3. The 802.11 standard defines which medium to be used in a distribution system (DS)?
   A. 802.3 Ethernet
   B. 802.15
   C. 802.5 token ring
   D. Star-bus topology
   E. None of the above

4. Which option is a wireless computer topology used for communication of computer devices within close proximity of a person?
   A. WLAN
   B. Bluetooth
   C. ZigBee
   D. WPAN
   E. WMAN

5. Support for roaming is required under the 802.11 standard. Which 802.11 service set may allow for roaming?
   A. ESS
   B. Basic service set
   C. Colocated APs
   D. IBSS
   E. Spread spectrum service set

6. What factors might affect the size of a BSA coverage area of an access point? (Choose all that apply.)
   - A. Antenna gain
   - B. CSMA/CA
   - C. Transmission power
   - D. Indoor/outdoor surroundings
   - E. Distribution system

7. What is the default configuration mode that allows an AP radio to operate in a basic service set?
   - A. Scanner
   - B. Repeater
   - C. Root
   - D. Access
   - E. Nonroot

8. Which terms describe an 802.11 topology involving STAs but no access points? (Choose all that apply.)
   - A. BSS
   - B. Ad hoc
   - C. DSSS
   - D. Infrastructure
   - E. IBSS
   - F. Peer-to-peer

9. STAs operating in infrastructure mode may communicate in which of the following scenarios? (Choose all that apply.)
   - A. 802.11 frame exchanges with other STAs via an AP
   - B. 802.11 frame exchanges with an AP in scanner mode
   - C. 802.11 frame peer-to-peer exchanges directly with other STAs
   - D. Frame exchanges with network devices on the DSM
   - E. All of the above

10. What are the only three topologies defined by the 802.11 standard? (Choose all that apply.)
    - A. Bridge mode
    - B. Extended service set
    - C. BSS
    - D. IBSS
    - E. FHSS

11. Which wireless topology provides citywide wireless coverage?
    A. WMAN
    B. WLAN
    C. WPAN
    D. WAN
    E. WWAN
12. At which layer of the OSI model will a BSSID address be used?
    A. Physical
    B. Network
    C. Session
    D. Data-Link
    E. Application
13. The basic service set identifier address can be found in which topologies? (Choose all that apply.)
    A. FHSS
    B. IBSS
    C. ESS
    D. DSSS
    E. BSS
14. 802.11 wireless networking will not scale appropriately to accommodate which wireless topology?
    A. WLAN
    B. WPAN
    C. WMAN
    D. WWAN
    E. VLAN
15. Which wired network hardware devices do autonomous access points most resemble? (Choose all that apply.)
    A. Switch
    B. Node
    C. Hub
    D. Router
    E. Server

16. What are some nonstandard modes in which an AP radio may be configured? (Choose all that apply.)
    - A. Scanner
    - B. Root
    - C. Bridge
    - D. Nonroot
    - E. Repeater

17. A network consisting of clients and one or more access points connected by an 802.3 Ethernet backbone is one example of which 802.11 topology? (Choose all that apply.)
    - A. ESS
    - B. Basic service set
    - C. Extended service set
    - D. IBSS
    - E. Ethernet service set

18. What term best describes two access points communicating with each other wirelessly while also allowing clients to communicate through the access point?
    - A. WDS
    - B. DS
    - C. DSS
    - D. DSSS
    - E. DSM

19. What components make up a distribution system? (Choose all that apply.)
    - A. HR-DSSS
    - B. Distribution system services
    - C. DSM
    - D. DSSS
    - E. Intrusion detection system

20. What type of wireless topology is defined by the 802.11 standard?
    - A. WAN
    - B. WLAN
    - C. WWAN
    - D. WMAN
    - E. WPAN

Answers to Review Questions

1. D, E. The service set identifier (SSID) is a 32-character, case-sensitive, logical name used to identify a wireless network. An extended service set identifier (ESSID) is the logical network name used in an extended service set. ESSID is often synonymous with SSID.

2. C, E. The 802.11 standard defines three service sets, or topologies. A basic service set (BSS) is defined as one AP and associated clients. An extended service set (ESS) is defined as one or more basic service sets connected by a distribution system medium. An independent basic service set (IBSS) does not use an AP and consists solely of client stations (STAs).

3. E. By design, the 802.11 standard does not specify a medium to be used in the distribution system. The distribution system medium (DSM) may be an 802.3 Ethernet backbone, an 802.5 token ring network, a wireless medium, or any other medium.

4. D. A wireless personal area network (WPAN) is a short-distance wireless topology. Bluetooth and ZigBee are technologies that are often used in WPANs.

5. A. The most common implementation of an extended service set (ESS) has access points with partially overlapping coverage cells. The purpose behind an ESS with partially overlapping coverage cells is seamless roaming.

6. A, C, D. The size and shape of a basic service area can depend on many variables, including AP transmit power, antenna gain, and physical surroundings.

7. C. The normal default setting of an access point is root mode, which allows the AP to transfer data back and forth between the DS and the 802.11 wireless medium. The default root configuration of an AP allows it to operate inside a basic service set (BSS).

8. B, E, F. The 802.11 standard defines an independent basic service set (IBSS) as a service set using client peer-to-peer communications without the use of an AP. Other names for an IBSS include ad hoc and peer-to-peer.

9. A, D. Clients that are configured in Infrastructure mode may communicate via the AP with other wireless client stations within a BSS. Clients may also communicate through the AP with other networking devices that exist on the distribution system medium, such as a server or a wired desktop.

10. B, C, D. The three topologies, or service sets, defined by the 802.11 standard are basic service set (BSS), extended service set (ESS), and independent basic service set (IBSS).

11. A. A wireless metropolitan area network (WMAN) provides coverage to a metropolitan area such as a city and the surrounding suburbs.

12. D. The basic service set identifier (BSSID) is a 48-bit (6-octet) MAC address. MAC addresses exist at the MAC sublayer of the Data-Link layer of the OSI model.

13. B, C, E. The BSSID is the layer 2 identifier of either a BSS or an IBSS service set. The 48-bit (6-octet) MAC address of an access point's radio card is the basic service set identifier (BSSID) within a BSS. An ESS topology utilizes multiple access points, thus the existence of multiple BSSIDs. In an IBSS network, the first station that powers up randomly generates a virtual BSSID in the MAC address format.

14. D. A wireless wide area network (WWAN) covers a vast geographical area. 802.11 solutions cannot scale to that magnitude.

15. A, C. An autonomous access point (AP) is a hybrid device that is half-duplex, much like a hub, yet possesses some switchlike intelligence. The integration service (IS) and distribution system services (DSS) reside in an autonomous AP. Lightweight access points do not have switchlike intelligence, and instead the IS and DSS reside in a WLAN controller.

16. A, C, E. The default standard mode for an access point is Root mode. Examples of non-standard modes include Bridge, Workgroup Bridge, Scanner, and Repeater modes.

17. A, C. An extended service set (ESS) is one or more basic service sets connected by a distribution system. An ESS is a collection of multiple access points and their associated client stations, all united by a single distribution system medium.

18. A. A wireless distribution system (WDS) can connect access points together using a wireless backhaul while allowing clients to also associate to the radio cards in the access point.

19. B, C. The distribution system consists of two main components. The distribution system medium (DSM) is a logical physical medium used to connect access points. Distribution system services (DSS) consist of services built inside an access point, usually in the form of software.

20. B. The 802.11 standard is considered a wireless local area networking (WLAN) standard. 802.11 hardware can, however, be utilized in other wireless topologies.

802.11 Medium Access

IN THIS CHAPTER, YOU WILL LEARN ABOUT THE FOLLOWING:

✓ **CSMA/CA vs. CSMA/CD**

- Collision detection

✓ **Distributed Coordination Function (DCF)**

- Interframe space (IFS)
- Duration/ID field
- Carrier sense
- Random back-off timer

✓ **Point Coordination Function (PCF)**

✓ **Hybrid Coordination Function (HCF)**

- Enhanced Distributed Channel Access (EDCA)
- HCF Controlled Channel Access (HCCA)

✓ **Block acknowledgment (BA)**

✓ **Wi-Fi Multimedia (WMM)**

One of the difficulties we had in writing this chapter was that in order for you to understand how a wireless station gains access to the media, we have to teach more than what is needed for the CWNA exam. The details are needed to understand the concepts; however, it is the concepts that you will be tested on. If you find the details of this chapter interesting, you should look into the Certified Wireless Network Expert (CWNE) certification, which gets into the nitty-gritty details of 802.11 communications. At that time, you will need to understand details far beyond what we have included in this chapter. However, at this time, take the details for what they are: a foundation for helping you understand the overall process of how a wireless station gains access to the half-duplex medium.

CSMA/CA vs. CSMA/CD

Network communication requires a set of rules to provide controlled and efficient access to the network medium. *Media access control (MAC)* is the generic term used when discussing the different methods of access. There are many ways of providing media access. The early mainframes used polling, which sequentially checked each terminal to see whether there was data to be processed. Later, token-passing and contention methods were used to provide access to the media. Two forms of contention that are heavily used in today's networks are *Carrier Sense Multiple Access with Collision Detection (CSMA/CD)* and *Carrier Sense Multiple Access with Collision Avoidance (CSMA/CA)*.

CSMA/CD is well known and is used by Ethernet networks. CSMA/CA is not as well known and is used by 802.11 networks. Stations using either access method must first listen to see whether any other device is transmitting. If another device is transmitting, the station must wait until the medium is available. The difference between CSMA/CD and CSMA/CA exists when a client wants to transmit and no other clients are presently transmitting. A CSMA/CD node can immediately begin transmitting. If a collision occurs while a CSMA/CD node is transmitting, the collision will be detected and the node will temporarily stop transmitting. 802.11 wireless stations are not capable of transmitting and receiving at the same time, so they are not capable of detecting a collision during their transmission. For this reason, 802.11 wireless networking uses CSMA/CA instead of CSMA/CD to try to avoid collisions.

When a CSMA/CA station has determined that no other stations are transmitting, the 802.11 radio will choose a random back-off value. The station will then wait an additional period of time, based on that back-off value, before transmitting. During this time, the station continues to monitor to make sure that no other stations begin transmitting.

Because of the half-duplex nature of the RF medium, it is necessary to ensure that at any given time only one radio card has control of the medium. CSMA/CA is a process used to ensure that only one radio card is transmitting at a time. Is this process perfect? Absolutely not! Collisions still do occur when two or more radios transmit at the same time. However, CSMA/CA defines a function called Distributed Coordination Function (DCF) as a medium access method that utilizes multiple checks and balances to try to minimize collisions. These checks and balances can also be thought of as several lines of defense. The various lines of defense are put in place to once again hopefully ensure that only one radio is transmitting while all other radios are listening. CSMA/CA minimizes the risk of collisions without excessive overhead.

CSMA/CA also defines an optional function called Point Coordination Function (PCF) that allows for the access point to poll client stations about their need to transmit data. Finally, CSMA/CA also encompasses a Hybrid Coordination Function (HCF) that specifies advanced *quality of service (QoS)* methods.

This entire process is covered in more detail in the next section of this chapter.

CSMA/CA Overview

Carrier sense determines whether the medium is busy. *Multiple access* ensures that every radio gets a fair shot at the medium (but only one at a time). *Collision avoidance* means only one radio gets on the medium at any given time, hopefully avoiding collisions.

Collision Detection

Earlier in this chapter, we mentioned that 802.11 radios were not able to transmit and receive at the same time and therefore cannot detect collisions. So if they cannot detect a collision, how do they know whether one occurred? The answer is simple. Every time an 802.11 radio transmits a unicast frame, if the frame is received properly, the 802.11 radio that received the frame will reply with an *acknowledgment (ACK)* frame.

If the ACK is received, the original station knows that the frame transfer was successful. All unicast 802.11 frames must be acknowledged. Broadcast and multicast frames do not require an acknowledgment. If any portion of a unicast frame is corrupted, the *cyclic redundancy check (CRC)* will fail and the receiving 802.11 radio will not send an ACK frame to the transmitting 802.11 radio. If an ACK frame is not received by the original transmitting radio, the unicast frame is not acknowledged and will have to be retransmitted.

This process does not specifically determine whether a collision occurs; in other words, there is no collision detection. However, if an ACK frame is not received by the original radio, there is collision assumption. Think of the ACK frame as a method of delivery verification. If no proof of delivery is provided, the original radio card assumes there was a delivery failure and retransmits the frame.

Distributed Coordination Function (DCF)

Distributed Coordination Function (DCF) is the fundamental access method of 802.11 communications. DCF is the mandatory access method of the 802.11 standard. The 802.11 standard also has an optional access method known as *Point Coordination Function (PCF)*, which is covered later in this chapter. With the addition of the 802.11e amendment, which is now part of the 802.11-2007 standard, a third coordination function known as the *Hybrid Coordination Function (HCF)* has been added, which also is covered later in this chapter. In this section, you will learn about some of the components that are part of the CSMA/CA process.

The four main components of DCF are interframe space, virtual carrier-sense, physical carrier-sense, and the random back-off timer. Think of these four components as checks and balances that work together at the same time to ensure that only one radio card is transmitting on the half-duplex medium.

Interframe Space (IFS)

Interframe space (IFS) is a period of time that exists between transmissions of wireless frames. There are five types of interframe spaces, which are listed here in order of shortest to longest:

- Short interframe space (SIFS), highest priority
- PCF interframe space (PIFS), middle priority
- DCF interframe space (DIFS), lowest priority
- Arbitration interframe space (AIFS), used by QoS stations
- Extended interframe space (EIFS), used with retransmissions

The actual length of time of each of the interframe spaces varies depending on the transmission speed of the network. Interframe spaces are one line of defense used by CSMA/CA to ensure that only certain types of 802.11 frames are transmitted following certain interframe spaces. For example, only ACK frames and clear-to-send (CTS) frames may follow a SIFS. The two most common interframe spaces used are the SIFS and the DIFS. As pictured in Figure 8.1, the ACK frame is the highest-priority frame, and the use of a SIFS ensures that it will be transmitted first, before any other type of 802.11 frame. Most other 802.11 frames follow a longer period of time called a DIFS.

Interframe spaces are all about what type of 802.11 traffic is allowed next. Interframe spacing also acts as a backup mechanism to virtual carrier-sense, which is discussed later in this section. The main thing that you need to understand at this time is that there are five interframe spaces of different durations of time, and the order is SIFS < PIFS < DIFS < AIFS < EIFS.

As you read further in this chapter, you will learn that timing is an important aspect of successful wireless communications. Interframe spaces are just one component of this tightly linked environment.

FIGURE 8.1 SIFS and DIFS

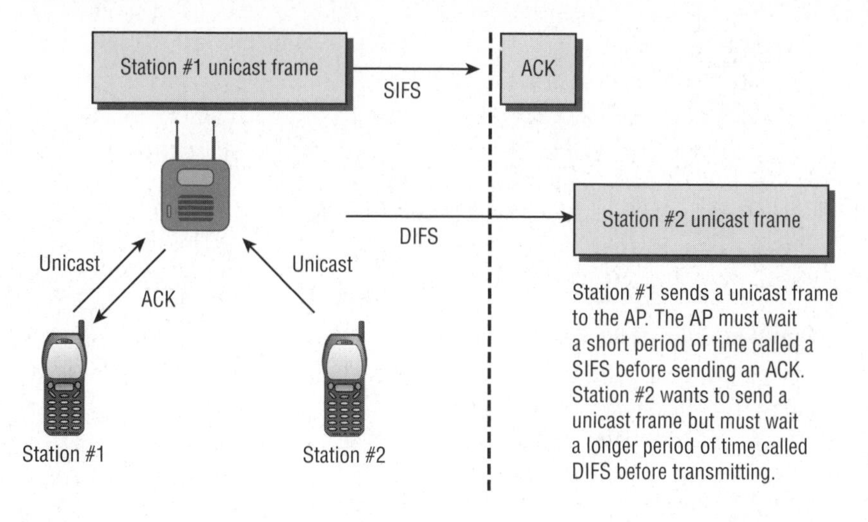

The current version of the CWNA test does not cover interframe spacing. This, however, may change in the future. The CWNE certification tests very heavily on interframe spacing.

Duration/ID Field

As pictured in Figure 8.2, one of the fields in the MAC header of an 802.11 frame is the *Duration/ID field*. When a client transmits a unicast frame, the Duration/ID field contains a value from 0 to 32,767. In this scenario, the Duration/ID value represents the time, in microseconds, that is required to transmit the ACK plus one SIFS interval, as illustrated in Figure 8.3. The client that is transmitting the data frame calculates how long it will take to receive an ACK frame and includes that length of time in the Duration/ID field in the MAC header of the transmitted unicast data frame. The value of the Duration/ID field in the MAC header of the ACK frame that follows is 0 (zero). To summarize, the value of the Duration/ID field indicates how long the RF medium will be busy before another station can contend for the medium.

The majority of the time, the Duration/ID field contains a Duration value that is used to reset other stations' network allocation vector (NAV) timers. In the rare case of a PS-Poll frame, the Duration/ID is used as an ID value of a client station using legacy power management. Power management is discussed in Chapter 9, "802/11 MAC Architecture."

FIGURE 8.2 Duration/ID field

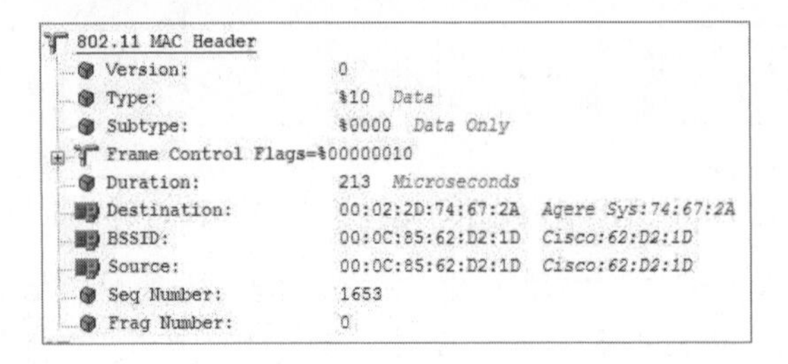

FIGURE 8.3 Duration value of SIFS + ACK

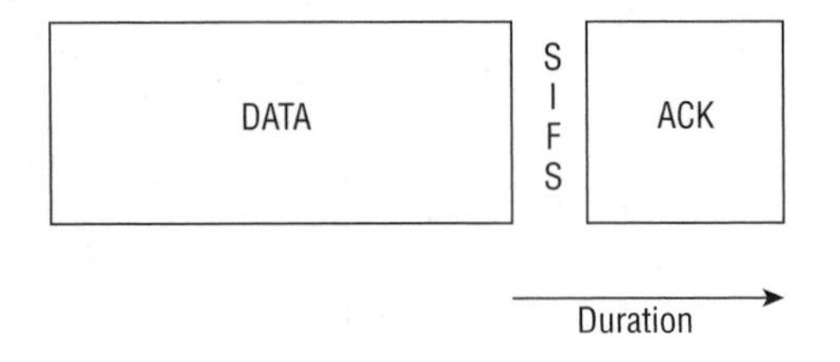

Carrier Sense

The first step that an 802.11 CSMA/CA device needs to do to begin transmitting is to perform a carrier sense. This is a check to see whether the medium is busy. Think of it like listening for a busy signal when you call someone on the phone. There are two ways that a carrier sense is performed: virtual carrier-sense and physical carrier-sense.

Virtual Carrier-Sense

Virtual carrier-sense uses a timer mechanism known as the *network allocation vector (NAV)*. The NAV timer maintains a prediction of future traffic on the medium based on Duration value information seen in a previous frame transmission. When an 802.11 radio is not transmitting, it is listening. As depicted in Figure 8.4, when the listening radio hears a frame transmission from another station, it looks at the header of the frame and determines whether the Duration/ID field contains a Duration value or an ID value. If the field contains a Duration value, the listening station will set its NAV timer to this value. The listening station will then use the NAV as a countdown timer, knowing that the RF medium should be busy until the countdown reaches 0.

This process essentially allows the transmitting 802.11 radio to notify the other stations that the medium will be busy for a period of time (Duration/ID value). The stations that are not transmitting listen and hear the Duration/ID, set a countdown timer (NAV), and wait until their timer hits 0 before they can contend for the medium and eventually transmit on the medium. A station cannot contend for the medium until its NAV timer is 0, nor can a station transmit on the medium if the NAV timer is set to a nonzero value. As stated earlier, there are several lines of defense used by CSMA/CA to prevent collisions, and the NAV timer is often considered the first line of defense.

FIGURE 8.4 Virtual carrier-sense

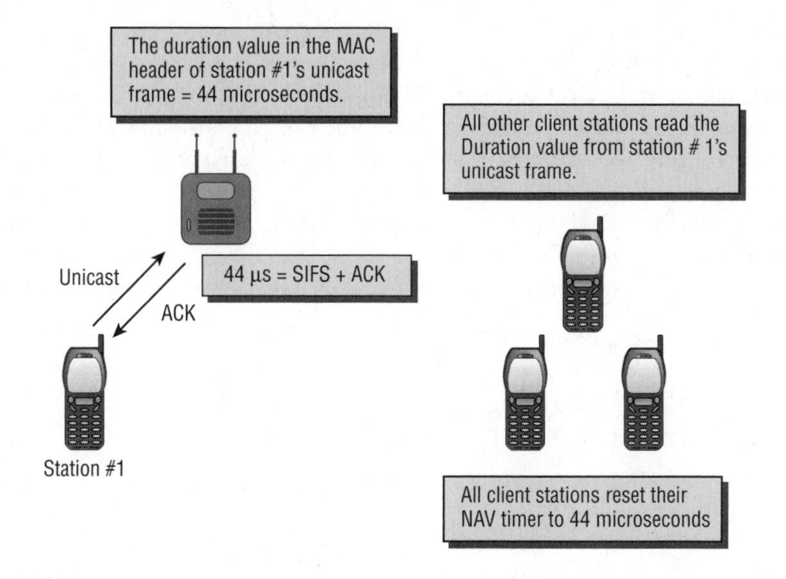

Physical Carrier-Sense

The virtual carrier-sense is one method of keeping other stations from transmitting while another radio has control of the RF medium. However, it is possible that a station did not hear the other radio transmitting so was unable to read the Duration/ID field and set its NAV timer. There could be numerous reasons why, but that's irrelevant at the moment. CSMA/CA utilizes another line of defense to ensure that a station does not transmit while another is already transmitting: The 802.11 standard defines a *physical carrier-sense.*

Physical carrier-sensing is performed constantly by all stations that are not transmitting or receiving. When a station performs a physical carrier-sense, it is actually listening to the channel to see whether any other transmitters are taking up the channel.

Physical carrier-sense has two purposes. The first purpose is to determine whether a frame transmission is inbound for a station to receive. If the medium is busy, the radio will attempt to synchronize with the transmission. The second purpose is to determine whether the medium is busy before transmitting. This is known as the *clear channel assessment (CCA).* The CCA involves listening for 802.11 RF transmissions at the Physical layer. The medium must be clear before a station can transmit.

It is important to understand that both virtual carrier-sense and physical carrier-sense are always happening at the same time. Virtual carrier-sense is a layer 2 line of defense, while physical carrier-sense is a layer 1 line of defense. If one line of defense fails, hopefully the other will prevent collisions from occurring.

Random Back-off Timer

An 802.11 station may contend for the medium during a window of time known as the *contention window*. At this point in the CSMA/CA process, the station selects a random back-off value. The random value is chosen from a range of 0 to the initial contention window value, as shown in Figure 8.5. The back-off value is then multiplied by the *slot time*, which is a period of time that differs among the different spread spectrum technologies. This starts a random back-off timer. The random back-off timer is the final timer used by a station before it transmits. The station's back-off timer begins to count down ticks of a clock known as slots. When the back-off time is equal to 0, the client can reassess the channel and if clear, begin transmitting. If the channel is still busy, it will restart this process.

FIGURE 8.5 Contention window length

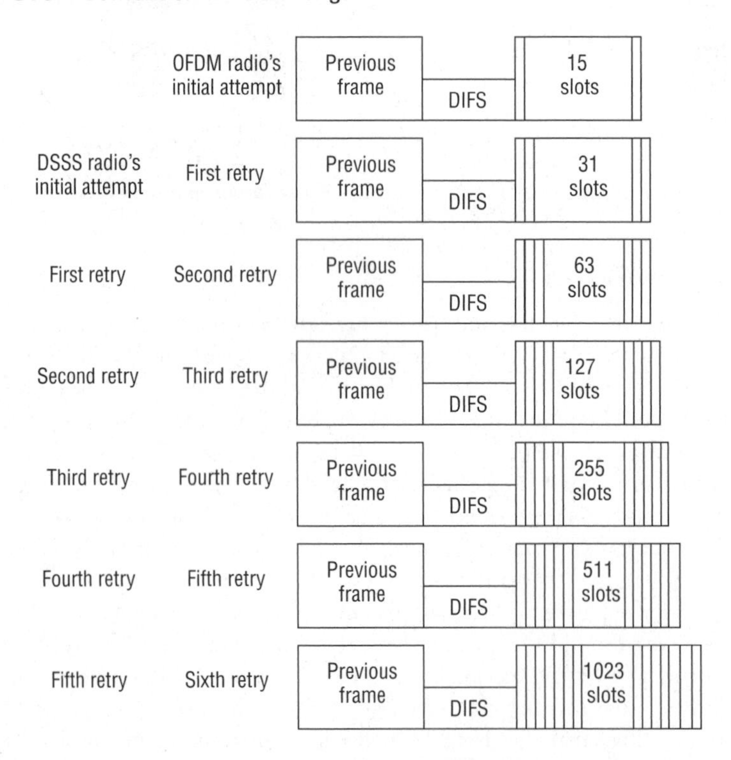

The random back-off timer is another line of defense and helps minimize the likelihood of two stations trying to communicate at the same time, although it does not fully prevent this from occurring. If a station does not receive an ACK, it starts the carrier-sense process over again.

Point Coordination Function (PCF)

In addition to DCF, the IEEE 802.11 standard defines an additional, optional medium access method known as *Point Coordination Function (PCF)*. This access method is a form of polling. The access point performs the function of the *point coordinator (PC)*. Because an access point is taking the role of the point coordinator, the PCF medium access method will work in only a basic service set (BSS). PCF cannot be utilized in an ad hoc network because no access point exists in an independent basic service set (IBSS). Because polling is performed from a central device, PCF provides managed access to the medium.

In order for PCF to be used, both the access point and the station must support it. If PCF is enabled, DCF will still function. The access point will alternate between PCF mode and DCF mode. When the access point is functioning in PCF mode, it is known as the *contention-free period (CFP)*. During the contention-free period, the access point polls only clients in PCF mode about their intention to send data. This is a method of prioritizing clients. When the access point is functioning in DCF mode, it is known as the *contention period (CP)*.

If you would like to learn more about PCF, we suggest that you read the 802.11-2007 standard document. As we stated earlier, PCF is an optional access method, and as this book is being written, we do not know of any vendor that has implemented it.

Hybrid Coordination Function (HCF)

The 802.11e quality of service amendment added a new coordination function to 802.11 medium contention, known as *Hybrid Coordination Function (HCF)*. The 802.11e amendment and HCF have since been incorporated into the 802.11-2007 standard. HCF combines capabilities from both DCF and PCF and adds enhancements to them to create two channel-access methods: Enhanced Distributed Channel Access (EDCA) and HCF Controlled Channel Access (HCCA).

DCF and PCF medium contention mechanisms discussed earlier allow for an 802.11 radio to transmit a single frame. After transmitting a frame, the 802.11 station must contend for the medium again before transmitting another frame. HCF defines the ability for an 802.11 radio to send multiple frames when transmitting on the RF medium. When an HCF-compliant radio contends for the medium, it receives an allotted amount of time to send frames. This period of time is called a *transmit opportunity (TXOP)*. During this TXOP, an 802.11 radio may send multiple frames in what is called a *frame burst*. During the frame burst, a short interframe space (SIFS) is used between each frame to ensure that no other radios transmit during the frame burst.

Enhanced Distributed Channel Access (EDCA)

Enhanced Distributed Channel Access (EDCA) is a wireless media access method that provides differentiated access for stations by using eight *user priority (UP)* levels. EDCA is an extension of DCF. The EDCA medium access method provides for the prioritization of traffic via priority tags that are identical to 802.1D priority tags. Priority tags provide a mechanism for implementing quality of service (QoS) at the MAC level.

Different classes of service are available, represented in a 3-bit user priority field in an IEEE 802.1Q header added to an Ethernet frame. 802.1D enables priority queuing (enabling some Ethernet frames to be forwarded ahead of others within a switched Ethernet network). Figure 8.6 depicts 802.1D priority tags from the Ethernet side that are used to direct traffic to access-category queues.

FIGURE 8.6 EDCA and 802.1D priority tags

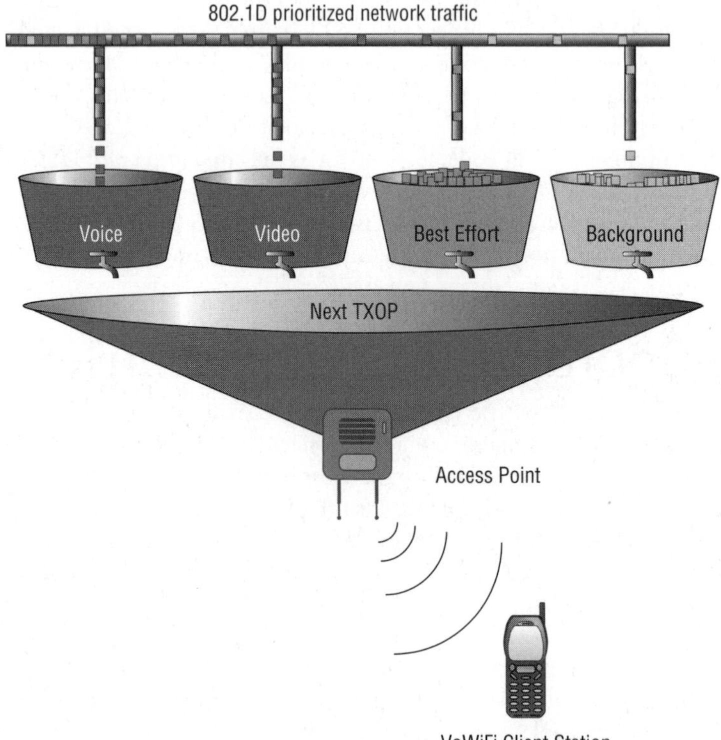

EDCA defines four access categories, based on the UPs. The four access categories from lowest priority to highest priority are AC_BK (Background), AC_BE (Best Effort), AC_VI (Video), and AC_VO (Voice). For each access category, an enhanced version of DCF known as *Enhanced Distributed Channel Access Function (EDCAF)* is used to contend for a TXOP. Frames with the highest-priority access category have the lowest back-off values and therefore are more likely to get a TXOP. The specific details of this process are beyond the scope of the CWNA exam.

HCF Controlled Channel Access (HCCA)

HCF Controlled Channel Access (HCCA) is a wireless media access method that uses a QoS-aware centralized coordinator known as a *hybrid coordinator (HC)*, which operates differently than the point coordinator in a PCF network. The HC is built into the access point and has a higher priority of access to the wireless medium. Using this higher priority level, it can allocate TXOPs to itself and other stations to provide a limited-duration controlled access phase (CAP), providing contention-free transfer of QoS data. The specific details of this process are beyond the scope of the CWNA exam.

Block Acknowledgment (BA)

The 802.11e amendment also introduced a *Block acknowledgment (BA)* mechanism that is now also defined by the 802.11-2007 standard. A Block ACK improves channel efficiency by aggregating several acknowledgments into one single acknowledgment frame. There are two types of Block ACK mechanisms: immediate and delayed. The immediate Block ACK is designed for use with low-latency traffic, and the delayed Block ACK is more suitable for latency-tolerant traffic. For the purposes of this book, we will discuss only the immediate Block ACK.

As pictured in Figure 8.7, an originator station sends a block of QoS data frames to a recipient station. The originator requests acknowledgment of all the outstanding QoS data frames by sending a BlockAckReq frame. Instead of acknowledging each unicast frame independently, the block of QoS data frames are all acknowledged by a single Block ACK. A bitmap in the Block ACK frame is used to indicate the status of all the received data frames. If only one of the frames is corrupted, only that frame will need to be retransmitted. The use of a Block ACK instead of a traditional ACK is a more efficient method that cuts down on medium contention overhead. Uses of Block ACK mechanisms are further defined in the 802.11n draft amendment.

FIGURE 8.7 Immediate Block acknowledgment

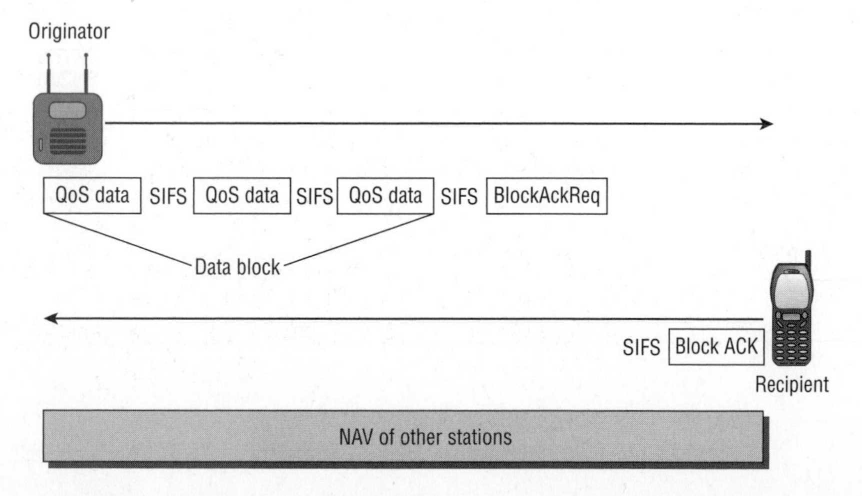

Wi-Fi Multimedia (WMM)

Since the adoption of the original 802.11 standard, there have not been any adequate QoS procedures defined for the use of time-sensitive applications such as *Voice over Wi-Fi (VoWiFi)*. Application traffic such as voice, audio, and video has a lower tolerance for latency and jitter, and requires priority before standard data traffic. The 802.11e amendment defined the layer 2 MAC methods needed to meet the QoS requirements for time-sensitive applications over IEEE 802.11 wireless LANs. The Wi-Fi Alliance introduced the *Wi-Fi Multimedia (WMM)* certification as a partial mirror of 802.11e amendment.

Real World Scenario

VoWiFi and QoS

Currently, WMM is based on EDCA mechanisms defined by the 802.11e amendment. The Wi-Fi Alliance had once proposed a certification called WMM-SA (Scheduled Access), which was to be based on HCCA mechanisms. However, the proposed WMM-SA certification no longer exists. Some VoWiFi vendors were interested in implementing HCCA mechanisms. However, other vendors were not. Support for HCCA mechanisms in the near future is doubtful. Additionally, some VoWiFi vendors still offer proprietary QoS solutions as well as WMM. Although not a recommended practice, it is not unusual to see a proprietary VoWiFi QoS vendor solution and another vendor using WMM within the same WLAN.

Because WMM is based on EDCA mechanisms, 802.1D priority tags from the Ethernet side are used to direct traffic to four access-category priority queues. The WMM certification provides for traffic prioritization via four access categories, as shown in Table 8.1.

TABLE 8.1 Wi-Fi Multimedia Access Categories

Access Category	Description	802.1D Tags
WMM Voice priority	This is the highest priority. It allows multiple and concurrent VoIP calls with low latency and toll voice quality.	7, 6
WMM Video priority	This supports prioritized video traffic before other data traffic. A single 802.11g or 802.11a channel can support three to four SDTV video streams or one HDTV video stream.	5, 4

TABLE 8.1 Wi-Fi Multimedia Access Categories *(continued)*

Access Category	Description	802.1D Tags
WMM Best Effort priority	This is traffic from applications or devices that cannot provide QoS capabilities, such as legacy devices. This traffic is not as sensitive to latency but is affected by long delays, such as Internet browsing.	0, 3
WMM Background priority	This is low-priority traffic that does not have strict throughput or latency requirements. This traffic includes file transfers and print jobs.	2, 1

The Wi-Fi Alliance also defined *WMM-PS (Power Save)*, which uses 802.11e power-saving mechanisms to increase the battery life via advanced power-saving mechanisms. More information about power management can be found in Chapter 9. WMM and HCF as defined by 802.11e is an extremely complicated medium access method, and entire books and classes will probably be created on the subject.

The Wi-Fi Alliance has two white papers we recommend you read about WMM. Both white papers from the Wi-Fi Alliance are included on the CD that accompanies this book:

- *Wi-Fi CERTIFIED for WMM—Support for Multimedia Applications with Quality of Service in Wi-Fi Networks*
- *WMM Power Save for Mobile and Portable Wi-Fi CERTIFIED Devices*

Summary

This chapter focused on 802.11 medium access. Every station has the right to communicate, and the management of access to the wireless medium is controlled through media access control. We discussed the difference between CSMA/CD and CSMA/CA as contention methods. CSMA/CA use a pseudorandom contention method called Distributed Coordination Function. DCF uses four lines of defense to ensure that only one 802.11 radio is transmitting on the half-duplex medium. We also discussed an optional contention-free method called Point Coordination Function. The 802.11e quality of service amendment added a new coordination function to 802.11 medium contention, known as Hybrid Coordination Function (HCF). The last topic covered was the Wi-Fi Multimedia (WMM) certification, which was introduced by the Wi-Fi Alliance as a partial mirror of the 802.11e amendment. WMM is designed to meet the QoS requirements for time-sensitive applications such as audio, video, and voice over IEEE 802.11.

Exam Essentials

Understand the similarities and differences between CSMA/CA and CSMA/CD. Understand both access methods and know what makes them similar and what makes them different.

Define the four checks and balances of CSMA/CA and DCF. Understand that virtual carrier-sense, physical carrier-sense, interframe spacing, and the random back-off timer all work together to ensure that only one 802.11 radio is transmitting on the half-duplex medium.

Define virtual and physical carrier-senses. Understand the purpose and basic mechanisms of the two carrier-senses.

Explain DCF and PCF. Define the basic operations of both Distributed Coordination Function and Point Coordination Function.

Define HCF quality of service mechanisms. Hybrid Coordination Function defines the use of TXOPs and access categories in EDCA as well as the use of TXOPs and polling during HCCA.

Understand Wi-Fi Multimedia (WMM) certification and its importance now and in the future. WMM is designed to provide quality of service capabilities to 802.11 wireless networks. WMM is a partial mirror of the 802.11e amendment. WMM currently provides for traffic priority via four access categories.

Key Terms

Before you take the exam, be certain you are familiar with the following terms:

acknowledgment (ACK)
Block acknowledgment (BA)
Carrier Sense Multiple Access with Collision Avoidance (CSMA/CA)
Carrier Sense Multiple Access with Collision Detection (CSMA/CD)
clear channel assessment (CCA)
contention period (CP)
contention window
contention-free period (CFP)
cyclic redundancy check (CRC)
Distributed Coordination Function (DCF)
Duration/ID field
Enhanced Distributed Channel Access (EDCA)
Enhanced Distributed Channel Access Function (EDCAF)
HCF Controlled Channel Access (HCCA)
hybrid coordinator (HC)
Hybrid Coordination Function (HCF)
interframe space (IFS)
media access control (MAC)
network allocation vector (NAV)
physical carrier-sense
Point Coordination Function (PCF)
point coordinator (PC)
quality of service (QoS)
slot time
transmit opportunity (TXOP)
user priority (UP)
virtual carrier-sense
Voice over Wi-Fi (VoWiFi)
Wi-Fi Multimedia (WMM)
WMM-PS (Power Save)

Review Questions

1. DCF is also known as what? (Choose all that apply.)
   - A. Carrier Sense Multiple Access with Collision Detection (CSMA/CD)
   - B. Carrier Sense Multiple Access with Collision Avoidance (CSMA/CA)
   - C. Data Control Function
   - D. Distributed Coordination Function
2. 802.11 collision detection is handled using which technology?
   - A. Network allocation vector (NAV).
   - B. Clear channel assessment (CCA).
   - C. Duration/ID value.
   - D. Receiving an ACK from the destination station.
   - E. Positive collision detection cannot be determined.
3. ACK and CTS frames follow which interframe space?
   - A. EIFS
   - B. DIFS
   - C. PIFS
   - D. SIFS
   - E. LIFS
4. The carrier sense portion of CSMA/CA is performed by using which of the following methods? (Choose all that apply.)
   - A. Virtual carrier-sense
   - B. Physical carrier-sense
   - C. Channel sense window
   - D. Clear channel assessment
5. After the station has performed the carrier sense and determined that no other devices are transmitting for a period of a DIFS interval, the next step is for the station to do what?
   - A. Wait the necessary number of slot times before transmitting.
   - B. Begin transmitting.
   - C. Select a random back-off value.
   - D. Begin the random back-off timer.

6. If PCF is implemented, it can function in which of the following network environments? (Choose all that apply.)
   - A. Ad Hoc mode
   - B. BSS
   - C. IBSS
   - D. Infrastructure mode
   - E. BSA

7. Which of the following terms are affiliated with the virtual carrier-sense mechanism? (Choose all that apply.)
   - A. Contention window
   - B. Network allocation vector
   - C. Random back-off time
   - D. Duration/ID field

8. Which of the following statements about PCF are true? (Choose all that apply.)
   - A. PCF will work in only a BSS.
   - B. PCF can be used in an IBSS.
   - C. Both the station and access point must support PCF for it to be used.
   - D. The access point will alternate between PCF and DCF mode.
   - E. PCF is a form of contention.

9. CSMA/CA and DCF define which mechanisms to ensure that only one 802.11 radio can transmit on the half-duplex RF medium? (Choose all that apply.)
   - A. Random back-off timer
   - B. NAV
   - C. CCMP
   - D. CCA
   - E. Interframe spacing

10. The Wi-Fi Alliance certification called Wi-Fi Multimedia (WMM) is based on which media access method defined by the 802.11-2007 standard?
    - A. DCF
    - B. PCF
    - C. EDCA
    - D. HCCA
    - E. HSRP

11. Hybrid Coordination Function (HCF) defines what allotted period of time in which a station can transmit multiple frames?
    A. Block acknowledgment
    B. Polling
    C. Virtual carrier-sense
    D. Physical carrier-sense
    E. TXOP

12. Currently, WMM is based on EDCA and provides for traffic prioritization via which of the following access categories? (Choose all that apply.)
    A. WMM Voice priority
    B. WMM Video priority
    C. WMM Audio priority
    D. WMM Best Effort priority
    E. WMM Background priority

13. The 802.11e amendment (now part of the 802.11-2007 standard) defines which of the following medium access methods to support QoS requirements? (Choose all that apply.)
    A. Distributed Coordination Function (DCF)
    B. Enhanced Distributed Channel Access (EDCA)
    C. Hybrid Coordination Function (HCF)
    D. Point Coordination Function (PCF)
    E. Hybrid Coordination Function Controlled Access (HCCA)

14. What information that comes from the wired side network is used to assign traffic into access categories on a WLAN controller? (Choose all that apply.)
    A. DSCP bits
    B. Duration/ID
    C. 802.1D priority tags
    D. Destination MAC address
    E. Source MAC address

15. What are the two reasons that 802.11 radios use physical carrier-sense? (Choose two.)
    A. Synchronize incoming transmissions.
    B. Synchronize outgoing transmissions.
    C. Reset the NAV.
    D. Start the random back-off timer.
    E. Assess the RF medium.

16. What CSMA/CA mechanism is used for medium contention? (Choose all that apply.)
    - A. NAV
    - B. Interframe spacing
    - C. CCA
    - D. Random back-off timer
    - E. Contention window

17. Which field in the MAC header of an 802.11 frame resets the NAV timer for all listening 802.11 stations?
    - A. NAV
    - B. Frame control
    - C. Duration/ID
    - D. Sequence number
    - E. Strictly ordered bit

18. The EDCA medium access method provides for the prioritization of traffic via priority tags that are identical to the 802.1D priority tags. What are the EDCA priority tags known as?
    - A. TXOP
    - B. UP levels
    - C. Priority levels
    - D. Priority bits
    - E. PT

19. ACKs are required for which of the following frames? (Choose all that apply.)
    - A. Unicast
    - B. Broadcast
    - C. Multicast
    - D. Anycast

20. What QoS mechanism can be used to reduce medium contention overhead during a frame burst of low-latency traffic?
    - A. Delayed Block ACK
    - B. Contention period
    - C. Contention window
    - D. Contention-free period
    - E. Immediate Block ACK

Answers to Review Questions

1. B, D. DCF is an abbreviation for Distributed Coordination Function. DCF is a CSMA/CA media access control method. CSMA/CD is used by 802.3, not 802.11. There is no such thing as Data Control Function.

2. E. 802.11 technology does not use collision detection. If an ACK frame is not received by the original transmitting radio, the unicast frame is not acknowledged and will have to be retransmitted. This process does not specifically determine whether a collision occurs. Failure to receive an ACK frame from the receiver means that a unicast frame was either not received by the destination station or the ACK frame was not received, but it cannot positively determine the cause. It may be due to collision or to other reasons such as high noise level. All of the other options are used to help prevent collisions.

3. D. Only ACK frames and CTS frames may follow a SIFS. LIFS do not exist. The other three interframe spaces are not covered on the CWNA test.

4. A, B, D. The NAV timer maintains a prediction of future traffic on the medium based on duration value information seen in a previous frame transmission. The virtual carrier-sense uses the NAV to determine medium availability. Physical carrier-sense checks the RF medium for carrier availability. Clear channel assessment is another name for physical carrier-sense. Channel sense window does not exist.

5. C. The first step is to select a random back-off timer. After the value is selected, it is multiplied by the slot time. The random back-off timer then begins counting down the number of slot times. When the number reaches 0, the station can begin transmitting.

6. B, D. PCF requires an access point. Ad Hoc mode and an independent basic service set (IBSS) are the same and do not use an access point. A basic service set (BSS) is a WLAN topology, where 802.11 client stations communicate through an access point. Infrastructure mode is the default client station mode that allows clients to communicate via an access point. BSA is basic service area, the area of coverage of a basic service set

7. B, D. The Duration/ID field is used to set the network allocation vector (NAV), which is a part of the virtual carrier-sense process. The contention window and random back-off time are part of the back-off process that is performed after the carrier sense process.

8. A, C, D. PCF will work only in a BSS because an access point is required, which also means that it cannot be used in an IBSS. Both the station and access point must support PCF if it is going to be used. The access point will alternate between PCF mode (contention-free period) and DCF mode (contention period). PCF is a form of polling, not contention.

9. A, B, D, E. DCF defines four checks and balances of CSMA/CA and DCF to ensure that only one 802.11 radio is transmitting on the half-duplex medium. Virtual carrier-sense (NAV), physical carrier-sense (CCA), interframe spacing, and the random back-off timer all work together.

10. C. Currently, WMM is based on EDCA mechanisms defined by the 802.11e amendment, which is now part of the 802.11-2007 standard. The WMM certification provides for traffic prioritization via four access categories. EDCA is a subfunction of Hybrid Coordination Function (HCF). The other subfunction of HCF is HCCA.

11. E. HCF defines the ability for an 802.11 radio to send multiple frames when transmitting on the RF medium. When an HCF-compliant radio contends for the medium, it receives an allotted amount of time to send frames called a transmit opportunity (TXOP). During this TXOP, an 802.11 radio may send multiple frames in what is called a *frame burst.*

12. A, B, D, E. WMM Audio priority does not exist. The WMM certification provides for traffic prioritization via the four access categories of voice, video, best effort, and background.

13. B, C, E. DCF and PCF were defined in the original 802.11 standard. The 802.11e quality of service amendment added a new coordination function to 802.11 medium contention, known as Hybrid Coordination Function (HCF). The 802.11e amendment and HCF have since been incorporated into the 802.11-2007 standard. HCF combines capabilities from both DCF and PCF and adds enhancements to them to create two channel access methods, HCF Controller Channel Access (HCCA) and Enhanced Distributed Channel Access (EDCA).

14. C. The EDCA medium access method provides for the prioritization of traffic via the use of 802.1D priority tags. 802.1D tags provide a mechanism for implementing quality of service (QoS) at the MAC level. Different classes of service are available, represented in a 3-bit user priority field in an IEEE 802.1Q header added to an Ethernet frame. 802.1D priority tags from the Ethernet side are used to direct traffic to different access-category queues.

15. A, E. The first purpose is to determine whether a frame transmission is inbound for a station to receive. If the medium is busy, the radio will attempt to synchronize with the transmission. The second purpose is to determine whether the medium is busy before transmitting. This is known as the clear channel assessment (CCA). The CCA involves listening for 802.11 RF transmissions at the Physical layer. The medium must be clear before a station can transmit.

16. D, E. An 802.11 radio uses a random back-off algorithm to contend for the medium during a window of time known as the contention window. The contention window is essentially a final countdown timer and is also known as the random back-off timer.

17. C. When the listening radio hears a frame transmission from another station, it looks at the header of the frame and determines whether the Duration/ID field contains a Duration value or an ID value. If the field contains a Duration value, the listening station will set its NAV timer to this value.

18. B. Enhanced Distributed Channel Access provides differentiated access for stations by using eight user priority (UP) levels. The EDCA medium access method provides for the prioritization of traffic via priority tags that are identical to the 802.1D priority tags.

19. A. All unicast 802.11 frames must be acknowledged. Broadcast and multicast frames do not require an acknowledgment. Anycast frames do not exist.

20. E. A Block ACK improves channel efficiency by aggregating several acknowledgments into one single acknowledgment frame. There are two types of Block ACK mechanisms: immediate and delayed. The immediate Block ACK is designed for use with low-latency traffic, while the delayed Block ACK is more suitable for latency-tolerant traffic.

802.11 MAC Architecture

IN THIS CHAPTER, YOU WILL LEARN ABOUT THE FOLLOWING:

✓ **Packets, frames, and bits**

✓ **Data-Link layer**

- MAC Service Data Unit (MSDU)
- MAC Protocol Data Unit (MPDU)

✓ **Physical layer**

- PLCP Service Data Unit (PSDU)
- PLCP Protocol Data Unit (PPDU)

✓ **802.11 and 802.3 interoperability**

✓ **Three 802.11 frame types**

- Management frames
- Control frames
- Data frames

✓ **Beacon management frame (beacon)**

✓ **Passive scanning**

✓ **Active scanning**

✓ **Authentication**

- Open System authentication
- Shared Key authentication

✓ **Association**

✓ **Authentication and association states**

✓ **Basic and supported rates**

- ✓ **Roaming**
- ✓ **Reassociation**
- ✓ **Disassociation**
- ✓ **Deauthentication**
- ✓ **ACK frame**
- ✓ **Fragmentation**
- ✓ **Protection mechanism**
- ✓ **RTS/CTS**
- ✓ **CTS-to-Self**
- ✓ **Data frames**
- ✓ **Power management**
    - Active mode
    - Power Save mode
    - Traffic indication map (TIM)
    - Delivery traffic indication message (DTIM)
    - Announcement traffic indication message (ATIM)
    - WMM Power Save (WMM-PS) and U-APSD
    - 802.11n power management

This chapter presents all of the components of the 802.11 MAC architecture. We discuss how upper-layer information is encapsulated within an 802.11 frame format. We cover the three major 802.11 frame types and a majority of the 802.11 frame subtypes. We discuss many MAC layer tasks such as active scanning and the specific 802.11 frames that are used to accomplish these tasks. An often misunderstood capability of 802.11 is the ERP protection mechanism. We describe exactly how 802.11b and 802.11g stations can coexist in the same BSS by using either the RTS/CTS or CTS-to-Self protection mechanism. The final section of the chapter discusses legacy 802.11 power management and enhanced WMM-PS power management, which are methods used to save battery life.

Packets, Frames, and Bits

When learning about any technology, it is important at times to step back and focus on the basics. If you have ever flown an airplane, it is important, when things get difficult, to refocus on the number one priority, the main objective—and that is to fly the airplane. Navigation and communications are secondary to flying the airplane. When dealing with any complex technology, it is easy to forget the main objective; this is as true with 802.11 communications as it is with flying. With 802.11 communications, the main objective is to transfer user data from one computing device to another.

As data is processed in a computer and prepared to be transferred from one computer to another, it starts at the upper layers of the OSI model and moves down until it reaches the Physical layer, where it is ultimately transferred to the other devices. Initially, a user may want to transfer a word processing document from their computer to a shared network disk on another computer. This document will start at the Application layer and work its way down to the Physical layer, get transmitted to the other computer, and then work its way back up the layers of the OSI model to the Application layer on the other computer.

As data travels down the OSI model for the purpose of being transmitted, each layer adds header information to that data. This enables the data to be reassembled when it is received by the other computer. At the Network layer, an IP header is added to the data that came from layers 4–7. A layer 3 IP *packet*, or datagram encapsulates the data from the higher layers. At the Data-Link layer, a MAC header is added and the IP packet is encapsulated inside a *frame*. Ultimately, when the frame reaches the Physical layer, a PHY header with more information is added to the frame.

Data is eventually transmitted as individual bits at the Physical layer. A *bit* is a binary digit, taking a value of either 0 or 1. Binary digits are a basic unit of communication in digital computing. A byte of information comprises 8 bits.

In this chapter, we discuss how upper-layer information moves down the OSI model through the Data-Link and Physical layers from an 802.11 perspective.

Data-Link Layer

The 802.11 *Data-Link layer* is divided into two sublayers. The upper portion is the IEEE 802.2 *Logical Link Control (LLC)* sublayer, which is identical for all 802-based networks, although not used by all IEEE 802 networks. The bottom portion of the Data-Link layer is the *Media Access Control (MAC) sublayer*, which is identical for all 802.11-based networks. The 802.11 standard defines operations at the MAC sublayer.

MAC Service Data Unit (MSDU)

When the Network layer (layer 3) sends data to the Data-Link layer, that data is handed off to the LLC and becomes known as the *MAC Service Data Unit (MSDU)*. The MSDU contains data from the LLC and layers 3–7. A simple definition of the MSDU is that it is the data payload that contains the IP packet plus some LLC data.

Later in this chapter, you will learn about the three major 802.11 frame formats. 802.11 management and control frames do not carry upper-layer information. Only 802.11 data frames carry an MSDU payload in the frame body. The 802.11-2007 standard states that the maximum size of the MSDU is 2,304 bytes. The maximum frame body size is determined by the maximum MSDU size (2,304 octets) plus any overhead from encryption.

MAC Protocol Data Unit (MPDU)

When the LLC sends the MSDU to the MAC sublayer, the MAC header information is added to the MSDU to identify it. The MSDU is now encapsulated in a *MAC Protocol Data Unit (MPDU)*. A simple definition of an MPDU is that it is an 802.11 frame. The 802.11 frame, as seen in Figure 9.1, contains a layer 2 header, a frame body, and a trailer, which is a 32-bit CRC known as the *frame check sequence (FCS)*. The 802.11 MAC header is discussed in more detail later in this chapter.

At this point, the frame is ready to be passed onto the Physical Layer, which will then further prepare the frame for transmission.

FIGURE 9.1 802.11 MPDU

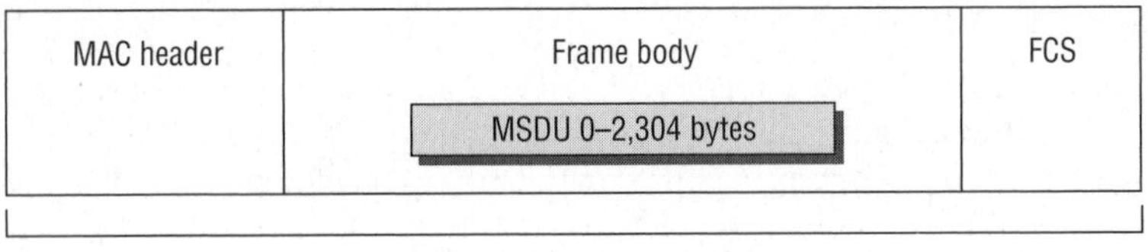

MAC header	Frame body	FCS
	MSDU 0–2,304 bytes	

MPDU—802.11 data frame

Physical Layer

Similar to the way the Data-Link layer is divided into two sublayers, the *Physical layer* is also divided into two sublayers. The upper portion of the Physical layer is known as the *Physical Layer Convergence Procedure (PLCP) sublayer*, and the lower portion is known as the *Physical Medium Dependent (PMD) sublayer*. The PLCP prepares the frame for transmission by taking the frame from the MAC sublayer and creating the PLCP Protocol Data Unit (PPDU). The PMD sublayer then modulates and transmits the data as bits.

PLCP Service Data Unit (PSDU)

When you are at a door, it could be the entrance or the exit. It depends on what side of the door you are on, but either way, it is the same door. The *PLCP Service Data Unit (PSDU)* is a view of the MPDU from the other side. The MAC layer refers to the frame as the MPDU, while the Physical layer refers to this same exact frame as the PSDU. The only difference is which side of the door you are on, or, in the OSI model, from which layer of the model you are looking at the frame.

PLCP Protocol Data Unit (PPDU)

When the PLCP receives the PSDU, it then prepares the PSDU to be transmitted and creates the *PLCP Protocol Data Unit (PPDU)*. The PLCP adds a preamble and PHY header to the PSDU. The preamble is used for synchronization between transmitting and receiving 802.11 radios. It is beyond the scope of this book and the CWNA exam to discuss all the details of the preamble and PHY header. When the PPDU is created, the PMD sublayer takes the PPDU and modulates the data bits and begins transmitting.

Figure 9.2 depicts a flowchart that shows the upper-layer information moving between the Data-Link and Physical layers.

FIGURE 9.2 Data-Link and Physical layers

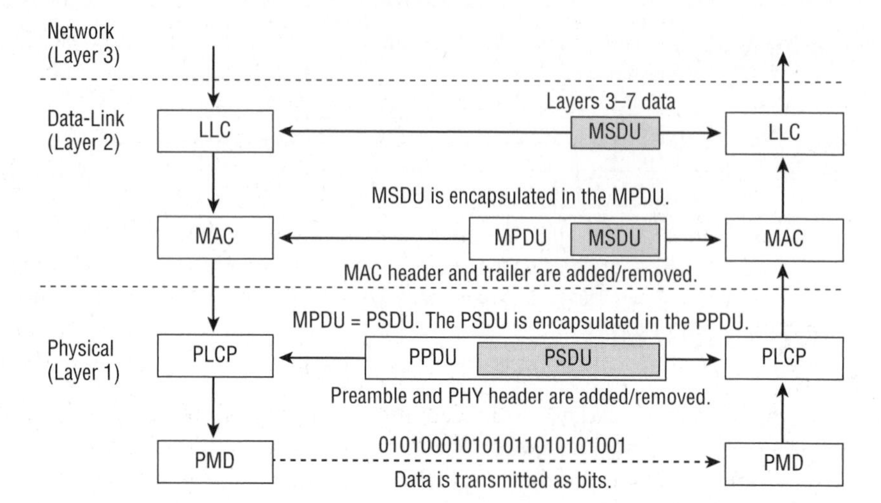

802.11 and 802.3 Interoperability

As you learned in Chapter 7, "Wireless LAN Topologies," the 802.11-2007 standard defines an *integration service (IS)* that enables delivery of MSDUs between the distribution system (DS) and a non-IEEE-802.11 local area network (LAN), via a portal. A simpler way of defining the integration service is to characterize it as a frame format transfer method. The portal is usually either an access point or a WLAN controller. As mentioned earlier, the payload of a wireless 802.11 data frame is the upper layer 3–7 information known as the MSDU. The eventual destination of this payload usually resides on a wired network infrastructure. Because the wired infrastructure is a different physical medium, an 802.11 data frame payload (MSDU) must be effectively transferred into an 802.3 Ethernet frame. It is beyond the scope of the 802.11-2007 standard to define how the integration service operates. Normally, the integration service transfers frame payloads between an 802.11 and 802.3 medium. However, the integration service could transfer an MSDU between the 802.11 medium and some sort of other medium such as 802.5 token ring.

All of the IEEE 802 frame formats share similar characteristics, and the 802.11 frame is no exception. Because the frames are similar, it makes it easier to translate the frames as they move from the 802.11 wireless network to the 802.3 wired network and vice versa.

One of the differences between 802.3 Ethernet and 802.11 wireless frames is the frame size. 802.3 frames have a maximum size of 1,518 bytes with a maximum MSDU of 1,500 bytes. As you have just learned, 802.11 frames are capable of transporting frames with a MSDU payload of 2,304 bytes of *upper-layer* data. This means that as the data moves between the wireless and the wired network, the access point may receive a data frame that is too large for the wired network. This is rarely a problem thanks to the TCP/IP protocol suite. TCP/IP, the most common communications protocol used on networks, typically has an IP *maximum transmission unit (MTU)* size of 1,500 bytes. IP packets are usually 1,500 bytes based on the MTUs. When the IP packets are passed down to 802.11, even though the maximum size of the MSDU is 2,304 bytes, the size will be limited to the 1,500 bytes of the IP packets.

Another difference between 802.3 and 802.11 frames is the MAC addressing fields. 802.3 frames have only a source address (SA) and destination address (DA) in the layer 2 header. As pictured in Figure 9.3, 802.11 frames have four address fields in the MAC header. 802.11 frames typically use only three of the MAC address fields. However, as we discussed in Chapter 7, an 802.11 frame sent within a wireless distribution system (WDS) requires four MAC addresses. The contents of these four fields can include the following MAC addresses: receiver address (RA), transmitter address (TA), basic service set identifier (BSSID), destination address (DA), and source address (SA). Certain frames may not contain some of the address fields. Even though the number of address fields is different, both 802.3 and 802.11 identify a source address and a destination address, and use the same MAC address format. The first three octets are known as the Organizationally Unique Identifier (OUI), and the last three octets are known as the extension identifier.

FIGURE 9.3 802.11 MAC header

Bytes	2	2	6	6	6	2	6	2
	Frame control	Duration/ ID	Address 1	Address 2	Address 3	Sequence control	Address 4	QoS control

It is beyond the scope of the CWNA exam to explain the purpose of every field in the 802.11 MAC header. However, a very important field that was discussed earlier in the book is the Duration/ID field. As you learned in Chapter 8, "802.11 Medium Access," the duration value in the MAC header of a transmitting station is used to reset to the NAV timer of other listening stations.

Three 802.11 Frame Types

Unlike many wired network standards such as IEEE 802.3, which uses a single data frame type, the IEEE 802.11 standard defines three major frame types: management, control, and data. These frame types are further subdivided into multiple subtypes. In Chapter 8, you learned about the optional media access method of Point Coordination Function (PCF) and the quality-of-service media access method called Hybrid Coordination Function (HCF). Some of the frame subtypes are defined to perform functions associated with PCF. PCF is optional and to date there are no known access points that support this technology. We have indicated any subtypes that are solely used for PCF by placing *PCF only* next to these subtypes but will not address or define them. It is also beyond the scope of this book to discuss all of the frame subtypes used for QoS in HCF. We have placed *HCF* next to the subtype but will not address or define them.

Management Frames

802.11 *management frames* make up a majority of the frame types in a WLAN. Management frames are used by wireless stations to join and leave the basic service set (BSS). They are not necessary on wired networks, since physically connecting or disconnecting the network cable performs this function. However, because wireless networking is an unbounded medium, it is necessary for the wireless station to first find a compatible WLAN, then authenticate to the WLAN (assuming they are allowed to connect), and then associate with the WLAN (typically with an access point) to gain access to the wired network (the distribution system).

Another name for an 802.11 management frame is a *Management MAC Protocol Data Unit (MMPDU)*. Management frames do not carry any upper-layer information. There is no MSDU encapsulated in the MMPDU frame body, which carries only layer 2 information fields and information elements. *Information fields* are fixed-length mandatory fields in the body of a management frame. *Information elements* are variable in length and are optional.

Following is a list of all 12 of the management frame subtypes as defined by the 802.11 standard:

- Association request
- Association response
- Reassociation request
- Reassociation response
- Probe request
- Probe response
- Beacon
- Announcement traffic indication message (ATIM)
- Disassociation
- Authentication
- Deauthentication
- Action

Control Frames

802.11 *control frames* assist with the delivery of the data frames. Control frames must be able to be heard by all stations; therefore, they must be transmitted at one of the basic rates. Control frames are also used to clear the channel, acquire the channel, and provide unicast frame acknowledgments. They contain only header information.

Following is a list of all eight of the control frame subtypes as defined by the 802.11 standard:

- Power Save (PS)-Poll
- Request to send (RTS)

- Clear to send (CTS)
- Acknowledgment (ACK)
- Contention-Free (CF)-End [PCF only]
- CF-End + CF-ACK [PCF only]
- Block ACK Request [HCF]
- Block ACK [HCF]

Data Frames

Most 802.11 *data frames* carry the actual data that is passed down from the higher-layer protocols. However, some 802.11 data frames carry no data at all but do have a specific purpose within the BSS. There are a total of 15 data frame subtypes; however, the CWNA exam should test only your knowledge of the data subtype and the null function subtype. The data subtype is usually referred to as the *simple data frame.* The simple data frame has MSDU upper-layer information encapsulated in the frame body. The integration service that resides in autonomous APs and WLAN controllers takes the MSDU payload of a simple data frame and transfers the MSDU into 802.3 Ethernet frames. Null function frames are sometimes used by client stations to inform the access point of changes in Power Save status.

Following is a list of all 15 of the data frame subtypes as defined by the 802.11 standard:

- Data (simple data frame)
- Null function (no data)
- Data + CF-ACK [PCF only]
- Data + CF-Poll [PCF only]
- Data + CF-ACK + CF-Poll [PCF only]
- CF-ACK (no data) [PCF only]
- CF-Poll (no data) [PCF only]
- CF-ACK + CF-Poll (no data) [PCF only]
- QoS Data [HCF]
- QoS Null (no data) [HCF]
- QoS Data + CF-ACK [HCF]
- QoS Data + CF-Poll [HCF]
- QoS Data + CF-ACK + CF-Poll [HCF]
- QoS CF-Poll (no data) [HCF]
- QoS CF-ACK + CF-Poll (no data) [HCF]

EXERCISE 9.1

802.11 Frame Analysis

The following directions should help you install and use the AirMagnet Wi-Fi Analyzer product demo software so that you can explore some of the topics covered in this chapter. To download the program and determine whether your wireless card is supported, go to www.airmagnet.com/cwna/sybex.

In this exercise, you will use a protocol analyzer to view 802.11 management, control, and data frames.

1. In your web browser, proceed to the following URL: www.airmagnet.com/cwna/sybex.
2. Fill out the product demo request forms. You will then be sent an email message with a private download URL.
3. Proceed to the private download URL. Verify that you have a supported Wi-Fi card. Review the system requirements and supported operating systems.
4. Download the Wi-Fi Analyzer demo to your desktop. This evaluation copy of Wi-Fi Analyzer will be licensed to work for seven days.
5. Double-click the file AirMagnet_WiFi_DemoInstaller.exe and follow the installation prompts. Install the proper driver for your Wi-Fi card.
6. You will now open a demo capture file. From Windows File Manager, browse to the capture file, which is located at C:\Program Files\AirMagnet Inc\AirMagnet Laptop Demo\Demo.amc. Double-click on the demo.amc capture file. The Wi-Fi Analyzer application should appear.
7. In the bottom-left corner of the application, click the Decodes button.

8. You can now view 802.11 frames in the main window. Click on frame #10. You are now looking at a Beacon management frame. Throughout this chapter, we will reference specific frame numbers for you to open in other exercises.

No	M
1	☐
2	☐
3	☐
4	☐
5	☐
6	☐
7	☐
8	☐
9	☐
10	☑
11	☐
12	☐

Beacon Management Frame (Beacon)

One of the most important frame types is the *beacon management frame*, commonly referred to as the beacon. Beacons are essentially the heartbeat of the wireless network. The access point of a basic service set sends the beacons while the clients listen for the beacon frames. Client stations only transmit beacons when participating in an independent basic service set (IBSS), also known as Ad Hoc mode. Each beacon contains a time stamp, which client stations use to keep their clocks synchronized with the access point. Because so much of successful wireless communications is based on timing, it is imperative that all stations are in synch with each other. Some of the information that can be found inside the body of a beacon frame includes the following:

Time stamp: Synchronization information

Spread spectrum parameter sets: FHSS-, DSSS-, or ERP-specific information

Channel information: Channel used by the AP or IBSS

Data rates: Basic and supported rates

Service set capabilities: Extra BSS or IBSS parameters

SSID: Logical WLAN name

Traffic indication map (TIM): A field used during the Power Save process

QoS capabilities: Quality of service and EDCA information

Security capabilities: TKIP or CCMP cipher information

Vendor proprietary information: Vendor-unique or vendor-specific information

The beacon frame contains all the necessary information for a client station to learn about the parameters of the basic service set before joining the BSS. Beacons are transmitted about 10 times per second. This interval can be configured on some APs but it cannot be disabled.

EXERCISE 9.2

Beacon Frame

1. From Windows File Manager, browse to the AirMagnet demo capture file, which is located at `C:\Program Files\AirMagnet Inc\AirMagnet Laptop Demo\Demo.amc`. Double-click on the `demo.amc` capture file. The Wi-Fi Analyzer application should appear.
2. In the bottom-left corner of the application, click the Decodes button.
3. From the AirMagnet Wi-Fi Analyzer main window, click on frame #24, which is a beacon frame.
4. In the lower section of the screen, browse through the information found inside the beacon frame body.

Passive Scanning

In order for a station to be able to connect to an access point, it must first discover an access point. A station discovers an access point by either listening for an AP (passive scanning) or searching for an AP (active scanning). In *passive scanning*, the client station listens for the beacon frames that are continuously being sent by the access points, as seen in Figure 9.4.

FIGURE 9.4 Passive scanning

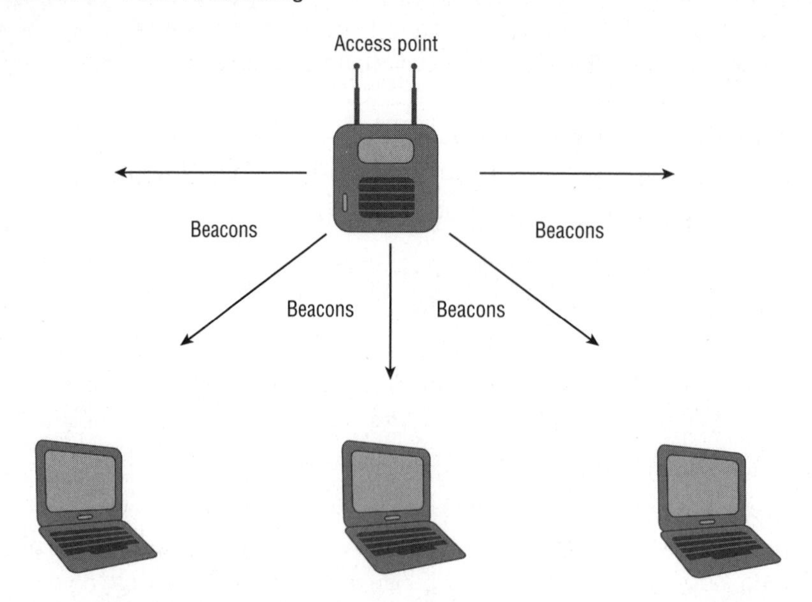

The client station will listen for the beacons that contain the same SSID that has been preconfigured in the client station's software utility. When the station hears one, it can then connect to that WLAN. If the client station hears beacons from multiple access points with the same SSID, it will determine which access point has the best signal, and it will attempt to connect to that AP.

It is important to understand that both active and passive scanning can be performed by different client stations at the same time. Also, when an independent basic service set is deployed, all of the stations in Ad Hoc mode take turns transmitting the beacons since there is no access point. Passive scanning occurs in an ad hoc environment just as it does in a basic service set.

Active Scanning

In addition to passively scanning for access points, client stations can actively scan for them. In *active scanning*, the client station transmits management frames known as probe requests. These probe requests either can contain the SSID of the specific WLAN that the client station is looking for or can look for any SSID. A client station that is looking for any SSID sends a probe request with the SSID field set to null. A probe request with the specific SSID information is known as a *directed probe request*. A probe request without the SSID information is known as a *null probe request*.

If a directed probe request is sent, all access points that support that specific SSID, and hear the request, should reply by sending a *probe response*. The information that is contained inside the body of a probe response frame is the exact same information that can be found in a beacon frame, with the exception of the traffic indication map (TIM). Just like the beacon frame, the probe response frame contains all of the necessary information for a client station to learn about the parameters of the basic service set before joining the BSS.

If a null probe request is sent, all access points that hear the request should reply by sending a probe response, as seen in Figure 9.5.

FIGURE 9.5 Active scanning

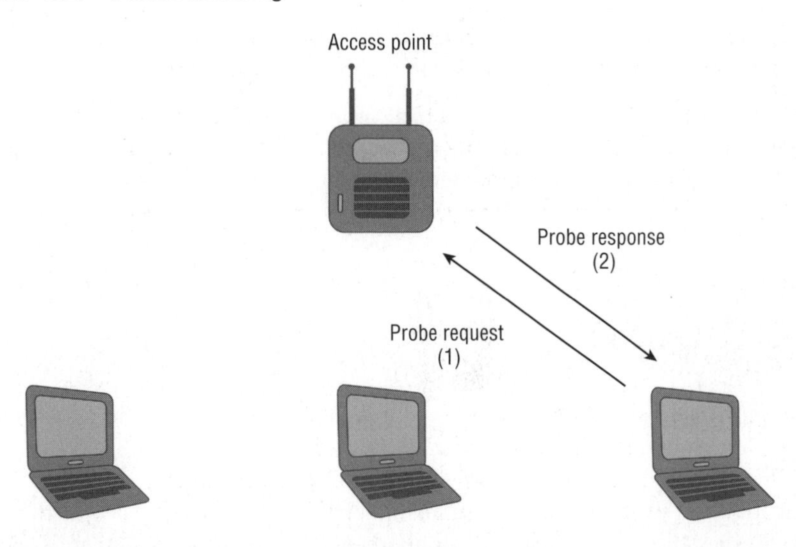

One drawback to passive scanning is that beacon management frames are broadcast only on the same channel as the access point. In contrast, active scanning uses probe request frames that are sent out across all available channels by the client station. If a client station receives probe responses from multiple access points, signal strength and quality characteristics are typically used by the client station to determine which access point has the best signal and thus which access point to connect to. The client station will sequentially send probe requests on each of the supported channels. In fact, it is common for a client station that is already associated to an access point and transmitting data to go off-channel and continue to send probe requests every few seconds across other channels. By continuing to actively scan, a client station can maintain and update a list of known access points, and if the client station needs to roam, it can typically do so faster and more efficiently. How often a client station goes off-channel for active scanning purposes is proprietary. For example, an 802.11 radio in a VoWiFi phone will probably send probe requests across all channels more frequently than an 802.11 radio in a laptop.

EXERCISE 9.3

Probe Requests and Probe Responses

1. From Windows File Manager, browse to the AirMagnet demo capture file, which is located at `C:\Program Files\AirMagnet Inc\AirMagnet Laptop Demo\Demo.amc`. Double-click on the `demo.amc` capture file. The Wi-Fi Analyzer application should appear.
2. In the bottom-left hand corner of the application, click the Decodes button.
3. From the AirMagnet Wi-Fi Analyzer main window, click on frame #12081, which is a probe request. Look at the SSID field in the frame body and notice that this is a directed probe request.
4. Click on frame #12083, which is a probe response.
5. In the lower section of the screen, browse through the information found inside the frame body and notice that the information is similar to a beacon frame.
6. Click on frame #33, which is a probe request. Look at the SSID field in the frame body and notice that this is a null probe request. Click on frame #34 and once again observe the information in the frame body of the probe response frame.

Authentication

Authentication is the first of two steps required to connect to the 802.11 basic service set. Both authentication and association must occur, in that order, before an 802.11 client can pass traffic through the access point to another device on the network.

Authentication is a process that is often misunderstood. When many people think of authentication, they think of what is commonly referred to as network authentication, entering a username and password in order to get access to the network. In this chapter, we are referring to 802.11 authentication. When an 802.3 device needs to communicate with other devices, the first step is to plug the Ethernet cable into the wall jack. When this cable is plugged in, the client creates a physical link to the wired switch and is now able to start transmitting frames. When an 802.11 device needs to communicate, it must first authenticate with the access point or with the other stations if it is configured for Ad Hoc mode. This authentication is not much more of a task than plugging the Ethernet cable into the wall jack. The 802.11 authentication merely establishes an initial connection between the client and the access point.

The 802.11-2007 standard specifies two different methods of authentication: Open System authentication and Shared Key authentication. The following two sections describe these two authentication methods.

Open System Authentication

Open System authentication is the simpler of the two authentication methods. It provides authentication without performing any type of client verification. It is essentially an exchange of hellos between the client and the access point. It is considered a null authentication because there is no exchange or verification of identity between the devices. Open System authentication occurs with an exchange of frames between the client and the access point, as seen in Exercise 9.4.

Wired Equivalent Privacy (WEP) security can be used with Open System authentication; however, WEP is used only to encrypt the upper-layer information of data frames and only after the client station is authenticated and associated. Because of its simplicity, Open System authentication is used in conjunction with more-advanced network security authentication methods such as 802.1X/EAP when implemented.

EXERCISE 9.4

Open System Authentication

1. From Windows File Manager, browse to the AirMagnet demo capture file located at `C:\Program Files\AirMagnet Inc\AirMagnet Laptop Demo\Demo.amc`. Double-click on the `demo.amc` capture file. The WiFi Analyzer application should appear.
2. In the bottom-left corner of the application, click the Decodes button.
3. From the AirMagnet Wi-Fi Analyzer main window, click on frame #12085, which is an authentication request. Look at the 802.11 MAC header and note the source address and destination address.
4. Click on frame #12087, which is an authentication response. Look at the 802.11 MAC header and note that the source address is the AP's BSSID and that the destination address is the MAC address of the client who sent the authentication request. Look at the frame body and note that authentication was successful.

Shared Key Authentication

Shared Key authentication uses WEP to authenticate client stations and requires that a static WEP key be configured on both the station and the access point. In addition to WEP being mandatory, authentication will not work if the static WEP keys do not match. The authentication process is similar to Open System authentication but includes a challenge and response between the AP and client station.

Shared Key authentication is a four-way authentication frame exchange. The client station sends an authentication request to the access point and then the access point sends a cleartext challenge to the client station in an authentication response. The client station then encrypts the cleartext challenge and sends it back to the access point in the body of another authentication request frame. The access point then decrypts the station's response and compares it to the challenge text. If they match, the access point will respond by sending a fourth and final authentication frame to the station, confirming the success. If they do not match, the access point will respond negatively. If the access point cannot decrypt the challenge, it will also respond negatively. If Shared Key authentication is successful, the same static WEP key that was used during the Shared Key authentication process will also be used to encrypt the 802.11 data frames.

Although it might seem that Shared Key authentication is a more secure solution than Open System authentication, in reality Shared Key could be the bigger security risk. Anyone who captures the cleartext challenge phrase and then captures the encrypted challenge phrase in the response frame could potentially derive the static WEP key. If the static WEP key is compromised, a whole new can of worms has been opened because now all the data frames can be decrypted. Neither of the legacy authentication methods is considered strong enough for enterprise security. The more secure 802.1X/EAP authentication method is discussed in Chapter 13, "802.11 Network Security Architecture."

Association

After the station has authenticated with the access point, the next step is for it to associate with the access point. When a client station associates, it becomes a member of a basic service set (BSS). *Association* means that the client station can send data through the access point and on to the distribution system medium. The client station sends an association request to the access point, seeking permission to join the BSS. The access point sends an association response to the client, either granting or denying permission to join the BSS. In the body of the association response frame is an association identifier (AID), a unique association number given to every associated client. You will learn later in this chapter that the AID is used during power management.

Association occurs after Shared Key or Open System authentication, as seen in Exercise 9.5. After a client station becomes a member of the BSS by completing association, the client will send a DHCP request and begin communications at upper layers when the DHCP response is received.

EXERCISE 9.5

Association

1. From Windows File Manager, browse to the AirMagnet demo capture file located at `C:\Program Files\AirMagnet Inc\AirMagnet Laptop Demo\Demo.amc`. Double-click on the `demo.amc` capture file. The Wi-Fi Analyzer application should appear.

2. In the bottom-left corner of the application, click the Decodes button.

3. From the AirMagnet Wi-Fi Analyzer main window, click on frame 12089, which is an association request. Look at the frame body.

4. Click on frame #12091, which is an association response. Look at the frame body and note that association was successful and that the client received an AID number.

Authentication and Association States

The 802.11 station keeps two variables for tracking the authentication state and the association state. The states that are tracked are as follows:

- Authentication state: unauthenticated or authenticated
- Association state: unassociated or associated

Together, these two variables create three possible states for the stations, as listed here and shown in Figure 9.6:

- State 1, initial start state, unauthenticated and unassociated
- State 2, authenticated and unassociated
- State 3, authenticated and associated

FIGURE 9.6 Authentication and association states

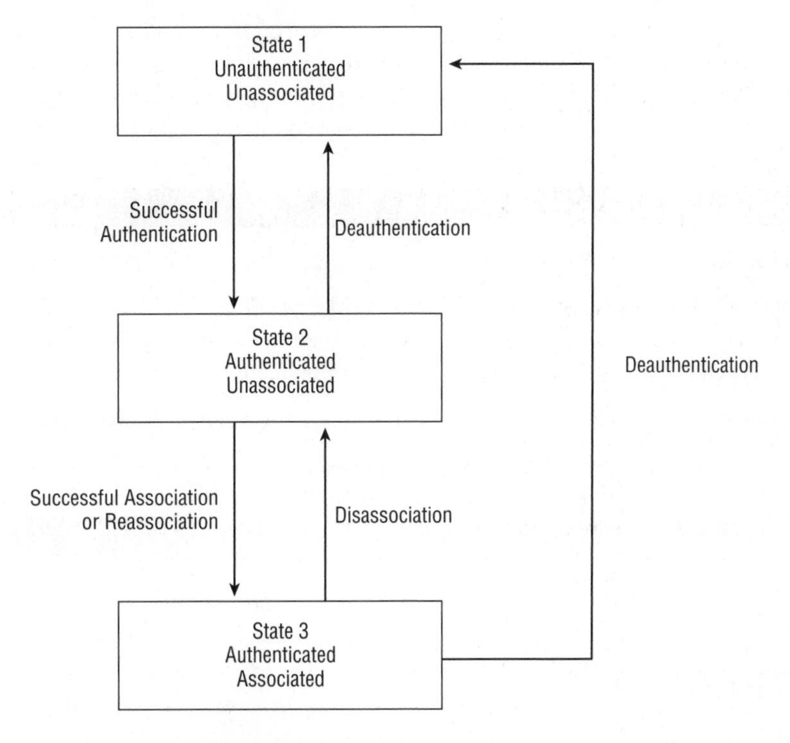

Because a station must authenticate before it can associate, it can never be unauthenticated and associated.

Basic and Supported Rates

As you have learned in earlier chapters, the 802.11-2007 standard defines supported rates for various RF technologies. For example, HR-DSSS (clause 18) radios are capable of supporting data rates of 1, 2, 5.5, and 11 Mbps. ERP (clause 19) radios are also capable of supporting the HR-DSSS data rates, but are also capable of supporting ERP-OFDM rates of 6, 9, 12, 18, 24, 36, 48, and 54 Mbps.

On any autonomous AP or WLAN controller, specific data rates can be configured as *required* rates. The 802.11-2007 standard defines required rates as *basic rates*. In order for a client station to successfully associate with an AP, the station must be capable of communicating by using the configured basic rates that the access point requires. As demonstrated in Exercise 9.6, if the client station is not capable of communicating with all of the basic rates, the client station will not be able to associate with the access point and will not be allowed to join the BSS.

In addition to the basic rates, the access point defines a set of supported rates. This set of supported rates is advertised by the access point in the beacon frame and is also in some of the other management frames. The supported rates are the group data rates that the access point will use when communicating with a station. After a station associates with an access point, it will use one of the advertised supported rates to communicate with the access point.

EXERCISE 9.6

Association and Basic Rates

1. From Windows File Manager, browse to the AirMagnet demo capture file located at `C:\Program Files\AirMagnet Inc\AirMagnet Laptop Demo\Demo.amc`. Double-click on the `demo.amc` capture file. The Wi-Fi Analyzer application should appear.
2. In the bottom-left corner of the application, click the Decodes button.
3. From the AirMagnet Wi-Fi Analyzer main window, click on frame #3147, which is an association response. Look at the frame body and look at the status code. Why was the client station denied association?

Roaming

As wireless LANs grew to multiple access points, the 802.11 standard provided the ability for the client stations to transition from one access point to another while maintaining network connectivity for the upper-layer applications. This ability is known as *roaming*, although the 802.11 standard does not define what roaming is.

The decision to roam is currently made by the client station. What actually causes the client station to roam is a set of proprietary rules determined by the manufacturer of the wireless card, usually determined by the signal strength, noise level, and bit-error rate. As the client station communicates on the network, it continues to look for other access points and will authenticate to those that are within range. Remember, a station can be authenticated to multiple access points but associated to only one AP. As the client station moves away from the access point that it is associated with and the signal drops below a predetermined threshold, the client station will attempt to connect to another access point and roam from its current BSS to a new BSS. As the station roams, the original access point and the new access point should communicate with each other across the distribution system medium and help provide a clean transition between the two. Many manufacturers provide this handoff, but it is not officially part of the 802.11 standard, so each vendor does it using its own method. In WLAN controller-based solutions, the roaming handoff mechanisms occur within the WLAN controller. The roaming handoffs that occur within WLAN controllers are much faster than roaming handoffs between autonomous access points.

Reassociation

When a client station decides to roam to a new access point, it will send a *reassociation* request frame to the new access point. It is called a reassociation not because you are reassociating to the access point but because you are reassociating to the SSID of the wireless network.

Reassociation occurs after the client and the access point have exchanged six frames, as described in the following steps:

1. In the first step, the client station sends a reassociation request frame to the new access point. As shown in Exercise 9.7, the reassociation request frame includes the BSSID (MAC address) of the access point it is currently connected to (we will refer to this as the original AP).
2. The new access point then replies to the station with an ACK.
3. The new access point attempts to communicate with the original AP by using the distribution system medium (DSM). The new access point attempts to notify the original AP about the roaming client and requests that the original AP forward any buffered data. Please remember that any communications between APs via the DSM are not defined by the 802.11-2007 standard and are proprietary. In a controller-based WLAN solution, the inter-access point communications occur within the controller.
4. If this communication is successful, the original access point will use the distribution system medium to forward any buffered data to the new access point.
5. The new access point then sends a reassociation response frame to the client via the wireless network.
6. The client sends an ACK to the new access point. The client does not need to send a disassociation frame to the original access point, because the client assumes that the two access points have communicated with each other across the distribution system medium.

If the reassociation is not successful, the client will retain its connection to the original AP and either continue to communicate with it or attempt to roam to another access point.

EXERCISE 9.7

Reassociation

1. From Windows File Manager, browse to the AirMagnet demo capture file located at `C:\Program Files\AirMagnet Inc\AirMagnet Laptop Demo\Demo.amc`. Double-click on the `demo.amc` capture file. The Wi-Fi Analyzer application should appear.
2. In the bottom-left corner of the application, click the Decodes button.
3. From the AirMagnet Wi-Fi Analyzer main window, click on frame #11659, which is a reassociation request. Look at the frame body and at the current AP field and note the BSSID of the original access point.

Disassociation

Disassociation is a notification, not a request. If a station wants to disassociate from an AP, or an AP wants to disassociate from stations, either device can send a disassociation frame. This is a polite way of terminating the association. A client will do so when you shut down the operating system. An AP will do so if it is being disconnected from the network for maintenance. Disassociation cannot be refused by either party. If the disassociation frame is not heard by the other party, MAC management is designed to accommodate loss of communications.

Deauthentication

Like disassociation, a *deauthentication* frame is a notification and not a request. If a station wants to deauthenticate from an AP, or an AP wants to deauthenticate from stations, either device can send a deauthentication frame. Because authentication is a prerequisite for association, a deauthentication frame will automatically cause a disassociation to occur. Deauthentication cannot be refused by either party.

ACK Frame

The *ACK frame* is one of the six control frames and one of the key components of the 802.11 CSMA/CA media access control method. Since 802.11 is a wireless medium that cannot guarantee successful data transmission, the only way for a station to know that a frame it transmitted was properly received is for the receiving station to notify the transmitting station. This notification is performed using an ACK.

The ACK is a simple frame consisting of 14 octets of information, as depicted in Figure 9.7. When a station receives data, it waits a short period of time known as a *short interframe space (SIFS)*. The receiving station copies the MAC address of the transmitting station from the data frame and places it in the Receiver Address (RA) field of the ACK frame. As seen in Exercise 9.8, the receiving station then replies by transmitting the ACK. If all goes well, the station that sent the data frame receives the ACK with its MAC address in the RA field and now knows that the frame was received and was not corrupted. The ACK frame is the highest-priority frame because of the half-duplex nature of the medium. The delivery of every unicast frame must be verified, or a retransmission must take place. The ACK frame is used for delivery verification.

FIGURE 9.7 ACK control frame

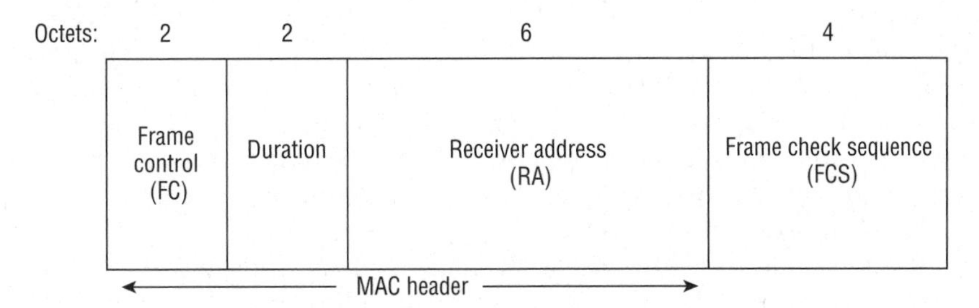

Every unicast frame must be followed by an ACK frame. If for any reason the unicast frame is corrupted, the 32-bit CRC known as the frame check sequence (FCS) will fail and the receiving station will not send an ACK. If a unicast frame is not followed by an ACK, it will be retransmitted. Broadcast and multicast frames do not require acknowledgment.

EXERCISE 9.8

Acknowledgment

1. From Windows File Manager, browse to the AirMagnet demo capture file located at `C:\Program Files\AirMagnet Inc\AirMagnet Laptop Demo\Demo.amc`. Double-click on the `demo.amc` capture file. The Wi-Fi Analyzer application should appear.
2. In the bottom-left corner of the application, click the Decodes button.
3. From the AirMagnet Wi-Fi Analyzer main window, observe the frame exchanges between frame #10683 and frame #10693. Notice that all the unicast frames are being acknowledged by the receiving station.

Fragmentation

The 802.11-2007 standard allows for fragmentation of frames. *Fragmentation* breaks an 802.11 frame into smaller pieces known as fragments, adds header information to each fragment, and transmits each fragment individually. Although the same amount of actual data is being transmitted, each fragment requires its own header, and the transmission of each fragment is followed by a SIFS and an ACK. In a properly functioning 802.11 network, smaller fragments will actually decrease data throughput because of the MAC sublayer overhead of the additional header, SIFS, and ACK of each fragment. On the other hand, if the network is experiencing a large amount of data corruption, lowering the 802.11 fragmentation setting may improve data throughput.

If an 802.11 frame is corrupted and needs to be retransmitted, the entire frame must be sent again. When the 802.11 frame is broken into multiple fragments, each fragment is smaller and transmits for a shorter period of time. If interference occurs, instead of an entire large frame becoming corrupted, it is likely that only one of the small fragments will become corrupted and only this one fragment will need to be retransmitted. Retransmitting the small fragment will take much less time than retransmitting the larger frame. If fragmentation is implemented, retransmission overhead may be reduced.

Figure 9.8 illustrates how smaller fragments reduce retransmission overhead. (Please note that this is a representation and not drawn to scale. Additionally, to simplify the illustration, ACKs were not included.) This illustration shows the transmission and retransmission of a large 1,500-byte frame above and the transmission and retransmission of smaller 500-byte fragments below. If there was no RF interference, only the solid-lined rectangles would need to be transmitted. Because of the additional headers (H) and the time between the fragments for each SIFS and ACK, the smaller fragments would take longer to transmit. However, if RF interference occurred, it would take less time to retransmit the smaller fragment than it would to retransmit the larger frame.

FIGURE 9.8 Frame fragmentation

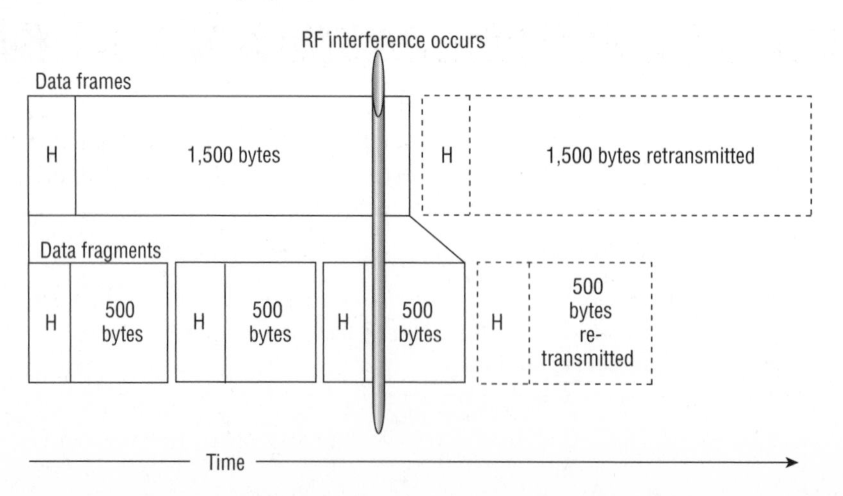

The transmission of a fragment is treated the same way as the transmission of a frame. Therefore, every fragment must participate in the CSMA/CA media access and must be followed by an ACK. If a fragment is not followed by an ACK, it will be retransmitted.

Not all wireless LAN adapters allow you to adjust the fragmentation settings. If you do set your wireless LAN adapter to use a smaller fragment size, you must realize that as you roam between access points and as you move between networks, all of your 802.11 frames will be fragmented using the setting you have configured. This means that if you roam to a location where there is no interference, your station will still be using the smaller frame fragments and will actually perform worse than if you had left the fragmentation value at its largest setting.

Real World Scenario

Will Fragmentation Increase Throughput?

Fragmentation may reduce retransmission overhead in an environment with a lot of data corruption. However, fragmentation always introduces more MAC sublayer overhead to the network. Usually, if fragmentation is used within a BSS, the additional MAC overhead will cause the network throughput to decrease. In some rare cases, the fragmentation threshold settings on an access point may be tweaked to improve throughput by reducing retransmission overhead caused by data corruption. Fragmentation is usually a temporary fix, and the better solution is to find the cause of the data corruption and permanently fix the problem. The many causes of layer 2 retransmissions and how to fix them are discussed in detail in Chapter 12, "WLAN Troubleshooting."

Protection Mechanism

The 802.11-2007 standard mandates support for both DSSS and OFDM technologies for clause 19 ERP radios (802.11g). When clause 18 HR-DSSS (802.11b) client stations need to communicate in a basic service set with an ERP (802.11g) access point and ERP (802.11g) client stations, the 802.11g devices need to provide compatibility for the slower 802.11b devices. ERP access points must also be backward compatible with legacy clause 15 DSSS client stations. This environment is often referred to as mixed-mode. Contrary to what some people believe, the 802.11g devices do not simply switch to 802.11b mode and communicate using 802.11b data rates. In order for 802.11g, 802.11b stations and legacy 802.11 DSSS stations to coexist within the same BSS, the 802.11g devices enable what is referred to as the protection mechanism, also known as 802.11g Protected mode.

In Chapter 5, "IEEE 802.11 Standards," you learned that vendors offer three configuration modes for an 802.11g access point:

802.11b-only mode When an 802.11g AP is running in this operational mode, support for DSSS technology is solely enabled. Effectively, the access point has been configured to be an 802.11b access point. Legacy 802.11 DSSS clients, 802.11b HR-DSSS clients, and 802.11g clients using ERP-DSSS will all be able to communicate with the AP at data rates of 1, 2, 5.5, and 11 Mbps. Aggregate throughput will be the same as achieved in an 802.11b network.

802.11g-only mode APs configured as G-only will communicate with only 802.11g client stations using ERP-OFDM technology. Support for DSSS and HR-DSSS is disabled; therefore, 802.11b HR-DSSS clients and legacy 802.11 DSSS clients will not be able to associate with the access point. Aggregate throughput will be equivalent to what can be achieved in an 802.11a network. For example, the aggregate throughput of an AP with a data rate of 54 Mbps might be about 19 to 20 Mbps. G-only wireless LANs are sometimes referred to as a *Pure* G networks.

802.11b/g mode This is the default operational mode of most 802.11g access points and is often called mixed-mode. Support for both DSSS and OFDM is enabled. Legacy 802.11 DSSS clients and 802.11b HR-DSSS clients will be able to communicate with the AP at data rates of 1, 2, 5.5, and 11 Mbps. The ERP (802.11g) clients will communicate with the AP by using the ERP-OFDM data rates of 6, 9, 12, 18, 24, 36, 48, and 54 Mbps.

You need to understand that these vendor configurations are not part of the 802.11-2007 standard. Although most vendors do indeed support these configurations, the standard mandates support for 802.11b clause 18 devices and 802.11g clause 19 devices within the ERP basic service set.

Real World Scenario

How Can You Make Sure That 802.11g Networks Are Transmitting at 802.11g Speeds?

Even if all of the wireless devices in your company support 802.11g, your WLAN will enable the protection mechanism if it sees even one 802.11b device. This 802.11b device could be a visitor to your company, someone driving past your building with an 802.11b wireless adapter enabled in their laptop, or a nearby business or home that also has a wireless network. If you want your network to always use the higher ERP-OFDM rates, you must configure the access points to support 802.11g clients only. Remember that if you do this, any 802.11b and legacy 802.11 DSSS devices will not be able to connect to your network.

In Chapter 8, you learned that one of the ways of preventing collisions is for the stations to set a countdown timer known as the network allocation vector (NAV). This notification is known as NAV distribution. NAV distribution is done through the Duration/ID field that is part of the data frame. When a data frame is transmitted by a station, the Duration/ID field is used by the listening stations to set their NAV timers. Unfortunately, this is not inherently possible in a mixed-mode environment. If an 802.11g device were to transmit a data frame, 802.11b devices would not be able to interpret the data frame or the Duration/ID value because the 802.11b HR-DSSS devices are not capable of understanding 802.11g ERP-OFDM transmissions. The 802.11b devices would not set their NAV timers and could incorrectly believe that the medium is available. To prevent this from happening, the 802.11g ERP stations switch into what is known as Protected mode.

In a mixed-mode environment, when an 802.11g device wants to transmit data, it will first perform a NAV distribution by transmitting a *request to send/clear to send (RTS/CTS)* or a CTS-to-Self using a data rate and modulation method that the 802.11b HR-DSSS stations can understand. The RTS/CTS or CTS-to-Self will be heard and understood by all of the 802.11b and 802.11g stations. The RTS/CTS or CTS-to-Self will contain a Duration/ID value that will be used by all of the listening stations to set their NAV timers. To put it simply, using a slow transmission that all stations can understand, the ERP (802.11g) device notifies all the stations to reset their NAV values. After the RTS/CTS or CTS-to-Self has been used to reserve the medium, the 802.11g station can transmit a data frame by using OFDM modulation without worrying about collisions with 802.11b HR-DSSS or legacy 802.11 DSSS stations.

Within an ERP basic service set, the HR-DSSS (802.11b) and legacy 802.11 DSSS stations are known as non-ERP stations. The purpose of the protection mechanism is that ERP stations (802.11g) can coexist with non-ERP stations (802.11b and 802.11 legacy) within the same BSS. This allows the ERP stations to use the higher ERP-OFDM data rates to transmit and receive data, yet still maintain backward compatibility with the older legacy non-ERP stations.

Included on the CD of this book is a white paper titled "Protection Ripple in ERP 802.11 WLANs" by Devin Akin. This white paper goes into much greater detail about the protection mechanism and is highly recommended extra reading for preparing for the CWNA exam.

So what exactly triggers the protection mechanism? When an ERP access point decides to enable the use of a protection mechanism, it needs to notify all of the ERP (802.11g) stations in the BSS that protection is required. It accomplishes this by setting the ERP information element in the beacon frame. Every time the access point transmits a beacon, any ERP station that hears the beacon will know that Protected mode is required. Three scenarios can trigger protection in an ERP basic service set:

- As shown in Exercise 9.9, if a non-ERP STA associates to an ERP AP, the ERP AP will enable the NonERP_Present bit in its own beacons, enabling protection mechanisms in its BSS. In other words, an HR-DSSS (802.11b) client association will trigger protection.
- If an ERP AP hears a beacon with an 802.11b or 802.11 DSSS supported rate set from another AP or an IBSS STA, it will enable the NonERP_Present bit in its own beacons, enabling protection mechanisms in its BSS. In simpler terms, if an 802.11g AP hears a

beacon frame from an 802.11b access point or an 802.11b ad hoc client, the protection mechanism will be triggered.

- If an ERP AP hears a beacon from another ERP access point with the NonERP_Present bit set to 1, it also will enable protection mechanisms in its BSS. In other words, an 802.11g access point will enable protection if it hears a beacon from another 802.11g access point that has already enabled the protection mechanism within its own BSS.

EXERCISE 9.9

Protection Mechanism

1. From Windows File Manager, browse to the AirMagnet demo capture file located at `C:\Program Files\AirMagnet Inc\AirMagnet Laptop Demo\Demo.amc`. Double-click on the `demo.amc` capture file. The Wi-Fi Analyzer application should appear.
2. In the bottom-left corner of the application, click the Decodes button.
3. From the AirMagnet Wi-Fi Analyzer main window, click on frame #7945, which is a beacon. Look at the frame body and the ERP information element. Under ERP flags, note that the Use Protection bit is set to 1.
4. Click on frame #7946, which is a CTS-to-Self frame. Under network media information, notice that the data frame was sent using an HR-DSSS rate of 11 Mbps.
5. Click on frame #7947, which is an encrypted data frame. Under network media information, notice that the data frame was sent using an ERP-OFDM rate of 24 Mbps.

Real World Scenario

How Does 802.11b Affect 802.11g Throughput?

A common misconception is that 802.11g radios revert to 802.11b data rates when the protection mechanism is used. In reality, ERP (802.11g) radios still transmit data at the higher ERP-OFDM rates. However, when an HR-DSSS (802.11b) station causes an ERP (802.11g) BSS to enable the protection mechanism, a large amount of RTS/CTS or CTS-to-Self overhead is added prior to every ERP-OFDM data transmission. The aggregate data throughput loss is caused by the extra overhead and not by using slower 802.11b rates. A data rate of 54 Mbps usually will provide about 18–20 Mbps of aggregate throughput when protection is not enabled. After protection is enabled, the overhead will reduce the aggregate data throughput to below 13 Mbps, and possibly as low as 9 Mbps.

RTS/CTS

In order for a client station to participate in a basic service set, it must be able to communicate with the access point. This is straightforward and logical; however, it is possible for the client station to be able to communicate with the access point but not be able to hear or be heard by any of the other client stations. This can be a problem because, as you may recall, a station performs collision avoidance by setting its NAV when it hears another station transmitting (virtual carrier sense) and by listening for RF (physical carrier sense). If a station cannot hear the other stations, or cannot be heard by the other stations, there is a greater likelihood that a collision can occur. *Request to send/clear to send (RTS/CTS)* is a mechanism that performs a NAV distribution and helps prevent collisions from occurring. This NAV distribution reserves the medium prior to the transmission of the data frame.

Now, let's look at the RTS/CTS from a slightly more technical perspective. This will be a basic explanation because an in-depth explanation is beyond the scope of the exam. When RTS/CTS is enabled on a station, every time the station wants to transmit a frame, it must perform an RTS/CTS exchange prior to the normal data transmissions. When the transmitting station goes to transmit data, it first sends an RTS frame. The duration value of the RTS frame resets the NAV timers of all listening stations so that they must wait until the CTS, DATA, and ACK have been transmitted. The receiving station then sends a CTS, which is also used for NAV distribution. The duration value of the CTS frame resets the NAV timer of all listening stations so that they must wait until the DATA and ACK have been transmitted.

As seen in Figure 9.9, the duration value of the RTS frame represents the time, in microseconds, that is required to transmit the CTS/DATA/ACK exchange plus three SIFS intervals. The duration value of the CTS frame represents the time, in microseconds, that is required to transmit the DATA/ACK exchange plus two SIFS intervals. If any station did not hear the RTS, it should hear the CTS. When a station hears either the RTS or the CTS, it will set its NAV to the value provided. At this point, all stations in the basic service set should have their NAV set, and the stations should wait until the entire data exchange is complete. Figure 9.10 depicts an RTS/CTS exchange between a client station and an access point.

RTS/CTS is used primarily in two situations. It can be used when a hidden node exists (this is covered in Chapter 12), or it can be used automatically as a protection mechanism for a mixed-mode environment in an ERP basic service set.

CTS-to-Self

CTS-to-Self is used strictly as a protection mechanism for mixed-mode environments. One of the benefits of using CTS-to-Self over RTS/CTS as a protection mechanism is that the throughput will be higher because fewer frames are being sent.

FIGURE 9.9 RTS/CTS duration values

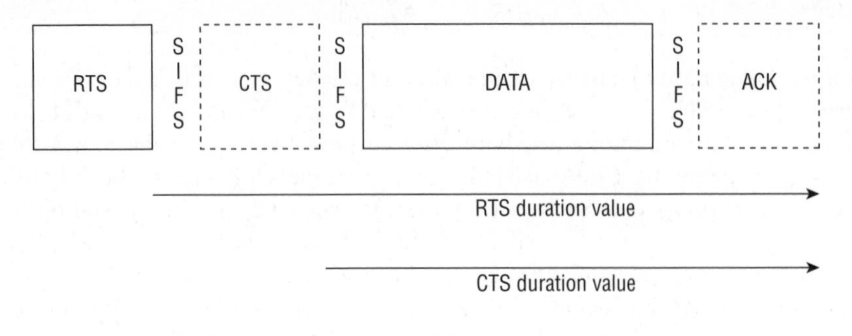

FIGURE 9.10 RTS/CTS frame exchange

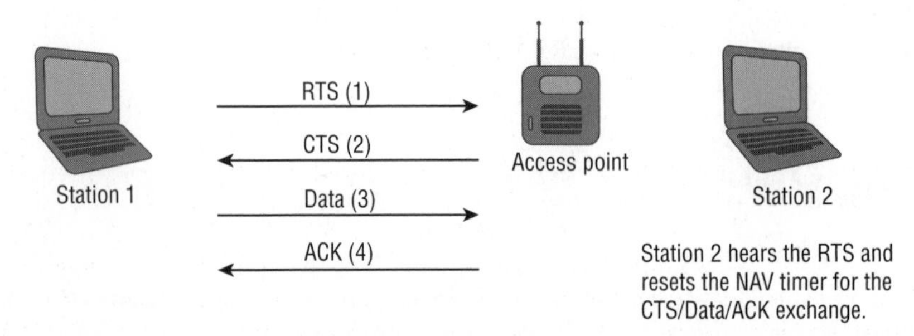

When a station using CTS-to-Self wants to transmit data, it performs a NAV distribution by sending a CTS frame. This CTS notifies all other stations that they must wait until the DATA and ACK have been transmitted. Any station that hears the CTS will set their NAV to the value provided.

Since the CTS-to-Self is used as a protection mechanism for mixed-mode environments, the ERP (802.11g) station will transmit the CTS by using DSSS technology that all stations can understand. Then the DATA and the ACK will be transmitted at a faster 802.11g speed by using (ERP-OFDM).

CTS-to-Self is better suited for use by an access point. It is important to make sure that all stations hear the CTS to reserve the medium, and this is most likely to occur if it is being sent by an access point. If a client station were to use CTS-to-Self, there is a chance that another client station on the opposite side of the BSS might be too far away from the CTS-to-Self and would not realize that the medium is busy.

Data Frames

As mentioned earlier, there are 15 types of data frames. The most common data frame is the *simple data frame*, which has MSDU upper-layer information encapsulated in the frame body. The integration service that resides in autonomous APs and WLAN controllers takes the MSDU payload of a simple data frame and transfers the MSDU into 802.3 Ethernet frames. For data privacy reasons, the MSDU data payload should usually be encrypted.

The *null function frame* is used by client stations to inform the access point of changes in Power Save status by changing the Power Management bit. When a client station decides to go off-channel for active scanning purposes, the client station will send a null function frame to the access point with the *Power Management bit* set to 1. As demonstrated in Exercise 9.10, when the Power Management bit is set to 1, the access point buffers all of that client's 802.11 frames. When the client station returns to the AP's channel, the station sends another null function frame with the Power Management bit set to 0. The AP then transmits the client's buffered frames. Some vendors also use the null function frame to implement proprietary power-management methods.

EXERCISE 9.10

Data Frames

1. From Windows File Manager, browse to the AirMagnet demo capture file located at `C:\Program Files\AirMagnet Inc\AirMagnet Laptop Demo\Demo.amc`. Double-click on the `demo.amc` capture file. The Wi-Fi Analyzer application should appear.

2. In the bottom-left corner of the application, click the Decodes button.

EXERCISE 9.10 *(continued)*

3. From the AirMagnet Wi-Fi Analyzer main window, click on frame #282, which is an unencrypted simple data frame. Look at the frame body and notice the upper-layer information such as IP addresses and UDP ports.
4. Click on frame #13, which is a null function frame. Look at the 802.11 MAC header. Look in the Frame Control field and note that the Power Management bit is set to 1. The AP will now buffer the client's traffic.

Power Management

One of the key uses of wireless networking is to provide mobility for the client station. Client mobility goes hand in hand with battery-operated client stations. When battery-operated devices are used, one of the biggest concerns is how long the battery will last until it needs to be recharged. To increase the battery time, a bigger, longer-lasting battery can be used or power consumption can be decreased. The 802.11 standard includes a power-management feature that can be enabled to help increase battery life. Battery life is extremely important for handheld scanners and VoWiFi phones. The battery life of handheld devices usually needs to last at least one 8-hour work shift. The two legacy power-management modes supported by the 802.11 standard are Active mode and Power Save mode. 802.11 power-management methods have also been enhanced by both the ratified 802.11e amendment and the 802.11n draft amendment.

Active Mode

Active mode is the default power-management mode for most 802.11 stations. When a station is set for Active mode, the wireless station is always ready to transmit or receive data. Active mode is sometimes referred to as *Continuous Aware mode*, and it provides no battery conservation. In the MAC header of an 802.11 frame, the Power Management field is 1 bit in length and is used to indicate the power-management mode of the station. A value of 0 indicates that the station is in Active mode. Stations running in Active mode will achieve higher throughput than stations running in Power Save mode, but the battery life will typically be much shorter.

Stations that are always connected to a power source should be configured to use Active mode.

Power Save Mode

Power Save mode is an optional mode for 802.11 stations. When a client station is set for Power Save mode, it will shut down some of the transceiver components for a period of time to conserve power. The wireless card basically takes a short nap. The station indicates that it is using Power Save mode by changing the value of the Power Management bit to 1. When the Power Management bit is set to 1, the access point is informed that the client station is using power management, and the access point buffers all of that client's 802.11 frames. Power Save mode functions differently when the station is part of an infrastructure network or an ad hoc network, which is covered later in this chapter.

Traffic Indication Map (TIM)

If a station is part of a basic service set, it will notify the access point that it is enabling Power Save mode by changing the Power Management field to 1. When the access point receives a frame from a station with this bit set to 1, the access point knows that the station is in Power Save mode. If the access point then receives any data that is destined for the station in Power Save mode, the AP will store the information in a buffer. Anytime a station associates to an access point, the station receives an *association identifier (AID)*. The access point uses this AID to keep track of the stations that are associated and the members of the BSS. If the access point is buffering data for a station in Power Save mode, when the access point transmits its next beacon, the AID of the station will be seen in a field of the beacon frame known as the *traffic indication map (TIM)*. As seen in Exercise 9.11, the TIM field is a list of all stations that have undelivered data buffered on the access point waiting to be delivered. Every beacon will include the AID of the station until the data is delivered.

After the station notifies the access point that it is in Power Save mode, the station shuts down part of its transceiver to conserve energy. A station can be in one of two states, either awake or doze. During the awake state, the client station can receive frames and transmit frames. During the doze state, the client station cannot receive or transmit any frames and operates in a very low power state to conserve power. Because beacons are transmitted at a consistent predetermined interval known as the *target beacon transmission time (TBTT)*, all stations know when beacons will occur. The station will remain asleep for a short period of time and awaken in time to hear a beacon frame. The station does not have to awaken for every beacon. To conserve more power, the station can sleep for a longer period of time and then awaken in time to hear an upcoming beacon. How often the client station awakens is based on a variable called the *listen interval* and is usually vendor specific.

When the station receives the beacon, it checks to see whether its AID is set in the TIM, indicating that a buffered unicast frame waits. If so, the station will remain awake and will send a PS-Poll frame to the access point. When the access point receives the PS-Poll frame, it will send the buffered unicast frame to the station. The station will stay awake while the access point transmits the buffered unicast frame. When the access point sends the data to the station, the station needs to know when all of the buffered unicast data has been received so that it can go back to sleep. Each unicast frame contains a 1-bit field called the More Data field. When the station receives a buffered unicast frame with the More Data field set to 1, the station knows that it cannot go back to sleep yet because there is some more buffered data that it has not yet received. When the More Data field is set to 1, the station knows that it needs to send another PS-Poll frame and wait to receive the next buffered unicast frame.

After all of the buffered unicast frames have been sent, the More Data field in the last buffered frame will be set to 0, indicating that there is currently no more buffered data, and the station will go back to sleep. The access point will remove the station's AID from the TIM, and when the next TBTT arrives, the access point will send a beacon. The station will remain asleep for a short period of time and again awaken in time to hear a beacon frame. When the station receives the beacon, it will again check to see whether its AID is set in the TIM. Assuming that there are no buffered unicast frames awaiting this station, the station's AID will not be set in the TIM and the station can simply go back to sleep until it is time to wake up and check again.

EXERCISE 9.11

Legacy Power Save

1. From Windows File Manager, browse to the AirMagnet demo capture file located at `C:\Program Files\AirMagnet Inc\AirMagnet Laptop Demo\Demo.amc`. Double-click on the `demo.amc` capture file. The Wi-Fi Analyzer application should appear.
2. In the bottom-left corner of the application, click the Decodes button.
3. From the AirMagnet Wi-Fi Analyzer main window, click on frame #7896, which is a beacon. Look at the frame body and the TIM field and note that the AID #30 has a buffered unicast frame.
4. Click on frame #7897, which is a PS-Poll. Look at the 802.11 MAC header and note that the client is identifying itself as AID #30.
5. Note that the AP sent the buffered unicast frame #7900 to the client station.

Delivery Traffic Indication Message (DTIM)

In addition to unicast traffic, network traffic includes multicast and broadcast traffic. Because multicast and broadcast traffic is directed to all stations, the BSS needs to provide a way to make sure that all stations are awake to receive these frames. A *delivery traffic indication message (DTIM)* is used to ensure that all stations using power management are awake when multicast or broadcast traffic is sent. DTIM is a special type of TIM. A TIM is transmitted as part of every beacon.

A configurable setting on the access point called the *DTIM interval* determines how often a DTIM is transmitted as part of the beacon. A DTIM interval of 3 means that every third beacon is a DTIM beacon, whereas a DTIM interval of 1 means that every beacon is a DTIM beacon. Every beacon contains DTIM information that informs the stations when the next DTIM will occur. All stations will wake up in time to receive the beacon with the DTIM. If the access point has multicast or broadcast traffic to be sent, it will transmit the beacon with the DTIM and then immediately send the multicast or broadcast data.

After the multicast or broadcast data is transmitted, if a station's AID was in the DTIM, the station will remain awake and will send a PS-Poll frame and proceed with retrieving its buffered unicast traffic from the access point. If a station did not see its AID in the DTIM, the station can go back to sleep.

The DTIM interval is important for any application that uses multicasting. For example, many VoWiFi vendors support *push-to-talk* capabilities that send VoIP traffic to a multicast address. A misconfigured DTIM interval would cause performance issues during a push-to-talk multicast.

Announcement Traffic Indication Message (ATIM)

If a station is part of an independent basic service set (IBSS), there is no central access point to buffer data while the stations are in Power Save mode. A station will notify the other stations that it is enabling Power Save mode by changing the Power Management field to 1. When the station transmits a frame with this field set to 1, the other stations know to buffer any data that they may have for this station because this station is now in Power Save mode.

Periodically, all stations must wake up and notify each other if any station has buffered data that needs to be delivered to another station. This recurring period of time when all devices must be awake to exchange this information is known as the *announcement traffic indication message (ATIM) window*. During the ATIM window, if a station has buffered data for another station, it will send a unicast frame known as an *ATIM frame* to the other station. This unicast frame informs the station that it must stay awake until the next ATIM window so that it can receive the buffered data. Any station that either has buffered data for another station or has received an ATIM will stay awake so that the buffered data can be exchanged. All of the other stations can go to sleep and wait until the next ATIM window to go through this process again.

When the ATIM window expires, the nodes that have stayed awake go through the usual CSMA/CA process to exchange the unsent data. If a station is unable to transmit the data during this time, it will simply send another ATIM frame during the next ATIM window and then attempt to send the data during the following CSMA/CA period.

Do not confuse the ATIM frame with the TIM field. The ATIM is a frame used for power management by ad hoc clients not communicating through an access point. The TIM is a field in the beacon frame that tells client stations in Power Management mode that the AP has buffered unicast frames for the clients.

WMM Power Save (WMM-PS) and U-APSD

The main focus of the 802.11e amendment, which is now part of the 802.11-2007 standard, is quality of service. However, the IEEE 802.11e amendment also introduced an enhanced power management method called *automatic power save delivery (APSD)*. The two APSD methods that are defined are *scheduled automatic power save delivery (S-APSD)* and *unscheduled automatic power save delivery (U-APSD)*. The S-APSD power management method is beyond the scope of this book. The Wi-Fi Alliance's *WMM Power Save (WMM-PS)* certification is based on U-APSD. WMM-PS is an enhancement over the legacy power saving mechanisms already discussed. The goal of WMM-PS is to have client devices spend more time in a doze state and consume less power. WMM-PS also is designed to minimize latency for time-sensitive applications such as voice during the power-management process.

The legacy power-management methods have several limitations. As pictured in Figure 9.11, a client using legacy power management must first wait for a beacon with a TIM before the client can request buffered unicast frames. The client must also send a unique PS-Poll frame to the AP to request every single buffered unicast frame. This ping-pong power-management method increases the latency of time-sensitive applications such as voice. The clients must also stay awake during the ping-pong process, which results in reduced battery life. In addition, the amount of time that the clients spend dozing is determined by the vendor's driver and not by the application traffic.

FIGURE 9.11 Legacy power management

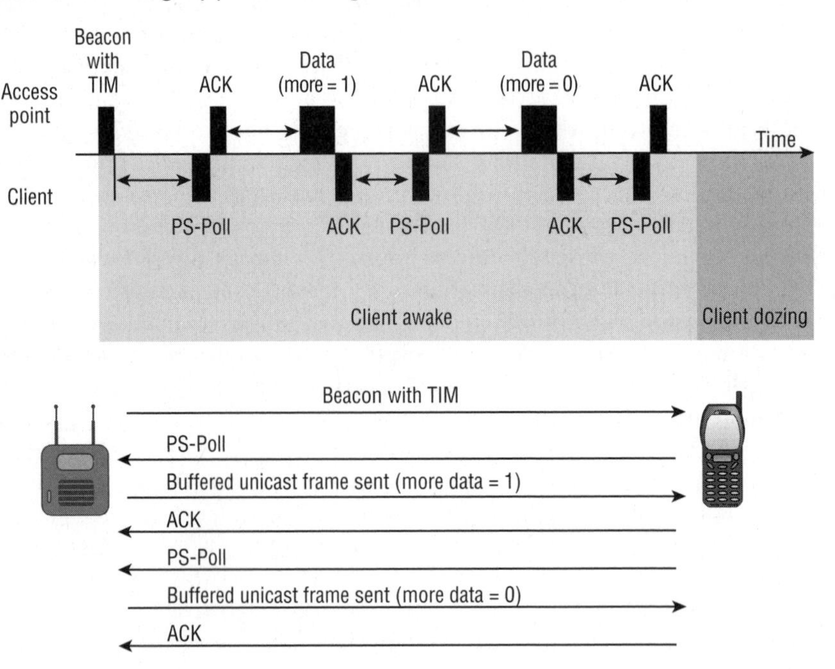

WMM-PS uses a trigger mechanism to received buffered unicast traffic based on WMM access categories. You learned in Chapter 8 that 802.1D priority tags from the Ethernet side are used to direct traffic to four different WMM access-category priority queues. The access category queues are voice, video, best effort, and background. As depicted in Figure 9.12, the client station sends a trigger frame related to a WMM access category to inform the AP that the client is awake and ready to download any frames that the access point may have buffered for that access category. The trigger frame can also be an 802.11 data frame, thus eliminating the need for a separate PS-Poll frame. The AP will then send an ACK to the client and proceed to send a "frame burst" of buffered application traffic during a transmit opportunity (TXOP).

FIGURE 9.12 WMM-PS

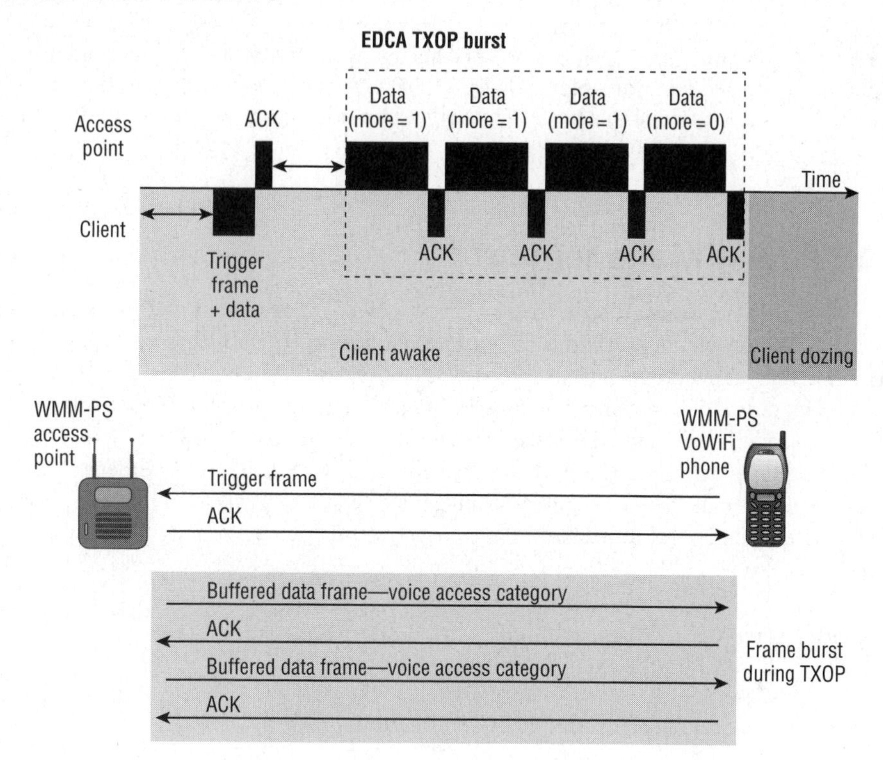

The advantages of this enhanced power-management method include the following:

- Applications now control the power-save behavior by setting doze periods and sending trigger frames. VoWiFi phones will obviously send triggers to the access point frequently during voice calls, while a laptop radio using a data application will have a longer doze period.
- The trigger and delivery method eliminates the need for PS-Poll frames.
- The client can request to download buffered traffic and does not have to wait for a beacon frame.
- All the downlink application traffic is sent in a faster frame burst during the AP's TXOP.

Three conditions have to be met for a Wi-Fi client to use the enhanced WMM-PS mechanisms:

- The client is Wi-Fi CERTIFIED for WMM-PS.
- The access point is Wi-Fi CERTIFIED for WMM-PS.
- Latency-sensitive applications must support WMM-PS.

It should be noted that applications that do not support WMM-PS can still coexist with WMM Power Save–enabled applications. The data from the other applications will be delivered with legacy power-save methods.

A white paper from the Wi-Fi Alliance called "WMM™ Power Save for Mobile and Portable Wi-Fi CERTIFIED Devices" is included on the CD that accompanies this book. This white paper is highly recommended extra reading for the CWNA exam.

802.11n Power Management

The 802.11n draft amendment also defines two new power-management methods. The first new power-management method is called *spatial multiplexing power save (SM power save)*. The purpose of SM power save is to enable a MIMO 802.11n device to power down all but one of its radios. The second new power-management method, *power save multi-poll (PSMP)*, has also been defined for use for HT clause 20 radios. PSMP is an extension of automatic power save delivery (APSD) that was defined by the 802.11e amendment. A more detailed discussion about these two power-management methods is covered in Chapter 18, "High Throughput (HT) and 802.11n."

Summary

This chapter covered key areas of the MAC architecture:

- 802.11 frame format
- Major 802.11 frame types
- 802.11 frame subtypes
- Fragmentation
- ERP protection mechanism
- Power management

It is important to understand the makeup of the three major 802.11 frame types and the purpose of each individual 802.11 frame and how they are used in scanning, authentication, association, and other MAC processes. It is important to understand the need for an ERP protection mechanism. Without one, mixed-mode networks would not be able to function. Both RTS/CTS and CTS-to-Self provide ERP (802.11g) protection mechanisms.

To help manage battery life, power management can be configured on a wireless station. Active mode provides no battery conservation of any kind, while Power Save mode can be invaluable for increasing the battery life of laptop and handheld computing devices. WMM and 802.11n have also enhanced power-management capabilities. We discussed the following power-management pieces in this chapter:

- Traffic indication map (TIM)
- Delivery traffic indication message (DTIM)
- Announcement traffic indication message (ATIM)
- WMM Power Save (WMM-PS)

Exam Essentials

Explain the differences between a PPDU, PSDU, MPDU, and MSDU. Understand at which layer of the OSI model each data unit operates and what each data unit comprises.

Understand the similarities and differences of 802.11 frames and 802.3 frames. The IEEE created both of these frame types. 802.11 and 802.3 frames share similar and different properties. Know how they compare to each other.

Know the three major 802.11 frame types. Make sure you know the function of the management, control, and data frames. Know what makes the major frame types different. Data frames contain an MSDU, whereas management and control frames do not. Understand the purpose of each individual frame subtype.

Know the media access control (MAC) process and all of the frames that are used during this process. Understand the function of each of the following: active scanning, passive scanning, beacon, probe request, probe response, authentication, association, reassociation, disassociation, deauthentication.

Know the importance of the ACK frame for determining that a unicast frame was received and uncorrupted. Understand that after a unicast frame is transmitted, there is a short interframe space (SIFS) and then the receiving station replies by transmitting an ACK. If this process is completed successfully, the transmitting station knows the frame was received and was not corrupted.

Know the benefits and detriments of fragmentation. By default, fragmentation adds overhead, and fragmented frames are inherently slower than unfragmented frames. If RF interference exists, fragmentation can reduce the amount of retransmitted overhead, thus actually increasing the data throughput. If fragmentation does increase throughput, this is a clear indication of a transmission problem such as multipath.

Understand the importance of ERP protection mechanisms and how they function. Protected mode allows ERP (802.11g), HR-DSSS (802.11b), and legacy DSSS devices to coexist within the same BSS. Protected mode can be provided by RTS/CTS or CTS-to-Self. CTS-to-Self is strictly a protection mechanism, but RTS/CTS can also be manually configured and used to identify or prevent hidden nodes.

Understand all of the technologies that make up power management. Power management can be enabled to decrease power usage and increase battery life. Understand how buffered unicast traffic is received in a different way than buffered broadcast and multicast traffic. Understand the power-management enhancements defined by WMM-PS.

Key Terms

Before you take the exam, be certain you are familiar with the following terms:

ACK frame
Active mode
active scanning
announcement traffic indication message (ATIM) window
association
association identifier (AID)
authentication
automatic power save delivery (APSD)
basic rates
beacon management frame
control frames
CTS-to-Self
data frames
deauthentication
delivery traffic indication message (DTIM)
directed probe request
disassociation
frame
frame check sequence (FCS)
information elements
information fields
integration service (IS)
Logical Link Control (LLC)
MAC Protocol Data Unit (MPDU)
MAC Service Data Unit (MSDU)
management frames
Management MAC Protocol Data Unit (MMPDU)
maximum transmission unit (MTU)
mixed-mode
null function frame
null probe request
Open System authentication
packet
passive scanning
Physical Layer Convergence Procedure (PLCP) sublayer
Physical Medium Dependent (PMD) sublayer
PLCP Protocol Data Unit (PPDU)
PLCP Service Data Unit (PSDU)
Power Management bit
Power Save mode
power save multi-poll (PSMP)
probe request
probe response
reassociation
request to send/clear to send (RTS/CTS)
roaming
scheduled automatic power save delivery (S-APSD)
Shared Key authentication
short interframe space (SIFS)
simple data frame
spatial multiplexing power save (SM power save)
traffic indication map (TIM)
unscheduled automatic power save delivery (U-APSD)
WMM Power Save (WMM-PS)

Review Questions

1. What is the difference between association frames and reassociation frames? (Choose all that apply.)
   - **A.** Association frames are management frames, whereas reassociation frames are control frames.
   - **B.** Reassociation frames are only used in fast BSS transition.
   - **C.** Association frames are used exclusively for roaming.
   - **D.** Reassociation frames contain the BSSID of the original AP.
   - **E.** Only association frames are used to join a BSS.
2. Which of the following contains the information found inside an IP packet?
   - **A.** MPDU
   - **B.** PPDU
   - **C.** PSDU
   - **D.** MSDU
   - **E.** MMPDU
3. Which of the following are protection mechanisms? (Choose all that apply.)
   - **A.** NAV back-off
   - **B.** RTS/CTS
   - **C.** RTS-to-Self
   - **D.** CTS-to-Self
   - **E.** WEP encryption
4. The presence of what type of transmissions can trigger the protection mechanism within an ERP basic service set? (Choose all that apply.)
   - **A.** Association of an HR-DSSS client
   - **B.** Association of an ERP-OFDM client
   - **C.** HR-DSSS beacon frame
   - **D.** ERP beacon frame with the NonERP_Present bit set to 1
   - **E.** Association of an FHSS client
5. Which of the following information is included in a probe response frame? (Choose all that apply.)
   - **A.** Time stamp
   - **B.** Supported data rates
   - **C.** Service set capabilities
   - **D.** SSID
   - **E.** Traffic indication map

6. Which of the following are true about beacon management frames? (Choose all that apply.)
   - **A.** Beacons can be disabled to hide the network from intruders.
   - **B.** Time-stamp information is used by the clients to synchronize their clocks.
   - **C.** In a BSS, clients share the responsibility of transmitting the beacons.
   - **D.** Beacons can contain vendor-proprietary information.

7. After a station sees its AID in the TIM, what typically is the next frame that the station transmits?
   - **A.** Data
   - **B.** PS-Poll
   - **C.** ATIM
   - **D.** ACK

8. When a station sends an RTS, the Duration/ID field notifies the other stations that they must set their NAV timers to which of the following values?
   - **A.** 213 microseconds
   - **B.** The time necessary to transmit the DATA and ACK frames
   - **C.** The time necessary to transmit the CTS frame
   - **D.** The time necessary to transmit the CTS, DATA, and ACK frames

9. How does a client station indicate that it is using Power Save mode?
   - **A.** It transmits a frame to the access point with the Sleep field set to 1.
   - **B.** It transmits a frame to the access point with the Power Management field set to 1.
   - **C.** Using DTIM, the access point determines when the client station uses Power Save mode.
   - **D.** It doesn't need to, because Power Save mode is the default.

10. What would cause an 802.11 station to retransmit a unicast frame? (Choose all that apply.)
    - **A.** The transmitted unicast frame was corrupted.
    - **B.** The ACK frame from the receiver was corrupted.
    - **C.** The receiver's buffer was full.
    - **D.** The transmitting station will attempt to retransmit the data frame.
    - **E.** The transmitting station will send a retransmit notification.

11. If a station is in Power Save mode, how does it know that the AP has buffered unicast frames waiting for it?
    - **A.** By examining the PS-Poll frame
    - **B.** By examining the TIM field
    - **C.** When it receives an ATIM
    - **D.** When the Power Management bit is set to 1
    - **E.** DTIM interval

12. When is an ERP (802.11g) access point required by the IEEE 802.11-2007 standard to respond to probe request frames from nearby HR-DSSS (802.11b) stations? (Choose all that apply.)
    A. When the probe request frames contain a null SSID value
    B. When the access point supports only ERP-OFDM data rates
    C. When the access point supports only HR/DSSS data rates
    D. When the Power Management bit is set to 1
    E. When the probe request frames contain the correct SSID value

13. Which of the following are true about scanning? (Choose all that apply.)
    A. There are two types of scanning: passive and active.
    B. Stations must transmit probe requests in order to learn about local access points.
    C. The 802.11 standard allows access points to ignore probe requests for security reasons.
    D. It is common for stations to continue to send probe requests after being associated to an access point.

14. Which of the following 802.11 frames carry an MSDU payload that may eventually be transferred by the integration service into an 802.3 Ethernet frame? (Choose all that apply.)
    A. 802.11 management frames
    B. 802.11 control frames
    C. 802.11 data frames
    D. 802.11 power-management frames
    E. 802.11 action frames

15. When a client station is first powered on, what is the order of frames generated by the client station and access point?
    A. Probe request, probe response, association request/response, authentication request/response
    B. Probe request, probe response, authentication request/response, association request/response
    C. Association request/response, authentication request/response, probe request, probe response
    D. Authentication request/response, association request/response, probe request, probe response

16. WLAN users have recently complained about gaps in audio and problems with the push-to-talk capabilities with the ACME Company's VoWiFi phones. What could be the cause of this problem?
    A. Misconfigured TIM setting
    B. Misconfigured DTIM setting
    C. Misconfigured ATIM setting
    D. Misconfigured BTIM setting

17. The WLAN help desk gets a call that all of the sudden, all of the HR-DSSS (802.11b) VoWiFi phones cannot connect to any of the ERP (802.11g) lightweight access points that are managed by a multiple-channel- architecture WLAN controller. All the laptops with ERP (802.11g) radios can still however connect. What are the possible causes of this problem? (Choose all that apply.)

A. The WLAN admin disabled the 1, 2, 5.5, and 11 Mbps data rates on the controller.

B. The WLAN admin disabled the 6 and 9 Mbps data rates on the controller.

C. The WLAN admin enabled the 6 and 9 Mbps data rates on the controller as basic rates.

D. The WLAN admin configured all the APs on channel 6.

18. In a multiple-channel architecture, roaming is controlled by the client station and occurs based on a set of proprietary rules determined by the manufacturer of the wireless radio. Which of the following parameters are often used when making the decision to roam? (Choose all that apply.)

A. Received signal level

B. Distance

C. SNR

D. WMM access categories

19. What are some of the advantages of using U-APSD and WMM-PS power management over legacy power-management methods? (Choose all that apply.)

A. Applications control doze time and trigger frames.

B. U-APSD access points transmit all voice and video data immediately.

C. The client does not have to wait for a beacon.

D. Downlink traffic is sent in a frame burst.

E. Data frames are used as trigger frames. PS-Poll frames are not used.

20. WMM-PS is based on which 802.11-2007 power-management method?

A. S-APSD

B. U-APSD

C. PSMP

D. SM Power Save

E. PS-Poll

Answers to Review Questions

1. B, D. Both frames are used to join a BSS. Reassociation frames are used during the roaming process. The reassociation frame contains an additional field called the Current AP Address. This address is the BSSID of the original AP that the client is leaving.

2. D. An IP packet comprises layer 3–7 information. The MAC Service Data Unit (MSDU) contains data from the LLC sublayer and/or any number of layers above the Data-Link layer. The MSDU is the payload found inside the body of 802.11 data frames.

3. B, D. RTS/CTS and CTS-to-Self provide 802.11g protection mechanisms, sometimes referred to as mixed-mode support. NAV back-off and RTS-to-Self do not exist. WEP encryption provides data security.

4. A, C, D. An ERP access point signals for the use of the protection mechanism in the ERP information element in the beacon frame. Three scenarios can trigger protection in an ERP basic service set. If a non-ERP STA associates to an ERP AP, the ERP AP will enable the NonERP_Present bit in its own beacons, enabling protection mechanisms in its BSS. In other words, an HR-DSSS (802.11b) client association will trigger protection. If an ERP AP hears a beacon with an 802.11b or 802.11 supported rate set from another AP or an IBSS STA, it will enable the NonERP_Present bit in its own beacons, enabling protection mechanisms in its BSS. If an ERP AP hears a beacon from another ERP access point with the NonERP_Present bit set to 1, it also will enable protection mechanisms in its BSS.

5. A, B, C, D. The probe response contains the same information as the beacon frame, with the exception of the traffic indication map.

6. B, D. Beacons cannot be disabled. Clients use the time-stamp information from the beacon to synchronize with the other stations on the wireless network. Only access points send beacons in a BSS; clients stations send beacons in an IBSS. Beacons can contain proprietary information

7. B. If a station finds its AID in the TIM, there is unicast data on the access point that the station needs to stay awake for and request to have downloaded. This request is performed by a PS-Poll frame.

8. D. When the RTS frame is sent, the value of the Duration/ID field is equal to the time necessary for the CTS, DATA, and ACK frames to be transmitted.

9. B. When the client station transmits a frame with the Power Management field set to 1, it is enabling Power Save mode. The DTIM does not enable Power Save mode; it only notifies clients to stay awake in preparation for a multicast or broadcast.

10. A, B. The receiving station may have received the data, but the returning ACK frame may have become corrupted and the original unicast frame will have to be retransmitted. If the unicast frame becomes corrupted for any reason, the receiving station will not send an ACK.

11. B. The PS-Poll frame is used by the station to request cached data. The ATIM is used to notify stations in an IBSS of cached data. The Power Management bit is used by the station to notify the AP that the station is going into Power Save mode. The DTIM is used to indicate to client stations how often to wake up to receive buffered broadcast and multicast frames. The traffic indication map (TIM) is a field in the beacon frame used by the AP to indicate that there are buffered unicast frames for clients in Power Save mode.

12. A, E. All 802.11 access points are required to respond to directed probe request frames that contain the correct SSID value. The access point must also respond to null probe request frames that contain a blank SSID value. Some vendors offer the capability to respond to null probe requests with a null probe response.

13. A, D. There are two types of scanning: passive, which occurs when a station listens to the beacons to discover an AP, and active, which occurs when a station sends probe requests looking for access points. Stations send probe requests only if they are performing an active scan. After a station is associated, it is common for the station to continue to learn about nearby access points. All client stations maintain a "known AP" list that is constantly updated by active scanning.

14. C. Only 802.11 data frames can carry an upper-layer payload (MSDU) within the body of the frame. The MSDU can be as large as 2,304 bytes and usually should be encrypted. 802.11 control frames do not have a body. 802.11 management frames have a body; however, the payload is strictly layer 2 information.

15. B. When the client first attempts to connect to an access point, it will first send a probe request and listen for a probe response. After it receives a probe response, it will attempt to authenticate to the access point and then associate to the network.

16. B. The delivery traffic indication message (DTIM) is used to ensure that all stations using power management are awake when multicast or broadcast traffic is sent. The DTIM interval is important for any application that uses multicasting. For example, many VoWiFi vendors support push-to-talk capabilities that send VoIP traffic to a multicast address. A misconfigured DTIM interval would cause performance issues during a push-to-talk multicast.

17. A, C. An ERP (802.11g) AP is backward compatible with HR-DSSS and supports the data rates of 1, 2, 5.5, and 11 Mbps as well as the ERP-OFDM data rates of 6, 9, 12, 18, 24, 36, 48, and 54 Mbps. If a WLAN admin disabled the 1, 2, 5.5, and 11 Mbps data rates, backward compatibility will effectively be disabled and the HR-DSSS clients will not be able to connect. The 802.11-2007 standard defines the use of *basic rates*, which are required rates. If a client station does not support any of the basic rates used by an AP, the client station will be denied association to the BSS. If a WLAN admin configured the ERP-OFDM data rates of 6 and 9 Mbps as basic rates, the HR-DSSS clients would be denied association because they do not support those rates.

18. A, C. The amplitude of the received signals from the access points is usually the main variable when clients make a roaming decision. Client roaming mechanisms are often based on RSSI values including received signal levels and signal-to-noise ratio (SNR). Distance and WMM access categories have nothing to do with the client's decision to roam to a new access point.

19. A, C, D, E. Applications now control the power-save behavior by setting doze periods and sending trigger frames. Clients using time-sensitive applications will send triggers to the access point frequently, while clients using more latency-tolerant applications will have a longer doze period. The trigger and delivery method eliminates the need for PS-Poll frames. The client can request to download buffered traffic and does not have to wait for a beacon frame. All the downlink application traffic is sent in a faster frame burst during the AP's TXOP.

20. B. The IEEE 802.11-2007 standard defines an enhanced power-management method called automatic power save delivery (APSD). The two APSD methods that are defined are scheduled automatic power save delivery (S-APSD) and unscheduled automatic power save delivery (U-APSD). The Wi-Fi Alliance's WMM Power Save (WMM-PS) certification is based on U-APSD.

Wireless Devices

IN THIS CHAPTER, YOU WILL LEARN ABOUT THE FOLLOWING:

✓ **Wireless LAN client devices**

- Radio card formats
- Radio card chipsets
- Client utilities

✓ **Progression of WLAN architecture**

- Autonomous access point—Intelligent edge architecture
- Wireless network management system (WNMS)
- Centralized WLAN architecture
- Lightweight access point
- WLAN controller
- Split MAC
- Remote office WLAN controller
- Distributed WLAN architecture
- Distributed WLAN hybrid
- Unified WLAN architecture

✓ **Specialty WLAN infrastructure**

- Wireless workgroup bridge
- Wireless LAN bridges
- Enterprise wireless gateway
- Residential wireless gateway
- VPN wireless router
- Wireless LAN mesh access points
- Enterprise encryption gateway
- WLAN array
- Cooperative control
- Virtual AP system
- Real-time location systems
- VoWiFi

In Chapter 7, "Wireless LAN Topologies," we discussed the various 802.11 WLAN topologies. You learned that both client and access point stations can be arranged in service sets to provide some sort of access to another medium. In this chapter, we discuss the multiple devices that can be used in both standard and nonstandard 802.11 topologies. Many choices exist for client station radio cards that can be used in desktops, laptops, PDAs, and so on. We also discuss the progression of WLAN infrastructure devices over the years. We cover the purpose of many WLAN specialty devices that exist in today's Wi-Fi marketplace. We explain how special solutions such as real-time location system (RTLS) and VoWiFi can be integrated with a WLAN. Finally we discuss other nontraditional WLAN solutions such as WLAN arrays, virtual APs, and cooperative control access points.

Wireless LAN Client Devices

The main hardware in a WLAN client adapter is a half-duplex radio transceiver, which can exist in many hardware formats and chipsets. All client adapters require a special driver to interface with the operating system, and software utilities to interface with the end user. Many cards can work with Windows, Linux, and Macintosh, though they require a different driver and client software for each operating system. The drivers for many manufacturers' cards may already be included in the operating system, but often newer cards require or can benefit from an updated driver installation. Most vendors will provide a CD with an automated driver installation wizard; however, some may require that the driver be installed manually in the operating system. Corrupted drivers are a common cause of a malfunctioning WLAN radio. A simple reinstallation of the 802.11 radio drivers usually fixes this problem.

With a software interface, the end user can configure a card to participate in a WLAN by using configuration settings that pertain to identification, security, and performance. These client utilities may be the manufacturer's own software utility or an internal software interface built into the operating system.

In the following sections, we discuss the various radio card formats, the chipsets that are used, and software client utilities.

Radio Card Formats

Radio cards are used in both client adapters and access points. This section focuses mainly on how radio cards can be used as client devices. 802.11 radios are manufactured in many *form factors*, meaning the radio card comes in different shapes and sizes.

For many years, the only option you had when purchasing an 802.11 client adapter was a standard *PCMCIA*-type adapter. A PCMCIA adapter, also known as a *PC Card*, is pictured in Figure 10.1. The PCMCIA radio card can be used in any laptop or handheld device that has a PC Card slot. Most PCMCIA cards have integrated antennas. Some cards have only internal integrated antennas, whereas others have both integrated antennas and external connectors.

FIGURE 10.1 PCMCIA adapter/PC card

The radio format that is becoming the most widely used is the *Mini PCI*. The Mini PCI is a variation of the Peripheral Component Interconnect (PCI) bus technology and was designed for use mainly in laptops. A Mini PCI radio is often used inside access points and is also the main type of radio used by manufacturers as the internal 802.11 wireless adapter inside laptops. It is almost impossible to buy a brand new laptop today that does not have an internal 802.11 Mini PCI radio card, as pictured in Figure 10.2. The Mini PCI card typically is installed from the bottom of the laptop and is connected to small diversity antennas that are mounted along the edges of the laptop's monitor.

Figure 10.3 shows another 802.11 radio card form factor that is gaining in popularity: the new *ExpressCard* format. ExpressCard is a hardware standard that is replacing PCMCIA cards. Many laptop manufacturers have replaced PCMCIA slots with ExpressCard slots. Although Mini PCI radios are removable from most laptops, there is no guarantee that an 802.11 Mini PCI radio card will work in another vendor's laptop. One advantage of using either WLAN ExpressCards or WLAN PCMCIA cards is that they can be moved and used in different laptops. Also, WLAN engineers usually use either a PCMCIA or ExpressCard 802.11 radio when running protocol analyzer software and/or site survey software applications.

FIGURE 10.2 Mini PCI radio

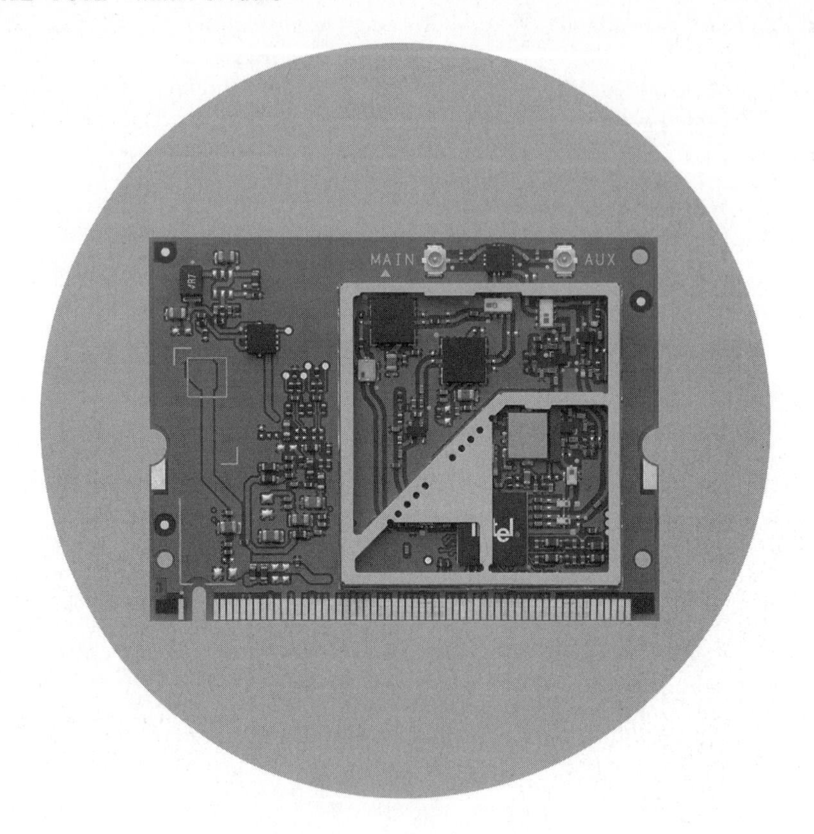

FIGURE 10.3 ExpressCard radio

COURTESY OF BELKIN.

Secure Digital (SD) and *CompactFlash (CF)* are two radio card formats that were often used with a handheld personal digital assistant (PDA). These cards typically require very low power and are smaller than the size of a matchbook. CompactFlash radio cards can sometimes be used in the PC slot of a laptop with the aid of a CF-to-PCMCIA adapter. The use of the SD and CF formats with PDAs has become less common because most handheld PDAs and handheld barcode scanners now have internal 802.11 radios using an embedded form factor.

So far we have discussed client adapters that are used with laptops or handheld devices. 802.11 client adapters also exist for desktops in the form of 802.11 PCI adapters or USB client adapters. Many 802.11 PCI adapters are simply a PCI peripheral card with a PCMCIA card attached or soldered onto the PCI card. Most desktop users place their computers underneath a desk. Therefore, the integrated antenna of an 802.11 PCI adapter is surrounded by the desk, resulting in poor communications and potentially a hidden node problem. Newer 802.11 PCI adapters often have an integrated radio card with a jack for an external antenna so that the user may place the antenna on top of the desk for better transmission and reception. Both types of 802.11 PCI adapters are shown in Figure 10.4.

FIGURE 10.4 Desktop 802.11 PCI adapters

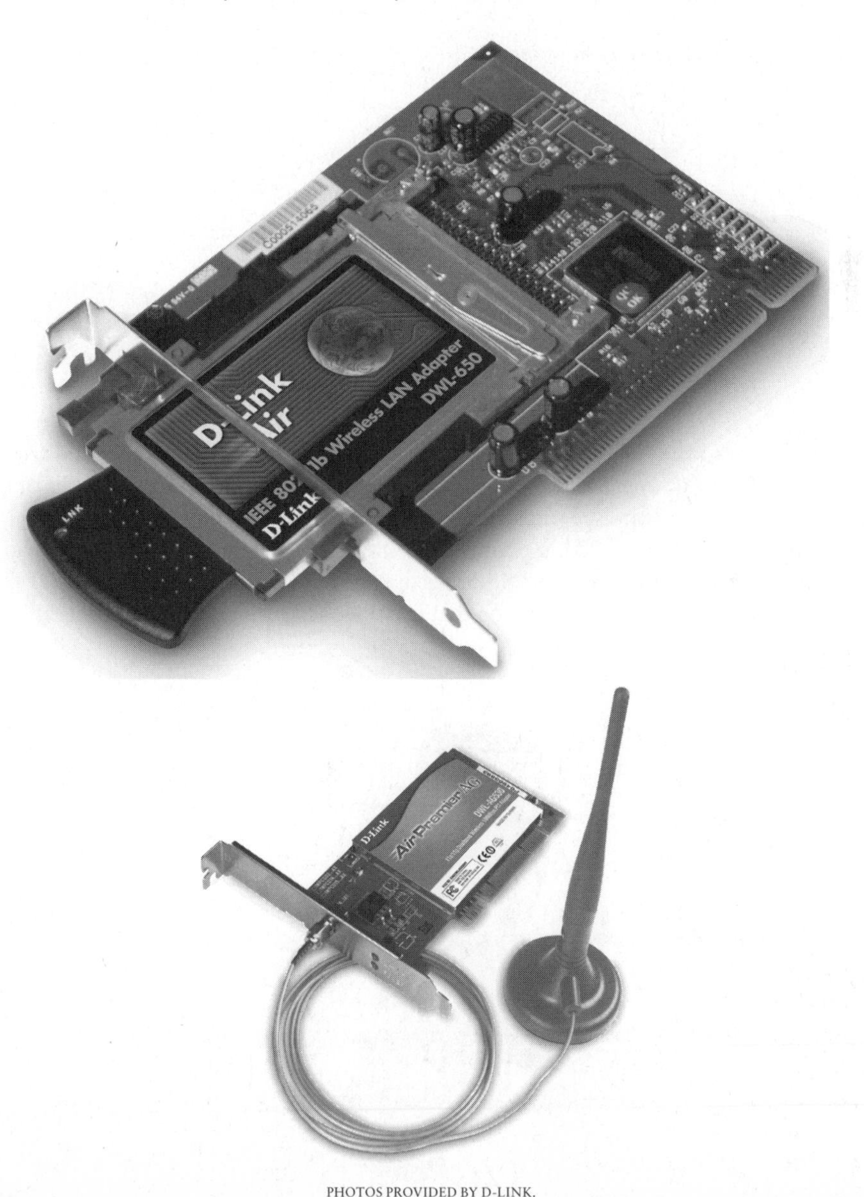

PHOTOS PROVIDED BY D-LINK.

The USB 802.11 radio adapter is a popular choice because almost all computers have USB ports. USB technology provides simplicity of setup and does not require an external power source. 802.11 USB radio adapters exist either in the form of a small dongle device (see Figure 10.5) or as an external wired USB device with a separate USB cable connector. The dongle devices are compact and portable for use with a laptop computer, while the external devices can be connected to a desktop computer with a USB extension cable and placed on top of a desk for better reception.

FIGURE 10.5 802.11 USB adapter

PHOTO PROVIDED BY D-LINK.

We have discussed the various types of 802.11 radio card formats that are used with laptops, PDAs, and desktops. 802.11 radio cards are also used in many other types of handheld devices, such as bar code scanners and VoWiFi phones. Barcode scanners, such as the Honeywell mobile device pictured in Figure 10.6, have made use of 802.11 radios for many years. Manufacturers of most handheld devices use an embedded form factor 802.11 radio (usually a Mini PCI or similar form factor that is embedded into the device's motherboard).

FIGURE 10.6: Barcode scanner

COURTESY OF HONEYWELL.

It should be noted that 802.11 radio cards used as client devices have begun to show up in many types of machines and solutions. Radio cards already exist in gaming devices, stereo systems, and video cameras. Appliance manufacturers are experimenting with putting Wi-Fi cards in washing machines and refrigerators. Because of the low cost of 802.11 radio cards, in the not-too-distant future, your entire house might be networked wirelessly, enabling you to control everything from remote locations.

Real World Scenario

Can I Use the Same Radio in Different Laptops?

The answer to this question depends entirely on two things: the type of radio card you are using and the operating system you are using. PCMCIA cards can be used in any laptop as long as the laptop has a PC Card slot. ExpressCards can be used in any laptop as long as the laptop has an ExpressCard slot. USB client adapters can be used by any laptop that has a USB port. Any laptop manufactured today will have a USB port and usually either a PC Card slot or an ExpressCard slot. Using the same Mini PCI radio card in different laptops might be a different story. Because Mini PCI radio cards are typically installed in laptop computers, they should not be inserted and removed too many times. Another potential problem is that laptop manufacturers may support only a specific Mini PCI radio chipset, which will limit your choice of laptops in which the card can be installed. Check with your laptop vendor before switching Mini PCI radios. Also make sure that you have appropriate drivers for the specific device you are using. Not all client devices are compatible with all operating systems and service packs.

Radio Card Chipsets

A group of integrated circuits designed to work together is often marketed as a *chipset*. Many 802.11 chipset manufacturers exist and sell their chipset technology to the various radio card manufacturers. Legacy chipsets will obviously not support all of the same features as newer chipset technologies. For example, a legacy chipset may support only DSSS technology, whereas newer chipsets will support both DSSS and OFDM technology.

Some chipsets may only support the ability to transmit on the 2.4 GHz ISM band, other chipsets can transmit on either the 2.4 GHz or 5 GHz unlicensed frequencies. Chipsets that support both frequencies are used in 802.11a/b/g client cards. The chipset manufacturers incorporate newer 802.11 technologies as they develop. Many proprietary technologies turn up in the individual chipsets, and some of these technologies become part of the standard in future 802.11 amendments. For example, the chipsets used by 802.11n draft 2.0 radios are vastly different from the chipsets used by 802.11a/b/g radios.

Although there are many chipset manufacturers, detailed information about some of the most widely used chipsets may be found at the following URLs: `www.atheros.com`, `www.broadcom.com`, and `www.intel.com`.

Client Utilities

An end user must have the ability to configure a wireless client card. Therefore, a software interface is needed in the form of *client utilities*. The software interface will usually have the ability to create multiple connection profiles. One profile may be used to connect to the wireless network at work, another for connecting at home, and a third for connecting at a hotspot.

Configuration settings for a client utility typically include the service set identifier (SSID), transmit power, WPA/WPA2 security settings, WMM quality-of-service capabilities, and power-management settings. As mentioned in Chapter 7, any client card can also be configured for either Infrastructure or Ad Hoc mode. Most good client utilities will typically have some sort of statistical information display along with some sort of received signal strength measurement indicator tool.

Four major types, or categories, of client utilities exist:

- Small office, home office (SOHO) client utilities
- Enterprise-class client utilities
- Integrated operating system client utilities
- Third-party client utilities

SOHO client utilities are usually simplistic in nature and are designed for ease of use for the average home user. Surprisingly, though, many of the SOHO utilities support some rather advanced features as 802.11 technologies progress. Enterprise-class client utilities provide the software interface for the more expensive enterprise-grade vendor cards. Typically, the enterprise-class utilities support more configuration features and have better statistical tools. Figure 10.7 depicts the Intel PROSet wireless client interface.

The most widely used client utility is an integrated operating system client utility, more specifically known as the *Wireless Zero Configuration (WZC) service*, that is enabled by default in Windows XP. The WZC is pictured in Figure 10.8. The main advantage of the WZC is that as an administrator, you have to support only one client utility even though your end users may have different radio cards.

It should be noted that the WZC has many published security risks, and therefore many government agencies and corporations ban the use of the integrated operating system utilities. Other disadvantages of the WZC are that it supports only a limited number of EAP protocols and does not have a built-in received signal measurement tool.

An added advantage of the WZC is that it utilizes a proprietary roaming method called *opportunistic PMK caching* that is supported by many of the WLAN controller vendors. It is beyond the scope of this book to discuss this roaming method; however, on the CD of this book is a white paper by Devin Akin, titled "Robust Security Network (RSN) Fast BSS Transition (FT)," that explains opportunistic PMK caching.

FIGURE 10.7 Enterprise-class client utility

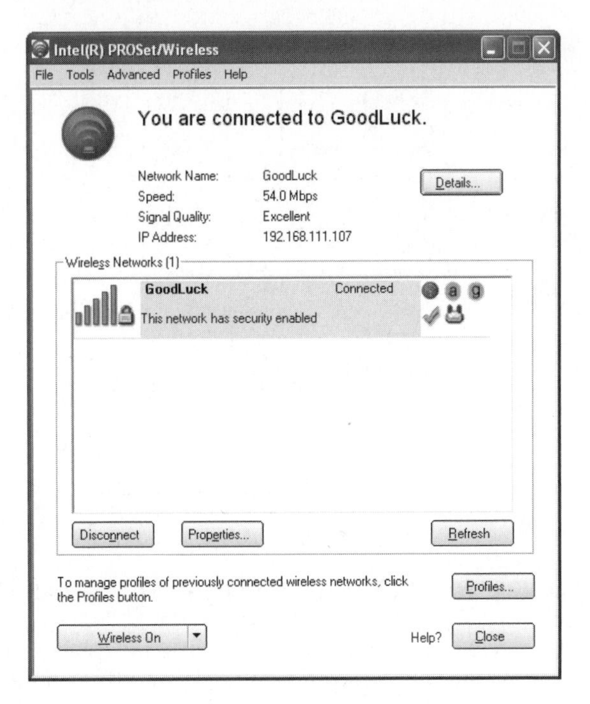

FIGURE 10.8 Wireless Zero Configuration service

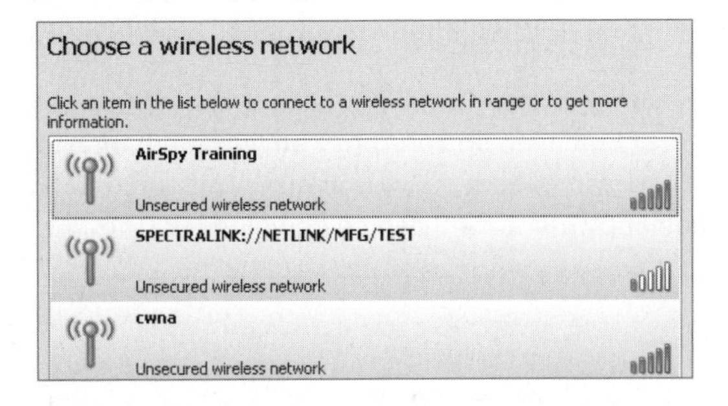

The last type of software interface for an 802.11 radio card is a third-party client utility, such as Juniper Networks Odyssey Access Client pictured in Figure 10.9. Much like any integrated OS client software, a third-party utility will work with radio cards from different vendors, making administrative support much easier. Third-party client utilities often bring the advantage of supporting many different EAP types, giving a WLAN administrator a wider range of security choices. The main disadvantage of third-party client utilities is that they cost extra money.

FIGURE 10.9 Third-party client utility

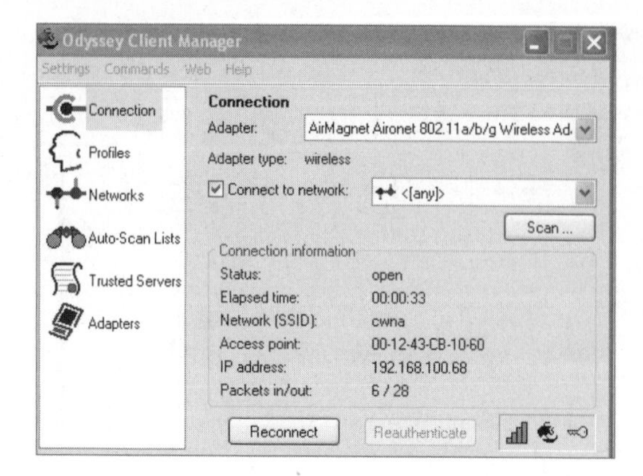

Real World Scenario

Which Wi-Fi Client Utility Should You Deploy in an Enterprise Environment?

This is a question that almost no one can agree on. Because of the ease of administration and the extended roaming capabilities, many of the WLAN controller vendors recommend using integrated OS client software such as the WZC that is built into Windows XP. Other administrators prefer to use a specific vendor's enterprise client utility that was designed to work with that vendor's card. This scenario, though, is feasible only in work environments where the administrator can mandate what type radio cards are deployed.

Because of some known security risks of the WZC, the authors of this book currently recommend using a third-party client utility in an enterprise environment or using only one vendor's solution. A popular third-party client utility is Juniper Networks' Odyssey Access Client (OAC). A 30-day evaluation copy of the OAC can be downloaded from `www.juniper.net`.

Progression of WLAN Architecture

While the acceptance of 802.11 technologies in the enterprise continues to grow, the evolution of WLAN architecture has kept an equivalent pace. In most cases, the main purpose of 802.11 technologies is to provide a wireless portal into a wired infrastructure network. How an 802.11 wireless portal is integrated into a typical 802.3 Ethernet infrastructure continues to change drastically.

Figure 10.10 depicts the progression of WLAN architecture that will be discussed in the following sections.

FIGURE 10.10 WLAN architecture progression

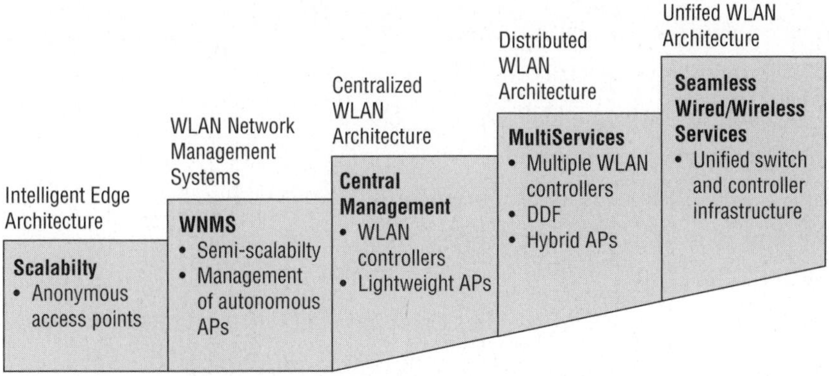

Intelligent Edge Architecture
Scalabilty
• Anonymous access points

WLAN Network Management Systems
WNMS
• Semi-scalabilty
• Management of autonomous APs

Centralized WLAN Architecture
Central Management
• WLAN controllers
• Lightweight APs

Distributed WLAN Architecture
MultiServices
• Multiple WLAN controllers
• DDF
• Hybrid APs

Unfifed WLAN Architecture
Seamless Wired/Wireless Services
• Unified switch and controller infrastructure

Autonomous Access Point—Intelligent Edge Architecture

For many years, the conventional access point has been thought of as a portal device where all the "brains" and horsepower exists inside the access point (AP) on the edge of the network architecture. Because all the intelligence exists inside each individual access point, these APs are often referred to as *fat APs*, *stand-alone APs*, or *intelligent edge APs*. However, the most common industry term for the traditional access point is *autonomous AP*.

All configuration settings exist in the autonomous access point itself, and, therefore, management and configuration occurs at the access layer. All encryption and decryption mechanisms and MAC layer mechanisms also operate within the autonomous AP. The distribution system service (DSS) and integration service (IS) that you learned about in Chapter 7 both function within an autonomous AP.

An autonomous access point, pictured in Figure 10.11, contains at least two physical interfaces: usually a radio frequency (RF) radio card and a 10/100BaseT port. The majority of the time, these physical interfaces are bridged together by a virtual interface known as a *Bridged Virtual Interface (BVI)*. The BVI is assigned an IP address that is shared by two or more physical interfaces.

FIGURE 10.11 Autonomous AP (Motorola AP-5131).

COURTESY OF MOTOROLA.

An autonomous access point will typically encompass both the 802.11 protocol stack and the 802.3 protocol stack. These APs might have some of the following features:

- Multiple management interfaces, such as command line, web GUI, and SNMP
- WEP, WPA, and WPA2 security capabilities
- WMM quality-of-service capabilities
- Fixed or detachable antennas
- Filtering options, such as MAC and protocol
- Connectivity modes, such as root, repeater, bridge, and scanner
- Removable radio cards
- Multiple radio card and dual-frequency capability: 2.4 GHz and 5 GHz
- Adjustable transmit power, which is used mostly for cell sizing
- VLAN support (VLANs are created on a managed wired switch.)
- IEEE standards support
- 802.3-2005, clause 33, Power over Ethernet (PoE) support

As pictured in Figure 10.12, autonomous APs are deployed at the access layer and typically are powered by a PoE-capable access layer switch. The integration service within an autonomous AP translates the 802.11 traffic into 802.3 traffic. The 802.11 traffic passes through a firewall before accessing network resources that exist at the core layer.

The autonomous AP that utilizes edge intelligence was the foundation that WLAN architects deployed for many years. However, most enterprise deployments of autonomous APs have been replaced by WLAN controller and lightweight AP solutions that are discussed later in this chapter. Autonomous access points are still very often deployed in the enterprise at small office or remote offices that require only a few APs.

Wireless Network Management System (WNMS)

One of the challenges for a WLAN administrator using a large WLAN intelligent edge architecture is management. As an administrator, would you want to configure 300 autonomous APs individually? One major disadvantage of using the traditional autonomous access point is that there is no central point of management. Any intelligent edge WLAN architecture with 25 or more autonomous access points is going to require some sort of *wireless network management system (WNMS)*.

A WNMS provides a central point of management to configure and maintain thousands of fat access points. A WNMS can be either a hardware appliance or a software solution. WNMS solutions can be either vender specific or vender neutral.

Because the main purpose of a WNMS is to provide a central point of management, both configuration settings and firmware upgrades can be pushed down to all the autonomous access points. The WMNS server is deployed at the core layer of the wired network and communicates via the 802.3 infrastructure with the autonomous APs that are deployed at the access layer. Although centralized management is the main goal, a WNMS has other capabilities as well, such as including RF spectrum planning and management. A WNMS can also be used to monitor an intelligent edge WLAN architecture with alarms and notifications centralized and integrated into a management console. Other capabilities include network reporting, trending, capacity planning, and policy enforcement. A WNMS might also be able to perform some rogue AP detection, but by no means should a WNMS be considered a wireless intrusion detection system (WIDS). One of the main disadvantages of a WNMS is that it will not assist in the roaming capabilities between access points, whereas the WLAN controller architecture has that capability. A WNMS system does not sit in the data path and as such has no knowledge of the traffic being passed through the APs. The WNMS is simply a utility on the sidelines that can be used to push configuration and firmware changes.

FIGURE 10.12 Autonomous AP topology

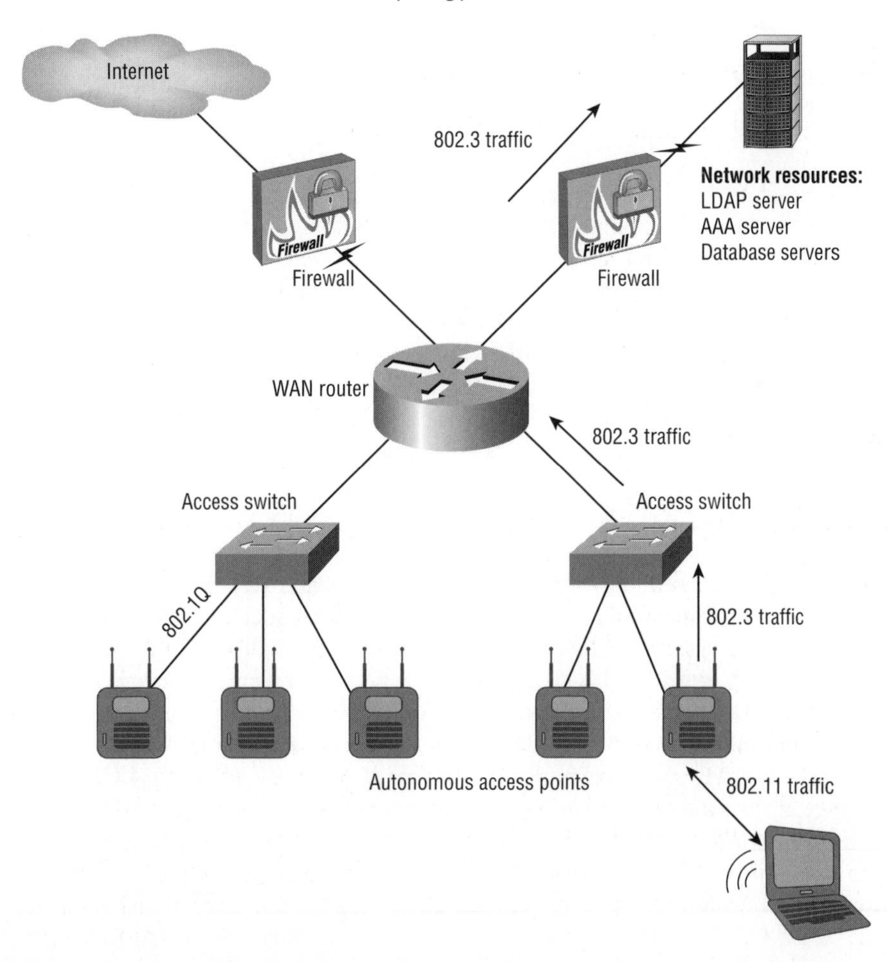

Currently, WNMSs are completely separate from any wired network management systems. A WNMS may also not recognize certain hardware, and the most current firmware updates from a vendor are not always immediately usable in a WNMS.

In the past, the whole point of a WNMS server was to provide a central point of management for autonomous access points. That definition has changed in recent years. Later in this chapter, you will learn about WLAN controllers, which are used as a central point of management for lightweight APs. WLAN controllers effectively replaced WNMS servers as a central point of management for access points. However, at times multiple WLAN controllers are needed in large-scale WLAN enterprise deployments. Currently, most WMNS servers are now used as a central point of management for multiple WLAN controllers in a large-scale WLAN enterprise. The current WNMS servers that are used to manage multiple WLAN controllers from a single vendor may also be used to manage other vendors' WLAN infrastructure, including autonomous APs.

Centralized WLAN Architecture

The next progression in the development of WLAN integration is the centralized WLAN architecture. This model uses a central WLAN controller that resides in the core of the network. In the centralized WLAN architecture, autonomous APs have been replaced with *lightweight access points*, also known as *thin APs*. A lightweight AP has minimal intelligence and is functionally just a radio card and an antenna. All the intelligence resides in the centralized WLAN controller, and all of the AP configuration settings such as channel and power are distributed to the lightweight APs from the WLAN controller and stored in the RAM of the lightweight AP. The encryption and decryption capabilities might reside in the centralized WLAN controller or may still be handled by the lightweight APs, depending on the vendor. The distribution system service (DSS) and integration service (IS) that you learned about in Chapter 7 both now typically function within the WLAN controller.

Lightweight AP

Many WLAN controller vendors only manufacture lightweight APs that have very limited software capabilities. Almost all of the configuration settings, such as transmission power, channel, security settings, and SSID are not configured on the lightweight AP but instead are configured on the centralized WLAN controller. Although security settings are configured on the WLAN controller, some lightweight access points still handle the encryption and decryption mechanisms of 802.11 data frames at the edge of the network.

Some lightweight APs may have multiple radio cards that provide dual 2.4 GHz and 5 GHz coverage at the same time. However, many lightweight access points have a single *software defined radio (SDR)*. An SDR can support a range of frequency bands, transmission techniques, and modulation schemes so that a single radio could replace multiple hardware designs. An 802.11 SDR in a lightweight access point has the ability to operate as either as a 2.4 GHz transceiver or as a 5 GHz transceiver, but it cannot operate on both frequency bands at the same time unless the AP is equipped with two radios. Please keep in mind that only the configuration of a lightweight AP with SDR capabilities takes place on the WLAN controller. Aruba Networks' AP-61, shown in Figure 10.13, is an example of a lightweight AP with a single radio that can operate either as a 2.4 GHz transceiver or as a 5 GHz transceiver.

FIGURE 10.13 Lightweight access point (Aruba AP-61)

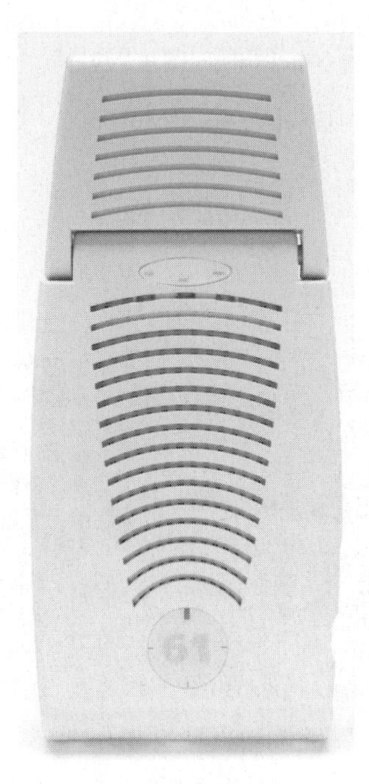

COURTESY OF ARUBA NETWORKS.

Other vendors manufacture access points that can be either an autonomous access point or a lightweight AP. A simple firmware update can convert a stand-alone autonomous AP into a lightweight AP. Siemens' HiPath Wireless AP2630 and AP2640, pictured in Figure 10.14, are examples of stand-alone autonomous APs that can be converted into lightweight APs that communicate directly with a WLAN controller.

FIGURE 10.14 Autonomous or lightweight access point (HiPath Wireless AP2630 and AP2640)

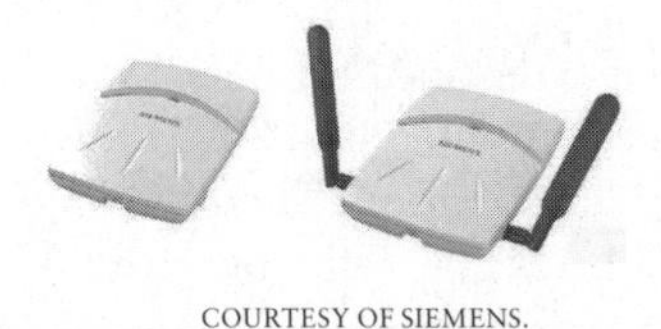

COURTESY OF SIEMENS.

WLAN Controller

At the heart of the centralized WLAN architecture model is the *WLAN controller* (see Figure 10.15). WLAN controllers are often referred to as *wireless switches* because they are indeed an Ethernet-managed switch that can process and route data at the Data-Link layer (layer 2) of the OSI model. Many of the WLAN controllers are multilayer switches that can also route traffic at the Network layer (layer 3). However, the phrase *wireless switch* does not adequately describe the many capabilities of a WLAN controller.

FIGURE 10.15 WLAN controller (Aruba MMC-6000)

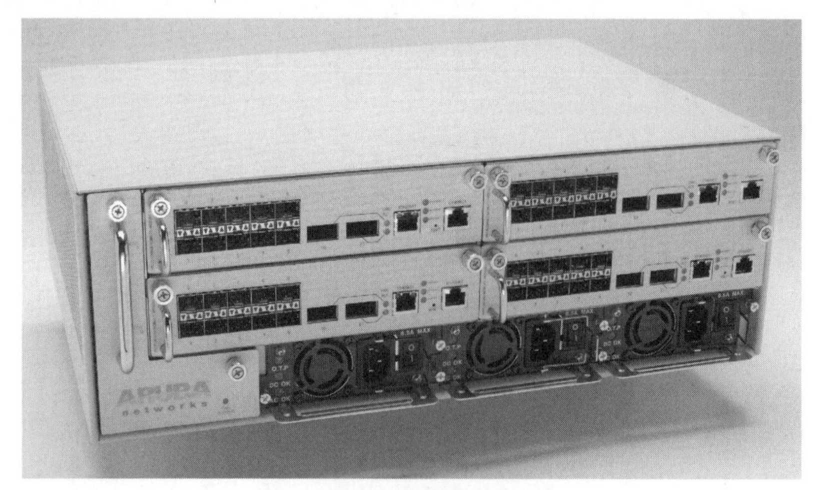

COURTESY OF ARUBA NETWORKS.

A WLAN controller may have some of these many features:

AP management As mentioned earlier, the majority of the lightweight access point functions such as power, channels, and supported data rates are configured on the WLAN controller. This allows for centralized management and configuration of lightweight APs. Most vendors use proprietary protocols for communications between the WLAN controller and their lightweight APs. These proprietary protocols can transfer configuration settings, update firmware, and maintain keep-alive traffic.

802.11 traffic tunneling A key feature of most WLAN controllers is that the integration service (IS) and distribution system service (DSS) operate within the WLAN controller. In other words, all 802.11 traffic that is destined for wired-side network resources must first pass through the controller and be translated into 802.3 traffic by the integration service before being sent to the final wired destination. Therefore, lightweight access points must send their 802.11 frames to the WLAN controller over an 802.3 wired connection.

As you learned in Chapter 9, "802.11 MAC Architecture," the 802.11 frame format is complex and is designed for a wireless medium and not a wired medium. An 802.11 frame cannot travel through an Ethernet 802.3 network by itself. So how can an 802.11 frame traverse between a lightweight AP and a WLAN controller? The answer is inside an IP-encapsulated tunnel. Each 802.11 frame is encapsulated entirely within the body of an IP packet. Many WLAN vendors use *Generic Routing Encapsulation (GRE)*, which is a commonly used network tunneling protocol. GRE can encapsulate an 802.11 frame inside an IP tunnel, creating a virtual point-to-point link between the lightweight AP and the WLAN controller. WLAN vendors that do not use GRE use other proprietary protocols for the IP tunneling.

As pictured in Figure 10.16, the WLAN controller is usually deployed close to network resources at the core layer. The lightweight access points are connected to third-party managed switches that provide PoE. The lightweight APs tunnel their 802.11 frames all the way back to the WLAN controller, from the access layer all the way back to the core layer. The distribution system service inside the controller directs the traffic, while the integration service translates an 802.11 data MSDU into an 802.3 frame. After 802.11 data frames have been translated into 802.3 frames, they are then sent to their final wired destination.

FIGURE 10.16 WLAN controller and IP tunneling—core layer

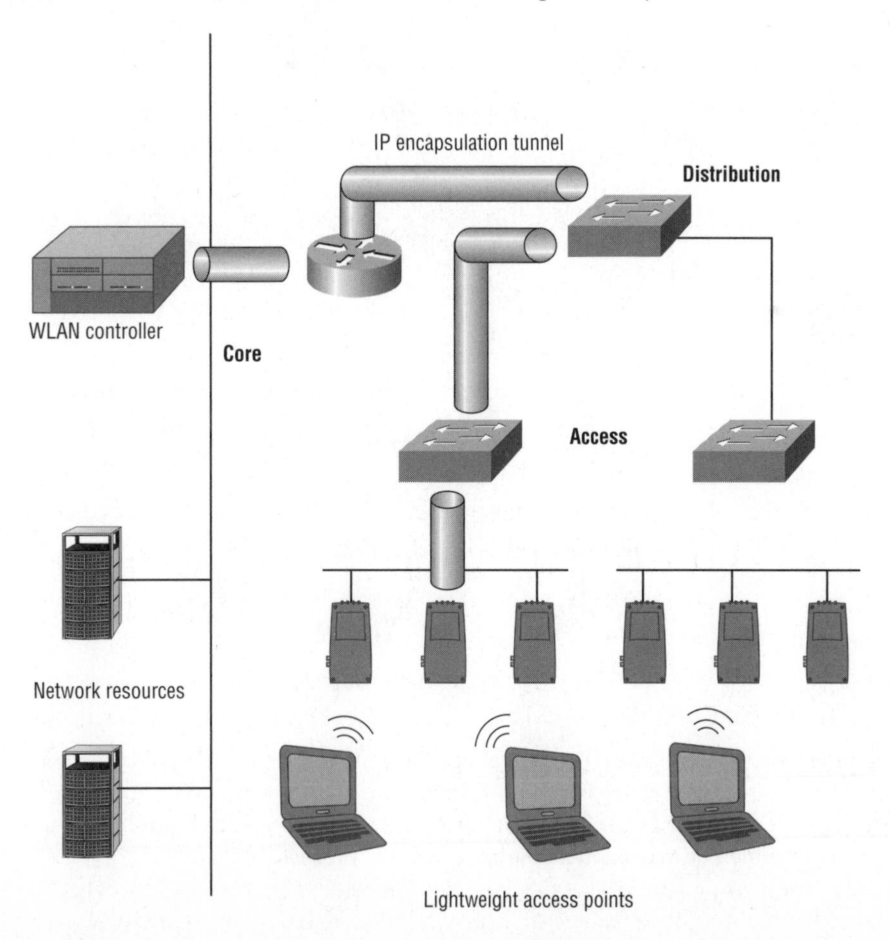

Most WLAN controllers are deployed at the core layer; however, they may also be deployed at either the distribution layer or even the access layer, as depicted in Figures 10.17 and 10.18. Exactly where a WLAN controller is deployed depends on the WLAN vendor's solution, and the intended wireless integration into the preexisting wired topology. Multiple WLAN controllers that communicate with each other may be deployed at different network layers, providing they can communicate with each other.

FIGURE 10.17 WLAN controller and IP tunneling—distribution layer

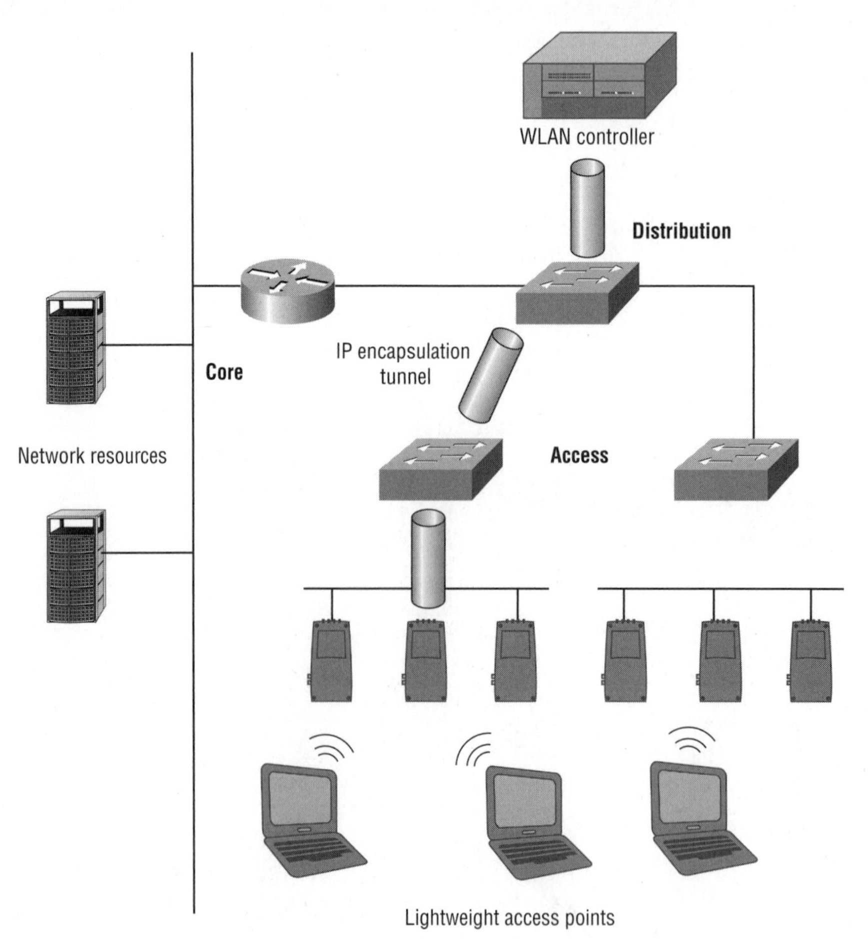

AP group profiles An AP group profile defines the configuration settings for a single AP or group of access points. Settings such as channel, transmit power, and supported data rates are examples of settings configured in an AP group profile. An AP can belong to only one AP group profile but may support multiple WLAN profiles.

FIGURE 10.18 WLAN controller and IP tunneling—access layer

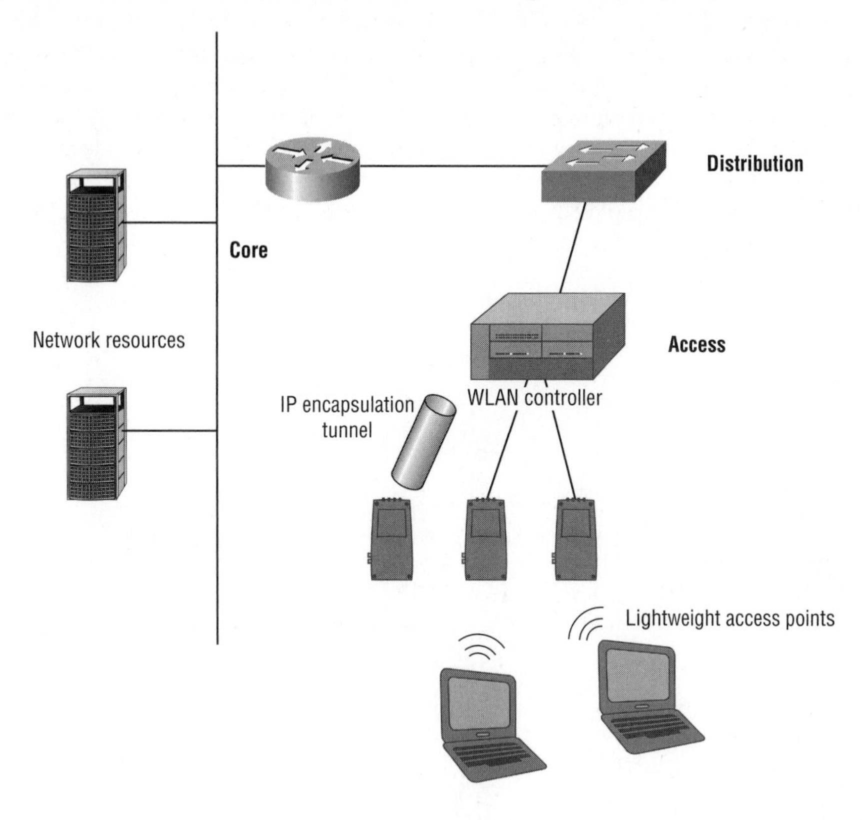

WLAN profiles WLAN controllers introduced the concept of virtual WLANs, which are often called *WLAN profiles*. Different groups of 802.11 clients exist in a virtual WLAN. The WLAN profile is a set of configuration parameters that are configured on the WLAN controller. The profile parameters can include the WLAN logical name (SSID), WLAN security settings, VLAN assignment, and quality-of-service (QoS) parameters.

WLAN profiles often work together with *role-based access control (RBAC)* mechanisms. When a user connects to a *virtual WLAN*, users are assigned to specific roles. Do not confuse the WLAN profile with an AP group profile. Multiple WLAN profiles can be supported by a single AP; however, an AP can alone belong to one AP group.

Virtual BSSIDs You learned in Chapter 7 that every WLAN has a logical name (SSID) and that each WLAN BSS has a unique layer 2 identifier, the *basic service set identifier (BSSID)*. The BSSID is typically the MAC address of the access point's radio card. WLAN controllers have the capability of creating multiple virtual BSSIDs. As you just learned, the WLAN controller allows for the creation of virtual WLANs, each with a unique logical identifier (SSID) that is also assigned to a specific VLAN. Because the BSSID is the MAC address of the AP, and because the WLAN controller can support many virtual WLANs on the same physical AP, each virtual WLAN is typically linked with a unique *virtual BSSID*. As shown in Figure 10.19, the virtual BSSIDs are usually increments of the original MAC address of the lightweight AP's radio. As depicted in Figure 10.19, within each lightweight AP's coverage area, multiple virtual WLANs can exist. Each virtual WLAN has a logical name (SSID) and a unique virtual layer 2 identifier (BSSID), and each WLAN is mapped to a unique layer 3 virtual local area network (VLAN). In other words, multiple layer 2/3 domains can exist within one layer 1 domain. Try to envision multiple basic service sets (BSSs) that are linked to multiple VLANs, yet they all exist within the same coverage area of a single access point.

FIGURE 10.19 Virtual WLANs, virtual BSSIDs, and VLANs

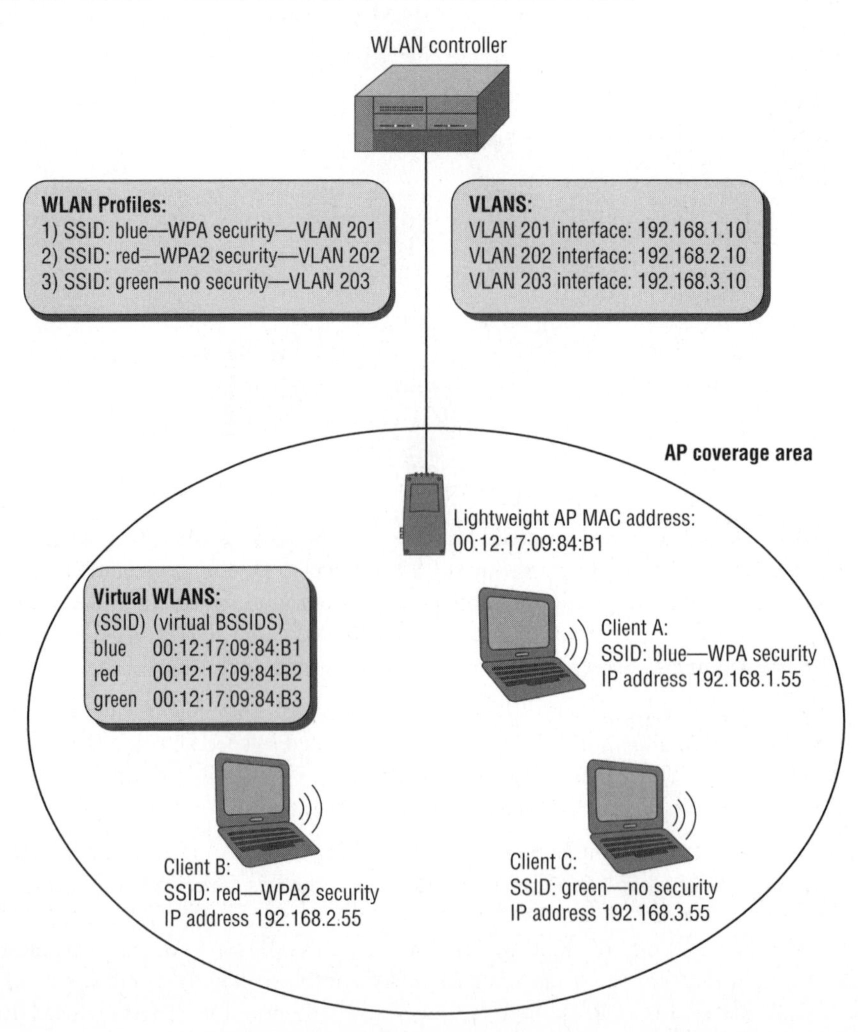

VLANs WLAN controllers fully support the creation of VLANs and 802.1Q VLAN tagging. Multiple wireless user VLANs can be created on the WLAN controller. The ability to create user VLANs is one of the main benefits of a WLAN controller, because they can provide for segmentation and security. VLANs may be assigned statically to WLAN profiles or may be assigned using a RADIUS attribute. A more detailed discussion of wireless VLANs can be found in Chapter 13, "802.11 Network Security Architecture."

User management WLAN controllers usually provide the ability to control the who, when, and where in terms of using role-based access control (RBAC) mechanisms. A more detailed discussion of RBAC can be found in Chapter 13.

Layer 2 security support WLAN controllers fully support layer 2 WEP, WPA, and WPA2 encryption. Authentication capabilities include internal databases as well as full integration with RADIUS and LDAP servers.

Layer 3 and 7 VPN concentrators Some WLAN controller vendors also offer VPN server capabilities within the controller. The controller can act as a VPN concentrator or end point for PPTP, IPsec, or SSL VPN tunnels.

Captive portal WLAN controllers have captive portal features that can be used with guest WLANs and guest WLAN profiles. Because the captive portal authenticates users but has very limited encryption capabilities, it is rarely used for anything other than guest access.

Automatic failover and load balancing WLAN controllers usually provide support for Virtual Router Redundancy Protocol (VRRP) for redundancy purposes. Most vendors also offer proprietary capabilities to load-balance wireless clients between multiple lightweight APs.

Internal Wireless Intrusion Detection Systems Some WLAN controllers have integrated WIDS capabilities for security monitoring. A more detailed discussion on WIDS can be found in Chapter 14, "Wireless Attacks, Intrusion Monitoring, and Policy."

Dynamic RF spectrum management The majority of WLAN controllers implement some type of *dynamic RF* capability. A WLAN controller is a centralized device that can dynamically change the configuration of the lightweight access points based on accumulated RF information gathered from the access points' radio cards. In a WLAN controller environment, the lightweight access points will monitor their respective channels as well as use off-channel scanning capabilities to monitor other frequencies. Any RF information heard by any of the access points is reported back to the WLAN controller. Based on all of the RF monitoring from multiple access points, the WLAN controller will make dynamic changes to the RF settings of the lightweight APs. Some lightweight access points may be told to change to a different channel, while other APs may be told to change their transmit power settings.

Dynamic RF is sometimes referred to as *radio frequency spectrum management (RFSM)*. However, the WLAN controller vendors all implement proprietary dynamic RF functionality. When implemented, RFSM provides automatic cell sizing, automatic monitoring, troubleshooting, and optimization of the RF environment, which can best be described as a self-organizing and self-healing wireless LAN.

Bandwidth management Bandwidth pipes can be restricted upstream or downstream.

Firewall capabilities Stateful packet inspection is available with an internal firewall in some WLAN controllers.

Layer 3 roaming support Capabilities to allow seamless roaming across layer 3 routed boundaries are fully supported. A more detailed discussion on layer 3 roaming and the Mobile IP standard can be found in Chapter 12, "WLAN Troubleshooting."

802.3-2005, clause 33—Power over Ethernet (PoE) When deployed at the access layer, WLAN controllers can provide direct power to lightweight APs via PoE. However, most lightweight APs are powered by third-party edge switches.

Management interfaces Many WLAN controllers offer full support for common management interfaces such as GUI, CLI, SSH, and so forth.

The most obvious advantages of the centralized architecture of a WLAN controller include AP management, user management, dynamic RF, and VLAN segmentation. Another major advantage of the WLAN controller model is that most of the controllers support some form of fast secure roaming, which can assist is resolving latency issues often associated with roaming across encrypted wireless networks.

One possible disadvantage of using a WLAN controller is that the WLAN controller might become a bottleneck because all data must be sent to and redirected from the WLAN controller. Most vendors are able to prevent this from occurring by providing a scalable hierarchical environment, which is discussed later in this chapter. QoS policies can also be enforced at the WLAN controller, which may cause latency issues if improperly configured. WLAN controllers and the lightweight APs might be separated by several hops, which can also introduce network latency. Some of the WLAN controllers have so many features and configuration settings that the user interface can be confusing for novice administrators.

Split MAC

The majority of WLAN controller vendors implement what is known as a *split MAC architecture*. With this type of WLAN architecture, some of the MAC services are handled by the WLAN controller, and some are handled by the lightweight access point. For example, the integration service and distribution system service are handled by the controller. WMM QoS methods are usually handled by the controller. Depending on the vendor, encryption and decryption of 802.11 data frames might be handled by the controller or by the AP.

You have already learned that 802.11 frames are tunneled between the lightweight APs and the WLAN controller. 802.11 data frames will always be tunneled to the controller because the controller's integration service transfers the layer 3–7 MSDU payload of the 802.11 data frames into 802.3 frames that are sent off to network resources. Effectively, the WLAN controller is needed to provide a gateway to network resources for the payload of 802.11 data frames. 802.11 management and control frames do not have an upper-layer payload and therefore are never translated into 802.3 frames. 802.11 management and control frames do not necessarily need to be tunneled to the WLAN controller because the controller does not need to provide a gateway to network resources for these types of 802.11 frames.

In a split MAC architecture, many of the 802.11 management and control frame exchanges occur only between the client station and the lightweight access point and are not tunneled back to the WLAN controller. For example, beacons, probe responses, and ACKs may be generated by the lightweight AP instead of the controller. It should be noted that most WLAN controller vendors implement split MAC architectures differently. The Internet Engineering Task Force (IETF) has proposed a set of standards for WLAN controller protocols called *Control and Provisioning of Wireless Access Points (CAPWAP)*. CAPWAP does define split MAC standards.

More information about the proposed Control and Provisioning of Wireless Access Points (CAPWAP) standards can be found at IETF's website, at `www.ietf.org/html.charters/capwap-charter.html`.

Remote Office WLAN Controller

Although WLAN controllers typically reside on the core of the network, they can also be deployed at the access layer, usually in the form of a remote office WLAN controller, such as that pictured in Figure 10.20. A remote office WLAN controller typically has much less processing power than a core WLAN controller and is also less expensive. The purpose of a remote office WLAN switch is to allow remote and branch offices to be managed from a single location. Remote WLAN controllers typically communicate with a central WLAN controller across a WAN link. Secure VPN tunneling capabilities are usually available between controllers across the WAN connection. Through the VPN tunnel, the central controller will download the network configuration settings to the remote WLAN controller, which will then control and manage the local APs. These remote controllers will allow for only a limited number of lightweight APs. Features typically include Power over Ethernet, internal firewalling, and an integrated router using NAT and DHCP for segmentation.

FIGURE 10.20 Remote office WLAN controller (Motorola WS-2000)

COURTESY OF MOTOROLA.

Distributed WLAN Architecture

Large enterprise WLAN deployments with hundreds or thousands of lightweight APs will require multiple WLAN controllers to manage them. Scalability with the WLAN controller/lightweight AP model is easily accomplished by simply adding more controllers at either the core or distribution layers. In order to scale a WLAN, more controllers will have to be added for two reasons.

First, controllers can support only a finite number of lightweight access points. As more access points are added, more controllers will be needed as a central point of management. Most WLAN vendors offer a distributed architecture that usually consists of one master or parent controller and multiple child controllers that communicate back to the parent controller. If numerous master or parent controllers are deployed across the enterprise, a wireless network management system (WNMS) server might be needed to provide a higher level of management to oversee or manage the multiple parent controllers.

The second reason for adding more controllers is to provide more data distribution gateways to network resources. As you have learned, lightweight access points tunnel their 802.11 data traffic to the controller with the eventual destination of the data payload being network resources. Because the integration service can exist in all controllers, each controller can act as an individual gateway for the data from the access points to be distributed onto the wired network at either the core or distribution layers. The use of multiple controllers as data distribution gateways onto the wired network is known as *distributed data forwarding (DDF)*. Figure 10.21 depicts multiple controllers acting as data-forwarding points at both the core and distribution layers.

FIGURE 10.21 Distributed WLAN architecture

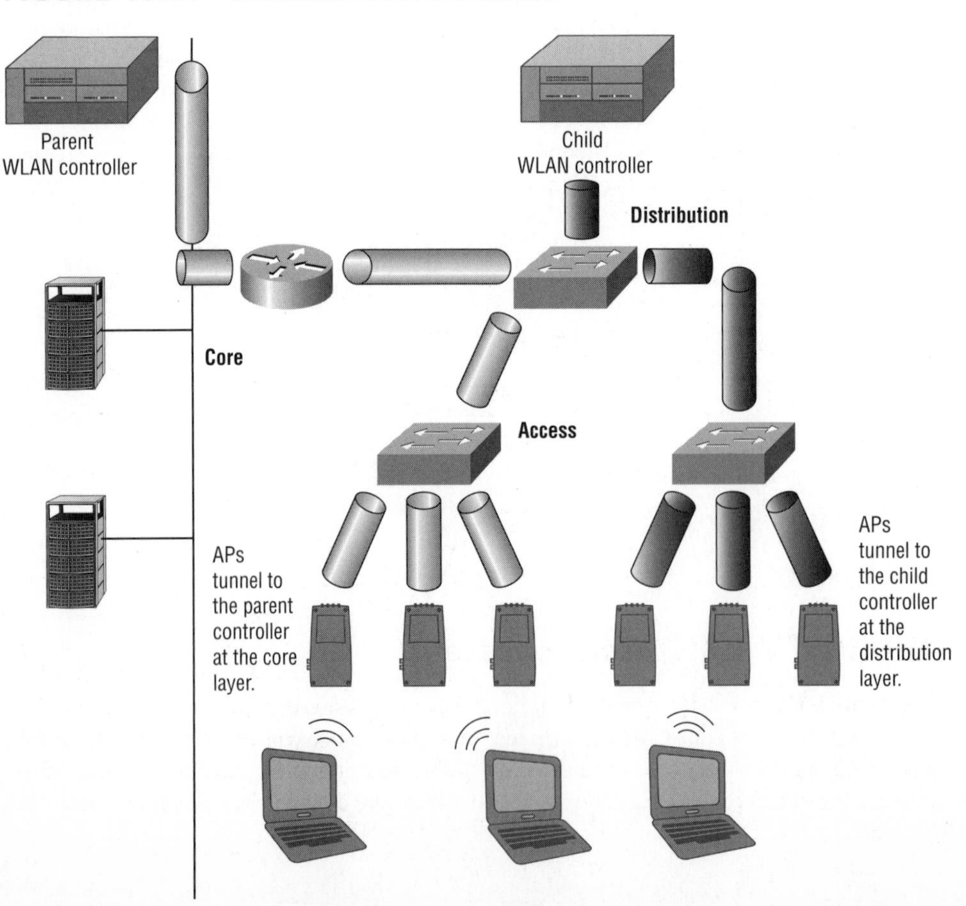

Distributed WLAN Hybrid

A few vendors have implemented a distributed WLAN architecture that uses a WLAN controller to manage hybrid fat/thin access points. The centralized controller acts as a central point of management for the hybrid access points. However, QoS policies and all of the 802.11 MAC data forwarding is handled at the edge of the network at the access points instead of back on the WLAN controller. The access points are the data distribution gateways to the wired backbone and are acting as distributed data forwarding (DDF) APs, while still being managed by the central WLAN controller.

The thinking behind these hybrid fat/thin solutions is that you maintain the centralized management but you eliminate the potential data bottlenecks and hopefully improve latency.

Unified WLAN Architecture

WLAN architecture could very well take another direction by fully integrating WLAN controller capabilities into wired network infrastructure devices. Wired switches and routers at both the core and the edge would also have WLAN controller capabilities, thereby allowing for the combined management of the wireless and wired networks. This unified architecture has already begun to be deployed by some vendors and will likely grow in acceptance as WLAN deployments become more commonplace and the need for fuller seamless integration continues to rise.

Specialty WLAN Infrastructure

In the previous sections, we discussed the progression of WLAN network infrastructure devices that are used to integrate an 802.11 wireless network into a wired network architecture. The Wi-Fi marketplace has also produced many specialty WLAN devices in addition to APs and WLAN controllers. Many of these devices, such as bridges and mesh networks, have become extremely popular, although they operate outside of the defined 802.11 standards. You will look at these devices in the following sections.

Wireless Workgroup Bridge

A *workgroup bridge (WGB)* is a wireless device that provides wireless connectivity for wired infrastructure devices that do not have radio cards. The radio card inside the WGB associates with an access point and joins the basic service set (BSS) as a client station. As depicted in Figure 10.22, multiple Ethernet devices are connected behind the wired side of the WGB. This provides fast wireless connectivity for wired devices through the association the WGB has with the access point. Because the WGB is an associated client of the access point, the WGB does not provide connectivity for other wireless clients. It is also important to understand that only the radio card inside the WGB can contend for the 802.11 wireless medium, and the wired cards behind the WGB cannot contend for the half-duplex RF medium.

FIGURE 10.22 Wireless workgroup bridge

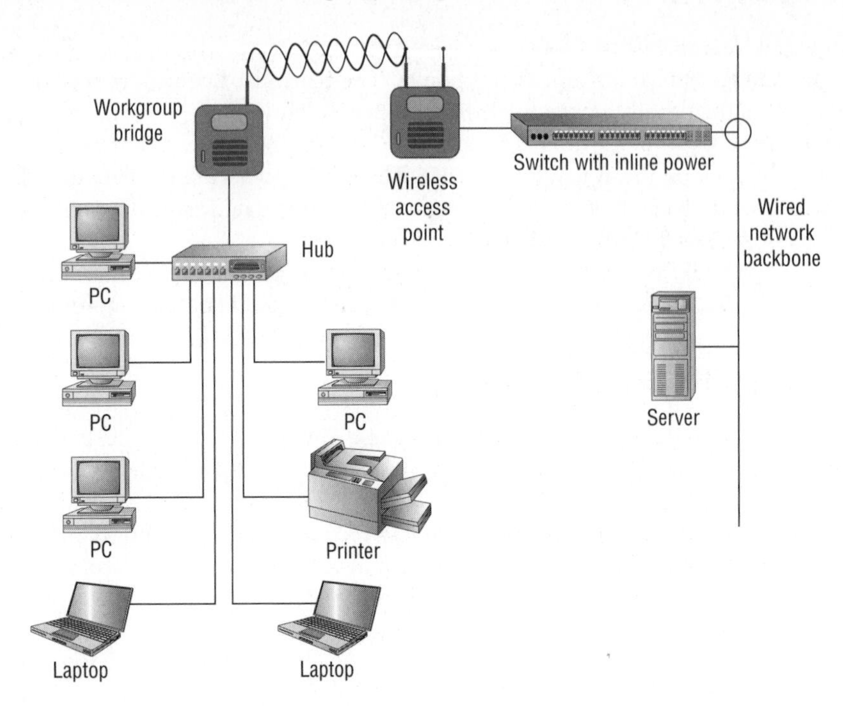

Most wireless workgroup bridges can provide connectivity for as many as eight wired devices, but it depends on the vendor. Some WGBs provide connectivity for only one wired device and are sometimes referred to as a *universal client*. The workgroup bridge can be very useful in providing wireless connectivity for small desktop workgroups, cash registers, network printers, and any other devices with Ethernet ports. The need for WGBs has greatly diminished because 802.11 radios are replacing Ethernet cards in many client devices.

Wireless LAN Bridges

A common nonstandard deployment of 802.11 technology is the *wireless LAN bridge*. The purpose of bridging is to provide wireless connectivity between two or more wired networks. A bridge generally supports all of the same features that an autonomous access point possesses, but the purpose is to connect wired networks and not to provide wireless connectivity to client stations. Although bridge links are sometimes used indoors, generally they are used outdoors to connect the wired networks inside two buildings. An outdoor bridge link is often used as a redundant backup to T1 or fiber connections between buildings. Outdoor wireless bridge links are even more commonly used as replacements to T1 or fiber connections between buildings because of their substantial cost savings.

Wireless bridges support two major configuration settings: *root* and *nonroot*. Bridges work in a parent/child-type relationship, so think of the root bridge as the parent and the nonroot bridge as the child.

A bridge link that connects only two wired networks is known as a *point-to-point (PtP)* bridge. Figure 10.23 shows a PtP connection between two wired networks using two 802.11 bridges and directional antennas. Note that one of the bridges must be configured as the parent root bridge while the other bridge is configured as the child nonroot bridge.

FIGURE 10.23 Point-to-point WLAN bridging

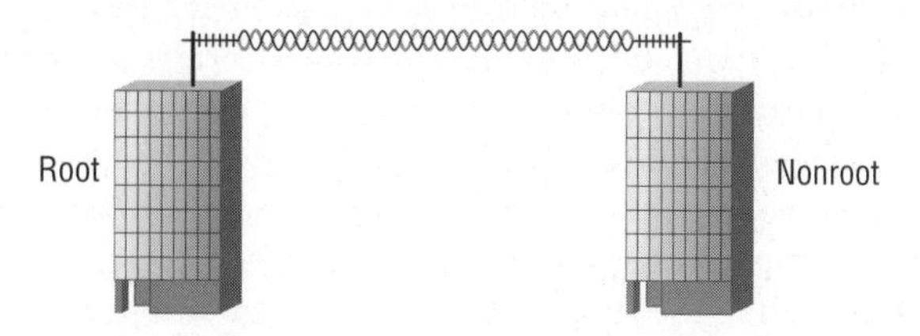

A *point-to-multipoint (PtMP)* bridge link connects multiple wired networks. The root bridge is the central bridge, and multiple nonroot bridges connect back to the root bridge. Figure 10.24 shows a PtMP bridge link between four buildings. Please note that the root bridge is using a high-gain omnidirectional antenna, while the nonroot bridges are all using unidirectional antennas pointing back to the antenna of the root bridge. Also notice that there is only one root bridge in a PtMP connection. There can never be more than one root bridge.

FIGURE 10.24 Point-to-multipoint WLAN bridging

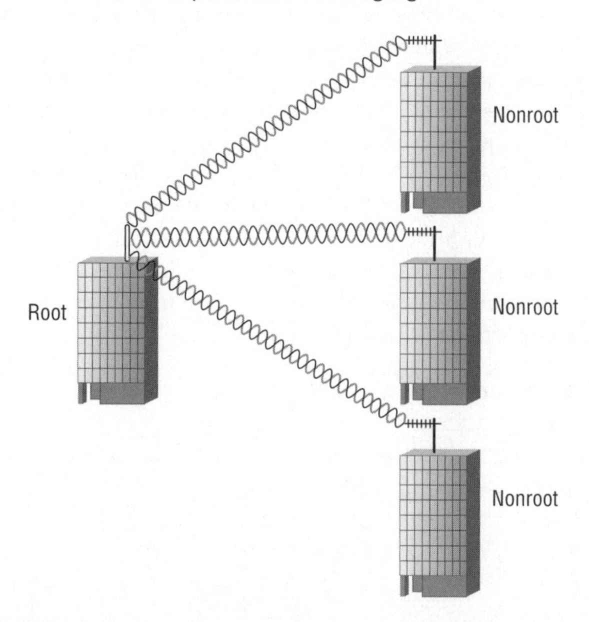

Besides the root and nonroot modes, bridges have other vendor configuration modes:

AP mode Converts a bridge into an access point

WGB mode Converts a bridge into a workgroup bridge

Repeater mode Repeats the cell of a root bridge to a nonroot bridge

Root with clients Root bridge that also allows clients to associate

Nonroot with clients Nonroot bridge that also allows clients to associate

The configuration settings that allow clients to associate, are highly discouraged. Allowing a client to associate to a bridge link is a security risk that can potentially leave network resources vulnerable. Clients can also affect the throughput of the bridge link, because the clients add medium contention overhead. Also, because of performance issues, the repeater mode is not a recommended mode for wireless bridging. If at all possible, a better bridge deployment practice is to use two separate bridge links as opposed to repeating the link of a root bridge to a nonroot bridge.

Considerations when deploying outdoor bridge links are numerous, including the Fresnel zone, earth bulge, free space path loss, link budget, and fade margin. There may be other considerations as well, including the IR and EIRP power regulations as defined by the regulatory body of your country.

Point-to-point links in the 2.4 GHz band can be as long as 24 miles. A problem that might occur over a very long distance link is an ACK time-out. Because of the half-duplex nature of the medium, every unicast frame must be acknowledged. Therefore, a unicast frame sent across a 24-mile link by one bridge must immediately receive an ACK frame from the opposite bridge, sent back across the same long-distance link. Even though RF travels at the speed of light, the ACK may not be received quickly enough. The original bridge will time-out after not receiving the ACK frame for a certain period of microseconds and will assume that a collision has occurred. The original bridge will then retransmit the unicast frame even though the ACK frame is on the way. Retransmitting unicast traffic that does not need to be resent can cause throughput degradation of as much as 50 percent. To resolve this problem, most bridges have an ACK time-out setting that can be adjusted to allow a longer period of time for a bridge to receive the ACK frame across the long-distance link.

A common problem with point-to-multipoint bridging is mounting the high-gain omnidirectional antenna of the root bridge too high, as pictured in Figure 10.25. The result is that the vertical line of sight with the directional antennas of the nonroot bridges is not adequate. The solution for this problem is to use a high-gain omnidirectional antenna that provides a certain amount of electrical downtilt or to use directional sector antennas aligned to provide omnidirectional coverage.

FIGURE 10.25 Common bridging challenge

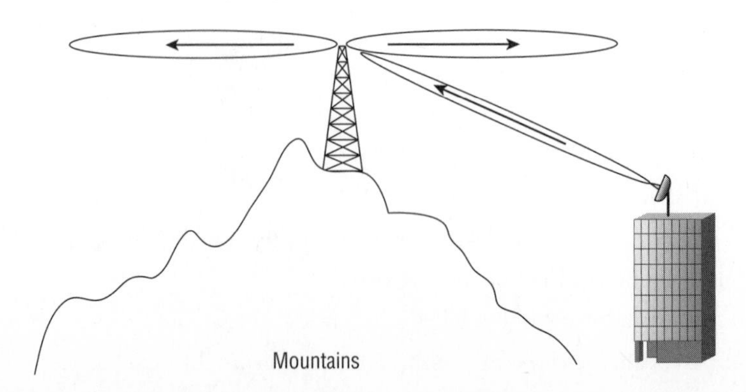

Enterprise Wireless Gateway

An *enterprise wireless gateway (EWG)* is a legacy middleware device that was used to segment autonomous access points from the protected wired network infrastructure. An EWG can segment the unprotected wireless network from the protected wired network by acting as a router, a VPN end point, and/or a firewall.

The EWG provided many of the same capabilities that a WLAN controller now provides, with some key differences. Because an EWG segmented fat access points and not thin access points, there was no AP management available within the EWG. The need still existed for a third-party WNMS to provide management of the autonomous APs from another central location. Also, unlike most WLAN controllers, enterprise wireless gateways did not have an integrated wireless intrusion detection system (WIDS), and the need for an overlay WIDS remained. An EWG also does not provide any RF spectrum management or control.

There are some similarities between an EWG and a WLAN controller, including layer 3 roaming capabilities, user management, role based access control, bandwidth throttling, redundancy support, layer 2 security support, and a captive portal. An EWG can also support VLANs that are created on a managed wired switch. Although enterprise wireless gateway devices still exist, they are a dying breed and are no longer manufactured. EWGs have been replaced by the various WLAN controller solutions.

Residential Wireless Gateway

Residential wireless gateway (RWG) is a fancy term for a home wireless router. The main function of a residential wireless gateway is to provide shared wireless access to a SOHO Internet connection while providing a level of security on the Internet. These SOHO Wi-Fi routers are generally inexpensive, yet they are surprisingly full featured.

The following features are supported by residential wireless gateways:

- Configurable 802.11 radio card
- Support for simple routing protocols such as RIP
- Network Address Translation (NAT)
- Port Address Translation (PAT)
- Port forwarding
- Firewall
- L2 security support (WEP or WPA-Personal or WPA2-Personal)
- DHCP server
- Multiport Ethernet switch for connecting wired clients

Keep in mind that any type of wireless router is a very different device than an access point. Unlike access points, which use a Bridged Virtual Interface (BVI), wireless routers have separate routed interfaces. The radio card exists on one subnet while the WAN Ethernet port exists on a different subnet.

Most CWNA candidates are already familiar with residential wireless gateways because more than likely they have one installed at home.

VPN Wireless Router

Much like the residential wireless gateway, enterprise-class wireless routers exist that can also act as an end point for a VPN tunnel. These enterprise *VPN wireless routers* have all of the same features that can be found in a SOHO wireless router, and they provide secure tunneling functionality in addition to 802.11 layer 2–defined security capabilities. Supported VPN protocols may include PPTP, IPsec, and SSL. VPN wireless routers are typically used as edge router solutions in remote or branch offices.

A short discussion of using VPN security with wireless networking can be found in Chapter 14.

Wireless LAN Mesh Access Points

Another specialty WLAN device gaining in popularity is the *WLAN mesh access point.* Wireless mesh APs communicate with each other by using proprietary layer 2 routing protocols, creating a self-forming and self-healing wireless infrastructure (a mesh) over which edge devices can communicate, as shown in Figure 10.26.

A self-forming WLAN mesh network automatically connects access points upon installation and dynamically updates routes as more clients are added. Because interference may occur, a self-healing WLAN mesh network will automatically reroute data traffic in a Wi-Fi mesh cell. Proprietary layer 2 intelligent routing protocols determine the dynamic routes based on measurement of traffic, signal strength, hops, and other parameters. Although a WLAN mesh network can be a mesh of repeater-like access points that all operate on one frequency, dual-band mesh APs are now much more common. With dual-band WLAN mesh APs, typically the 5 GHz radios are used for the mesh infrastructure and to provide backhaul while the 2.4 GHz radios are used to provide access to the client stations.

Although the 802.11s Task Group is currently working on standardizing WLAN mesh networking, all current vendor solutions are proprietary.

FIGURE 10.26 Wireless LAN mesh network

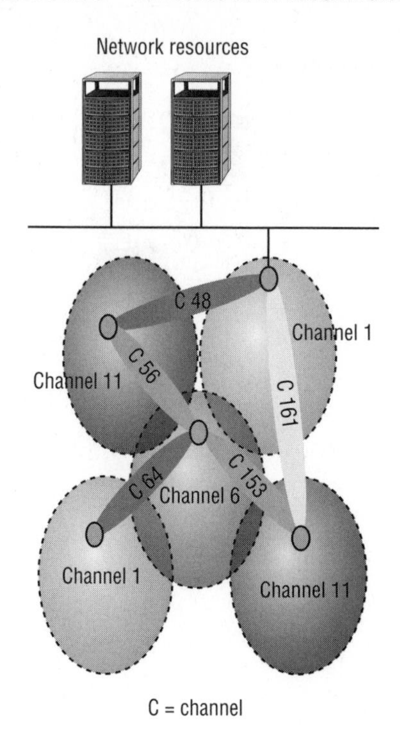

Enterprise Encryption Gateway

An *enterprise encryption gateway (EEG)* is an 802.11 middleware device that provides for segmentation and encryption. The main purpose of an EEG is to provide an overlay encryption solution. An EEG solution could be characterized as a proprietary layer 2 VPN solution, as pictured in Figure 10.27. The EEG typically sits behind several autonomous access points and segments the wireless network from the protected wired network infrastructure. Proprietary encryption technology using the AES algorithm at layer 2 is provided by the enterprise encryption gateway. Standard WPA2-compliant WLAN devices use Advanced Encryption Standard (AES) 128-bit dynamic encryption keys. EEG solutions can also provide 192-bit and 256-bit AES encryption. Figure 10.28 shows a picture of an EEG hardware device.

All the access points are managed from the unencrypted side of each gateway, and special client software is required for the end-user client stations. EEGs can also offer data compression and are typically certified to meet government security regulations such as FIPS 140-2. A central management server is used so that user and device authentication methods are also provided.

FIGURE 10.27 EEG topology

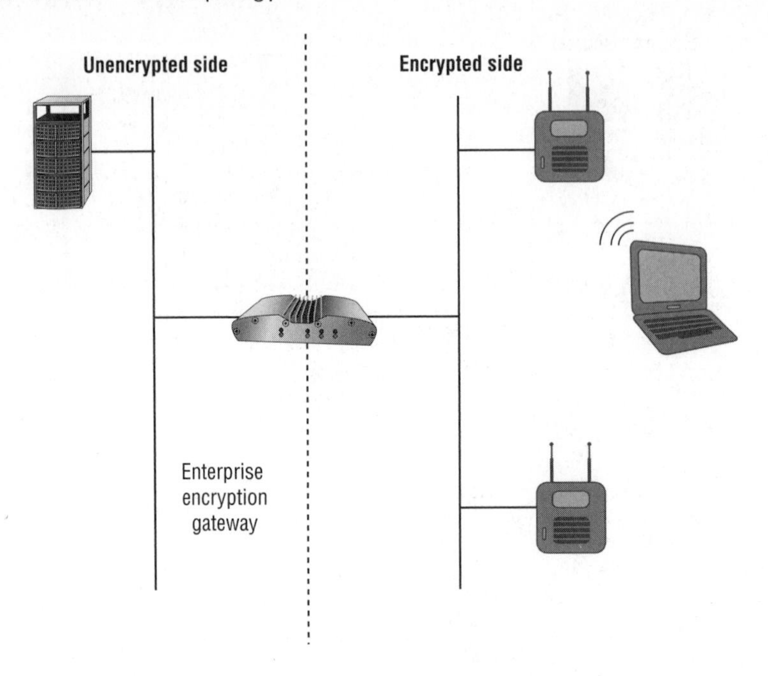

FIGURE 10.28 Enterprise encryption gateway

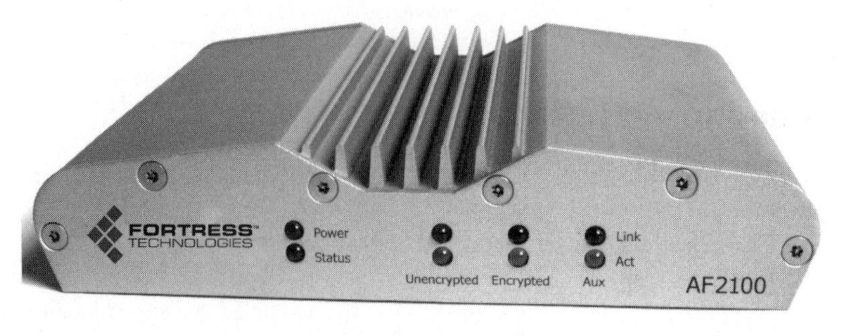

COURTESY OF FORTRESS TECHNOLOGIES.

WLAN Array

A company called Xirrus offers a proprietary solution that combines a WLAN controller and multiple access points in a single hardware device known as a Wi-Fi Array. The CWNP program uses the generic term *WLAN array* to describe this technology. As pictured in Figure 10.29, up to 16 access-point radios using sector antennas and an embedded WLAN controller all reside in one device. The WLAN controller is obviously deployed at the access layer because the device is mounted on the ceiling. The embedded WLAN controller offers many of the same features and capabilities found in more-traditional WLAN controllers.

FIGURE 10.29 WLAN array

COURTESY OF XIRRUS.

One of the key points of a WLAN array is that each individual AP has a sector antenna providing unidirectional coverage. Each AP therefore provides a sector of coverage. The WLAN array is simply an indoor sectorized array solution that provides 360 degrees of horizontal coverage by combining the unidirectional coverage of all the sector APs. The unidirectional coverage of each individual AP increases the range much like an outdoor sectorized array. The number of radios that are in a WLAN array often depends on the model and configuration. A WLAN array may have four AP radios, eight AP radios, or even as many as sixteen AP radios. A 16 access-point WLAN array would consist of four 2.4 GHz radios and twelve 5 GHz radios. One of the radios can be used as a full-time sensor device for the WIDS that is embedded with the controller.

One major advantage of the WLAN array solution is that there is much less physical equipment that needs to be deployed; therefore, the number of devices that need to be installed and managed is drastically reduced.

Cooperative Control

Aerohive Networks offers a WLAN solution using a *cooperative control access point (CC-AP)*. A CC-AP combines an autonomous access point with a suite of *cooperative control* protocols, without requiring a WLAN controller.

As pictured in Figure 10.30, the cooperative control protocols enable multiple CC-APs to be organized into groups, or *hives*, that share control information between CC-APs to provide functions such as layer 2 roaming, layer 3 roaming, cooperative RF management, security, and mesh networking. The best way to describe a cooperative control architecture is to think of it as an architecture that consists of groups of autonomous access points with WLAN controller intelligence and capabilities.

FIGURE 10.30 Cooperative control topology

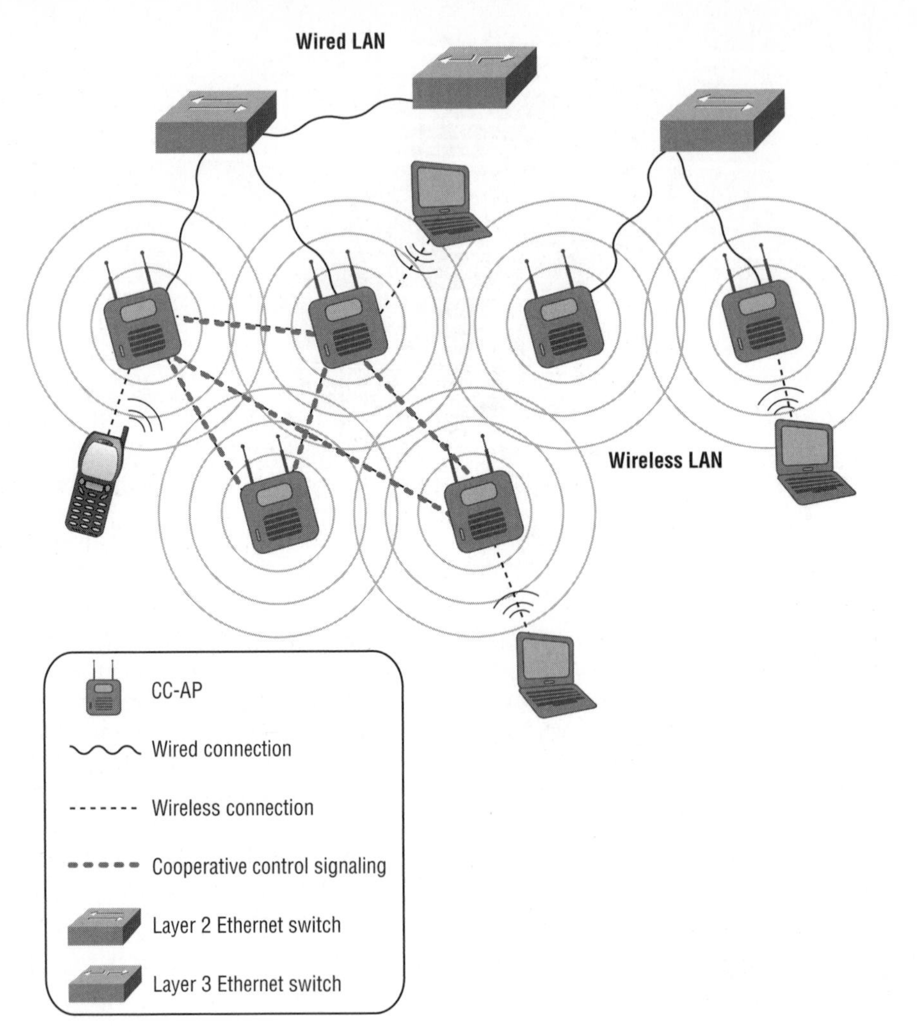

Virtual AP System

Several WLAN vendors offer solutions know as *virtual APs*. A virtual access-point solution uses multiple access points that all share a single basic service set identifier (BSSID). Because the multiple access points advertise only one single virtual MAC address (BSSID), client stations believe they are connected to only a single access point, although they may be roaming across multiple physical APs. The main advantage is that clients experience a *zero handoff* time and many of the latency issues associated with roaming are resolved. All the handoff and management is handled by a central WLAN switch. A virtual AP solution also uses a unique WLAN topology called *single channel architecture (SCA)*. All of the access points in an SCA transmit on the same channel yet do not interfere with each other. WLAN vendors such as Meru Networks and Extricom use very creative proprietary methods within the constraints of the 802.11-2007 standard to provide virtual AP and SCA topologies.

A more detailed discussion about single channel architecture (SCA) and multiple channel architecture (MCA) can be found in Chapter 12.

Real-Time Location Systems

WLAN controllers and WIDS solutions have some integrated capabilities to track 802.11 clients by using the access points as sensors. However, the tracking capabilities are not necessarily real-time and may be accurate to within only about 25 feet. The tracking capabilities in WLAN controllers and WIDS solutions provide a *near-time* solution and cannot track Wi-Fi RFID tags. Several companies such as AeroScout and Ekahau provide a WLAN *real-time location system (RTLS)*, which can track the location of any 802.11 radio device as well as active Wi-Fi RFID tags with much greater accuracy. The components of an overlay WLAN RTLS solution include the preexisting WLAN infrastructure, preexisting WLAN clients, Wi-Fi RFID tags, and an RTLS server. Additional RTLS WLAN sensors can also be added to supplement the preexisting WLAN APs.

Active RFID tags and/or standard Wi-Fi devices transmit a brief signal at a regular interval, adding status or sensor data if appropriate. Figure 10.31 shows an active RFID tag attached to a hospital IV pump. The signal is received by standard wireless APs (or RTLS sensors), without any infrastructure changes needed, and is sent to a processing engine that resides in the RTLS server at the core of the network. The RTLS server uses signal strength and/or time-of-arrival algorithms to determine location coordinates.

FIGURE 10.31 Active 802.11 RFID tag

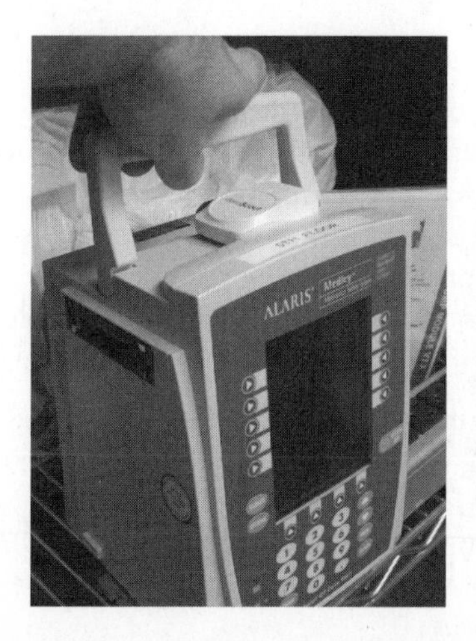

COURTESY OF AEROSCOUT.

As pictured in Figure 10.32, a software application interface is then used to see location and status data on a display map of the building's floor plan. The RTLS application can display maps, enable searches, automate alerts, manage assets, and interact with third-party applications.

FIGURE 10.32 RTLS application

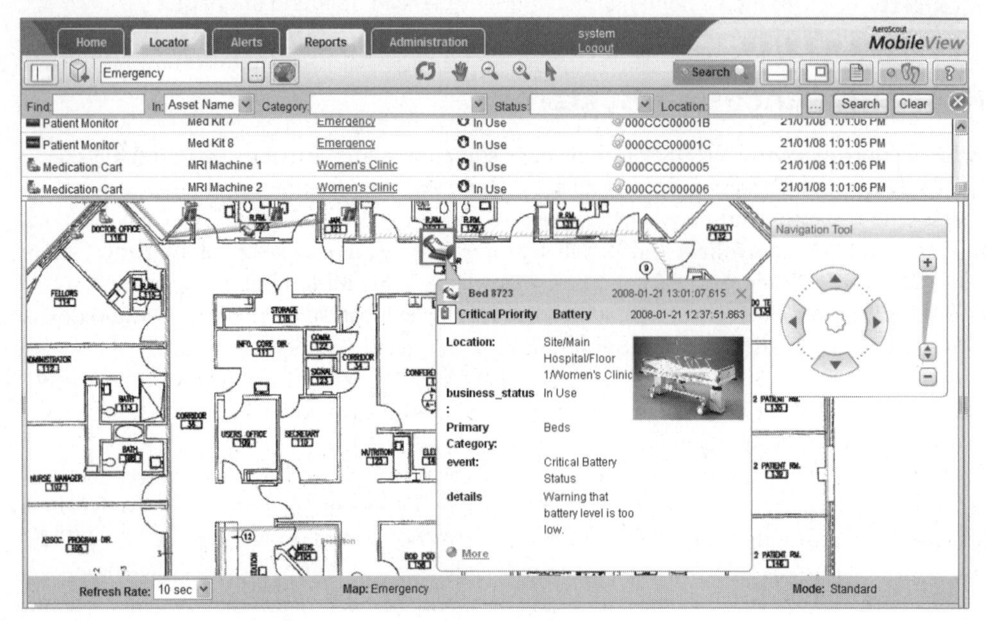

COURTESY OF AEROSCOUT.

VoWiFi

VoIP communications have been around for many years on wired networks. However, using VoIP on an 802.11 wireless LAN presents many challenges due to the RF environment and QoS considerations. In recent years, the demand for *Voice over Wi-Fi (VoWiFi)* solutions has grown considerably. The WLAN can be used to provide communications for all data applications while at the same time providing for voice communications using the same WLAN infrastructure, The components needed to deploy a VoWiFi solution include the following:

VoWiFi telephones A VoWiFi phone is similar to a cell phone except the radio is an 802.11 radio instead of a cellular radio. VoWiFi phones are 802.11 client stations that communicate through an access point. They fully support WEP, WPA, and WPA2 encryption, and WMM quality-of-service capabilities. Most VoWiFi phones use 802.11b radios using HR-DSSS and transmit in only the 2.4 GHz ISM band. However, in recent years,

VoWiFi phones that can also use OFDM technology and operate in the 5 GHz UNII bands have become available. Figure 10.33 is a picture of Polycom's SpectraLink 8030 VoWiFi phone, which has an 802.11a/b/g radio and can operate in either the 2.4 GHz band or the 5 GHz bands. VoWiFi technology can also reside in form factors other than a telephone. As pictured in Figure 10.34, VoWiFi vendor, Vocera, sells an 802.11 communications badge which is a wearable device that weighs less than two ounces. The Vocera badge is a fully-functional VoWiFi phone that also uses speech recognition and voiceprint verification software.

FIGURE 10.33 VoWiFi phone (SpectraLink 8030)

COURTESY OF POLYCOM.

FIGURE 10.34 Vocera Communications Badge

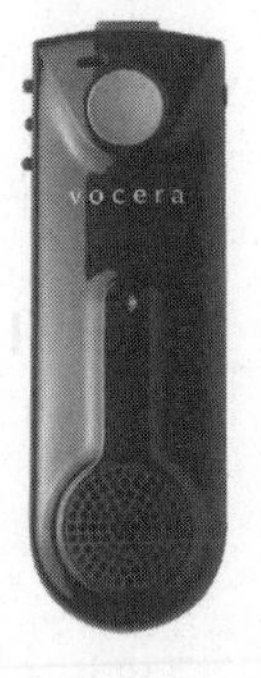

COURTESY OF VOCERA.

802.11 infrastructure (APs and controllers) An existing WLAN infrastructure is used for 802.11 communications between the VoWiFi and access points. Autonomous APs and/or WLAN controller solutions can both be used.

PBX A *private branch exchange (PBX)* is a telephone exchange that serves a particular business or office. PBXs make connections among the internal telephones of a private company and also connect them to the *public switched telephone network (PSTN)* via trunk lines. The PBX provides dial tone and may provide other features such as voicemail.

QoS server (optional) Before the ratification of the 802.11e QoS amendment and WMM QoS certification testing, quality of service capabilities were practically nonexistent within an 802.11 WLAN. However, VoWiFi companies such as Polycom offer proprietary QoS capabilities in the form of a QoS server. As pictured in Figure 10.35, the QoS server is a gateway between the PBX and the WLAN infrastructure. The QoS server ensures that the VoIP packets are prioritized and are transmitted in a timely fashion within the 802.11 infrastructure. The QoS server also has call admission control and load-balances the active calls between the access points. It is beyond the scope of this book to discuss how these proprietary QoS mechanisms operate. Many customers of these VoWiFi vendors still use these proprietary QoS server solutions; however, as acceptance of WMM (standardized 802.11 QoS) continues to grow, the need for the extra QoS servers will diminish.

FIGURE 10.35 VoWiFi topology

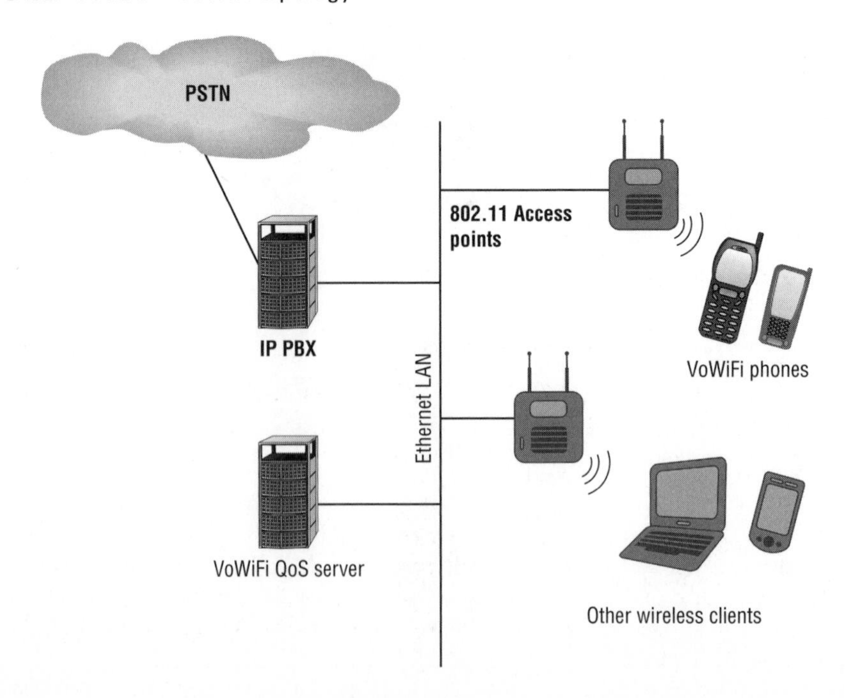

Summary

In this chapter, we discussed the different types of radio card formats, their chipsets, and the software interfaces needed for client station configuration. We also showed you the logical progression that WLAN devices have made, starting from autonomous access points, moving to WLAN controllers, and then moving along a path toward fully integrating wireless capabilities into wired network infrastructure devices. In addition, we covered specialty WLAN infrastructure devices that often meet needs that may not be met by more-traditional WLAN architecture.

The authors of this book recommend that before you take the CWNA exam, you get some hands-on experience with some WLAN infrastructure devices. We understand that most individuals cannot afford a $10,000 core WLAN controller; however, we do recommend that you purchase at least one 802.11a/b/g client adapter and either an autonomous access point or a SOHO wireless router. Hands-on experience will solidify much of what you have learned in this chapter as well as in many of the other chapters in this book.

Exam Essentials

Know the major radio card formats. The 802.11 standard does not mandate what type of format can be used by an 802.11 radio. 802.11 radios exist in multiple formats.

Understand the need for client adapters to have an operating system interface and a user interface. A client adapter requires a special driver to communicate with the operating system and a software client utility for user configuration.

Identify the four major types of client utilities. The four types of client utilities are SOHO, enterprise, integrated, and third party.

Explain the progression of WLAN architecture. Be able to explain the differences and similarities of autonomous AP solutions and WLAN controller solutions.

Identify the capabilities of all WLAN legacy infrastructure devices. Understand the capabilities of autonomous APs. Explain the differences between autonomous APs and lightweight APs.

Identify the capabilities of a WLAN controller solution. Understand all the features and functionality that a WLAN controller solution provides. Be able to explain IP tunneling, split MAC architecture, virtual BSSIDs, WLAN profiles, and dynamic RF. Be able to explain the various ways that a WLAN controller solution can be scaled. Explain the concept of distributed data forwarding.

Explain the role and configuration of WLAN bridges and workgroup bridges. The CWNA test covers bridging quite extensively. Know all of the different types of bridges and the difference between root and nonroot bridges. Be able to explain the differences between point-to-point and point-to-multipoint bridging. Understand bridging problems such as ACK time-out, and study other bridging considerations that are covered in other chapters, such as the Fresnel zone and system operating margin.

Define WLAN mesh networking. Be able to explain that WLAN mesh routers use self-healing and self-forming methods and proprietary layer 2 routing protocols. Understand the difference between single-band and dual-band mesh networks.

Explain other WLAN specialty infrastructure. Be able to explain how EEG solutions, RTLS solutions, and VoWiFi solutions can all be integrated with a WLAN. Explain other nontraditional WLAN solutions such as WLAN arrays, virtual APs, and cooperative control APs.

Key Terms

Before you take the exam, be certain you are familiar with the following terms:

autonomous AP
basic service set identifier (BSSID)
Bridged Virtual Interface (BVI)
chipset
client utilities
CompactFlash (CF)
Control and Provisioning of Wireless Access Points (CAPWAP)
cooperative control
cooperative control access point (CC-AP)
distributed data forwarding (DDF)
dynamic RF
enterprise encryption gateway (EEG)
enterprise wireless gateway (EWG)
ExpressCard
Generic Routing Encapsulation (GRE)
lightweight access points
Mini PCI
nonroot
PC Card
PCMCIA
point-to-multipoint (PtMP)
point-to-point (PtP)
private branch exchange (PBX)
public switched telephone network (PSTN)
radio frequency spectrum management (RFSM)
real-time location system (RTLS)
residential wireless gateway (RWG)
role-based access control (RBAC)
root
Secure Digital (SD)
single channel architecture (SCA)
software defined radio (SDR)
split MAC architecture
virtual AP
virtual BSSID
virtual WLAN
Voice over Wi-Fi (VoWiFi)
VPN wireless routers
wireless LAN bridge
wireless network management system (WNMS)
wireless workgroup bridge (WGB)
Wireless Zero Configuration (WZC) service
WLAN array
WLAN controller
WLAN mesh access points
WLAN profiles
workgroup bridge (WGB)

Review Questions

1. A WNMS server can be used to manage what kind of devices? (Choose all that apply.)
   - A. Autonomous APs
   - B. Virtual APs
   - C. Residential wireless gateways
   - D. Enterprise wireless gateways
   - E. WLAN controllers

2. How many root bridges exist in a point-to-multipoint bridge link?
   - A. None
   - B. One
   - C. Two
   - D. Multiple
   - E. All of the above

3. Which radio formats can be used by 802.11 technology? (Choose all that apply.)
   - A. CF
   - B. Secure Digital
   - C. PCMCIA
   - D. Mini PCI
   - E. ExpressCard
   - F. Proprietary
   - G. All of the above

4. What 802.11 infrastructure devices are capable of participating in a WLAN by using distributed data forwarding (DDF) at various network layers?
   - A. Autonomous AP
   - B. Core layer WLAN controller
   - C. Distribution layer WLAN controller
   - D. WLAN array
   - E. Hybrid access point
   - F. All of the above

5. What capabilities can be found on a WLAN controller? (Choose all that apply.)
   A. VRRP
   B. Captive portal
   C. HSRP
   D. RBAC
   E. Wireless intrusion detection system

6. Which protocols are used to transport 802.11 frames between a lightweight AP and a WLAN controller? (Choose all that apply.)
   A. GRE
   B. CCMP
   C. Mobile IP
   D. PPTP
   E. Proprietary IP tunneling protocols
   F. All of the above

7. What devices can be used to manage multiple autonomous access points from a centralized location?
   A. WLAN controller
   B. WNMS
   C. WLAN switch
   D. Enterprise wireless gateway
   E. Wireless IDS

8. What are some of the common capabilities of a WLAN controller architecture? (Choose all that apply.)
   A. Dynamic RF
   B. AP management
   C. Layer 3 roaming support
   D. Bandwidth throttling
   E. Firewall
   F. All of the above

9. Which option best describes a layer 3 device used to secure an Internet connection for a small number of wireless users?
   A. VPN router
   B. Wireless workgroup bridge
   C. Wireless mesh router
   D. Residential wireless gateway
   E. Wireless hub

10. Which option best describes a device used to provide wireless connectivity for a small number of wired clients?

    A. VPN router
    B. Wireless workgroup bridge
    C. Wireless mesh router
    D. Wireless Ethernet repeater
    E. Wireless bridge

11. Which terms best describe components of a centralized WLAN architecture in which all the intelligence resides in a centralized device and pushes the configuration settings down to the access points? (Choose all that apply.)

    A. WLAN controller
    B. Wireless network management system
    C. Enterprise wireless gateway
    D. Cooperative control AP
    E. Lightweight AP

12. A ________________ uses a proprietary layer 2 roaming protocol, and a ____________ utilizes a proprietary layer 2 encryption.

    A. Wireless mesh access point, enterprise encryption gateway
    B. Enterprise wireless gateway, WLAN controller
    C. WLAN controller, CC-AP
    D. CC-AP, enterprise wireless gateway
    E. WLAN array, CC-AP

13. A network administrator is having a hard time getting two WLAN bridges to associate with one another in a PtP link. The bridge in building A is on the 172.16.1.0/24 network, while the bridge in building B resides on the 172.16.2.0/24 network. What is the most likely cause?

    A. The bridges are on different subnets.
    B. The bridges are both configured as nonroot.
    C. The gateway address is incorrect.
    D. The ACK time-out setting is short.
    E. There is impedance overflow.

14. Billy must connect building A via a WLAN bridge link to building C, which is 30 miles away. He cannot make a direct connection of that distance because of regulatory power restrictions in his country. Building B sits between the two remote buildings. What is the best way for Billy to link the two buildings together using WLAN bridges?

A. Place a root bridge on building A with a highly directional antenna, a nonroot bridge on building B with an omnidirectional antenna, and a root bridge on building C with a highly directional antenna.

B. Place a root bridge on building A with a highly directional antenna, a repeater bridge on building B with an omnidirectional antenna, and a root bridge on building C with a highly directional antenna.

C. Place a nonroot bridge on building A with a highly directional antenna, a root bridge on building B with an omnidirectional antenna, and a nonroot bridge on building C with a highly directional antenna.

D. Place a root bridge on building A with a highly-directional antenna, and a nonroot bridge on building B with a highly directional antenna. Set up another root bridge on building B with a highly directional antenna, and a nonroot bridge on building C with a highly directional antenna. Connect the two bridges on building B via a switch or router.

E. None of the above.

15. On which device can you create and configure VLANs? (Choose all that apply.)

A. WIPS server

B. Lightweight AP

C. Ethernet switch

D. WLAN controller

E. All of the above

16. What term best describes a WLAN architecture where the integration service (IS) and distribution system services (DSS) are handled by a WLAN controller while generation of certain 802.11 management and control frames are handled by a lightweight AP?

A. Cooperative control

B. Distributed data forwarding

C. Distributed hybrid architecture

D. Distributed WLAN architecture

E. Split MAC

17. What options best describe components of a WLAN intelligent edge architecture? (Choose all that apply.)

A. Lightweight access point

B. Autonomous AP

C. WLAN controller

D. VPN router

E. WNMS

18. What would be needed for multiple basic service sets (BSSs) to exist within the same coverage area of a single access point and in which all the client stations would be segmented in separate layer 2/3 domains but all communicate within a single layer 1 RF domain?

A. Virtual BSSIDs

B. SSIDs

C. VLANs

D. Autonomous APs

E. None of the above

19. What are some of the parameters of a WLAN profile that can be configured on a WLAN controller? (Choose all that apply.)

A. SSID

B. Channel

C. VLAN

D. WMM

E. WPA-2

20. What are the required components of a VoWiFi architecture using WMM quality-of-service mechanisms? (Choose all that apply.)

A. VoWiFi phone

B. WLAN controller

C. QoS server

D. Lightweight access point

E. PBX

Answers to Review Questions

1. A, E. The main purpose of a WNMS is to provide a central point of management for autonomous APs. Configuration settings and firmware upgrades can be pushed down to the autonomous access points from the WNMS server. However, WLAN controllers effectively have replaced the WNMS as a central point of management for access points. Currently, most WMNS servers are now used as a central point of management for multiple WLAN controllers in a large-scale WLAN enterprise.

2. B. All bridge links can have only one root bridge. A PtP link will have only one root bridge, and a PtMP link will also have only one root bridge.

3. G. The 802.11 standard does not mandate what type of form factor must used by an 802.11 radio. Although PCMCIA and Mini PCI client adapters are the most common, 802.11 radios exist in many other formats, such as CompactFlash cards, Secure Digital cards, USB dongles, ExpressCards, and other proprietary formats.

4. B, C, E. Because the integration service can exist in all controllers, each controller can act as an individual gateway for the data to be distributed onto the wired network at either the core or distribution layers. Hybrid thin/fat APs handle all of the 802.11 MAC data forwarding at the edge of the network instead of back on the WLAN controller. Hybrid APs are the data distribution gateways to the wired backbone and are acting as distributed data forwarding (DDF) access points while still being managed by the central WLAN controller. Autonomous APs and a WLAN array also handle MAC data forwarding, but not in a distributed method.

5. A, B, D, E. WLAN controllers support the VRRP redundancy protocol. HSRP is a proprietary redundancy protocol. WLAN controllers have a captive portal option and support user management via role-based access control. WLAN controllers may also have an integrated IDS server.

6. A, E. An IP-encapsulated tunnel is needed for 802.11 frames to be able to traverse between a lightweight AP and a WLAN controller over a wired medium. Each 802.11 frame is encapsulated entirely within the body of an IP packet. Many WLAN vendors use Generic Routing Encapsulation (GRE), a commonly used network tunneling protocol. WLAN vendors that do not use GRE use other proprietary protocols for the IP tunneling.

7. B. One major disadvantage of using the traditional autonomous access point is that there is no central point of management. Any intelligent edge WLAN architecture with 25 or more access points is going to require some sort of wireless network management system (WNMS).

8. F. WLAN controllers support layer 3 roaming capabilities, bandwidth policies, and stateful packet inspection. Dynamic RF and AP management are also supported on a controller.

9. D. The main function of a residential wireless gateway is to provide shared wireless access to a SOHO Internet connection.

10. B. A wireless workgroup bridge (WGB) is a wireless device that provides wireless connectivity for wired infrastructure devices that do not have radio cards.

11. A, E. In the centralized WLAN architecture, autonomous APs have been replaced with lightweight access points. All the intelligence resides on the centralized device known as a WLAN controller.

12. A. A wireless mesh access point uses proprietary layer 2 roaming protocols. Proprietary encryption technology using the AES algorithm at layer 2 is provided by an enterprise encryption gateway.

13. B. In a point-to-point bridge link, one bridge must be the root bridge while the other must be a nonroot bridge. Although they are on separate subnets, this does not come into account during the association process. Typically, the IP address of the bridges is purely for management purposes and has no impact on the traffic being passed.

14. D. Because of performance issues, repeater mode is not a recommended mode for wireless bridging. If at all possible, a better bridge deployment practice is to use two separate bridge links as opposed to repeating the link of a root bridge to a nonroot bridge.

15. C, D. All of these devices support VLANs, but VLANs must be created and configured on either a managed Ethernet switch or a WLAN controller.

16. E. The majority of WLAN controller vendors implement what is known as a split MAC architecture. With this type of WLAN architecture, some of the MAC services are handled by the WLAN controller and some are handled by the lightweight access point.

17. B, D. VPN routers and autonomous APs are both devices with intelligence that reside on the edge of network architecture.

18. A. WLAN controllers have the capability to create multiple virtual BSSIDs. Within each lightweight AP's coverage area, multiple virtual WLANs can exist. Each virtual WLAN has a logical name (SSID) and a unique virtual layer 2 identifier (BSSID), and each WLAN is mapped to a unique layer 3 virtual local area network (VLAN). Multiple layer 2/3 domains can exist within one layer 1 domain.

19. A, C, D, E. WLAN controllers introduced the concept of virtual WLANs, which are often called WLAN profiles. Different groups of 802.11 clients exist in a virtual WLAN. The WLAN profile is a set of configuration parameters that are configured on the WLAN controller. The profile parameters can include the WLAN logical name (SSID), WLAN security settings, VLAN assignment, and QoS parameters. Do not confuse the WLAN profile with an AP group profile. Multiple WLAN profiles can be supported by a single AP; however, an AP can alone belong to one AP group. An AP group profile defines the configuration settings for a single AP or group of access points. Settings such as channel, transmit power, and supported data rates are examples of settings configured in an AP group profile.

20. A, B, D, E. VoWiFi phones are 802.11 client stations that communicate through most WLAN architecture, including a WLAN controller/lightweight AP solution. The PBX is needed to make connections among the internal telephones of a private company and also connect them to the public switched telephone network (PSTN) via trunk lines. Many customers of the VoWiFi vendors still use the proprietary QoS server solutions. If WMM quality-of-service capabilities are supported by both the VoWiFi phone and WLAN controller, a QoS server is not needed.

WLAN Deployment and Vertical Markets

IN THIS CHAPTER, YOU WILL LEARN ABOUT THE FOLLOWING:

- ✓ **Corporate data access and end-user mobility**
- ✓ **Network extension to remote areas**
- ✓ **Bridging—building-to-building connectivity**
- ✓ **Wireless ISP (WISP)—last-mile data delivery**
- ✓ **Small office/home office (SOHO)**
- ✓ **Mobile office networking**
- ✓ **Educational/classroom use**
- ✓ **Industrial—warehousing and manufacturing**
- ✓ **Healthcare—hospitals and offices**
- ✓ **Municipal networks**
- ✓ **Hotspots—public network access**
- ✓ **Transportation networks**
- ✓ **Law-enforcement networks**
- ✓ **First-responder networks**
- ✓ **Fixed mobile convergence**
- ✓ **WLAN and health**
- ✓ **WLAN vendors**

In this chapter, you will learn about some of the environments where wireless networks are commonly deployed. At times, some of the pros and cons of wireless in the different environments will be looked at along with some of the areas of concern. Finally, we discuss the major commercial WLAN vendors and provide links to their websites.

Corporate Data Access and End-User Mobility

As corporations decide whether to install wireless networking, they are typically looking to add two capabilities to their existing network. The first is the ability to easily add network access in areas where installation of wired connections is difficult or expensive. The second is to provide easy mobility for the wireless user within the corporate building or campus environment.

The installation of wired network jacks is very expensive, often costing as much as or even more than $200 (in U.S. dollars) per jack. As companies reorganize workers and departments, network infrastructure typically needs to be changed as well. Other areas such as warehouses, conference rooms, manufacturing lines, research labs, and cafeterias are often difficult places to effectively install wired network connections. In these and other environments, the installation of wireless networks can save the company money and provide consistent network access to all users.

Another key reason for companies to install wireless networking is to provide continuous access and availability throughout the facility. With computer access and data becoming critical components of many people's jobs, it is important for the networks to be continuously available and to be able to provide up-to-the-moment information. By installing a wireless network throughout the building or campus, the company makes it easier for employees to meet and discuss or brainstorm while maintaining access to corporate data, email, and the Internet from their laptops, no matter where they are in the building or on the campus.

Whatever the reason for installing wireless networking, companies must remember its benefits and its flaws. Wireless networking is typically slower than wired networking and therefore cannot always provide a direct replacement to wired networking. Wireless provides mobility, accessibility, and convenience, but can lack in performance and throughput. Wireless is an access technology, providing connectivity to end-user stations. Wireless should rarely be considered for distribution or core roles, except for building-to-building bridging. Even in these scenarios, make sure that the wireless bridge will be capable of handling the traffic load and throughput needs.

Network Extension to Remote Areas

If you think about it carefully, network extension to remote areas was one of the driving forces of home wireless networking, which also helped drive the demands for wireless in the corporate environment. As households connected to the Internet and as more households purchased additional computers, there was a need to connect all of the computers in the house to the Internet. Although many people installed Ethernet cabling to connect their computers, this was typically too costly, impractical because of accessibility, or beyond the capabilities of the average homeowner.

Around this time, 802.11b wireless devices were becoming more affordable. The same reasons for installing wireless networking in a home are also valid reasons for installing wireless in offices, warehouses, and just about any other environment. The cost of installing network cabling for each computer is expensive, and in many environments, running cable or fiber is difficult because of building design or aesthetic restrictions. When wireless networking equipment is installed, far fewer cables are required, and equipment placement can often be performed without affecting the aesthetics of a building.

Bridging—Building-to-Building Connectivity

To provide network connectivity between two buildings, you can install an underground cable or fiber between the two buildings, you can pay for a high-speed leased telephone connection, or you can use a building-to-building wireless bridge. All three are very capable solutions, each with its benefits and downfalls.

Although a copper or fiber connection between two buildings will potentially provide you with the highest throughput, installing copper or fiber between two buildings can be expensive. If the buildings are separated by a long distance or by someone else's property, this may not even be an option. After the cable is installed, there are no monthly service fees since you own the cable.

Leasing a high-speed telephone connection can provide flexibility and convenience, but because you do not own the connection, you will pay monthly service fees. Depending on the type of service that you are paying for, you may or may not be able to easily increase the speed of the link.

A wireless building-to-building bridge requires that the two buildings have a clear RF line of sight between them. After this has been determined or created, a point-to-point (PTP) or point-to-multipoint (PTMP) transceiver and antenna can be installed. The installation is typically easy for trained professionals to perform, and there are no monthly service fees after installation, because you own the equipment.

In addition to connecting two buildings via a point-to-point bridge, three or more buildings can be networked together by using a point-to-multipoint solution. In a point-to-multipoint installation, the building that is most centrally located will be the central communication point, with the other devices communicating directly to the central building. This is known as a *hub and spoke* or star configuration. A potential problem with the point-to-multipoint solution is that the central communication point becomes a single point of failure for all of the buildings. To prevent a single point of failure and to provide higher data throughput, it is not uncommon to install multiple point-to-point bridges.

Wireless ISP (WISP)—Last-Mile Data Delivery

The term *last mile* is often used by the telephone and cable companies to refer to the last segment of their service that connects a home subscriber to their network. The last mile of service can often be the most difficult and costly to run because at this point a cable must be run individually to every subscriber. This is particularly true in rural areas where there are very few subscribers and they are separated by large distances. In many instances, even if a subscriber is connected, the subscriber may not be able to receive some services such as high-speed Internet because services such as xDSL have a maximum distance limitation of 18,000 feet (5.7 km) from the central office.

Wireless Internet Service Providers (WISPs) deliver Internet services via wireless networking. Instead of directly cabling each subscriber, a WISP can provide services via RF communications from central transmitters. WISPs often use wireless technology other than 802.11, enabling them to provide wireless coverage to much greater areas. Some small towns have had limited success using 802.11 mesh networks as the infrastructure for a WISP. However, 802.11 technology generally is not intended to scale to the size needed for citywide WISP deployments. Service from WISPs is not without its own problems. As with any RF technology, the signal can be degraded or corrupted by obstacles such as roofs, mountains, trees, and other buildings. Proper designs and professional installations can ensure a properly working system.

Small Office/Home Office (SOHO)

One common theme of a small or home office is that your job description includes everything from janitor to IT staff and everything in between. Small-business owners and home-office employees are typically required to be very self-sufficient because there are usually few, if any, other people around to help them. Wireless networking has helped to make it easier for a small- or home-office employee to connect the office computers and peripheral devices together and also to the Internet. The main purpose of a SOHO 802.11 network is typically to provide wireless access to an Internet gateway.

Most *small office/home office (SOHO)* wireless routers provide fairly easy-to-follow installation instructions and offer performance and security near what their corporate counterparts provide. They are generally not as flexible or feature rich as comparable corporate products, but most SOHO environments do not need all of the additional capabilities. What the small- or home-office person gets is a capable device at a quarter of the price paid by their corporate counterparts. Dozens of devices are available to provide the SOHO worker with the ability to install and configure their own secure Internet-connected network without spending a fortune.

Mobile Office Networking

Mobile home offices are used for many purposes: as temporary offices during construction or after a disaster or as temporary classrooms to accommodate unplanned changes in student population, for example. Mobile offices are simply an extension of the office environment. These structures are usually buildings on wheels that can be easily deployed for short- or long-term use on an as-needed basis. Since these structures are not permanent, it is usually easier to extend the corporate or school network to these offices by using wireless networking.

A wireless bridge can be used to distribute wireless networking to the mobile office. If needed, an AP can then be used to provide wireless network access to multiple occupants of the office. By providing networking via wireless communications, you can alleviate the cost of running wired cables and installing jacks. Additional users can connect and disconnect from the network without having to make any changes to the networking infrastructure. When the mobile office is no longer needed, the wireless equipment can simply be unplugged and removed.

Moveable wireless networks are used in many environments, including military maneuvers, disaster relief, concerts, flea markets, and construction sites. Because of the ease of installation and removal, mobile wireless networking can be an ideal networking solution.

Educational/Classroom Use

Wireless networking can be used to provide a safe and easy way of connecting students to a school network. Because the layout of most classrooms is flexible (with no permanently installed furniture), installing a wired network jack for each student is not possible. Because students would be constantly connecting and disconnecting to the network at the beginning and end of class, the jacks would not last long even if they were installed. Prior to wireless networking, in classrooms that were wired with Ethernet, usually all of the computers were placed on tables along the classroom walls, with the students typically facing away from the instructor. Wireless networking enables any classroom seating arrangement to be used, without the safety risk of networking cables being strung across the floor. A wireless network also enables students to connect to the network and work on schoolwork anywhere in the building without having to worry about whether a wired network jack is nearby or whether someone else is already using it. The use of wireless bridging is also very prevalent in campus environments. Many universities and colleges use many types of wireless bridge links, including 802.11, to connect buildings campuswide.

Industrial—Warehousing and Manufacturing

Warehouses and manufacturing facilities are two environments in which wireless networking has been used for years, even before the 802.11 standard was created. Because of the vast space and the mobile nature of the employees in these environments, companies saw the need to provide mobile network access to their employees so they could more effectively perform their jobs. Warehouse and manufacturing environments often deploy wireless handheld devices such as bar code scanners, which are used for inventory control. Most 802.11 networks deployed in either a warehouse or manufacturing environment are designed for coverage rather than capacity. Handheld devices typically do not require much bandwidth, but large coverage areas are needed to provide true mobility. Most early deployments of 802.11 frequency hopping technology were in manufacturing and warehouse environments. Some legacy 802.11 FHSS deployments still exist today. Wireless networks are able to provide the coverage and mobility required in a warehouse environment and provide it cost-effectively.

Healthcare—Hospitals and Offices

Although healthcare facilities such as hospitals, clinics, and doctors' offices may seem very different from other businesses, they essentially have the same networking needs as other companies: corporate data access and end-user mobility. Healthcare providers need quick, secure, and accurate access to their data so they can react and make decisions. Wireless networks can provide mobility, giving healthcare providers faster access to important data by delivering the data directly to a handheld device that the doctor or nurse carries with them. Medical carts used to enter and monitor patient information often have wireless connections back to the nursing station. Some companies have even integrated 802.11 wireless adapters directly into the equipment that is used to monitor and track the patient's vital signs.

VoWiFi is another common use of 802.11 technology in a medical environment, providing immediate access to personnel no matter where they are in the hospital. Real-time location system (RTLS) solutions using 802.11 RFID tags for inventory control are also commonplace.

Hospitals rely on many forms of proprietary and industry-standard wireless communications that may have the potential of causing RF interference with 802.11 wireless networks. Many hospitals have designated a person or department to help avoid RF conflicts by keeping track of the frequencies and biomedical equipment that are used within the hospital.

Advanced security is often required for hospitals to meet government regulations on privacy.

Municipal Networks

Over the past few years, municipal networks have received much attention. Cities and towns announced their intentions of providing wireless networking access to their citizens throughout the area. Many municipalities viewed this as a way of providing service to some of their residents who could not necessarily afford Internet access. Although this is a well-intentioned idea, communities typically underestimated the scale and cost of these projects, and many taxpayers did not want their taxes spent on what they considered to be an unnecessary service. Although most of these plans for municipal 802.11 networks have been scrapped, there are many downtown areas where limited 802.11 coverage and services are being offered. Some of these are provided by the municipality, and others are provided by individuals or business groups.

Hotspots—Public Network Access

The term *hotspot* typically refers to a free or pay-for-use wireless network that is provided as a service by a business. When people think of hotspots, they typically associate them with cafes, bookstores, or a hospitality-type business such as a hotel or convention center. Hotspots can be used effectively by businesses to attract customers. Business travelers often frequent restaurants or cafes that are known to provide free Internet access. Many of these establishments benefit from the increased business generated by offering a hotspot. Free hotspots have drawn much attention to the 802.11 wireless industry, helping to make more people aware of the benefits of the technology.

Other hotspot providers have had difficulty convincing people to pay upward of $40 per month for a subscription. Many airports and hotel chains have installed pay-for-use hotspots; however, there are many providers, each one offering a separate subscription, which is often not practical for the consumer.

Most hotspot providers perform network authentication by using a special type of web page known as a *captive portal*. When a user connects to the hotspot, the user must open up a web browser. No matter what web page the user attempts to go to, a logon web page will be displayed instead. This is the captive portal page. If the hotspot provider is a paid service, the user must enter either their subscription information if they are a subscriber to the service, or credit card information if they are paying for hourly or daily usage. Many free hotspots also use captive portals as a method for requiring users to agree to a usage policy before they are allowed access to the Internet. If the user agrees to the terms of the policy, they are required to either enter some basic information or click a button, validating their agreement with the usage policy. Many corporations also use captive portals to authenticate guest users onto their corporate networks.

Real World Scenario

Do Hotspots Provide Data Security?

It is important to remember that hotspot providers (free or pay-for-use) do not care about the security of your data. The free provider typically offers you Internet access as a way of encouraging you to visit their location, such as a cafe, and buy some of whatever it is they sell. The pay-for-use hotspot provider performs authentication to make sure you are a paid subscriber, and after you have proven that, they will provide you with access to the Internet. Except for rare occasions, neither of these hotspot providers performs any data encryption. Because of this, business users often use VPN client software to provide a secure encrypted tunnel back to their corporate network whenever they are using a hotspot. Many companies require employees to use a VPN during any connection to a public hotspot. Further discussion of security issues related to hotspot use is found in Chapter 14, "Wireless Attacks, Intrusion Monitoring, and Policy."

Transportation Networks

In discussing Wi-Fi transportation networks, the three main modes of transportation—trains, planes, and automobiles—are typically mentioned. In addition to these three primary methods of transportation, two others need to be mentioned. The first is boats, both cruise ships and commuter ferries, and the second is buses, similar to but different from automobiles.

Providing Wi-Fi service to any of the transportation methods is easy. Simply install one or more access points in the vehicle. Except for the cruise ship and large ferries, most of these methods of transportation would require only a few access points to provide Wi-Fi coverage. The primary use of these networks is to provide hotspot services for end users so that they can gain access to the Internet. The difference between a transportation network and a typical hotspot is that the network is continually moving, making it necessary for the transportation network to use some type of mobile uplink services.

To provide an uplink for a train, which is bound to the same path of travel for every trip, a metropolitan wireless networking technology such WiMAX could be used along the path of the tracks. With the other transportation networks, for which the path of travel is less bounded, the more likely uplink method would be via some type of cellular network connection. However, if WiMAX begins to be deployed in larger areas, either could be an acceptable uplink method for trains, buses, or automobiles.

Commuter ferries are likely to provide uplink services via cellular or WiMAX, because they are likely within range of these services. For ferries that travel farther distances away from shore and cruise ships, a satellite link is likely to be used.

Many airlines are in the process of installing Wi-Fi on their planes. The Wi-Fi service inside the airplane consists of one or more access points connected to a cellular router, which communicates to the cellular towers on the ground. This service is typically offered for a nominal fee and is available only while the airplane is flying, and typically when the airplane is at cruising altitude. Bandwidth metering is used to prevent any one user from monopolizing the connection.

Law-Enforcement Networks

Although Wi-Fi networks cannot provide the wide area coverage necessary to provide continuous wireless communications needed by law-enforcement personnel, they can still provide a major role in fighting crime. Many law-enforcement agencies are using Wi-Fi as a supplement to their public-safety wireless networks.

In addition to the obvious mobility benefits of using Wi-Fi inside police stations, many municipalities have installed Wi-Fi in the parking lots outside the police station and other municipal buildings as a supplement to their wireless metropolitan networks. These outdoor networks are sometimes viewed as secured hotspots. Unlike public hotspots, these networks provide both authentication and high levels of encryption. In addition to municipalities incorporating wireless technology into law enforcement, many are also adding non Wi-Fi based automation to utilities through the use of supervisory control and data acquisition (SCADA) equipment. Because of this growth in the use of different wireless technologies, we are starting to see municipalities designate a person or department to keep track of the frequencies and technologies that are being used.

Municipal Wi-Fi hotspots typically provide high-speed communications between networking equipment in the police cars and the police department's internal network. An interesting example of a good use of this network is the uploading of vehicle video files. With many police cars being equipped with video surveillance, and with these surveillance videos often being used as evidence, it is important to not only transfer these video files to a central server for cataloging and storage, but to also do it with the least amount of interaction by the police officer to preserve the chain of evidence.

When a police car arrives at one of these municipal Wi-Fi hotspots, the computer in the car automatically uploads the video files from the data storage in the car to the central video library. Automating this process minimizes the risk of data corruption and frees up the officer to do other, more-important tasks.

First-Responder Networks

When medical and fire rescue personnel arrive at the scene of an emergency, it is important for them to have fast and easy access to the necessary resources to handle the emergency at hand. Many rescue vehicles are being equipped with either permanently mounted Wi-Fi access points or easily deployed, self-contained portable access points that can quickly and easily blanket a rescue scene with a Wi-Fi bridge to the emergency personnel's data network. In a disaster, when public service communications systems such as cellular telephone networks may not be working because of system overload or outages, a Wi-Fi first-responder network may be able to provide communications between local personnel and possibly shared access to central resources.

During a disaster, assessing the scene and triaging the victims (grouping victims based on the severity of their injuries) is one of the first tasks. Historically, the task of triage included paper tags that listed the medical information and status of the victim. Some companies have created electronic triage tags that can hold patient information electronically and transmit it via Wi-Fi communications.

Fixed Mobile Convergence

One of the hottest topics currently relating to Wi-Fi is known as *fixed mobile convergence (FMC)*. The goal of FMC systems is to provide a single device, with a single telephone number that is capable of switching between networks and always using the lowest-cost network.

With the flexibility and mobility of cellular telephones, it is common for people to use them even in environments (home or work) where they are stationary and have access to other telephone systems that are frequently less costly. FMC devices typically are capable of communicating via either a cellular telephone network or a VoWiFi network. If you had an FMC telephone and were at your office or home, where a Wi-Fi network is available, the telephone would use the Wi-Fi network for any incoming or outgoing telephone calls. If you were outside either of these locations and did not have access to a Wi-Fi network, the telephone would use the cellular network for any incoming or outgoing telephone calls. FMC devices would also allow you to roam across networks. So you could initiate a telephone call from within your company by using the Wi-Fi network. As you walked outside, the FMC telephone would roam from the Wi-Fi network to the cellular network and seamlessly transition between the two networks. With fixed mobile convergence, you would be able to have one device and one telephone number that would work wherever you were, using the least costly network that was available at the time.

There is an 802.11 amendment that may well address seamless handoff and session persistence with other external networks such as cellular networks. The IEEE 802.11u amendment, which is often referred to as the Wireless InterWorking with External Networks (WIEN) amendment, could possibly standardize FMC communications between 802.11 and cellular networks.

WLAN and Health

Over the years, there has always been a concern about adverse health effects from the exposure of humans and animals to radio waves. The World Health Organization and government agencies set standards that establish exposure limits to radio waves, to which RF products must comply. Tests performed on WLANs have shown that they operate substantially below the required safety limits set by these organizations. Also Wi-Fi signals, as compared to other RF signals, are much lower in power. The World Health Organization has also concluded that there is no convincing scientific evidence that weak radio-frequency signals, such as those found in 802.11 communications, cause adverse health effects.

You can read more about some of these findings at the following websites:

- U.S. Federal Communications Commission

  www.fcc.gov/oet/rfsafety/rf-faqs.html

- World Health Organization

  www.who.int/peh-emf

- Wi-Fi Alliance

  www.wi-fi.org

WLAN Vendors

There are many vendors in the 802.11 WLAN marketplace, including established companies such as Cisco and Aruba, along with startup WLAN companies such as Xirrus. The following is a list of some of the major WLAN vendors. Please note that each vendor is listed in only one category, even if they offer products and services that cover multiple categories. This is most notable with the infrastructure vendors, who often offer additional capabilities as features of their products, such as security and troubleshooting.

WLAN Infrastructure

These 802.11 enterprise equipment vendors manufacture and sell WLAN controllers and access points:

- Aerohive Networks—www.aerohive.com
- Aruba Networks—www.arubanetworks.com
- Bluesocket—www.bluesocket.com

- Cisco—www.cisco.com
- Colubris Networks—www.colubris.com
- Extricom—www.extricom.com
- Meru Networks—www.merunetworks.com
- Motorola—www.motorola.com
- Proxim Wireless Corporation—www.proxim.com
- Ruckus Wireless—www.ruckuswireless.com
- Siemens—www.siemens.com
- Trapeze Networks—www.trapezenetworks.com
- Xirrus—www.xirrus.com

WLAN Mesh Infrastructure

These WLAN vendors specialize in 802.11 mesh networking:

- BelAir Networks—www.belairnetworks.com
- Firetide—www.firetide.com
- Meraki—www.meraki.com
- MeshDynamics —www.meshdynamics.com
- Strix Systems —www.strixsystems.com
- Tropos Networks —www.tropos.com

WLAN Troubleshooting and Design Solutions

These are some companies that sell 802.11 protocol analyzers, spectrum analyzers, site survey software, and other WLAN analysis solutions:

- AirMagnet—www.airmagnet.com
- Berkeley Varitronics Systems —www.bvsystems.com
- CACE Technologies—www.cacetech.com
- Ekahau—www.ekahau.com
- Fluke Networks—www.flukenetworks.com
- MetaGeek—www.metageek.net
- TamoSoft—www.tamos.com
- WildPackets—www.wildpackets.com
- Wireshark—www.wireshark.org

WLAN Management

These companies provide wireless network management system (WNMS) solutions:

- AirWave—www.airwave.com
- Wavelink—www.wavelink.com

WLAN Security Solutions

These WLAN companies offer overlay encryption solutions, WLAN IDS solutions, or 802.1X/EAP supplicant/server solutions:

- AirDefense—www.airdefense.net
- AirTight Networks—www.airtightnetworks.com
- Fortress Technologies—www.fortresstech.com
- Juniper Networks—www.juniper.net

VoWiFi Solutions

Manufacturers of 802.11 VoWiFi phones and VoIP gateway solutions include the following:

- Ascom—www.ascom.com
- Polycom—www.polycom.com
- Vocera—www.vocera.com

WLAN Fixed Mobile Convergence

Manufacturers of 802.11 and cellular convergence solutions include the following:

- Agito Networks—www.agitonetworks.com
- DiVitas Networks—www.divitas.com

WLAN RTLS Solutions

Manufacturers of 802.11 real-time location system (RTLS) solutions include the following:

- AeroScout—www.aeroscout.com
- Newbury Networks—www.newburynetworks.com

WLAN SOHO Vendors

These are some of the many WLAN vendors selling SOHO solutions that can provide Wi-Fi for the average home user:

- Apple —`www.apple.com`
- Buffalo Technology—`www.buffalotech.com`
- Belkin International—`www.belkin.com`
- D-Link —`www.dlink.com`
- Hawking Technology—`www.hawkingtech.com`
- Linksys—`www.linksys.com`
- Netgear—`www.netgear.com`
- SMC Networks—`www.smc.com`

Summary

This chapter covered some of the design, implementation, and management environments in which wireless networking is used. Although many of these environments are similar, each has unique characteristics. It is important to understand these similarities and differences and how wireless networking is commonly deployed.

Exam Essentials

Know the different WLAN vertical markets. Wireless networking can be used in many environments, with each vertical market having a different primary reason or focus for installing the wireless network. Know these environments and their main reasons for deploying 802.11 wireless networking.

Know fixed mobile convergence With cellular networking and Wi-Fi networking so common, telephone vendors are beginning to provide phones that are capable of communicating over both networks and provide roaming between them. Know what FMC is and the reasons and benefits of deploying it.

Key Terms

Before you take the exam, be certain you are familiar with the following terms:

fixed mobile convergence (FMC)
hotspot
last mile
small office/home office (SOHO)
Wireless Internet Service Providers (WISPs)

Review Questions

1. Which of the following are objectives of fixed mobile convergence? (Choose all that apply.)
   - A. Single telephone number
   - B. Single device
   - C. Always use the best-performing network
   - D. Always use the lowest-cost network
2. Which of the following is another form of a public hotspot network? (Choose all that apply.)
   - A. Law-enforcement network
   - B. First-responder network
   - C. Transportation network
   - D. Municipal network
3. Which type of organization often has a person responsible for keeping track of frequency usage inside the organization?
   - A. Law enforcement
   - B. Hotspot
   - C. Hospital
   - D. Cruise ship
4. On which of these transportation networks is satellite a functional solution for providing uplink to the Internet? (Choose all that apply.)
   - A. Bus
   - B. Automobile
   - C. Train
   - D. Cruise ship
5. Fixed mobile convergence provides roaming across which of the following wireless technologies? (Choose all that apply.)
   - A. Bluetooth
   - B. Wi-Fi
   - C. WiMAX
   - D. Cellular telephone
6. Which of the following is typically the most important design criteria when designing a warehouse WLAN?
   - A. Capacity
   - B. Throughput
   - C. RF interference
   - D. Coverage

7. Corporations typically install wireless networks to provide which of the following capabilities? (Choose all that apply.)
   - **A.** Easy mobility for the wireless user within the corporate building or campus environment
   - **B.** High-speed network access comparable to wired networking
   - **C.** Secure access for employees from their homes or on the road
   - **D.** The ability to easily add network access in areas where installation of wired connections is difficult or expensive

8. Last-mile Internet service is provided by which of the following? (Choose all that apply.)
   - **A.** Telephone company
   - **B.** Long-distance carrier
   - **C.** Cable provider
   - **D.** WISPs

9. Which of the following is the main purpose of a SOHO 802.11 network?
   - **A.** Shared networking
   - **B.** Internet gateway
   - **C.** Network security
   - **D.** Print sharing

10. Which of the following are examples of mobile office networking? (Choose all that apply.)
    - **A.** Construction-site offices
    - **B.** Temporary disaster-assistance office
    - **C.** Remote sales office
    - **D.** Temporary classrooms

11. Warehousing and manufacturing environments typically have which of the following requirements? (Choose all that apply.)
    - **A.** Mobility
    - **B.** High-speed access
    - **C.** High capacity
    - **D.** High coverage

12. Which of the following is least likely to be offered by a hotspot provider?
    - **A.** Free access
    - **B.** Paid access
    - **C.** Network authentication
    - **D.** Data encryption

13. Which of the following are good uses for mobile networks? (Choose all that apply.)
    A. Military maneuvers
    B. Disaster relief
    C. Construction sites
    D. Manufacturing plants

14. Which of the following terms refer to the same network design? (Choose all that apply.)
    A. PTP
    B. PTMP
    C. Hub and spoke
    D. Star

15. Most early deployments of 802.11 FHSS were used in which type of environment?
    A. Mobile office networking
    B. Educational/classroom use
    C. Industrial (warehousing and manufacturing)
    D. Healthcare (hospitals and offices)

16. When using a hotspot, you should do which of the following to ensure security back to your corporate network?
    A. Enable WEP.
    B. Enable 802.1X/EAP.
    C. Use an IPsec VPN.
    D. Security cannot be provided, because you do not control the access point.

17. What are some popular 802.11 applications used in the healthcare industry? (Choose all that apply.)
    A. VoWiFi
    B. Bridging
    C. RTLS
    D. Patient monitoring

18. Multiple point-to-point bridges between the same locations are often installed for which of the following reasons? (Choose all that apply.)
    A. To provide higher throughput
    B. To prevent channel overlap
    C. To prevent single point of failure
    D. To enable support for VLANs

19. What are some of the key concerns of healthcare providers when installing a wireless network? (Choose all that apply.)

A. RF interference

B. Faster access to patient data

C. Secure and accurate access

D. Faster speed

20. Public hotspots typically provide which of the following security features?

A. Authentication.

B. Encryption.

C. TKIP.

D. No security is available.

Answers to Review Questions

1. A, B, D. The goal of fixed mobile convergence is to enable the user to have a single device with a single telephone number and to enable the user to roam between different networks, taking advantage of the least expensive network that is available.

2. C, D. Municipal and transportation networks are both specific types of public hotspots. Law-enforcement and first-responder networks are hotspot-type networks, but they are not intended for public use.

3. C. Because of the potential for interference and the importance of preventing it, hospitals often have a person responsible for keeping track of frequencies used within the organization. Some municipalities are starting to do this as well—not just for law enforcement, but for all of their wireless needs, because they often use wireless technologies for SCADA networks, traffic cameras, traffic lights, two-way radios, point to point bridging, hotspots, and more.

4. D. Since cruise ships are often not near land where cellular or WiMAX uplink is available, it is necessary to use a satellite uplink to connect the ship to the Internet.

5. B, D. Fixed mobile convergence allows roaming between Wi-Fi networks and cellular telephone networks, choosing the available network that is least expensive.

6. D. When designing a warehouse network, the networking devices are often barcode scanners that do not capture much data, so high capacity and throughput are not typically needed. Because the data-transfer requirements are so low, these networks are typically designed to provide coverage for large areas. Security is always a concern; however, it is not usually a design criteria.

7. A, D. Corporations typically install a WLAN to provide easy mobility and/or access to areas that are difficult or extremely expensive to connect via wired networks. Although providing connectivity to the Internet is a service that the corporate wireless network offers, it is not the driving reason for installing the wireless network.

8. A, C, D. The telephone company, cable providers, and WISPs are all examples of companies that provide last-mile services to users and businesses.

9. B. The main purpose of SOHO networks is to provide a gateway to the Internet.

10. A, B, D. Mobile office networking solutions are temporary solutions that include all of the options listed except for the remote sales office, which would more likely be classified as a SOHO installation.

11. A, D. Warehousing and manufacturing environments typically have a need for mobility, but their data transfers are typically very small. Therefore, their networks are often designed for high coverage rather than high capacity.

12. D. Hotspot providers are not likely to provide data encryption. It is more difficult to deploy, and there is no benefit or business reason for them to provide it.

13. A, B, C. Manufacturing plants are typically fixed environments and are better served by installing permanent access points.

14. B, C, D. Point-to-multipoint, hub and spoke, and star all describe the same communication technology, which connects multiple devices by using a central device. Point-to-point communications connects two devices.

15. C. Most of the 802.11 implementations used FHSS, with industrial (warehousing and manufacturing) companies being some of the biggest implementers. Their requirement of mobility with low data-transfer speeds was ideal for using the technology.

16. C. To make wireless access easy for the subscriber, hotspot vendors typically deploy authentication methods that are easy to use but that do not provide data encryption. Therefore, to ensure security back to your corporate network, the use of an IPsec VPN is necessary.

17. A, C, D. VoWiFi is a common use of 802.11 technology in a medical environment, providing immediate access to personnel no matter where they are in the hospital. Real-time location system (RTLS) solutions using 802.11 RFID tags for inventory control are also commonplace. WLAN medical carts are used to monitor patient information and vital signs.

18. A, C. The installation of multiple point-to-point bridges is either to provide higher throughput or to prevent a single point of failure. Care must be taken in arranging channel and antenna installations to prevent self-inflicted interference.

19. A, B, C. Healthcare providers often have many other devices that use RF communications, and therefore, RF interference is a concern. Fast access along with secure and accurate access is critical in healthcare environments. Faster access can be performed without faster speed. The mobility of the technology will satisfy the faster access that is typically needed.

20. A. Public hotspots are most concerned about ensuring that only valid users are allowed access to the hotspot. This is performed using authentication.

WLAN Troubleshooting

IN THIS CHAPTER, YOU WILL LEARN ABOUT THE FOLLOWING:

✓ **Layer 2 retransmissions**

- RF interference
- Multipath
- Adjacent cell interference
- Low SNR
- Mismatched power settings
- Near/far
- Hidden node

✓ **802.11 coverage considerations**

- Dynamic rate switching
- Roaming
- Layer 3 roaming
- Co-channel interference
- Channel reuse/multiple channel architecture
- Single channel architecture
- Capacity vs. coverage
- Oversized coverage cells
- Physical environment

✓ **Voice vs. data**

✓ **Performance**

✓ **Weather**

Diagnostic methods that are used to troubleshoot wired 802.3 networks should also be applied when troubleshooting a wireless local area network (WLAN). A bottoms-up approach to analyzing the OSI reference model layers also applies to wireless networking. A wireless networking administrator should always try to first determine whether problems exist at layer 1 and layer 2.

As with most networking technologies, most problems usually exist at the Physical layer. Simple layer 1 problems such as nonpowered access points or client card driver problems are often the root cause of connectivity or performance issues. Because WLANs use radio frequencies to deliver data, troubleshooting a WLAN offers many unique layer 1 challenges not found in a typical wired environment. The bulk of this chapter discusses the numerous potential problems that can occur at layer 1 and the solutions that might be implemented to prevent or rectify the layer 1 problems. A spectrum analyzer is often a useful tool when diagnosing layer 1 issues.

After eliminating layer 1 as a source of possible troubles, a WLAN administrator should try to determine whether the problem exists at the Data-Link layer. Authentication and association problems often occur because of improperly configured security and administrative settings on access points, WLAN controllers, and client utility software. A WLAN protocol analyzer is often an invaluable tool for troubleshooting layer 2 problems.

In this chapter, we discuss many coverage considerations and troubleshooting issues that may develop when deploying an 802.11 wireless network. RF propagation behaviors and RF interference will affect both the performance and coverage of your WLAN. Because mobility is usually required in a WLAN environment, many roaming problems often occur and must be addressed. The half-duplex nature of the medium also brings unique challenges typically not seen in a full-duplex environment. Different considerations also need to be given to outdoor 802.11 deployments due to weather conditions. In this chapter, we discuss how to identify, troubleshoot, prevent, and fix instances of potential WLAN problems.

Layer 2 Retransmissions

The mortal enemy of WLAN performance is layer 2 retransmissions that occur at the MAC sublayer. As you have learned, all unicast 802.11 frames must be acknowledged. If a collision occurs or any portion of a unicast frame is corrupted, the *cyclic redundancy check (CRC)* will fail and the receiving 802.11 radio will not return an ACK frame to the transmitting 802.11 radio. If an ACK frame is not received by the original transmitting radio, the unicast frame is not acknowledged and will have to be retransmitted.

Excessive layer 2 retransmissions adversely affect the WLAN in two ways. First, layer 2 retransmissions increase overhead and therefore decrease throughput. Many different factors can affect throughput, including a WLAN environment with abundant layer 2 retransmissions.

Second, if application data has to be retransmitted at layer 2, the timely delivery of application traffic becomes delayed or inconsistent. Applications such as VoIP depend on the timely and consistent delivery of the IP packet. Excessive layer 2 retransmissions usually result in latency and jitter problems for time-sensitive applications such as voice and video. When discussing VoIP, latency and jitter often get confused. *Latency* is the time it takes to deliver a VoIP packet from the source device to the destination device. A delay in the delivery (increased latency) of a VoIP packet due to layer 2 retransmissions can result in echo problems. *Jitter* is a variation of latency. Jitter measures how much the latency of each packet varies from the average. If all packets travel at exactly the same speed through the network, jitter will be zero. A high variance in the latency (jitter) is the more common result of 802.11 layer 2 retransmissions. Jitter will result in choppy audio communications and reduced battery life for VoWiFi phones.

Most data applications in a Wi-Fi network can handle a layer 2 retransmission rate of up to 10 percent without any noticeable degradation in performance. However, time-sensitive applications such as VoIP require that higher-layer IP packet loss be no greater than 2 percent. Therefore, Voice over Wi-Fi (VoWiFi) networks need to limit layer 2 retransmissions to 5 percent or less to guarantee the timely and consistent delivery of VoIP packets.

How can you measure layer 2 retransmissions? As shown in Figure 12.1, any good 802.11 protocol analyzer can track layer 2 retry statistics for the entire WLAN. 802.11 protocol analyzers can also track retry statistics for each individual WLAN access point and client station.

Unfortunately, layer 2 retransmissions are a result of many possible problems. Multipath, RF interference, and low SNR are problems that exist at layer 1 yet result in layer 2 retransmissions. Other causes of layer 2 retransmissions include hidden node, near/far, mismatched power settings, and adjacent cell interference, which are all usually a symptom of improper WLAN design.

RF Interference

Various types of RF interference can greatly affect the performance of an 802.11 WLAN. Interfering devices may prevent an 802.11 radio from transmitting, thereby causing a denial of service. If another RF source is transmitting with strong amplitude, 802.11 radios can sense the energy during the clear channel assessment (CCA) and defer transmission entirely. The other typical result of RF interference is that 802.11 frame transmissions become corrupted. If frames are corrupted due to RF interference, excessive retransmissions will occur and therefore throughput will be reduced significantly. There are several different types of interference:

Narrowband interference A narrowband RF signal occupies a smaller and finite frequency space and will not cause a denial of service (DoS) for an entire band such as the 2.4 GHz ISM band. A narrowband signal is usually very high amplitude and will absolutely disrupt communications in the frequency space in which it is being transmitted.

Narrowband signals can disrupt one or several 802.11 channels. Narrowband RF interference can also result in corrupted frames and layer 2 retransmissions. The only way to eliminate narrowband interference is to locate the source of the interfering device with a spectrum analyzer. To work around interference, use a spectrum analyzer to determine the affected channels and then design the channel reuse plan around the interfering narrowband signal. Figure 12.2 shows a spectrum analyzer capture of a narrowband signal close to channel 11 in the 2.4 GHz ISM band.

FIGURE 12.1 Layer 2 retransmission statistics

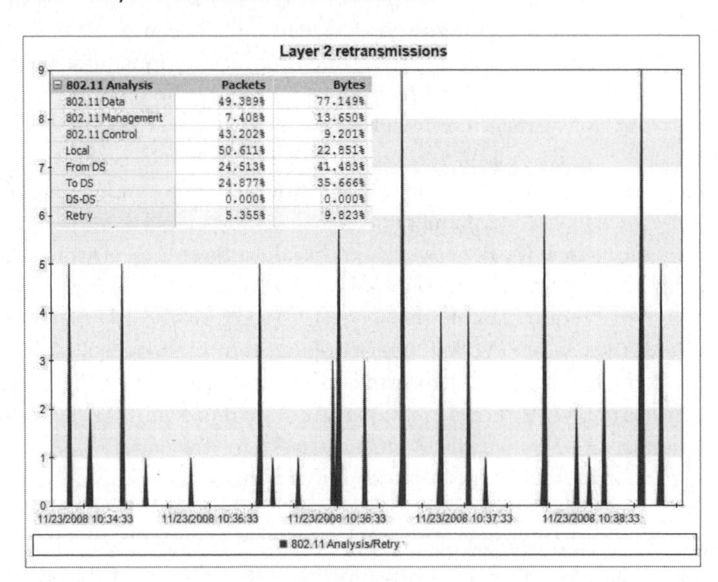

Wideband interference A source of interference is typically considered wideband if the transmitting signal has the capability to disrupt the communications of an entire frequency band. Wideband jammers exist that can create a complete DoS for the 2.4 GHz ISM band. The only way to eliminate wideband interference is to locate the source of the interfering device with a spectrum analyzer and remove the interfering device. Figure 12.3 shows a spectrum analyzer capture of a wideband signal in the 2.4 GHz ISM band with average amplitude of –60 dBm.

All-band interference The term *all-band interference* is typically associated with frequency hopping spread spectrum (FHSS) communications that usually disrupt HR-DSSS and/or ERP-OFDM channel communications at 2.4 GHz. As you learned in earlier chapters, FHSS constantly hops across an entire band, intermittingly transmitting on very small subcarriers of frequency space. A legacy 802.11 FHSS radio, for example, transmits on hops that are 1 MHz wide. While hopping and dwelling, an FHSS device will transmit in sections of the frequency space occupied by an HR-DSSS or ERP-OFDM channel. Although an FHSS device will not typically cause a denial of service, the frame transmissions from the HR-DSSS and ERP-OFDM devices can be corrupted from the all-band transmissions of the FHSS interfering radio.

FIGURE 12.2 Narrowband RF interference

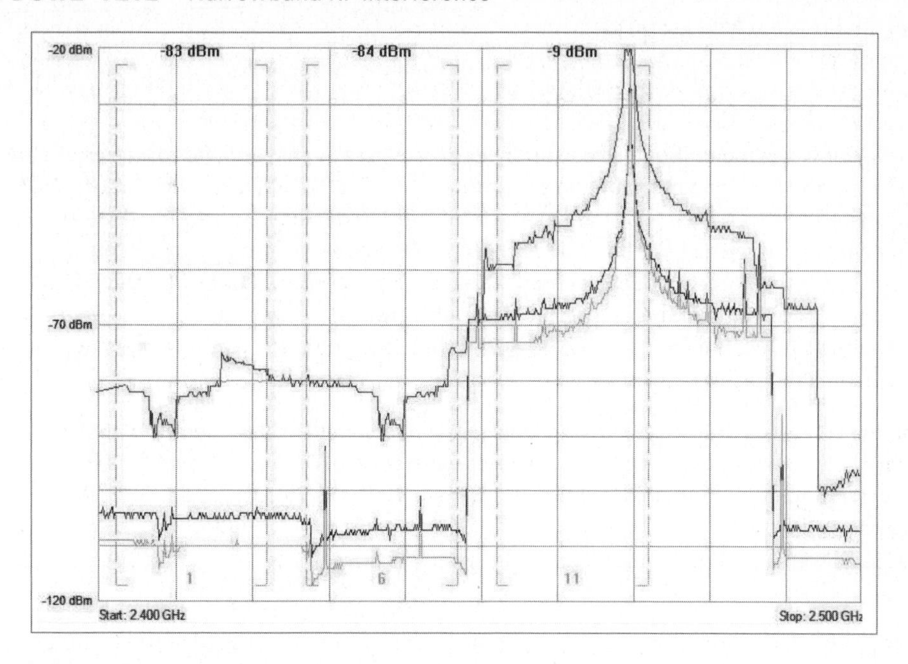

FIGURE 12.3 Wideband RF interference

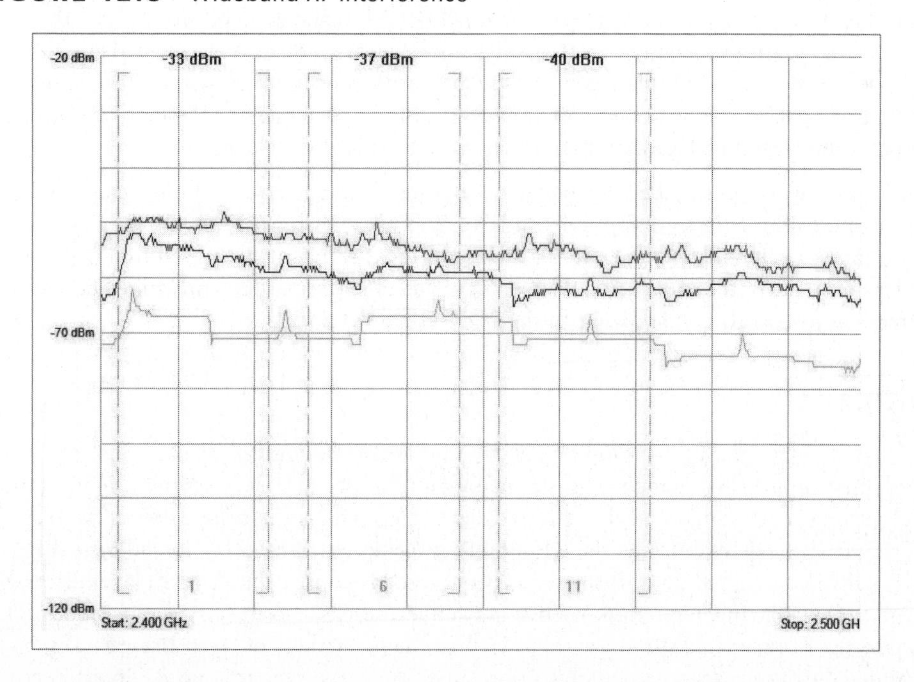

Real World Scenario

What Devices Cause RF Interference?

Numerous devices, including cordless phones, microwave ovens, and video cameras, can cause RF interference and degrade the performance of an 802.11 WLAN. The 2.4 GHz ISM band is extremely crowded, with many known interfering devices. Interfering devices also transmit in the 5 GHz UNII bands, but the 2.4 GHz frequency space is much more crowded. Often the biggest source of interference is signals from nearby WLANs. If RF interference cannot be eliminated at 2.4 GHz, special consideration should be given to deploying the WLAN in the less-crowded 5 GHz frequency bands. The tool that is necessary to locate sources of interference is a spectrum analyzer.

In Chapter 16, "Site Survey System and Devices," we discuss proper spectrum analysis techniques that should be part of every wireless site survey. Chapter 16 also lists the many interfering devices that can cause problems in both the 2.4 GHz and 5 GHz frequency ranges.

Bluetooth (BT) is a short-distance RF technology defined by the 802.15 standard. Bluetooth uses FHSS and hops across the 2.4 GHz ISM band at 1,600 hops per second. Older Bluetooth devices were known to cause severe all-band interference. Newer Bluetooth devices utilize adaptive mechanisms to avoid interfering with 802.11 WLANs. Digital Enhanced Cordless Telecommunications (DECT) cordless telephones also use frequency hopping transmissions. A now-defunct WLAN technology known as HomeRF also used FHSS; therefore, HomeRF devices can potentially cause all-band interference.

The existence of a high number of frequency-hopping transmitters in a finite space will result in some 802.11 data corruption and layer 2 retransmissions. The only way to eliminate all-band interference is to locate the source of the interfering device with a spectrum analyzer and remove the interfering device. Figure 12.4 shows a spectrum analyzer capture of a frequency hopping transmission in the 2.4 GHz ISM band.

Multipath

As discussed in Chapter 2, "Radio Frequency Fundamentals," *multipath* can cause *intersymbol interference (ISI)*, which causes data corruption. Because of the difference in time between the primary signal and the reflected signals, known as the *delay spread*, the receiver can have problems demodulating the RF signal's information. The delay spread time differential results in corrupted data. If the data is corrupted because of multipath, layer 2 retransmissions will result. In Chapter 16, we discuss active and passive site survey techniques. The main purpose of the active site survey is to look at the percentage of layer 2 retries. If it is determined during the spectrum analysis portion of the site survey that no RF interference occurred, the most likely cause of the layer 2 retransmissions will be multipath. As pictured in Figure 12.5, WLAN vendor Berkeley Varitronics Systems makes a line of WLAN troubleshooting tools that can detect and then visualize occurrences of multipath and the delay spread into a useful graphical display.

FIGURE 12.4 All-band RF interference

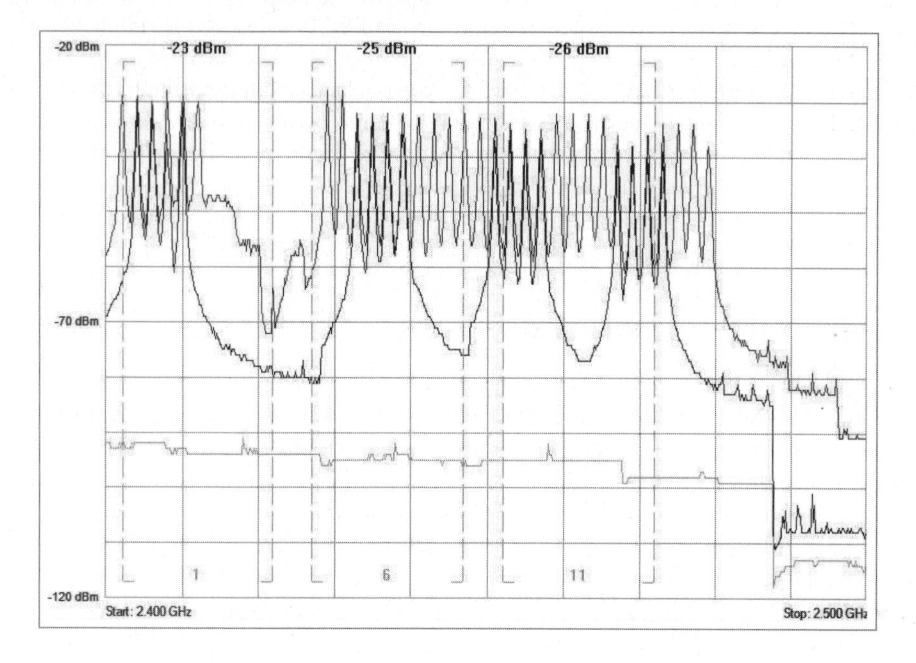

FIGURE 12.5 Multipath analysis troubleshooting tool

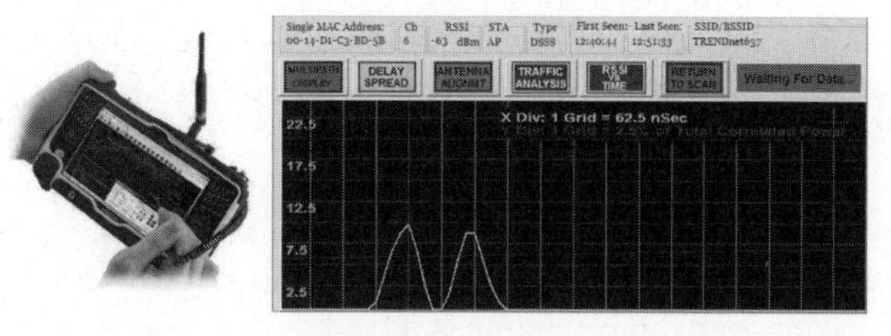

PHOTOS COURTESY OF BERKELEY VARITRONICS SYSTEMS.

There is no way to "fix" multipath indoors because some reflection will always occur, and thus there will always be multiple paths of the same signal. However, many of the negative effects of multipath, including intersymbol interference, can be compensated for with the use of antenna diversity, which is covered in Chapter 4, "Radio Frequency Signal and Antenna Concepts." High-multipath environments exist indoors in areas such as long corridors and anywhere metal is located (for example, warehouses with metal shelving or metal racks). The use of indoor diversity patch antennas is highly recommended in high-multipath environments. Using a unidirectional antenna will cut down on reflections and thereby decrease data corruption and layer 2 retransmissions.

Some RF technologies also compensate for multipath better than other RF technologies. Any 802.11 radio that uses OFDM technology will be more resilient to multipath than any radio using DSSS technology. Therefore, 802.11a and 802.11g radios that both use OFDM will handle multipath better than older legacy 802.11b radios. In Chapter 18, "High Throughput (HT) and 802.11n," we discuss the 802.11n amendment, which defines the use of High Throughput (HT) clause 20 radios. 802.11n radios use *multiple-input multiple-output (MIMO)* technology, which actually takes advantage of multipath. It should be noted that 802.11n clause 20 radios (HT) are required to be backward compatible with older clause 18 radios (HR-DSSS), clause 17 radios (OFDM), and clause 19 radios (ERP). 802.11n access points will not solve the problems that multipath creates for WLANs because older client devices will still be negatively affected by multipath.

Adjacent Cell Interference

Most Wi-Fi vendors use the term *adjacent channel interference* to refer to degradation of performance resulting from overlapping frequency space that occurs due to an improper channel reuse design. In the WLAN industry, an adjacent channel is considered to be the next or previous numbered channel. For example, channel 3 is adjacent to channel 2.

As you learned in Chapter 6, "Wireless Networks and Spread Spectrum Technologies," the 802.11-2007 standard requires 25 MHz of separation between the center frequencies of HR-DSSS and ERP-OFDM channels in order for them to be considered nonoverlapping. As pictured in Figure 12.6, only channels 1, 6, and 11 can meet these IEEE requirements in the 2.4 GHz ISM band in the United States if three channels are needed. Channels 2 and 7 are nonoverlapping, as well as 3 and 8, 4 and 9, and 5 and 10. The important thing to remember is that there must be five channels of separation in adjacent coverage cells. Some countries allow the use of all fourteen IEEE 802.11-defined channels in the 2.4 GHz ISM band, but because of the positioning of the center frequencies, no more than three channels can be used while avoiding frequency overlap. Even if all fourteen channels are available, most vendors and end users still choose to use channels 1, 6, and 11.

When designing a wireless LAN, you need overlapping coverage cells in order to provide for roaming. However, the overlapping cells should not have overlapping frequencies, and in the United States only channels 1, 6, and 11 should be used in the 2.4 GHz ISM band to get the most available, nonoverlapping channels. Overlapping coverage cells with overlapping frequencies cause what is known as *adjacent cell interference*. Although 802.11 radios might have overlapping frequency space, they do have different center frequencies and can therefore transmit at the same time. If overlapping coverage cells also have frequency overlap, the transmitted frames will become corrupted, the receivers will not send ACKs, and layer 2 retransmissions will significantly increase. Later in this chapter, we discuss channel reuse patterns that are used to mitigate adjacent cell interference.

FIGURE 12.6 2.4 GHz nonoverlapping channels

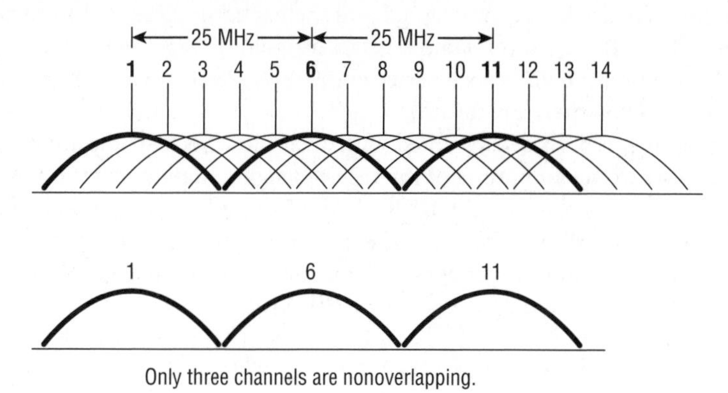

As defined by the IEEE, 23 channels are currently available in the 5 GHz UNII bands, as pictured in Figure 12.7. These 23 channels are technically considered nonoverlapping channels because there is 20 MHz of separation between the center frequencies. In reality, there will be some frequency overlap of the sidebands of each OFDM channel. The good news is that you are not limited to only three channels, and all twenty-three channels of the 5 GHz UNII bands can be used in a channel reuse pattern, which is discussed later in this chapter.

FIGURE 12.7 5 GHz nonoverlapping channels

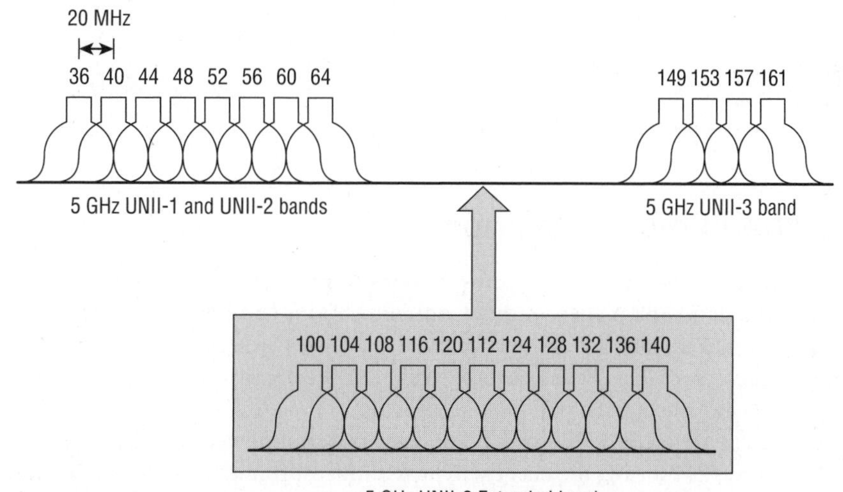

Low SNR

The *signal-to-noise ratio (SNR)* is an important value, because if the background noise is too close to the received signal or the received signal level is too low, data can get corrupted and retransmissions will increase. The SNR is not actually a ratio. It is simply the difference in decibels between the received signal and the background noise (noise floor), as depicted in Figure 12.8. If an 802.11 radio receives a signal of –85 dBm and the noise floor is measured at –100 dBm, the difference between the received signal and the background noise is 15 dB. The SNR is therefore 15 dB.

Data transmissions can become corrupted with a very low SNR. If the amplitude of the noise floor is too close to the amplitude of the received signal, data corruption will occur and result in layer 2 retransmissions. An SNR of 25 dB or greater is considered good signal quality, and an SNR of 10 dB or lower is considered poor signal quality. To ensure that frames do not get corrupted, many vendors recommend a minimum SNR of 18 dB for data WLANs and a minimum SNR of 25 dB for voice WLANs.

FIGURE 12.8 Signal-to-noise ratio

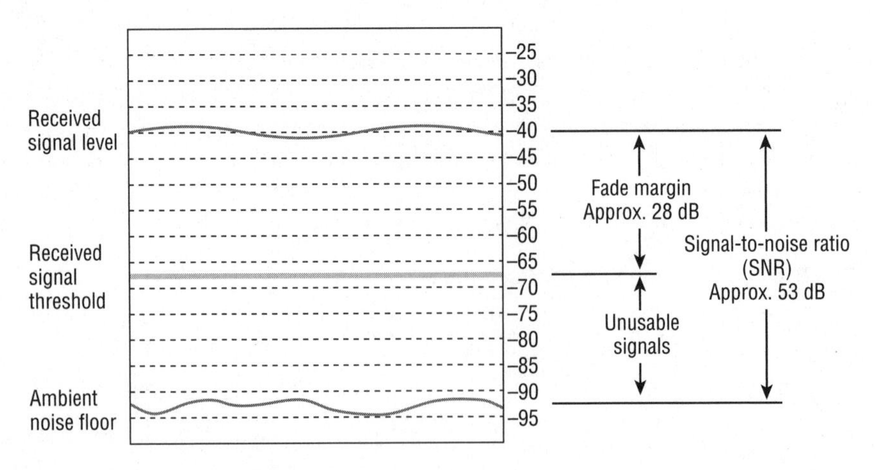

Mismatched Power Settings

An often overlooked cause of layer 2 retransmissions is mismatched transmit power settings between an access point and a client radio. Communications can break down if a client station's transmit power level is less than the transmit power level of the access point. As a client moves to the outer edges of the coverage cell, the client can "hear" the AP; however, the AP cannot "hear" the client.

As depicted in Figure 12.9, if an access point has a transmit power of 100 mW and a client has a transmit power of 20 mW, the client will hear a unicast frame from the AP because the received signal is within the client station's receive sensitivity capabilities. However, when the client sends an ACK frame back to the AP, the amplitude of the client's transmitted signal has dropped well below the receive sensitivity threshold of the AP's radio. The ACK frame is not "heard" by the AP, which then must retransmit the unicast frame. All of the client's transmissions are effectively seen as noise by the AP, and layer 2 retransmissions are the result.

How do you prevent layer 2 retries that are caused by mismatched power settings between the AP and clients? The best solution is to ensure that all of the client transmit power settings match the access point's transmit power. If this is not possible, the access point's power should never be set to more than the lowest-powered client station.

One way to test whether the mismatched AP/client power problem exists is to listen with a protocol analyzer. An AP/client power problem exists if the frame transmissions of the client station are corrupted when you listen near the access point, but are not corrupted when you listen near the client station.

AP/client power problems usually occur because APs are often deployed at full power to increase range. Increasing the power of an access point is the wrong way to increase range. If you want to increase the range for the clients, the best solution is to increase the antenna gain of the access point. Most people do not understand the simple concept of *antenna reciprocity*, which means that antennas amplify received signals just as they amplify transmitted signals. A high-gain antenna on an access point will amplify the AP's transmitted signal and extend the range at which the client is capable of hearing the signal. The AP's high-gain antenna will also amplify the received signal from a distant client station.

FIGURE 12.9 Mismatched AP and client power

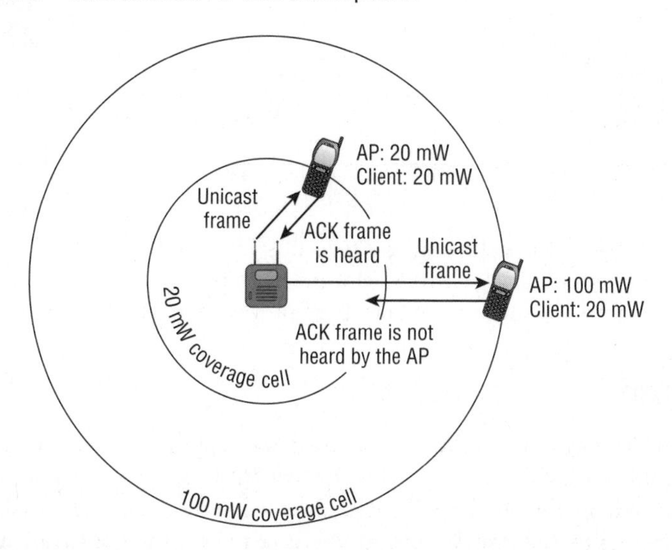

It should be noted that dynamic RF capabilities used by many WLAN controller vendors are notorious for causing mismatched power settings between the lightweight APs and client stations. A WLAN controller might dynamically increase a lightweight AP's power to a level above the client's transmit power. Dynamic changes of AP transmit power are well known to cause problems with VoWiFi phones. If the AP cannot hear the phone because of mismatched power, choppy audio may occur or phone conversations may drop entirely. The ratified 802.11k amendment does make it possible for an AP to inform clients to use transmit power control (TPC) capabilities to change their transmit amplitude dynamically to match the AP's power.

Near/Far

Disproportionate transmit power settings between multiple clients may also cause communication problems within a basic service set (BSS). A low-powered client station that is at a great distance from the access point could become an unheard client if other high-powered stations are very close to that access point. The transmissions of the high-powered stations could raise the noise floor near the AP to a higher level. The higher noise floor would corrupt the far station's incoming frame transmissions and would prevent this lower-powered station from being heard, as seen in Figure 12.10. This scenario is referred to as the *near/far* problem.

The half-duplex nature of the medium usually prevents most near/far occurrences. You can troubleshoot near/far problems with a protocol analyzer the same way you would troubleshoot the mismatched AP/client power problem.

FIGURE 12.10 The near/far problem

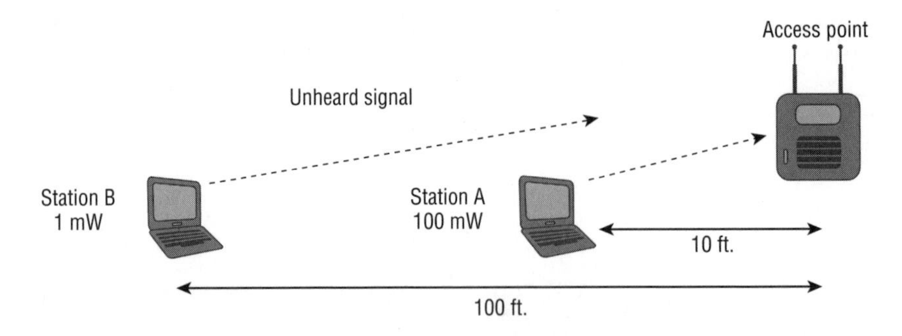

Please understand that the medium access methods employed by Carrier Sense Multiple Access with Collision Avoidance (CSMA/CA) usually averts the near/far problem, and that it is not as common a problem as, say, mismatched power between an AP and a client.

Hidden Node

In Chapter 8, "802.11 Medium Access," you learned about physical carrier-sense and clear channel assessment (CCA). CCA involves listening for 802.11 RF transmissions at the Physical layer, and the medium must be clear before a station can transmit. The problem with physical carrier-sense is that all stations may not be able to hear each other. Remember that the medium is half-duplex and, at any given time, only one radio card can be transmitting. What would happen, however, if one client station that was about to transmit performed a CCA but did not hear another station that was already transmitting? If the station that was about to transmit did not detect any RF energy during its CCA, it would transmit. The problem is that you then have two stations transmitting at the same time. The end result is a collision, and the frames will become corrupted. The frames will have to be retransmitted.

The *hidden node* problem occurs when one client station's transmissions are heard by the access point, but are not heard by any or all of the other client stations in the basic service set (BSS). The clients would not hear each other and therefore could transmit at the same time. The access point would hear both transmissions, which would be interfering with each other, and the incoming transmissions would be corrupted.

Figure 12.11 shows the coverage area of an access point. Note that a thick block wall resides between one client station and all of the other client stations that are associated to the access point. The RF transmissions of the lone station on the other side of the wall cannot be heard by all of the other 802.11 client stations, even though all the stations can hear the AP. That unheard station is the hidden node. What keeps occurring is that every time the hidden node transmits, another station is also transmitting, and a collision occurs. The hidden node continues to have collisions with the transmissions from all the other stations that cannot hear it during the clear channel assessment. The collisions continue on a regular basis and so do the layer 2 retransmissions, with the final result being a decrease in throughput. A hidden node can drive retransmission rates above 15 to 20 percent or even higher. Retransmissions, of course, will affect throughput, latency, and jitter.

FIGURE 12.11 Hidden node—obstruction

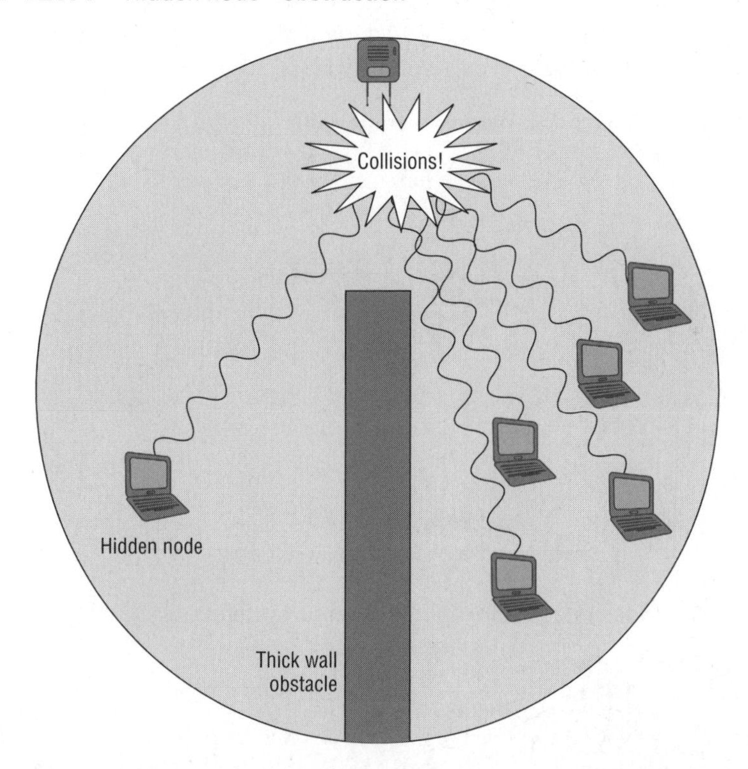

The hidden node problem may exist for several reasons—for example, poor WLAN design or obstructions such as a newly constructed wall or a newly installed bookcase. A user moving behind some sort of obstruction can cause a hidden node problem. VoWiFi phones often become hidden nodes because users take the phone into quiet corners or areas where the RF signal of the phone cannot be heard by other client stations. Users with wireless desktops often place their radio card underneath a metal desk and effectively transform that radio card into an unheard hidden node.

The hidden node problem can also occur when two client stations are at opposite ends of an RF coverage cell and they cannot hear each other, as seen in Figure 12.12. This often happens when coverage cells are too large as a result of the access point's radio transmitting at too much power. Later in this chapter, you will learn that it is a recommended practice to disable the data rates of 1 and 2 Mbps on an 802.11b/g access point for capacity purposes. Another reason for disabling those data rates is that a 1 and 2 Mbps coverage cell at 2.4 GHz can be quite large and often results in hidden nodes. If hidden node problems occur in a network planned for coverage, then RTS/CTS may be needed. This is discussed in detail later in this chapter.

Another cause of the hidden node problem is distributed antenna systems. Some manufacturers design distributed systems, which are basically made up of a long coaxial cable with multiple antenna elements. Each antenna in the distributed system has its own coverage area. Many companies purchase distributed antenna systems for cost-saving purposes, but a hidden node problem as pictured in Figure 12.13 will almost always occur. Distributed antenna systems and leaky cable systems should always be avoided.

FIGURE 12.12 Hidden node—large coverage cell

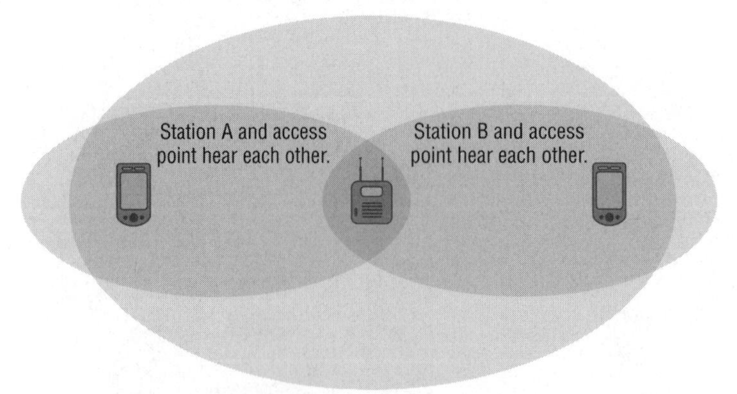

FIGURE 12.13 Hidden node—distributed antenna system

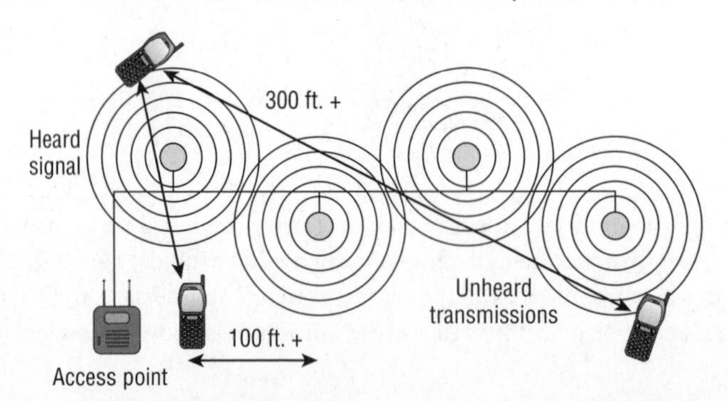

So how do you troubleshoot a hidden node problem? If your end users complain of a degradation of throughput, one possible cause is a hidden node. A protocol analyzer is a useful tool in determining hidden node issues. If the protocol analyzer indicates a higher retransmission rate for the MAC address of one station when compared to the other client stations, chances are a hidden node has been found. Some protocol analyzers even have hidden node alarms based on retransmission thresholds.

Another way is to use request to send/clear to send (RTS/CTS) to diagnose the problem. Try lowering the RTS/CTS threshold on a suspected hidden node to about 500 bytes. This level may need to be adjusted depending on the type of traffic being used. For instance, let's say you have deployed a terminal emulation application in a warehouse environment and a hidden node problem exists. In this case, the RTS/CTS threshold should be set for a much lower size, such as 30 bytes. Use a protocol analyzer to determine the appropriate size. As you learned in Chapter 9, "802.11 MAC Architecture," RTS/CTS is a method in which client stations can reserve the medium. In Figure 12.14, you see a hidden node initiating an RTS/CTS exchange.

FIGURE 12.14 Hidden node and RTS/CTS

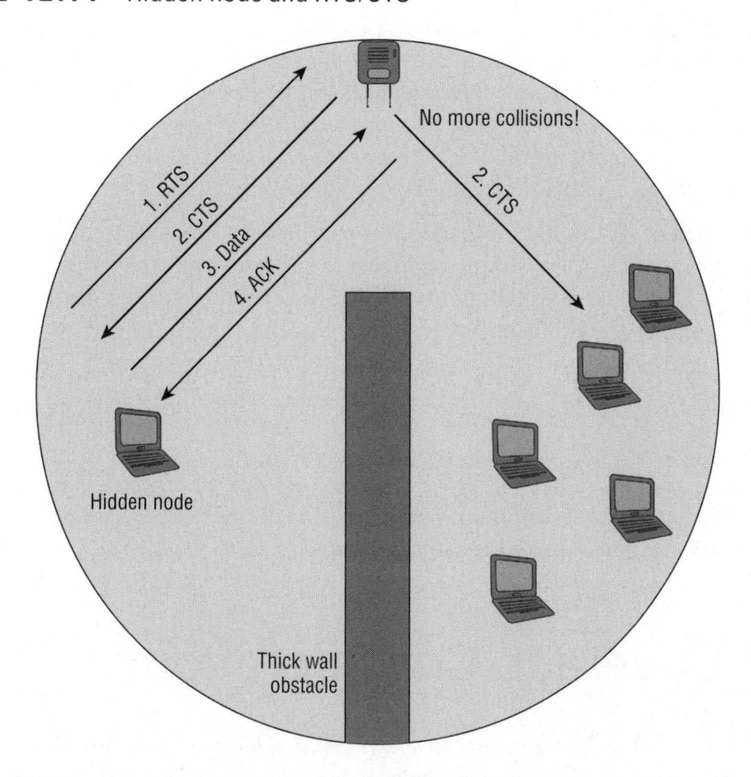

The stations on the other side of the obstacle may not hear the RTS frame from the hidden node, but they will hear the CTS frame sent by the access point. The stations that hear the CTS frame will reset their NAV for the period of time necessary for the hidden node to transmit the data frame and receive its ACK frame. Implementing RTS/CTS on a hidden node will reserve the medium and force all other stations to pause; thus the collisions and retransmissions will stop.

Collisions and retransmissions as a result of a hidden node will cause throughput to decrease. RTS/CTS usually decreases throughput as well. However, if RTS/CTS is implemented on a suspected hidden node, throughput will probably *increase* due to the stoppage of the collisions and retransmissions. If you implement RTS/CTS on a suspected hidden node and throughput increases, you have confirmed the existence of a hidden node.

RTS/CTS typically should not be viewed as a mechanism to fix the hidden node problem. RTS/CTS can be a temporary fix for the hidden node problem but should usually be used for only diagnostic purposes. One exception to that rule is point-to-multipoint (PtMP) bridging. The nonroot bridges in a PtMP scenario will not be able to hear each other because they are miles apart. RTS/CTS should be implemented on nonroot PtMP bridges to eliminate collisions caused by hidden node bridges that cannot hear each other.

The following methods can be used to fix a hidden node problem:

Use RTS/CTS to diagnose. Use either a protocol analyzer or RTS/CTS to diagnose the hidden node problem. RTS/CTS can also be used as a temporary fix to the hidden node problem.

Increase power to all stations. Most client stations have a fixed transmission power output. However, if power output is adjustable on the client side, increasing the transmission power of client stations will increase the transmission range of each station. If the transmission range of all stations is increased, the likelihood of the stations hearing each other also increases. This is not a recommended fix because, as you learned earlier, best practice dictates that client stations use the same transmit power used by all other radios in the BSS.

Remove the obstacles. If it is determined that some sort of obstacle is preventing client stations from hearing each other, simply removing the obstacle will solve the problem. Obviously, you cannot remove a wall, but if a metal desk or file cabinet is the obstacle, it can be moved to resolve the problem.

Move the hidden node station. If one or two stations are in an area where they become unheard, simply moving them within transmission range of the other stations will solve the problem.

Add another access point. The best fix for a continuous hidden problem is to add another AP. If moving the hidden nodes is not an option, adding another access point in the hidden area to provide coverage will also rectify the problem.

802.11 Coverage Considerations

Providing for both coverage and capacity in a WLAN design solves many problems. Roaming problems and interference issues will often be mitigated in advance if proper WLAN design techniques are performed as well as a thorough site survey. In the following sections, we discuss many considerations that should be addressed to provide proper coverage, capacity, and performance within an 802.11 coverage zone.

Dynamic Rate Switching

As client station radios move away from an access point, they will shift down to lower-bandwidth capabilities by using a process known as *dynamic rate switching (DRS)*. Access points can support multiple data rates depending on the spread spectrum technology used by the AP's radio card. For example, an HR-DSSS (802.11b) radio supports data rates of 11, 5.5, 2, and 1 Mbps. Data rate transmissions between the access point and the client stations will shift down or up depending on the quality of the signal between the two radio cards, as pictured in Figure 12.15. There is a correlation between signal quality and distance from the AP. As a result, transmissions between two 802.11b radio cards may be at 11 Mbps at 30 feet, but 2 Mbps at 150 feet.

FIGURE 12.15 Dynamic rate switching

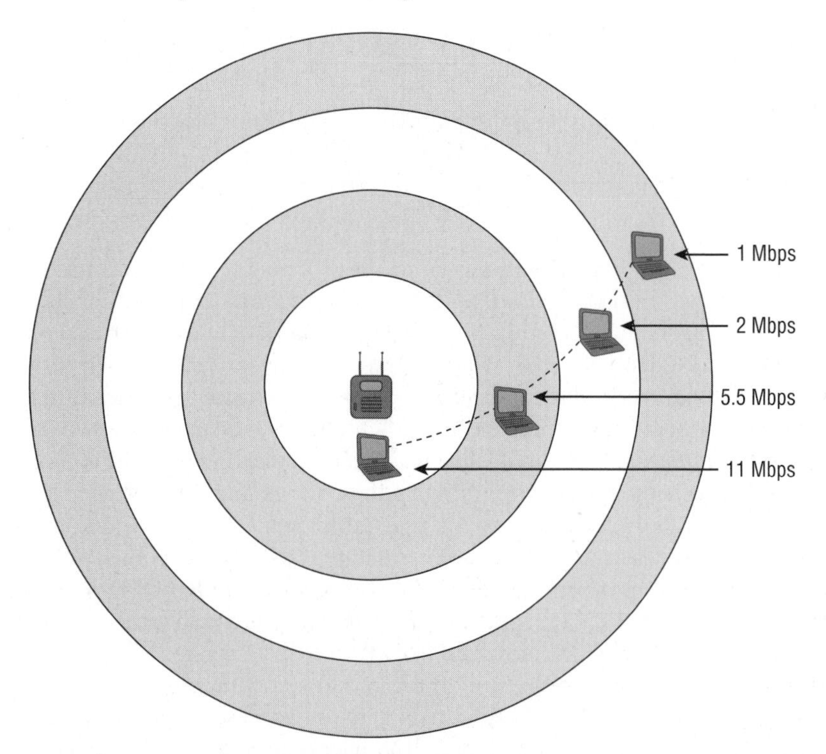

DRS is also referred to as *dynamic rate shifting*, *adaptive rate selection*, and *automatic rate selection*. All these terms refer to a method of speed fallback on a wireless LAN client as signal quality from the access point decreases. The objective of DRS is upshifting and downshifting for rate optimization and improved performance. Effectively, the lower data rates will have larger concentric zones of coverage than the higher data rates, as pictured in Figure 12.16.

FIGURE 12.16 Data rate coverage zones

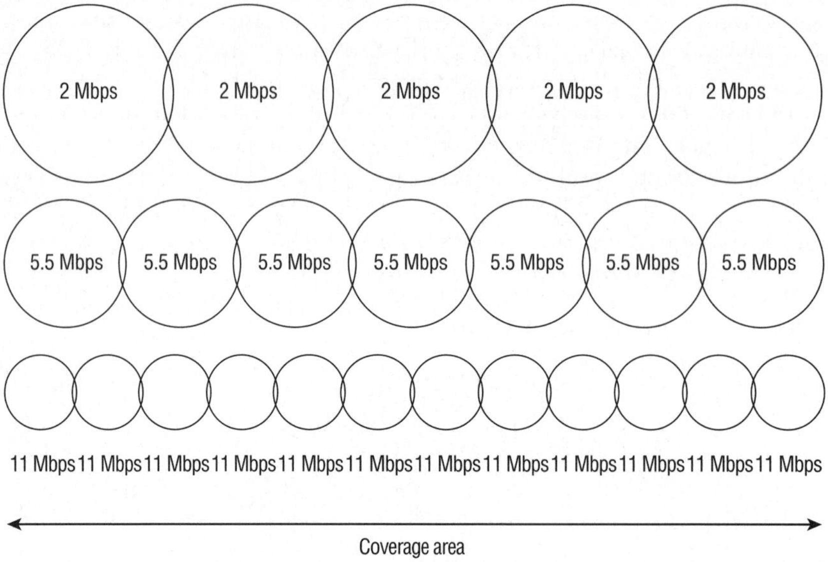

The algorithms used for dynamic rate switching are proprietary and are defined by radio card manufacturers. Most vendors base DRS on receive signal strength indicator (RSSI) thresholds, packet error rates, and retransmissions. RSSI metrics are usually based on signal strength and signal quality. In other words, a station might shift up or down between data rates based on both received signal strength in dBm and possibly on a signal-to-noise ratio (SNR) value. Because vendors implement DRS differently, you may have two different vendor client cards at the same location, while one is communicating with the access point at 5.5 Mbps and the other is communicating at 1 Mbps. For example, one vendor might shift down from data rate 11 Mbps to 5 Mbps at –70 dBm while another vendor might shift between the same two rates at –75 dBm. Keep in mind that DRS works with all 802.11 PHYs. For example, the same shifting of rates will also occur with ERP-OFDM radios shifting between 54, 48, 36, 24, 18, 12, 9, and 6 Mbps data rates. As a result, there is a correlation between signal quality and distance from the AP.

It is often a recommend practice to turn off the two lowest data rates of 1 and 2 Mbps when designing an 802.11b/g network. A WLAN network administrator might want to consider disabling the two lowest rates on an 802.11b/g access point for two reasons: medium contention and the hidden node problem. In Figure 12.17, you will see that multiple client stations are in the 1 Mbps zone, and only one lone client is in the 11 Mbps zone. Remember that wireless is a half-duplex medium and only one radio card can transmit on the medium at a time.

All radio cards access the medium in a pseudorandom fashion as defined by CSMA/CA. A radio transmitting a 1,500-byte data frame at 11 Mbps might occupy the medium for 300 microseconds. Another radio transmitting at 1 Mbps per second may take 3,300 microseconds to deliver that same 1,500 bytes. Radio cards transmitting at slower data rates will occupy the medium much longer, while faster radios have to wait. If multiple radio cards get on the outer cell edges and transmit at slower rates consistently, the perceived throughput for the cards transmitting at higher rates is much slower because of having to wait for slower transmissions to finish. For this reason, too many radios on outer 1 and 2 Mbps cells can adversely affect throughput. Another reason to consider turning off the lower data rates is the hidden node problem, which was explained earlier in this chapter.

FIGURE 12.17 Frame transmission time

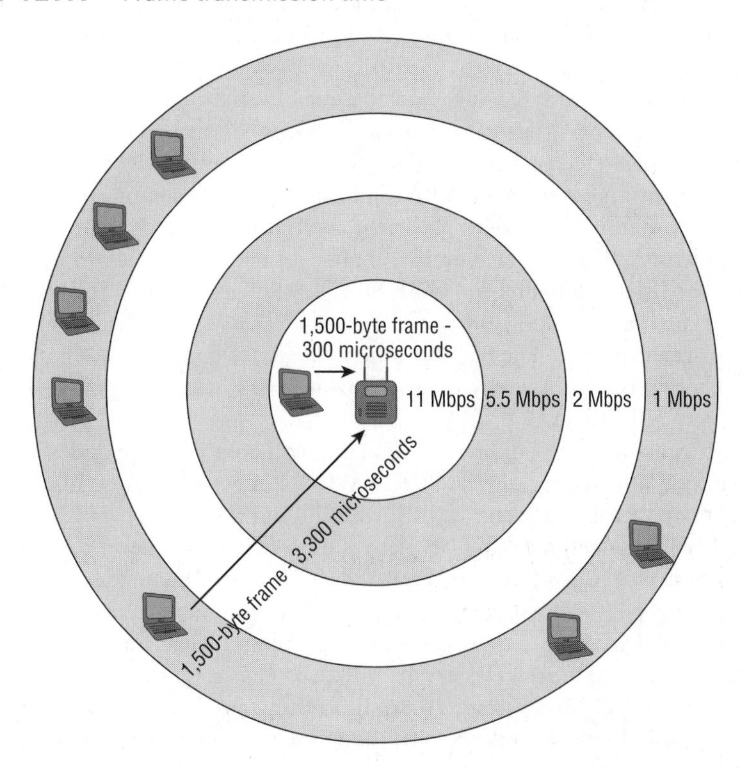

Roaming

As you have learned throughout this book, *roaming* is the method by which client stations move between RF coverage cells in a seamless manner. Client stations switch communications through different access points. Seamless communications for stations moving between the coverage zones within an extended service set (ESS) is vital for uninterrupted mobility. One of the most common issues you'll need to troubleshoot is problems with roaming. Roaming problems are usually caused by poor network design. Because of the proprietary nature of roaming, problems can also occur when radio cards from multiple vendors are deployed. Changes in the WLAN environment can also cause roaming hiccups.

Client stations, and not the access point, make the decision on whether or not to roam between access points. Some vendors may involve the access point or WLAN controller in the roaming decision, but ultimately, the client station initiates the roaming process with a reassociation request frame. The method in which client stations decide how to roam is entirely proprietary. All vendor client stations use roaming algorithms that can be based on multiple variables. The variable of most importance will always be received signal strength: As the received signal from the original AP grows weaker and a station hears a stronger signal from another known access point, the station will initiate the roaming process. However, other variables such as SNR, error rates, and retransmissions may also have a part in the roaming decision.

Because roaming is proprietary, a specific vendor client station may roam sooner than a second vendor client station as they move through various coverage cells. Some vendors like to encourage roaming, whereas others use algorithms that roam at lower received signal thresholds. In an environment where a WLAN administrator must support multiple vendor radios, different roaming behaviors will most assuredly be seen. For the time being, a WLAN administrator will always face unique challenges because of the proprietary nature of roaming. As discussed in Chapter 5, "IEEE 802.11 Standards," the 802.11k amendment has defined the use of *radio resource measurement (RRM)* and *neighbor reports* to enhance roaming performance. The 802.11r amendment also defines faster handoffs when roaming occurs between cells in a wireless LAN using the strong security defined in a robust security network (RSN).

The best way to ensure that seamless roaming will commence is proper design and a thorough site survey. When designing an 802.11 WLAN, most vendors recommend 15 to 25 percent overlap in coverage cells at the lowest desired signal level. The only way to determine whether proper cell overlap is in place is by conducting a coverage analysis site survey. Proper site survey procedures are discussed in detail in Chapter 16.

Roaming problems will occur if there is not enough overlap in cell coverage. Too little overlap will effectively create a roaming dead zone, and connectivity may even temporarily be lost. On the flip side, too much cell overlap will also cause roaming problems. For example, if two cells have 60 percent overlap, a station may stay associated with its original AP and not connect to a second access point even though the station is directly underneath the second access point. This can also create a situation in which the client device is constantly switching back and forth between the two or more APs. This often presents itself when a client device is directly under an AP and there are constant dropped frames.

Another design issue of great importance is latency. The 802.11-2007 standard defines the use of an 802.1X/EAP security solution in the enterprise. The average time involved during the authentication process can be 700 milliseconds or longer. Every time a client station roams to a new access point, reauthentication is required when an 802.1X/EAP security solution has been deployed. The time delay that is a result of the authentication process can cause serious interruptions with time-sensitive applications. VoWiFi requires a roaming handoff of 150 milliseconds or less when roaming. A *fast secure roaming (FSR)* solution is needed if 802.1X/EAP security and time-sensitive applications are used together in a wireless network. Currently, FSR solutions are proprietary, although the 802.11i amendment defined optional FSR. The fast secure roaming mechanisms defined by the ratified 802.11r amendment will hopefully standardize fast secure roaming in the enterprise.

The 802.1X standard and Extensible Authentication Protocol (EAP) are discussed in detail in Chapter 13, "802.11 Network Security Architecture." Included on the CD of this book is a white paper titled "Robust Security Network (RSN) Fast BSS Transition (FT)" by Devin Akin. The CWNA exam will not test you on the details of this paper. However, it discusses FSR solutions and is recommended extra reading for the CWSP exam.

Changes in the WLAN environment can also cause roaming headaches. RF interference will always affect the performance of a wireless network and can make roaming problematic as well. Very often new construction in a building will affect the coverage of a WLAN and create new dead zones. If the physical environment where the WLAN is deployed changes, the coverage design may have to change as well. It is always a good idea to periodically conduct a coverage survey to monitor changes in coverage patterns.

Troubleshooting roaming by using a protocol analyzer is tricky because the reassociation roaming exchanges occur on multiple channels. To troubleshoot a client roaming between channels 1, 6, and 11, you would need three separate protocol analyzers on three separate laptops that would produce three separate frame captures. CACE Technologies offers a product called AirPcap that is a USB 802.11 radio. As pictured in Figure 12.18, three AirPcap USB radios can be configured to capture frames on channels 1, 6, and 11 simultaneously. All three radios are connected to a USB hub and save the frame captures of all three channels into a single time-stamped capture file. The AirPcap solution allows for multichannel monitoring with a single protocol analyzer.

FIGURE 12.18 AirPcap provides multichannel monitoring and roaming analysis.

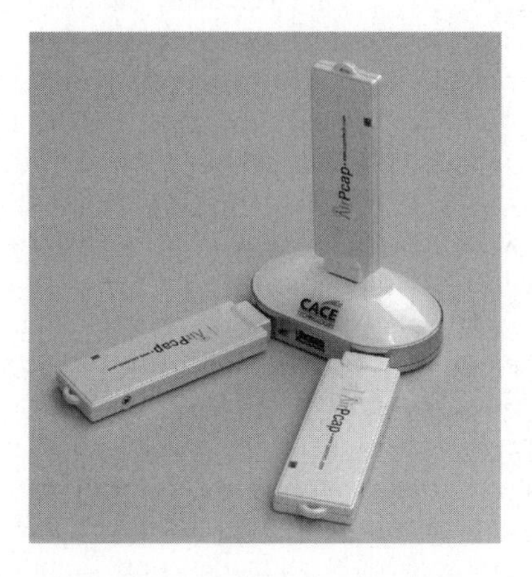

Layer 3 Roaming

One major consideration when designing a WLAN is what happens when client stations roam across layer 3 boundaries. As pictured in Figure 12.19, the client station is roaming between two access points. The roam is seamless at layer 2, but a router sits between the two access points, and each access point resides in a separate subnet. In other words, the client station will lose layer 3 connectivity and must acquire a new IP address. Any connection-oriented applications that are running when the client reestablishes layer 3 connectivity will have to be restarted. For example, a VoIP phone conversation would disconnect in this scenario, and the call would have to be reestablished.

FIGURE 12.19 Layer 3 roaming boundaries

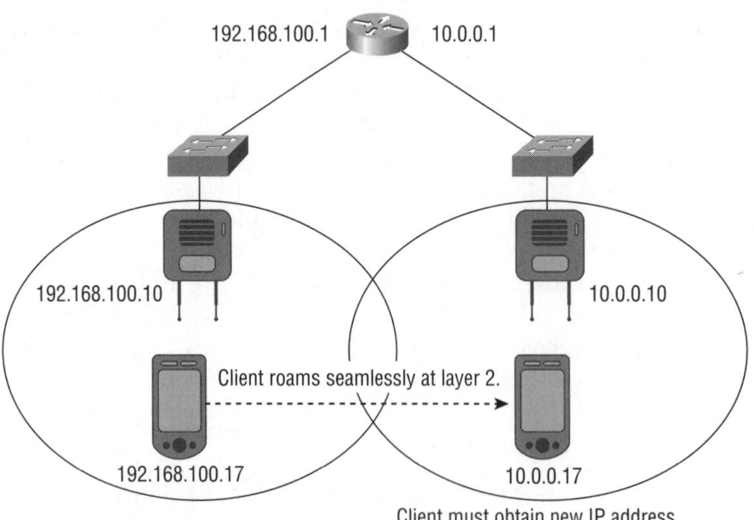

The preferred method when designing a WLAN is to have overlapping Wi-Fi cells that exist in only the same layer 3 domains through the use of VLANs. However, because 802.11 wireless networks are usually integrated into preexisting wired topologies, crossing layer 3 boundaries is often a necessity, especially in large deployments. The only way to maintain upper-layer communications when crossing layer 3 subnets is to provide either a Mobile IP solution or a proprietary *layer 3 roaming* solution. *Mobile IP* is an Internet Engineering Task Force (IETF) standard protocol that allows mobile device users to move from one layer 3 network to another while maintaining their original IP address. Mobile IP is defined in IETF request for comment (RFC) 3344. Mobile IP and proprietary solutions both use some type of tunneling method and IP header encapsulation to allow packets to traverse between separate layer 3 domains with the goal of maintaining upper-layer communications. It is beyond the scope of this book to explain either the standards-based Mobile IP or proprietary layer 3 roaming solutions; however, most WLAN controllers now support some type of layer 3 roaming solution. Although maintaining upper-layer connectivity is possible with these layer 3 roaming solutions, increased latency is often an issue. Additionally, layer 3 roaming may not be a requirement for your network. Even if there are layer 3 boundaries, your users may not need to seamlessly roam between subnets. Before you go to all the hassle of building a roaming solution, be sure to properly define your requirements.

Co-channel Interference

One of the most common mistakes many businesses make when first deploying a WLAN is to configure multiple access points all on the same channel. If all of the APs are on the same channel, unnecessary medium contention overhead occurs. As you have learned, CSMA/CA dictates half-duplex communications, and only one radio can transmit on the same channel at any given time.

As pictured in Figure 12.20, if an AP on channel 1 is transmitting, all nearby access points and clients on the same channel will defer transmissions. The result is that throughput is adversely affected: Nearby APs and clients have to wait much longer to transmit because they have to take their turn. The unnecessary medium contention overhead that occurs because all the APs are on the same channel is called *co-channel interference (CCI)*. In reality, the 802.11 radios are operating exactly as defined by the CSMA/CA mechanisms, and this behavior should really be called *co-channel cooperation*. The unnecessary medium contention overhead caused by co-channel interference is a result of improper channel reuse design, which is discussed in the next section of this chapter.

FIGURE 12.20 Co-channel interference

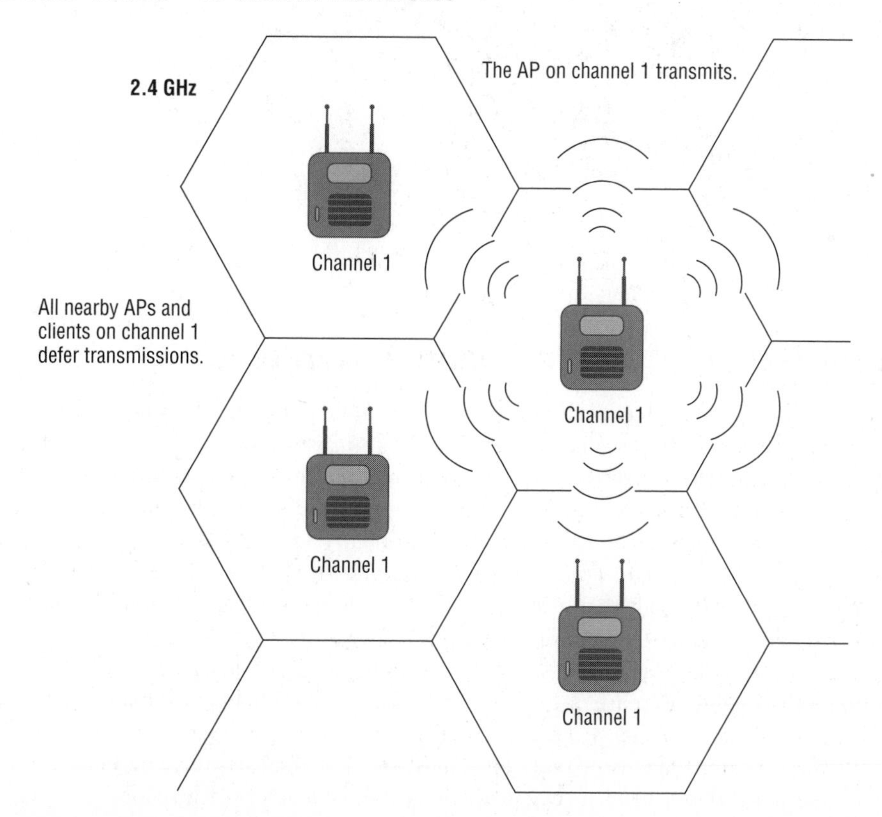

Please do not confuse adjacent cell interference with co-channel interference. However, adjacent cell interference is also a result of improper channel reuse design. As pictured in Figure 12.21, overlapping coverage cells that also have overlapping frequency space from adjacent cells will result in corrupted data and layer 2 retransmissions. Please refer back to Figure 12.6 and you will see that channels 1 and 4, channels 4 and 7, and channels 7 and 11 all have overlapping frequency space. Adjacent cell interference is a much more serious problem than co-channel interference because of the corrupted data and layer 2 retries. Proper channel reuse design is the answer to both co-channel and adjacent cell interference.

FIGURE 12.21 Adjacent cell interference: Access points with overlapping coverage cells transmit at the same time. The overlapping frequencies cause data corruption.

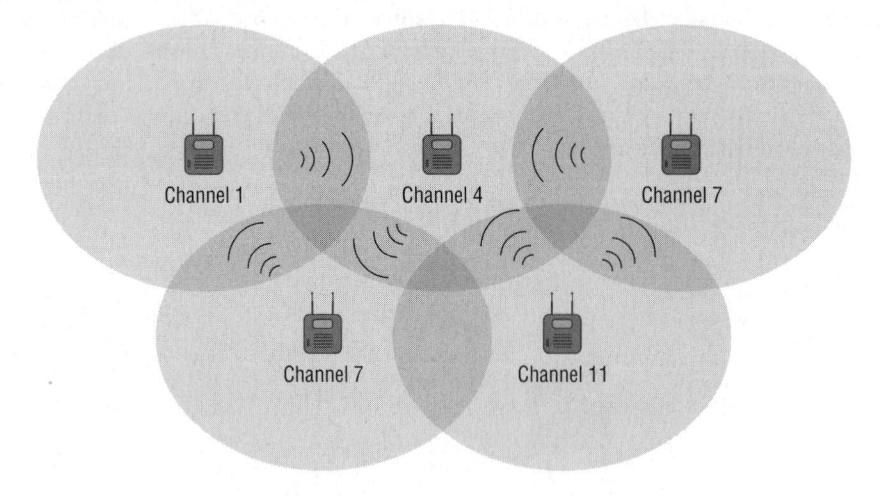

Channel Reuse/Multiple Channel Architecture

To avoid co-channel and adjacent cell interference, a channel reuse design is necessary. Once again, overlapping RF coverage cells are needed for roaming, but overlap frequencies must be avoided. The only three channels that meet these criteria in the 2.4 GHz ISM band are channels 1, 6, and 11 in the United States. Overlapping coverage cells therefore should be placed in a *channel reuse* pattern similar to the one pictured in Figure 12.22. A WLAN channel reuse pattern also goes by the name of *multiple channel architecture (MCA)*. WLAN architecture with overlapping coverage cells that utilizes three channels at 2.4 GHz, or numerous channels at 5 GHz, would be considered a multiple channel architecture.

It should be noted that it is impossible to avoid all instances of co-channel interference when using a three-channel reuse pattern at 2.4 GHz, because clients also cause co-channel interference. As pictured in Figure 12.23, if a client is at the outer edges of a coverage cell, the client's transmissions may propagate into another cell using the same channel. All of the radios in the other cell will defer if they hear the original client's transmissions.

Channel reuse patterns should also be used in the 5 GHz UNII bands. If all UNII bands are legally available for transmissions, a total of 23 channels may be used in a channel reuse pattern at 5 GHz. It may not be necessary to use all 23 channels; however, the more channels, the better.

FIGURE 12.22 2.4 GHz multiple channel architecture

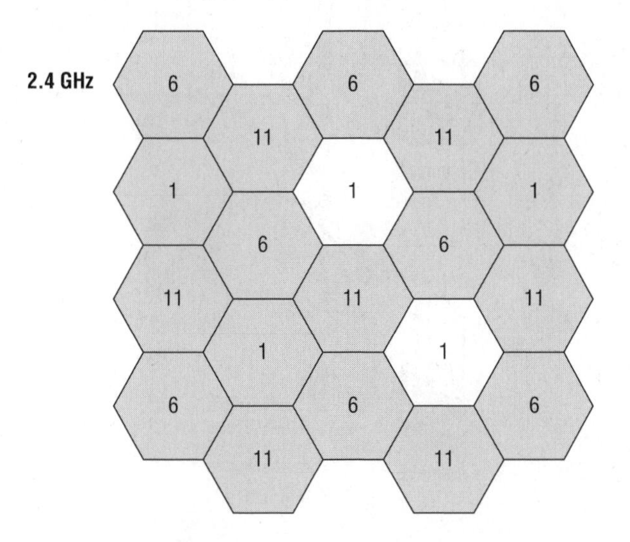

Figure 12.24 depicts a 5 GHz channel reuse pattern using the 12 channels available in UNII-1, UNII-2, and UNII-3. Although by the IEEE's definition, all 5 GHz OFDM channels are considered nonoverlapping, in reality there is some frequency sideband overlap from adjacent channels. It is a recommend practice that any adjacent coverage cells use a frequency that is at least two channels apart and not use an adjacent frequency. Following this simple rule will prevent adjacent cell interference from the sideband overlap.

As shown in Figure 12.24, the second recommended practice for 5 GHz channel reuse design dictates that there are always at least two cells of coverage space distance between any two access points transmitting on the same channel. Following this rule will prevent co-channel interference from APs and most likely also from clients. The client's signal will have to propagate a greater distance and should attenuate to an amplitude below the noise floor before the signal reaches another coverage cell using the same channel.

It is necessary to always think three-dimensionally when designing a multiple channel architecture reuse pattern. If access points are deployed on multiple floors in the same building, a reuse pattern will be necessary, such as the one pictured in Figure 12.25. A common mistake is to deploy a cookie-cutter design by performing a site survey on only one floor and then placing the access points on the same channels and same locations on each floor. A site survey must be performed on all floors, and the access points often need to be staggered to allow for a three-dimensional reuse pattern. Also, the coverage cells of each access point should not extend beyond more than one floor above and below the floor on which the access point is mounted. It is inappropriate to always assume that the coverage bleed over to other floors will provide sufficient signal strength and quality. In some cases, the floors are concrete or steel and allow very little, if any, signal coverage through. As a result, a survey is absolutely required.

FIGURE 12.23 Clients and co-channel interference

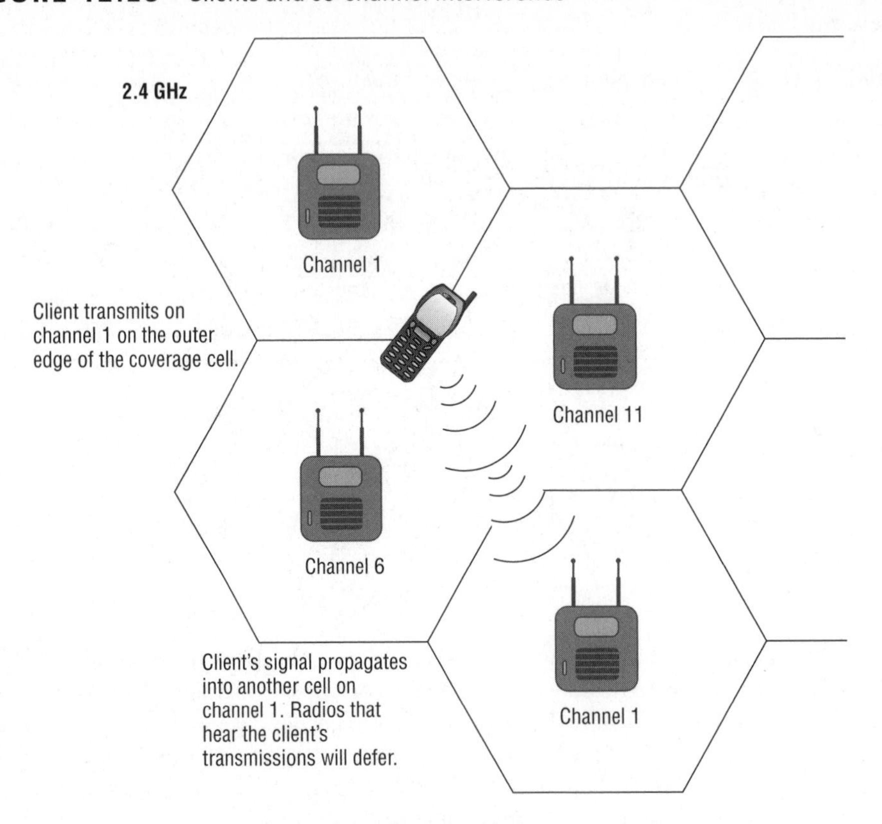

FIGURE 12.24 5 GHz multiple channel architecture

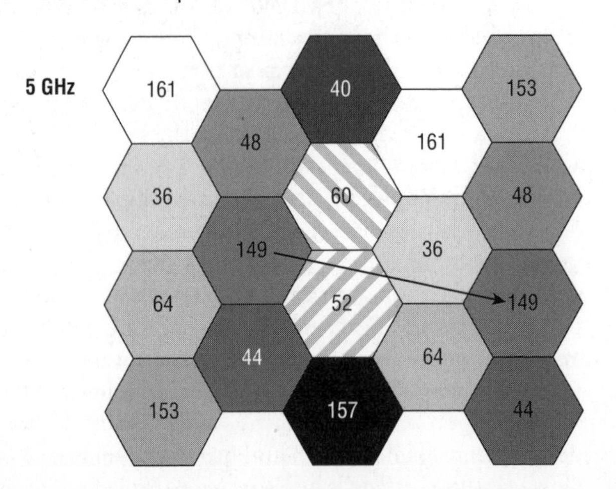

Many enterprise access points currently have dual radio card capabilities, allowing for both 2.4 GHz and 5 GHz wireless networks to be deployed in the same area. The 802.11a radio in an access point transmits at 5 GHz, and the signal will attenuate faster than the signal that is being transmitted at 2.4 GHz from the 802.11b/g radio card. Therefore, when performing a site survey for deploying dual-frequency WLANs, it is a recommended practice to perform the 5 GHz site survey first and determine the placement of the access points. After those locations are identified, channel reuse patterns will have to be used for each respective frequency.

FIGURE 12.25 Three-dimensional channel reuse

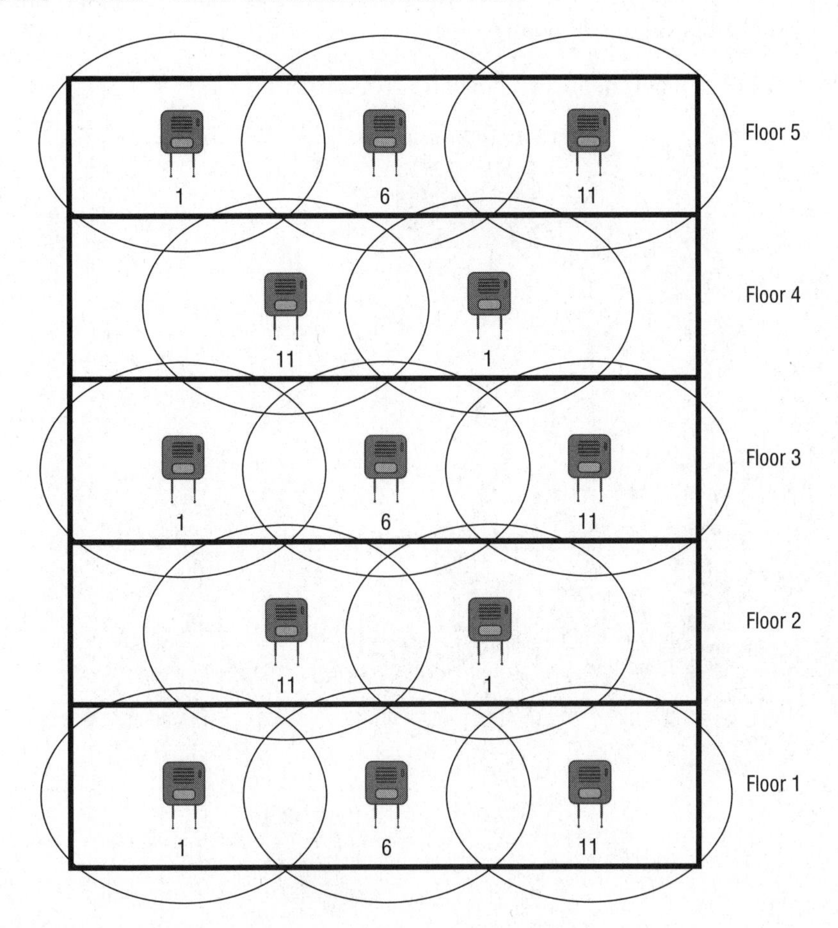

Single Channel Architecture

At the time of writing this book, two vendors, Meru Networks and Extricom, are offering an alternative WLAN channel design solution known as the *single channel architecture (SCA)*. Imagine a WLAN network with multiple access points all transmitting on the same channel and all sharing the same BSSID. A single channel architecture is exactly what you have just imagined! The client stations see transmissions on only a single channel with one SSID (logical WLAN identifier) and one BSSID (layer 2 identifier). From the perspective of the client station, only one access point exists. In this type of WLAN architecture, all access points in the network can be deployed on one channel in 2.4 GHz or 5 GHz frequency bands. Uplink and downlink transmissions are coordinated by a WLAN controller on a single 802.11 channel in such a manner that the effects of co-channel interference are minimized.

Let us first discuss the single BSSID. Single channel architecture consists of a WLAN controller and multiple lightweight access points. As shown in Figure 12.26, each AP has its own radio card with its own MAC address; however, they all share a *virtual BSSID* that is broadcast from all of the access points.

FIGURE 12.26 Single channel architecture

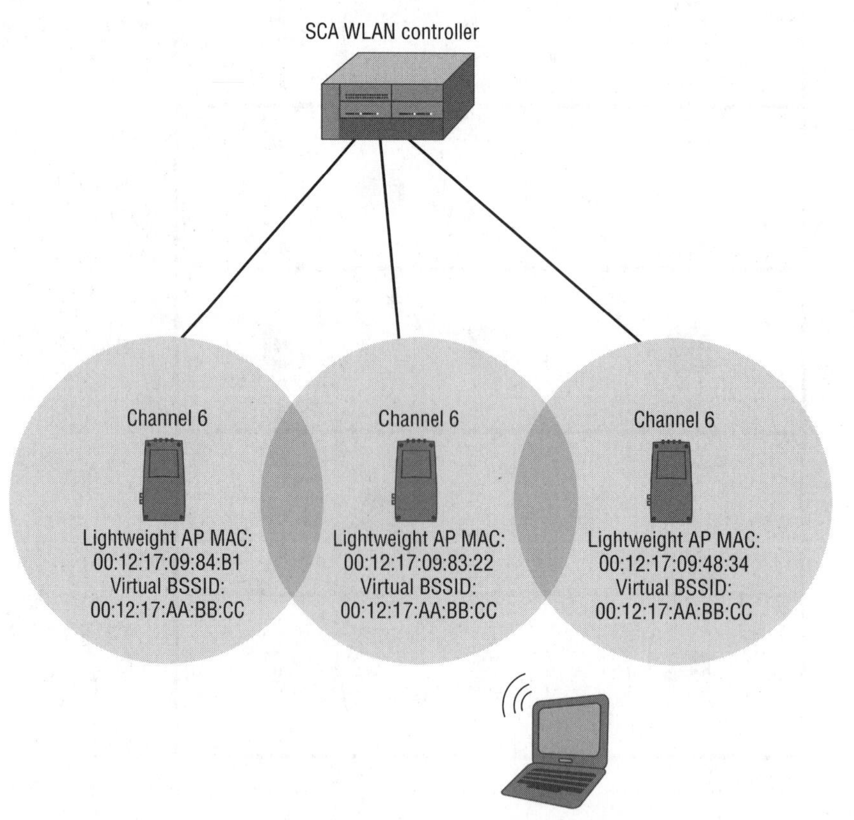

Because the multiple access points advertise only one single virtual MAC address (BSSID), client stations believe they are connected to only a single access point, although they may be roaming across multiple physical APs. You have learned that clients make the roaming decisions. In an single channel architecture (SCA) system, the clients think they are associated to only one AP, so they never initiate a layer 2 roaming exchange. All of the roaming hand-offs are handled by a central WLAN controller.

As pictured in Figure 12.27, the main advantage is that clients experience a *zero handoff* time, and the latency issues associated with roaming times are resolved. The *virtual AP* used by SCA solutions is potentially an excellent marriage for VoWiFi phones and 802.1X/EAP solutions. As we discussed earlier, the average time involved during the EAP authentication process can be 700 milliseconds or longer. Every time a client station roams to a new access point, reauthentication is required when an 802.1X/EAP security solution has been deployed. VoWiFi requires a roaming handoff of 150 ms or less. The virtual BSSID eliminates the need for reauthentication while physically roaming within a single channel architecture and thus a zero handoff time.

FIGURE 12.27 Zero handoff time

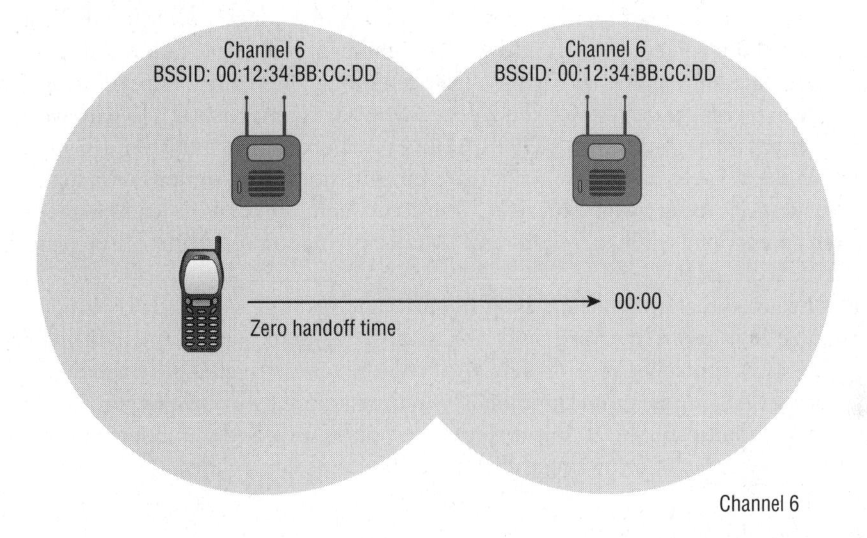

You have learned that client stations make the roaming decision in an MCA environment. However, client stations do not know that they roam in a SCA environment. The clients must still be mobile and transfer layer 2 communications between physical access points. All the client-roaming mechanisms are now handled back on the WLAN controller, and client-side-roaming decisions have been eliminated. All station associations are maintained at the SCA WLAN controller, and the SCA controller manages all the lightweight APs. The SCA controller assigns a unique lightweight access point the responsibility of handling downlink transmissions for an individual client station. When the controller receives the incoming transmissions of a client, the SCA controller evaluates the RSSI values of the client's transmissions. Based on incoming RSSI measurements, the SCA controller can allocate a specific AP for downlink transmissions. The client believes that it is associated to a single AP. However, the client moves between different physical APs based on RSSI measurements evaluated by the controller.

One big advantage of the single channel architecture is that adjacent cell interference is no longer an issue. If all the access points are on the same channel, there can be no adjacent cell interference, which is caused by frequency overlap. However, a legitimate question about a SCA WLAN solution is, Why doesn't co-channel interference occur if all of the channels are on the same channel? If all of the APs are on the same channel in an MCA wireless network, unnecessary medium contention overhead occurs. In a typical MCA environment, each access point has a unique BSSID and a separate channel, and each AP's coverage cell is a single-collision domain. In an SCA wireless environment, the collision domains are managed dynamically by the SCA controller based on RSSI algorithms. The controller ensures that nearby devices on the same channel are not transmitting at the same time. Most of the mechanisms used by SCA vendors to centrally manage co-channel interference is proprietary and beyond the scope of this book.

SCA solutions also use processes to limit medium contention overhead created by client stations transmitting at very low data rates. As we discussed earlier in this chapter, client stations transmitting at slower data rates will occupy the medium much longer, while faster clients have to wait. Medium contention overhead will be significant if multiple radio cards get on the outer cell edges transmitting at 1 or 2 Mbps. That is why it is often a recommend practice to turn off the two lowest data rates of 1 and 2 Mbps when designing an 802.11b/g network. An SCA controller has *airtime fairness (ATF)* capabilities that use layer 1 and layer 4–7 mechanisms back on the WLAN controller to prioritize transmissions from stations with higher data rates over the stations using lower data rates. One component of ATF is to load-balance the clients between access points at higher data rates. A full explanation of airtime fairness processes are beyond the scope of this book.

It should be noted that airtime fairness is not just used by SCA wireless LAN vendors. Several multiple channel architecture vendors have also begun to implement airtime fairness processes. The SCA controller vendors may also have the option to configure and deploy all of the lightweight APS in the more common MCA architecture by using a multiple channel reuse pattern and multiple BSSIDs. In the future, do not be surprised to see some of the MCA architecture vendors also offering single channel architecture capabilities.

Capacity vs. Coverage

When a wireless network is designed, two concepts that typically compete with each other are *capacity* and *coverage*. In the early days of wireless networks, it was common to install an access point with the power set to the maximum level to provide the largest coverage area possible. This was typically acceptable because there were very few wireless devices. Because the access points were also very expensive, companies tried to provide the most coverage while using the fewest access points. Figure 12.28 shows the outline of a building along with the coverage area that is provided by three APs in a multiple channel architecture. If there are just a few client stations, this type of wireless design is quite acceptable.

With the proliferation of wireless devices, network design has changed drastically from the early days. Proper network design now entails providing necessary coverage while trying to limit the number of devices connected to any single access point at the same time. This is what is meant by *capacity vs. coverage*. As you know, all of the client stations that connect to a single access point share the throughput capabilities of that access point. Therefore, it is important to design the network to try to limit the number of stations that are simultaneously connected to a single access point. This is performed by first determining the maximum number of stations that you want connected to an access point at the same time (this will vary from company to company depending on network usage). In an MCA environment, you need to determine how big the cell size needs to be to provide the proper capacity, and then you need to adjust the power level of the access point in order to create a cell of the desired size.

FIGURE 12.28 RF coverage of a building using three APs with few wireless stations

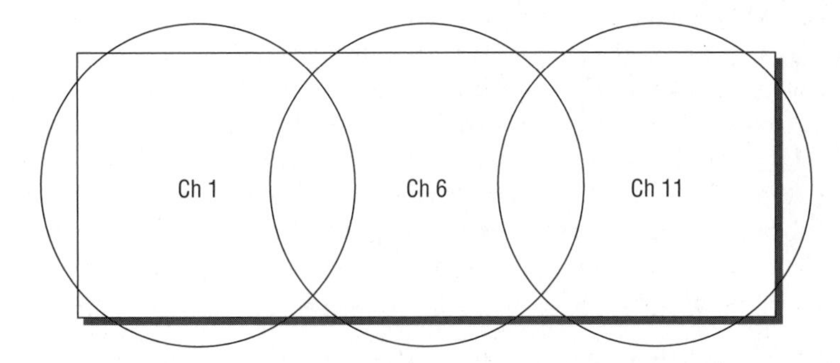

Figure 12.29 shows the outline of the same building, but because there are many more wireless stations, the cell sizes have been decreased while the number of cells has been increased. Adjusting the transmit power to limit the coverage area is known as *cell sizing* and is the most common method of meeting capacity needs in an MCA environment.

FIGURE 12.29 Cell sizing—multiple channel architecture

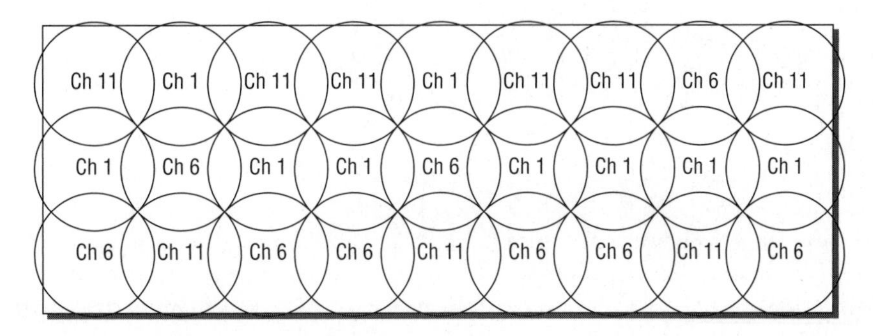

Another way of providing wireless support for a large capacity of users is by access point colocation. *Colocation* refers to placing multiple access points in the same physical space to provide for greater capacity.

802.11b and 802.11g APs are capable of having only three access points in the same area without causing interference. The three APs would need at least a five-channel separation to prevent RF interference. By colocating three APs, theoretically the potential cumulative speed is three times the speed of a single AP (assuming the three APs are equal). For example, three colocated 802.11g APs would provide a cumulative maximum speed of 162 Mbps (remember that actual throughput will be significantly less).

When colocating APs, it is important to make sure that the APs are physically separated from each other by at least 15 feet (5 meters) to prevent possible interference by the sideband signals that the APs generate. In an MCA environment, access point colocation is recommended only when the concentration of users is so dense that even when the cell size is at its smallest, there are still more stations per cell than desired. This often will occur in large meeting halls or university lecture halls. When colocated, end users can be load-balanced and segmented by MAC filters or by separate SSIDs.

Cell sizing is almost always the preferable method for meeting capacity needs in an MCA environment. Colocation can provide for capacity with MCA wireless LANs in a single predefined area. However, colocation does not scale in a WLAN that uses multiple channel architecture. Colocation does, however, scale within a single channel architecture design. As pictured in Figure 12.30, in a 2.4 GHz SCA deployment, multiple APs can be colocated by using three channels and three virtual BSSIDs. Colocation design in a single channel architecture is often referred to as *channel stacking*. Each layer of multiple APs on a single channel and using the same virtual BSSID is known as a *channel blanket* or *channel span*.

FIGURE 12.30 2.4 GHz channel stacking—single channel architecture

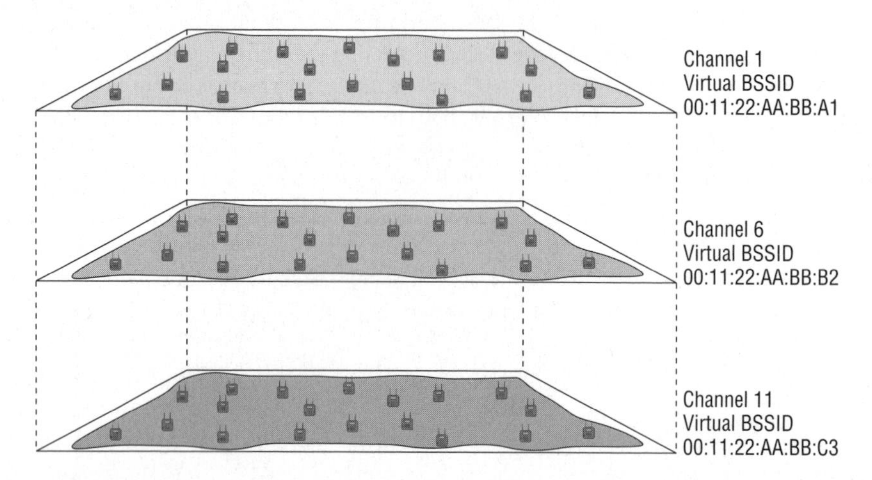

In Chapter 18, we discuss the 40 MHz channels that can be used by 802.11n. You will learn that using 40 MHz channels at 2.4 GHz is not possible in an MCA design due to adjacent cell interference caused by the frequency overlap of the 40 MHz channels. However, a single 40 MHz channel blanket could be deployed at 2.4 GHz with a single channel architecture. 40 MHz 802.11n channels do not scale in the 2.4 GHz ISM band within an MCA design, but a single 40 MHz channel blanket can scale at 2.4 GHz with a SCA design.

Oversized Coverage Cells

A mistake often made when deploying access points is to have the APs transmitting at full power. Effectively, this extends the range of the access point but causes many problems that have been discussed throughout this chapter. Oversized coverage usually will not meet your capacity needs. Oversized coverage cells can cause hidden node problems. Access points at full power may not be able to hear the transmissions of client stations with lower transmit power. Access points at full power will most likely also increase the odds of co-channel interference due to bleed-over transmissions. If the access point's coverage and range is a concern, the best method of extending range is to increase the AP's antenna gain instead of increasing transmit power.

Physical Environment

Although physical environment does not cause RF interference, physical obstructions can indeed disrupt and corrupt an 802.11 signal. An example of this is the scattering effect caused by a chain-link fence or safety glass with wire mesh. The signal is scattered and rendered useless. The only way to eliminate physical interference is to remove the obstruction or add more APs. Keep in mind that the physical environment of every building and floor is different, and the shape and size of coverage cells will widely vary. The best method of dealing with the physical environment is to perform a proper site survey as described in detail in Chapter 16.

Voice vs. Data

As you have already learned, most data applications in a Wi-Fi network can handle a layer 2 retransmission rate of up to 10 percent without any noticeable degradation in performance. However, time-sensitive applications such as VoIP require that higher-layer IP packet loss be no greater than 2 percent. Therefore, Voice over Wi-Fi (VoWiFi) networks need to limit layer 2 retransmissions to 5 percent or less to guarantee the timely and consistent delivery of VoIP packets. When layer 2 retransmissions exceed 5 percent, latency problems may develop and jitter problems will most likely surface.

The canary-and-coal-mine analogy is often used to describe the difference between voice traffic and other data application traffic within a WLAN environment. Early coal mines did not have ventilation systems installed in them, so miners would bring a caged canary into new coal shafts. Canaries are more sensitive to methane and carbon monoxide than humans, which made them ideal for detecting dangerous gas buildups. As long as the canary (voice traffic) was singing, the miners knew their air supply was safe. A dead canary signaled the presence of deadly gases, and the miners would evacuate or put on respirator masks. However, some species, such as a cockroach (data traffic) can still survive within the coal mine despite the existence of the deadly gases.

All too often, WLANs are deployed in the enterprise without any type of site survey. Also, many WLANs are initially designed to provide coverage only for data applications and not for voice. Most enterprise data applications will operate within a poorly designed WLAN, but are not running optimally due to the lack of a site survey or improper survey. Many companies decide to add a VoWiFi solution to their WLAN at a later date and quickly discover that the WLAN has many problems. The VoWiFi phones may have choppy audio or echo problems. The VoWiFi phones may disconnect or "die" like a canary. Adding voice to the WLAN often exposes existing problems: Voice is the canary, and the WLAN is the coal mine. Because data applications can withstand a much higher layer 2 retransmission rate, problems that existed within the WLAN may have gone unnoticed. The data applications are analogous to the cockroach that can still live in the coal mine but probably would have had a better life if the poor conditions did not exist. As shown in Table 12.1, IP voice traffic is more susceptible to late or inconsistent packet delivery due to layer 2 retransmissions.

TABLE 12.1 IP Voice and IP Data Comparison

IP Voice	IP Data
Small, uniform-size packets	Variable-size packets
Even, predictable delivery	Bursty delivery
Highly affected by late or inconsistent packet delivery	Minimally affected by late or inconsistent packet delivery
"Better never than late"	"Better late than never"

Optimizing the WLAN to support voice traffic will optimize the network for all wireless clients, including the clients running data applications other than voice. A proper site survey will reduce lower layer 2 retransmissions and provide an environment with seamless coverage that is required for VoWiFi networks.

Performance

When designing and deploying a WLAN, you will always be concerned about both coverage and capacity. Various factors can affect the coverage range of a wireless cell, and just as many factors can affect the aggregate throughput in an 802.11 WLAN. The following variables can affect the *range* of a WLAN:

Transmission power rates The original transmission amplitude (power) will have an impact on the range of an RF cell. An access point transmitting at 30 mW will have a larger coverage zone than an access point transmitting at 1 mW, assuming that the same antenna is used. APs with too much transmission amplitude can cause many problems already discussed in this chapter.

Antenna gain Antennas are passive-gain devices that focus the original signal. An access point transmitting at 30 mW with a 6 dBi antenna will have greater range than it would if it used only a 3 dBi antenna. If you want to increase the range for the clients, the best solution is to increase the antenna gain of the access point.

Antenna type Antennas have different coverage patterns. Using the right antenna will give the proper coverage and reduce multipath and nearby interference.

Wavelength Higher frequency signals have a smaller wavelength property and will attenuate faster than a lower-frequency signal with a larger wavelength. All things being equal, 2.4 GHz access points have a greater range than 5 GHz access points, due to the difference in the length of their waves.

Free space path loss In any RF environment, free space path loss (FSPL) attenuates the signal as a function of distance and frequency.

Physical environment Walls and other obstacles will attenuate an RF signal because of absorption and other RF propagation behaviors. A building with concrete walls will require more access points than a building with drywall because concrete is denser and attenuates the signal faster than drywall.

As you have learned in earlier chapters, proper WLAN design must take into account both coverage and capacity. The variables just mentioned all affect coverage and range. Capacity performance considerations are equally as important as range considerations. Please remember that 802.11 data rates are considered data bandwidth and not throughput. The following are among the many variables that can affect the *throughput* of a WLAN:

Carrier Sense Multiple Access with Collision Avoidance (CSMA/CA) The medium access method that uses interframe spacing, physical carrier-sense, virtual carrier-sense, and the random back-off timer creates overhead and consumes bandwidth. The overhead due to medium contention usually is 50 percent or greater.

Encryption Extra overhead is added to the body of an 802.11 data frame whenever encryption is implemented. WEP/RC4 encryption adds an extra 8 bytes of overhead per frame, TKIP/RC4 encryption adds an extra 20 bytes of overhead per frame, and CCMP/AES encryption adds an extra 16 bytes of overhead per frame. Layer 3 VPNs often use DES or 3DES encryption, both of which also consume significant bandwidth.

Application use Different types of applications will have variable affects on bandwidth consumption. VoWiFi and data collection scanning typically do not require a lot of bandwidth. Other applications that require file transfers or database access are often more bandwidth intensive.

Number of clients Remember that the WLAN is a shared medium. All throughput is aggregate, and all available bandwidth is shared.

Layer 2 retransmissions As we have discussed throughout this chapter, various problems can cause frames to become corrupted. If frames are corrupted, they will need to be retransmitted and throughput will be affected.

Weather

When deploying a wireless mesh network outdoors or perhaps an outdoor bridge link, a WLAN administrator must take into account the adverse affect of weather conditions. The following weather conditions must be considered:

Lightning Direct and indirect lightning strikes can damage WLAN equipment. Lightning arrestors should be used for protection against transient currents. Solutions such as lightning rods or copper/fiber transceivers may offer protection against lightning strikes.

Wind Because of the long distances and narrow beamwidths, highly directional antennas are susceptible to movement or shifting caused by wind. Even slight movement of a highly directional antenna can cause the RF beam to be aimed away from the receiving antenna, interrupting the communications. In high-wind environments, a grid antenna will typically remain more stable than a parabolic dish. Other mounting options may be necessary to stabilize the antenna from movement.

Water Conditions such as rain, snow, and fog present two unique challenges. First, all outdoor equipment must be protected from damage caused by exposure to water. Water damage is often a serious problem with cabling and connectors. Connectors should be protected with drip loops and coax seals to prevent water damage. Cables and connectors should be checked on a regular basis for damage. A radome (weatherproof protective cover) should be used to protect antennas from water damage. Outdoor bridges, access points, and mesh routers should be protected from the weather elements by using appropriate National Electrical Manufacturers Association (NEMA) enclosure units. Precipitation can also cause an RF signal to attenuate. A torrential downpour can attenuate a signal as much as 0.08 dB per mile (0.05 dB per kilometer) in both the 2.4 GHz and 5 GHz frequency ranges. Over long-distance bridge links, a system operating margin (SOM) of 20 dB is usually recommended to compensate for attenuation due to rain, fog, or snow.

Air stratification A change in air temperature at high altitudes is known as *air stratification* (layering). Changes in air temperature can cause refraction. Bending of RF signals over long-distance point-to-point links can cause misalignment and performance issues. K-factor calculations may be necessary to compensate for refraction over long-distance links.

UV/sun UV rays and ambient heat from rooftops can damage cables over time if proper cable types are not used.

Summary

In this chapter, we discussed numerous 802.11 coverage considerations. Troubleshooting for coverage, capacity, and performance problems can quite often be avoided with proper network design and comprehensive site surveys. We discussed the many causes of layer 2 retransmissions and the negative effects on the WLAN because of retries. Because wireless should always be considered an ever-changing environment, problems such as roaming, hidden nodes, and interference are bound to surface. Tools such as protocol analyzers and spectrum analyzers are invaluable when troubleshooting both layer 2 and layer 1 problems. We discussed and compared the differences between multiple channel architecture and single channel architecture. We also discussed the many performance variables that can affect both range and throughput. We discussed the challenges that are unique to both voice and data WLAN deployments. Finally, we discussed weather conditions that can impact outdoor RF communications and the steps that might be necessary for protection against Mother Nature.

Exam Essentials

Explain the causes and effect of Layer 2 retransmissions. Understand that layer 2 retransmissions can be caused by multipath, hidden nodes, mismatched power settings, RF interference, low SNR, near/far problems, and adjacent cell interference. Layer 2 retransmissions affect throughput, latency, and jitter.

Define dynamic rate switching. Understand the process of stations shifting between data rates. Know that dynamic rate switching is also referred to as dynamic rate shifting, adaptive rate selection, and automatic rate selection. Explain why disabling the two lower 802.11b/g data rates is often recommended.

Explain the various aspects of roaming. Understand that roaming is proprietary in nature. Know the variables that client stations may use when initiating the roaming process. Understand the importance of proper coverage cell overlap. Describe latency issues that can occur with roaming. Understand why crossing layer 3 boundaries can cause problems and what solutions might exist.

Define the differences between adjacent channel interference and co-channel interference. Understand the negative effects of both adjacent cell interference and co-channel interference. Explain why channel reuse patterns minimize the problems. Know what to consider when designing channel reuse patterns at both 2.4 GHz and 5 GHz in a multiple channel architecture.

Explain the differences between MCA and SCA wireless LAN design. Understand that MCA uses cell sizing to meet capacity needs, whereas SCA uses channel stacking to meet capacity needs. Explain the virtual BSSID and other aspects of an SCA design.

Identify the various types of interference. Know the differences between all-band, narrowband, wideband, physical, and intersymbol interference. Understand that a spectrum analyzer is your best interference-troubleshooting tool.

Explain the hidden node problem. Identify all the potential causes of the hidden node problem. Explain how to troubleshoot hidden nodes as well as how to fix the hidden node problem.

Define the near/far problem. Explain what causes near/far and how the problem can be rectified.

Identify performance variables. Explain all the variables that affect both the range of RF coverage and the throughput that can result within a basic service set.

Understand the consequences of weather conditions. Explain the problems that might arise due to water conditions, wind, lightning, and air stratification. Explain how these problems might be solved.

Key Terms

Before you take the exam, be certain you are familiar with the following terms:

adjacent cell interference
airtime fairness (ATF)
all-band interference
antenna reciprocity
Bluetooth (BT)
capacity
cell sizing
channel blanket
channel reuse
channel span
co-channel interference (CCI)
colocation
coverage
cyclic redundancy check (CRC)
delay spread
dynamic rate switching (DRS)
fast secure roaming (FSR)
hidden node
inter-symbol interference (ISI)
layer 3 roaming
Mobile IP
multipath
multiple channel architecture (MCA)
multiple-input multiple-output (MIMO)
near/far
radio resource measurement (RRM)
range
roaming
signal-to-noise ratio (SNR)
single channel architecture (SCA).
virtual BSSID

Review Questions

1. What type of solution must be deployed to provide continuous connectivity when a client station roams across layer 3 boundaries? (Choose all that apply.)
   A. Nomadic roaming solution
   B. Proprietary layer 3 roaming solution
   C. Seamless roaming solution
   D. Mobile IP solution
   E. Fast secure roaming solution

2. If the access points transmit on the same frequency channel in an MCA architecture, what type of interference is caused by overlapping coverage cells?
   A. Intersymbol interference
   B. Adjacent cell interference
   C. All-band interference
   D. Narrowband interference
   E. Co-channel interference

3. What variables might affect range in an 802.11 WLAN? (Choose all that apply.)
   A. Transmission power
   B. CSMA/CA
   C. Encryption
   D. Antenna gain
   E. Physical environment

4. What can be done to fix the hidden node problem? (Choose all that apply.)
   A. Increase the power on the access point.
   B. Move the hidden node station.
   C. Increase power on all client stations.
   D. Remove the obstacle.
   E. Decrease power on the hidden node station.

5. Layer 2 retransmissions occur when frames become corrupted. What are some of the causes of layer 2 retries? (Choose all that apply.)
   A. Multipath
   B. Low SNR
   C. Co-channel interference
   D. RF interference
   E. Adjacent cell interference

6. What scenarios might result in a hidden node problem? (Choose all that apply.)
   A. Distributed antenna system
   B. Too large coverage cell
   C. Too small coverage cell
   D. Physical obstruction
   E. Co-channel interference

7. What are some of the negative effects of layer 2 retransmissions? (Choose all that apply.)
   A. Decreased range
   B. Excessive MAC sublayer overhead
   C. Decreased latency
   D. Increased latency
   E. Jitter

8. Several users are complaining that their VoWiFi phones keep losing connectivity. The WLAN administrator notices that the frame transmissions of the VoWiFi phones are corrupted when listened to with a protocol analyzer near the access point, but are not corrupted when listened to with the protocol analyzer near the VoWiFi phone. What is the most likely cause of this problem?
   A. RF interference
   B. Multipath
   C. Hidden node
   D. Adjacent cell interference
   E. Mismatched power settings

9. A single user is complaining that her VoWiFi phone has choppy audio. The WLAN administrator notices that the user's MAC address has a retry rate of 25 percent when observed with a protocol analyzer. However, all the other users have a retry rate of about 5 percent when also observed with the protocol analyzer. What is the most likely cause of this problem?
   A. Near/far
   B. Multipath
   C. Co-channel interference
   D. Hidden node
   E. Low SNR

10. What type of interference is caused by overlapping cover cells with overlapping frequencies?
   A. Inter-symbol interference
   B. Adjacent cell interference
   C. All-band interference
   D. Narrowband interference
   E. Co-channel interference

11. Based on RSSI metrics, concentric zones of variable data rate coverage exist around an access point due to the upshifting and downshifting of client stations between data rates. What is the correct name of this process, according to the IEEE 802.11-2007 standard? (Choose all that apply.)
    - **A.** Dynamic rate shifting
    - **B.** Dynamic rate switching
    - **C.** Automatic rate selection
    - **D.** Adaptive rate selection
    - **E.** All of the above

12. Which of these weather conditions is a concern when deploying a long-distance point-to-point bridge link? (Choose all that apply.)
    - **A.** Wind
    - **B.** Rain
    - **C.** Fog
    - **D.** Changes in air temperature
    - **E.** All of the above

13. What variables might affect range in an 802.11 WLAN? (Choose all that apply.)
    - **A.** Wavelength
    - **B.** Free space path loss
    - **C.** Brick walls
    - **D.** Trees
    - **E.** All of the above

14. Which WLAN architecture can use the 40 MHz OFDM channel capabilities of an 802.11n access point in the 2.4 GHz ISM band?
    - **A.** Multiple channel architecture
    - **B.** Single channel architecture
    - **C.** Distributed WLAN architecture
    - **D.** Unified architecture
    - **E.** None of the above

15. Which of the following can cause roaming problems? (Choose all that apply.)
    - **A.** Too little cell coverage overlap
    - **B.** Too much cell coverage overlap
    - **C.** Free space path loss
    - **D.** CSMA/CA
    - **E.** Hidden node

16. What are some problems that can occur when an access point is transmitting at full power? (Choose all that apply.)

A. Hidden node

B. Adjacent cell interference

C. Co-channel interference

D. Mismatched power between the AP and the clients

E. Intersymbol interference

17. Why would a WLAN network administrator consider disabling the two lowest rates on an 802.11b/g access point? (Choose all that apply.)

A. Medium contention

B. Adjacent cell interference

C. Hidden node

D. Co-channel interference

E. All of the above

18. Which type of interference is caused by multipath?

A. Intersymbol interference

B. All-band interference

C. Narrowband interference

D. Wideband interference

E. Physical interference

19. In a multiple channel architecture (MCA) design, what is the greatest number of nonoverlapping channels that can be deployed in the 2.4 GHz ISM band?

A. 3

B. 12

C. 11

D. 14

E. 4

20. Colocating access points in the same physical space is one method of meeting capacity needs. Colocation scales best in the enterprise using which WLAN architecture?

A. Multiple channel architecture

B. Single channel architecture

C. Distributed WLAN architecture

D. Unified architecture

E. None of the above

Answers to Review Questions

1. B, D. The only way to maintain upper-layer communications when crossing layer 3 subnets is to provide either a Mobile IP solution or a proprietary layer 3 roaming solution.

2. E. In an MCA architecture, if all the access points are mistakenly configured on the same channel, unnecessary medium contention overhead is the result. If an AP is transmitting, all nearby access points and clients on the same channel will defer transmissions. The result is that throughput is adversely affected: Nearby APs and clients have to wait much longer to transmit because they have to take their turn. The unnecessary medium contention overhead that occurs because all the APs are on the same channel is called *co-channel interference (CCI)*. In reality, the 802.11 radios are operating exactly as defined by the CSMA/CA mechanisms, and this behavior should really be called *co-channel cooperation*.

3. A, D, E. The original transmission amplitude will have an impact on the range of an RF cell. Antennas amplify signal strength and can increase range. Walls and other obstacles will attenuate an RF signal and affect range. CSMA/CA and encryption do not affect range but do affect throughput.

4. B, C, D. The hidden node problem arises when client stations cannot hear the RF transmissions of another client station. Increasing the transmission power of client stations will increase the transmission range of each station, resulting in increased likelihood of all the stations hearing each other. Increasing client power is not a recommended fix because best practice dictates that client stations use the same transmit power used by all other radios in the BSS, including the AP. Moving the hidden node station within transmission range of the other stations also results in stations hearing each other. Removing an obstacle that prevents stations from hearing each other also fixes the problem. The best fix to the hidden node problem is to add another access point in the area that the hidden node resides.

5. A, B, D, E. If any portion of a unicast frame is corrupted, the cyclic redundancy check (CRC) will fail and the receiving 802.11 radio will not return an ACK frame to the transmitting 802.11 radio. If an ACK frame is not received by the original transmitting radio, the unicast frame is not acknowledged and will have to be retransmitted. Multipath, RF interference, low SNR, hidden nodes, mismatched power settings, near/far problems, and adjacent cell interference may all cause layer 2 retransmissions. Co-channel interference does not cause retries but does add unnecessary medium contention overhead.

6. A, B, D. The hidden node problem arises when client stations cannot hear the RF transmissions of another client station. Distributed antenna systems with multiple antenna elements are notorious for causing the hidden node problem. When coverage cells are too large as a result of the access point's radio transmitting at too much power, client stations at opposite ends of an RF coverage cell often cannot hear each other. Obstructions such as a newly constructed wall can also result in stations not hearing each other.

7. B, D, E. Excessive layer 2 retransmissions adversely affect the WLAN in two ways. First, layer 2 retransmissions increase MAC overhead and therefore decrease throughput. Second, if application data has to be retransmitted at layer 2, the timely delivery of application

traffic becomes delayed or inconsistent. Applications such as VoIP depend on the timely and consistent delivery of the IP packet. Excessive layer 2 retransmissions usually result in increased latency and jitter problems for time-sensitive applications such as voice and video.

8. E. An often overlooked cause of layer 2 retransmissions is mismatched transmit power settings between an access point and a client radio. Communications can break down if a client station's transmit power level is less than the transmit power level of the access point. As a client moves to the outer edges of the coverage cell, the client can "hear" the AP; however, the AP cannot "hear" the client. If the client station's frames are corrupted near the AP but not near the client, the most likely cause is mismatched power settings.

9. D. If an end user complains of a degradation of throughput, one possible cause is a hidden node. A protocol analyzer is a useful tool in determining hidden node issues. If the protocol analyzer indicates a higher retransmission rate for the MAC address of one station when compared to the other client stations, chances are a hidden node has been found. Some protocol analyzers even have hidden node alarms based on retransmission thresholds. Another way is to use request to send/clear to send (RTS/CTS) to diagnose the problem.

10. B. Overlapping coverage cells with overlapping frequencies cause adjacent cell interference, which causes a severe degradation in latency, jitter, and throughput. If overlapping coverage cells also have frequency overlap, frames will become corrupt, retransmissions will increase, and performance will suffer significantly.

11. B. As client station radios move away from an access point, they will shift down to lower bandwidth capabilities by using a process known as dynamic rate switching (DRS). The objective of DRS is upshifting and downshifting for rate optimization and improved performance. Although dynamic rate switching is the proper name for this process, all these terms refer to the method of speed fallback that a wireless LAN client uses as distance increases from the access point.

12. E. Highly directional antennas are susceptible to what is known as *antenna wind loading*, which is antenna movement or shifting caused by wind. Grid antennas may be needed to alleviate the problem. Rain and fog can attenuate an RF signal; therefore, a system operating margin (also known as fade margin) of 20 dB is necessary. A change in air temperature is also known as air stratification, which causes refraction. K-factor calculations may also be necessary to compensate for refraction.

13. E. Higher-frequency signals have a smaller wavelength property and will attenuate faster than a lower-frequency signal with a larger wavelength. Higher-frequency signals therefore will have shorter range. In any RF environment, free space path loss (FSPL) attenuates the signal as a function of distance. Loss in signal strength affects range. Brick walls exist in an indoor physical environment, while trees exist in an outdoor physical environment. Both will attenuate an RF signal, thereby affecting range.

14. B. The 802.11n draft amendment defines the use of 40 MHz OFDM channels. Using 40 MHz channels at 2.4 GHz is not possible in an MCA design due to adjacent cell interference caused by the frequency overlap of the 40 MHz channels. However, a single 40 MHz channel blanket could be deployed at 2.4 GHz with a single channel architecture. 40 MHz 802.11n channels do not scale in the 2.4 GHz ISM band within an MCA design, but a single 40 MHz channel blanket can scale at 2.4 GHz with a SCA design.

15. A, B. Roaming problems will occur if there is not enough overlap in cell coverage. Too little overlap will effectively create a roaming dead zone, and connectivity may even temporarily be lost. If two RF cells have too much overlap, a station may stay associated with its original AP and not connect to a second access point even though the station is directly underneath the second access point.

16. A, C, D. A mistake often made when deploying access points is to have the APs transmitting at full power. Effectively, this extends the range of the access point but causes many problems that have been discussed throughout this chapter. Oversized coverage usually will not meet your capacity needs. Oversized coverage cells can cause hidden node problems. Access points at full power may not be able to hear the transmissions of client stations with lower transmit power. Access points at full power will most likely also increase the odds of co-channel interference due to bleed-over transmissions. If the access point's coverage and range is a concern, the best method of extending range is to increase the AP's antenna gain instead of increasing transmit power.

17. A, C. Medium contention, also known as CSMA/CA, requires that all radios access the medium in a pseudorandom fashion. Radio cards transmitting at slower data rates will occupy the medium much longer, while faster radios have to wait. Data rates of 1 and 2 MBPS can create very large coverage cells, which may prevent a hidden node station at one edge of the cell from being heard by other client stations at the opposite side of the coverage cell.

18. A. Multipath can cause intersymbol interference (ISI), which causes data corruption. Because of the difference in time between the primary signal and the reflected signals, known as the *delay spread*, the receiver can have problems demodulating the RF signal's information. The delay spread time differential results in corrupted data and therefore layer 2 retransmissions.

19. A. HR-DSSS (802.11b) and ERP (802.11g) channels require 25 MHz of separation between the center frequencies to be considered nonoverlapping. The three channels of 1, 6, and 11 meet these requirements in the United States. In other countries, 2, 7, and 12; and 3, 8, and 13; and 4, 9, and 14 would work as well. Traditionally, 1, 6, and 11 are chosen almost universally.

20. B. Cell sizing is almost always the preferable method for meeting capacity needs in an MCA environment. Colocation can provide for capacity with MCA wireless LANs in a single predefined area. However, colocation does not scale in a WLAN that uses multiple channel architecture. Colocation does, however, scale within a single channel architecture design. Colocation design in a single channel architecture is often referred to as *channel stacking*. Each layer of multiple APs on a single channel and using the same virtual BSSID is known as a *channel blanket* or *channel span*.

802.11 Network Security Architecture

IN THIS CHAPTER, YOU WILL LEARN ABOUT THE FOLLOWING:

✓ **802.11 security basics**

- Data privacy
- Authentication, authorization, and accounting (AAA)
- Segmentation
- Monitoring and policy

✓ **Legacy 802.11 security**

- Legacy authentication
- Static WEP encryption
- MAC filters
- SSID cloaking

✓ **Robust security**

- Robust security network (RSN)
- Authentication and authorization
- 802.1X/EAP framework
- EAP types
- Dynamic encryption-key generation
- 4-Way Handshake
- WPA/WPA2-Personal
- TKIP encryption
- CCMP encryption

- ✓ **Segmentation**
  - VLANs
  - RBAC
- ✓ **Infrastructure security**
  - Physical security
  - Interface security
- ✓ **VPN wireless security**
  - Layer 3 VPNs

In the next two chapters, you will learn about what is probably the most often discussed topic in terms of 802.11 wireless networks: security. In this chapter, we discuss legacy 802.11 security solutions as well as more-robust solutions that are now defined by the 802.11-2007 standard. Although there is no such thing as 100 percent security, solutions do exist that can help fortify and protect your wireless network.

Numerous wireless security risks exist, and in Chapter 14, "Wireless Attacks, Intrusion Monitoring, and Policy," you will learn about many of the potential attacks that can be attempted against an 802.11 wireless network and how these attacks can be monitored.

Many of the attacks against an 802.11 network can be defended against with proper implementation of the security architectures that are discussed in this chapter. However, many attacks cannot be mitigated and can merely be monitored and hopefully responded to.

Although 10 percent of the CWNA exam covers 802.11 security, the CWNP program also offers another certification, titled Certified Wireless Security Professional (CWSP), which focuses on just the topic of wireless security. The CWSP certification exam requires a more in-depth understanding of 802.11 security. However, the next two chapters will give you a foundation of wireless security that should help you pass the security portions of the CWNA exam as well as give you a head start in the knowledge you will need to implement proper wireless security.

802.11 Security Basics

When you're securing a wireless 802.11 network, five major components are typically required:

- Data privacy
- Authentication, authorization, and accounting (AAA)
- Segmentation
- Monitoring
- Policy

Because data is transmitted freely and openly in the air, proper protection is needed to ensure data privacy, so strong encryption is needed. The function of most wireless networks is to provide a portal into some other network infrastructure, such as an 802.3 Ethernet backbone. The wireless portal must be protected, and therefore an authentication solution is needed to ensure that only authorized users can pass through the portal via a wireless access point. After users have been authorized to pass through the wireless portal, VLANs and identity-based mechanisms are needed to further restrict access to network resources. 802.11 wireless networks can be further protected with continuous monitoring by a wireless intrusion detection system. All of these security components should also be cemented with policy enforcement.

Network security should never be taken lightly for wired or wireless networks. WLAN security still has a bad reputation with some people because of the weak legacy 802.11 security mechanisms that were deployed. In 2004, the 802.11i amendment was ratified, defining stronger encryption and better authentication methods. The 802.11i amendment is now part of the 802.11-2007 standard and fully defines a robust security network (RSN), which is discussed later in this chapter. If proper encryption and authentication solutions are deployed, a wireless network can be just as secure, if not more secure, than the wired segments of a network. If properly implemented, the five components of 802.11 security discussed in this chapter and the next chapter will lay a solid foundation for protecting your WLAN.

Data Privacy

802.11 wireless networks operate in license-free frequency bands, and all data transmissions travel in the open air. Protecting data privacy in a wired network is much easier because physical access to the wired medium is more restricted. However, physical access to wireless transmissions is available to anyone in listening range. Therefore, using cipher encryption technologies to obscure information is mandatory to provide proper data privacy. A *cipher* is an algorithm used to perform encryption.

The two most common algorithms used to protect data are the RC4 algorithm (RC stands for Ron's code or Rivest cipher) and the Advanced Encryption Standard (AES) algorithm. Some ciphers encrypt data in a continuous stream, while others encrypt data in blocks.

The *RC4 algorithm* is a streaming cipher used in technologies that are often used to protect Internet traffic, such as Secure Sockets Layer (SSL). The RC4 algorithm is used to protect 802.11 wireless data and is incorporated into two encryption methods known as WEP and TKIP, both of which are discussed later in this chapter.

The *AES algorithm*, originally named the Rijndael algorithm, is a block cipher that offers much stronger protection than the RC4 streaming cipher. AES is used to encrypt 802.11 wireless data by using an encryption method known as Counter Mode with Cipher Block Chaining Message Authentication Code Protocol (CCMP), which will also be discussed later in this chapter. The AES algorithm encrypts data in fixed data blocks with choices in encryption key strength of 128, 192, or 256 bits. The AES cipher is the mandated algorithm of the U.S. government for protecting both sensitive and classified information.

In Chapter 9, "802.11 MAC Architecture," you learned about the three major types of 802.11 wireless frames. Inside the body of a management frame is layer 2 information necessary for the operation of the 802.11 network, and, therefore, 802.11 management frames are not encrypted. Control frames have no body and also are not encrypted. The information that needs to be protected is the upper-layer information inside the body of 802.11 data frames. If data encryption is enabled, the *MAC Service Data Unit (MSDU)* inside the body of any 802.11 data frame is protected by layer 2 encryption. Most of the encryption methods discussed in this chapter use layer 2 encryption, which is used to protect the layer 3 through layer 7 information found inside the body of an 802.11 data frame.

EXERCISE 13.1

Unencrypted and Encrypted Data Frames

In this exercise, you will use a protocol analyzer to view encrypted and unencrypted 802.11 data frames. The following directions should assist you with the installation and use of the AirMagnet Wi-Fi Analyzer product demo software. If you have already installed AirMagnet Wi-Fi Analyzer, you can skip steps 1–5.

1. In your web browser, proceed to the following URL: `www.airmagnet.com/cwna/sybex`.
2. Fill out the product demo request forms. You will then be sent an email message with a private download URL.
3. Proceed to the private download URL. Verify that you have a supported Wi-Fi card. Review the system requirements and supported operating systems.
4. Download the Wi-Fi Analyzer demo to your desktop. This evaluation copy of Wi-Fi Analyzer will be licensed to work for seven days.
5. Double-click the file `AirMagnet_WiFi_DemoInstaller.exe` and follow the installation prompts. Install the proper driver for your Wi-Fi card.
6. From Windows, choose Start ➢ Programs ➢ AirMagnet and then click the demo icon. The Wi-Fi Analyzer application should appear.
7. From the AirMagnet Wi-Fi Analyzer main window, click on frame #282, which is an unencrypted simple data frame. Look at the frame body and notice the upper-layer information such as IP addresses and UDP ports.
8. Click on frame #334, which is an encrypted simple data frame. Look at the frame body and notice that WEP encryption is being used and that the upper-layer information cannot be seen.

Authentication, Authorization, and Accounting (AAA)

Authentication, authorization, and accounting (AAA) is a common computer security concept that defines the protection of network resources.

Authentication is the verification of user identity and credentials. Users must identify themselves and present credentials such as usernames and passwords or digital certificates. More-secure authentication systems use multifactor authentication, which requires at least two sets of different credentials to be presented.

Authorization involves granting access to network resources and services. Before authorization to network resources can be granted, proper authentication must occur.

Accounting is tracking the use of network resources by users. It is an important aspect of network security, used to keep a paper trail of who used what resource, when and where. A record is kept of user identity, which resource was accessed, and at what time. Keeping an accounting trail is often a requirement of many industry regulations such as the U.S. Health Insurance Portability and Accountability Act of 1996 (HIPAA).

Remember that the usual purpose of an 802.11 wireless network is to act as a portal into an 802.3 wired network. It is therefore necessary to protect that portal with very strong authentication methods so that only legitimate users with the proper credentials will be authorized onto network resources.

Segmentation

Although it is of the utmost importance to secure an enterprise wireless network by utilizing both strong encryption and an AAA solution, an equally important aspect of wireless security is segmentation. Prior to the introduction of stronger authentication and encryption techniques, wireless was viewed as an untrusted network segment. Therefore, before the ratification of the 802.11i security amendment, the entire wireless segment of a network was commonly treated as the untrusted segment while the wired 802.3 network was considered the trusted segment.

Now that better security solutions exist, properly secured WLANs are more seamlessly and securely integrated into the wired infrastructure. It is still important to separate users into proper groups, much like what is done on any traditional network. Once authorized onto network resources, users can be further restricted as to what resources may be accessed and where they can go. Segmentation can be achieved through a variety of means, including firewalls, routers, VPNs, and VLANs. The most common wireless segmentation strategy used in 802.11 enterprise WLANs is layer 3 segmentation using virtual LANs (VLANs). Segmentation is also intertwined with role-based access control (RBAC), which is discussed later in this chapter.

Monitoring and Policy

Encryption, AAA, and segmentation security components will provide data privacy and secure network resources. However, a full-time monitoring solution is still needed to protect against possible attacks against a WLAN. Numerous layer 1 and layer 2 attacks are possible, and, in Chapter 14, “Wireless Attacks, Intrusion Monitoring, and Policy,” you will learn about many of the potential attacks that can be attempted against an 802.11 wireless network and how these attacks can be monitored by a wireless intrusion detection system (WIDS). Chapter 14 also discusses some of the fundamental components of a wireless security policy that are needed to cement a foundation of Wi-Fi security.

Legacy 802.11 Security

The original 802.11 standard defined very little in terms of security. The authentication methods first outlined in 1997 basically provided an open door into the network infrastructure. The encryption method defined in the original 802.11 standard has long been cracked and is considered inadequate for data privacy. In the following sections, you will learn about the legacy authentication and encryption methods that were the only defined standards for 802.11 wireless security from 1997 until 2004. Later in this chapter, you will learn about the more robust security that was defined in the *802.11i* security amendment that is now part of the current 802.11-2007 standard.

Legacy Authentication

You already learned about legacy authentication in Chapter 9. The original 802.11 standard specified two methods of authentication: Open System authentication and Shared Key authentication.

Open System authentication provides authentication without performing any type of client verification. It is essentially a two-way exchange between the client and the access point. The client sends an authentication request, and the access point then sends an authentication response. Because Open System authentication does not require the use of any credentials, every client gets authenticated and therefore authorized onto network resources after they have been associated. Static WEP encryption is optional with Open System authentication and may be used to encrypt the data frames after Open System authentication and association occur.

As you learned in Chapter 9, Shared Key authentication uses Wired Equivalent Privacy (WEP) to authenticate client stations and requires that a static WEP key be configured on both the station and the access point. In addition to WEP being mandatory, authentication will not work if the static WEP keys do not match. The authentication process is similar to Open System authentication but includes a challenge and response between the radio cards. Shared Key authentication is a four-way authentication frame handshake. The client station sends an authentication request to the access point, and then the access point sends a cleartext challenge to the client station in an authentication response. The client station encrypts the cleartext challenge and sends it back to the access point in the body of another authentication request frame. The access point decrypts the station's response and compares it to the challenge text. If they match, the access point will respond by sending a fourth and final authentication frame to the station confirming the success. If they do not match, the access point will respond negatively. If the access point cannot decrypt the challenge, it will also respond negatively. If Shared Key authentication is successful, the same static WEP key that was used during the Shared Key authentication process will also be used to encrypt the 802.11 data frames.

Although it might seem that Shared Key authentication is a more secure solution than Open System authentication, in reality Shared Key could be the bigger security risk. Anyone who captures the cleartext challenge phrase and then captures the encrypted challenge phrase in the response frame could potentially derive the static WEP key. If the static WEP key is compromised, a whole new can of worms has been opened because now all the data frames can be decrypted. Neither of the legacy authentication methods is considered strong enough for enterprise security. More-secure 802.1X/EAP authentication methods are discussed later in this chapter.

Static WEP Encryption

Wired Equivalent Privacy (WEP) is a layer 2 encryption method that uses the RC4 streaming cipher. The original 802.11 standard defined both 64-bit WEP and 128-bit WEP as supported encryption methods. The three main intended goals of WEP encryption include confidentiality, access control, and data integrity. The primary goal of confidentiality was to provide data privacy by encrypting the data before transmission. WEP also provides access control, which is basically a crude form of authorization. Client stations that do not have the same matching static WEP key as an access point are refused access to network resources. A data integrity checksum known as the Integrity Check Value (ICV) is computed on data before encryption and used to prevent data from being modified.

Although both 64-bit and 128-bit WEP were defined by the original 802.11 standard, initially the U.S. government allowed the export of only 64-bit technology. After the U.S. government loosened export restrictions on key size, radio card manufacturers began to produce equipment that supported 128-bit WEP encryption. As pictured in Figure 13.1, 64-bit WEP uses a secret 40-bit static key, which is combined with a 24-bit number selected by the card's device drivers. This 24-bit number, known as the *Initialization Vector (IV)*, is sent in cleartext, and is different on every frame. Although the IV is said to be different on every frame, there are only 16,777,216 different IV combinations; therefore, you are forced to reuse the IV values. The effective key strength of combining the IV with the 40-bit static key is 64-bit encryption. 128-bit WEP encryption uses a 104-bit secret static key that is also combined with a 24-bit IV.

FIGURE 13.1 Static WEP encryption key and Initialization Vector

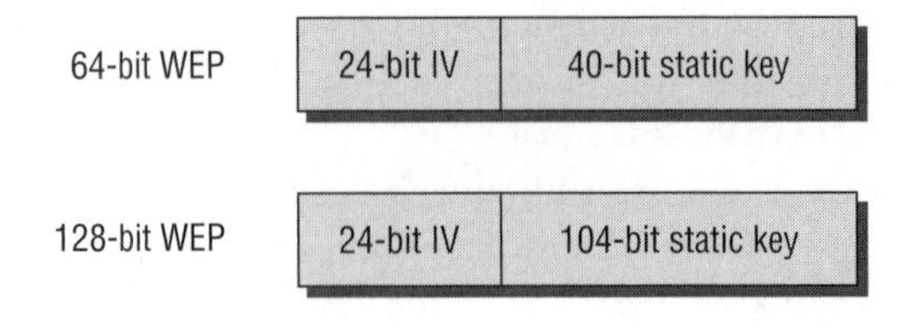

A static WEP key can usually be entered as hexadecimal (hex) characters (0–9 and A–F) or ASCII characters. The static key must match on both the access point and the client device. A 40-bit static key consists of 10 hex characters or 5 ASCII characters, while a 104-bit static key consists of 26 hex characters or 13 ASCII characters. Not all client stations or access points support both hex and ASCII. Most clients and access points support the use of up to four separate static WEP keys from which a user can choose one as the default transmission key (Figure 13.2 shows an example). The transmission key is the static key that is used to encrypt data by the transmitting radio. A client or access point may use one key to encrypt outbound traffic and a different key to decrypt received traffic. However, all keys must match exactly on both sides of a link for encryption/decryption to work properly.

FIGURE 13.2 Transmission key

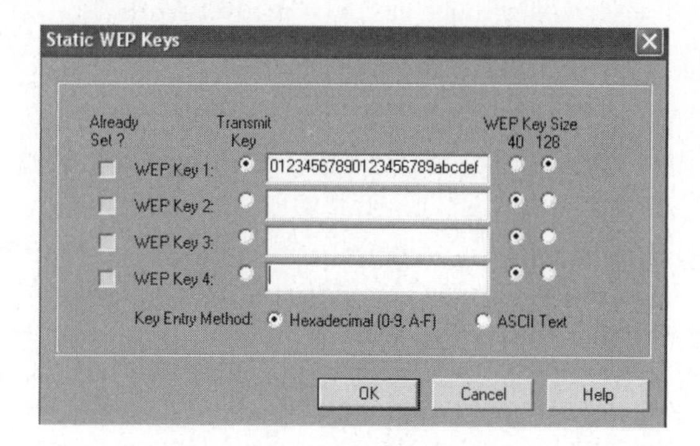

How does WEP work? WEP runs a cyclic redundancy check (CRC) on the plaintext data that is to be encrypted and then appends the Integrity Check Value (ICV) to the end of the plaintext data. A 24-bit cleartext Initialization Vector (IV) is then generated and combined with the static secret key. WEP then uses both the static key and the IV as seeding material through a pseudorandom algorithm that generates random bits of data known as a keystream. These pseudorandom bits are equal in length to the plaintext data that is to be encrypted. The pseudorandom bits in the keystream are then combined with the plaintext data bits by using a Boolean XOR process. The end result is the WEP ciphertext, which is the encrypted data. The encrypted data is then prefixed with the cleartext IV. Figure 13.3 illustrates this process.

FIGURE 13.3 WEP encryption process

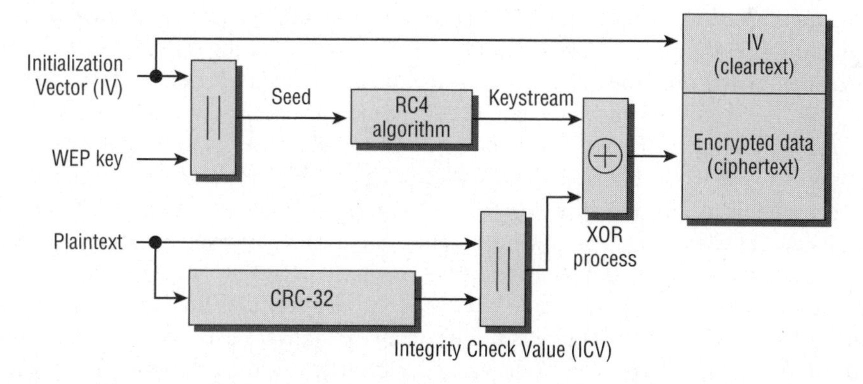

Unfortunately, WEP has quite a few weaknesses, including the following four main attacks:

IV collisions attack Because the 24-bit Initialization Vector is in cleartext and is different in every frame, all 16 million IVs will eventually repeat themselves in a busy WEP encrypted network. Because of the limited size of the IV space, IV collisions occur, and an attacker can recover the secret key much easier when IV collisions occur in wireless networks.

Weak key attack Because of the RC4 key-scheduling algorithm, weak IV keys are generated. An attacker can recover the secret key much easier by recovering the known weak IV keys.

Reinjection attack Hacker tools exist that implement a packet reinjection attack to accelerate the collection of weak IVs on a network with little traffic.

Bit-flipping attack The ICV data integrity check is considered weak. WEP encrypted packets can be tampered with.

Current WEP cracking tools may use a combination of the first three mentioned attacks and can crack WEP in less than 5 minutes. After an attacker has compromised the static WEP key, any data frame can be decrypted with the newly discovered key. Later in this chapter, we discuss TKIP, which is an enhancement of WEP and has not been cracked. CCMP encryption uses the AES algorithm and is an even stronger encryption method. As defined by the original 802.11 standard, WEP encryption is considered optional. Although WEP encryption has indeed been cracked and is viewed as unacceptable in the enterprise, it is still better than using no encryption at all.

MAC Filters

Every network card has a physical address known as a MAC address. This address is a 12-digit hexadecimal number. 802.11 client stations each have unique MAC addresses, and as you have already learned, 802.11 access points use MAC addresses to direct frame traffic. Most vendors provide MAC filtering capabilities on their access points. MAC filters can be configured to either allow or deny traffic from specific MAC addresses.

Most MAC filters apply restrictions that will allow traffic only from specific client stations to pass through based on their unique MAC addresses. Any other client stations whose MAC addresses are not on the allowed list will not be able to pass traffic through the virtual port of the access point and onto the distribution system medium. It should be noted that MAC addresses can be *spoofed*, or impersonated, and any amateur hacker can easily bypass any MAC filter by spoofing an allowed client station's address. Because of spoofing and because of all the administrative work that is involved with setting up MAC filters, MAC filtering is not considered a reliable means of security for wireless enterprise networks. The 802.11 standard does not define MAC filtering, and any implementation of MAC filtering is vendor specific.

MAC filters are often used as a security measure to protect legacy radios that do not support stronger security. For example, older handheld barcode scanners may use 802.11 radios that support only static WEP. Best practices dictate an extra layer of security by segmenting the handheld devices in a separate VLAN with a MAC filter based on the manufacturer's OUI address (the first three octets of the MAC address that are manufacturer specific).

SSID Cloaking

Remember in *Star Trek* when the Romulans "cloaked" their spaceship but somehow Captain Kirk always found the ship anyway? Well there is a way to "cloak" your service set identifier (SSID). Access points typically have a setting called *Closed Network* or *Broadcast SSID*. By either enabling a closed network or disabling the broadcast SSID feature, you can hide, or cloak, your wireless network name.

When you implement a closed network, the SSID field in the beacon frame is null (empty), and therefore passive scanning will not reveal the SSID to client stations that are listening to beacons. The SSID, which is also often called the ESSID, is the logical identifier of a WLAN. The idea behind cloaking the SSID is that any client station that does not know the SSID of the WLAN will not be able to associate.

Many wireless client software utilities transmit probe requests with null SSID fields when actively scanning for access points. Additionally, there is a popular and freely available software program called NetStumbler that is used by individuals to discover wireless networks. NetStumbler also sends out null probe requests actively scanning for access points. When you implement a closed network, the access point responds to null probe requests with probe responses; however, as in the beacon frame, the SSID field is null, and therefore the SSID is hidden to client stations that are using active scanning. Effectively, your wireless network is temporarily invisible, or cloaked. It should be noted that an access point in a closed network will respond to any configured client station that transmits directed probe requests with the properly configured SSID. This ensures that legitimate end users will be able to authenticate and associate to the AP. However, any stations that are not configured with the correct SSID will not be able to authenticate or associate. Although implementing a closed network will indeed hide your SSID from NetStumbler and other WLAN discovery tools, anyone with a layer 2 wireless protocol analyzer can capture the frames transmitted by any legitimate end user and discover the SSID, which is transmitted in cleartext. In other words, a hidden SSID can be found usually in seconds with the proper tools. Many wireless professionals will argue that hiding the SSID is a waste of time, while others view a closed network as just another layer of security.

Although you can hide your SSID to cloak the identity of your wireless network from novice hackers (often referred to as *script kiddies*) and nonhackers, it should be clearly understood that SSID cloaking is by no means an end-all wireless security solution. The 802.11 standard does not define SSID cloaking, and therefore all implementations of a closed network are vendor specific. As a result, incompatibility can potentially cause connectivity problems with older legacy cards or when using cards from mixed vendors on your own network. Be sure to know the capabilities of your devices before implementing a closed network. Cloaking the SSID can also become an administration and support issue. Requiring end users to configure the SSID in the radio software interface often results in more calls to the help desk because of misconfigured SSIDs.

Robust Security

In 2004, the 802.11i security amendment was ratified and is now part of the 802.11-2007 standard. The 802.11-2007 standard defines an enterprise authentication method as well as a method of authentication for home use. The current standard requires the use of an 802.1X/EAP authentication method in the enterprise and the use of a preshared key or a passphrase in a SOHO environment. The 802.11-2007 standard also requires the use of strong, dynamic encryption-key generation methods. CCMP/AES encryption is the default encryption method, while TKIP/RC4 is an optional encryption method.

Prior to the ratification of the 802.11i amendment, the Wi-Fi Alliance introduced the *Wi-Fi Protected Access (WPA)* certification as a snapshot of the not-yet-released 802.11i amendment, supporting only TKIP/RC4 dynamic encryption-key generation. 802.1X/EAP authentication was required in the enterprise, and passphrase authentication was required in a SOHO environment.

After 802.11i was ratified, the Wi-Fi Alliance introduced the WPA2 certification. *WPA2* is a more complete implementation of the 802.11i amendment and supports both CCMP/AES and TKIP/RC4 dynamic encryption-key generation. 802.1X/EAP authentication is required in the enterprise, and passphrase authentication is required in a SOHO environment. Table 13.1 offers a valuable comparison of the various security standards and certifications.

TABLE 13.1 Security Standards and Certifications Comparison

802.11 Standard	Wi-Fi Alliance Certification	Authentication Method	Encryption Method	Cipher	Key Generation
802.11 legacy		Open System or Shared Key	WEP	RC4	Static
	WPA-Personal	WPA Passphrase (also known as WPA PSK and WPA Preshared Key)	TKIP	RC4	Dynamic
	WPA-Enterprise	802.1X/EAP	TKIP	RC4	Dynamic
802.11- 2007 (RSN)	WPA2-Personal	WPA2 Passphrase (also known as WPA2 PSK and WPA2 Preshared Key)	CCMP (mandatory) TKIP (optional)	AES (mandatory) RC4 (optional)	Dynamic
802.11-2007 (RSN)	WPA2-Enterprise	802.1X/EAP	CCMP (mandatory) TKIP (optional)	AES (mandatory) RC4 (optional)	Dynamic

Robust Security Network (RSN)

The 802.11-2007 standard defines what is known as a *robust security network (RSN)* and *robust security network associations (RSNAs)*. Two stations (STAs) must establish a procedure to authenticate and associate with each other as well as create dynamic encryption keys through a process known as the 4-Way Handshake. This association between two stations is referred to as an RSNA. In other words, any two radios must share dynamic encryption keys that are unique between those two radios. CCMP/AES encryption is the mandated encryption method, while TKIP/RC4 is an optional encryption method. A robust security network (RSN) is a network that allows for the creation of only robust security network associations (RSNAs). An RSN can be identified by a field found in beacons, probe response frames, association request frames, and reassociation request frames. This field is known as the *RSN Information Element (IE)*. This field may identify the cipher suite capabilities of each station. The 802.11-2007 standard does allow for the creation of pre–robust security network associations (pre-RSNAs) as well as RSNAs. In other words, legacy security measures can be supported in the same basic service set (BSS) along with RSN-security-defined mechanisms. A *transition security network (TSN)* supports RSN-defined security as well as legacy security such as WEP within the same BSS.

Authentication and Authorization

As you learned earlier in this chapter, authentication is the verification of user identity and credentials. Users must identify themselves and present credentials such as passwords or digital certificates. Authorization involves granting access to network resources and services. Before authorization to network resources can be granted, proper authentication must occur.

The following sections detail more-advanced authentication and authorization defenses. You will also learn that dynamic encryption capabilities are also possible as a by-product of these stronger authentication solutions.

802.1X/EAP Framework

The IEEE *802.1X* standard is not specifically a wireless standard and is often mistakenly referred to as 802.11x. The 802.1X standard is a *port-based access control* standard. 802.1X provides an authorization framework that allows or disallows traffic to pass through a port and thereby access network resources. An 802.1X framework may be implemented in either a wireless or wired environment. The 802.1X framework consists of three main components:

Supplicant A host with software that is requesting authentication and access to network resources. Each supplicant has unique authentication credentials that are verified by the authentication server.

Authenticator A device that blocks or allows traffic to pass through its port entity. Authentication traffic is normally allowed to pass through the authenticator, while all other traffic is blocked until the identity of the supplicant has been verified. The authenticator maintains two virtual ports: an uncontrolled port and a controlled port. The uncontrolled port allows EAP authentication traffic to pass through, while the controlled port blocks all other traffic until the supplicant has been authenticated.

Authentication server (AS) A server that validates the credentials of the supplicant that is requesting access and notifies the authenticator that the supplicant has been authorized. The authentication server will maintain a user database or may proxy with an external user database to authenticate user credentials.

Within an 802.3 Ethernet network, the supplicant would be a desktop host, the authenticator would be a managed switch, and the authentication server would typically be a Remote Authentication Dial-In User Service (RADIUS) server. In an 802.11 wireless environment, the supplicant would be a client station requesting access to network resources. As seen in Figure 13.4, an autonomous access point would be the authenticator, blocking access via virtual ports, and the AS is typically a RADIUS server. Figure 13.4 also shows that when an 802.1X security solution is used with a WLAN controller solution, the WLAN controller is the authenticator— and not the lightweight access points.

As depicted in Figure 13.5, the root bridge would be the authenticator and the nonroot bridge would be the supplicant if 802.1X security is used in a WLAN bridged network.

FIGURE 13.4 802.1X comparison—autonomous access point and WLAN controller

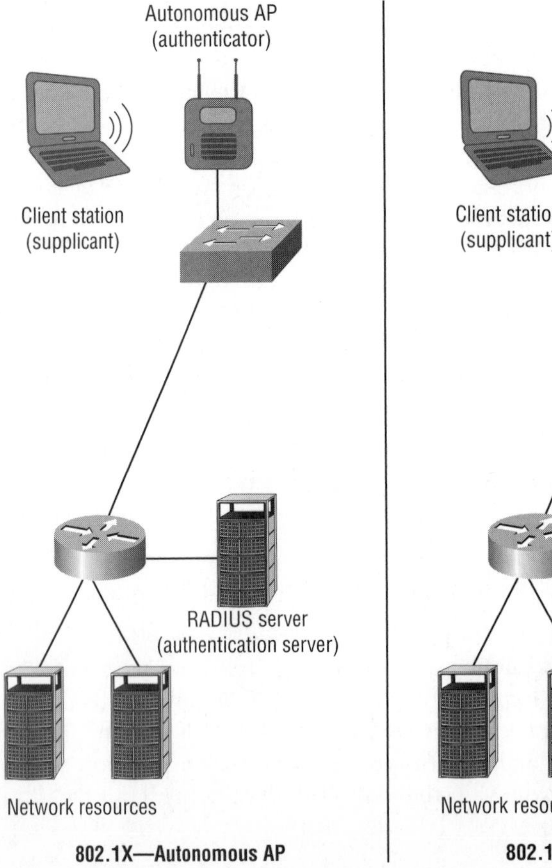

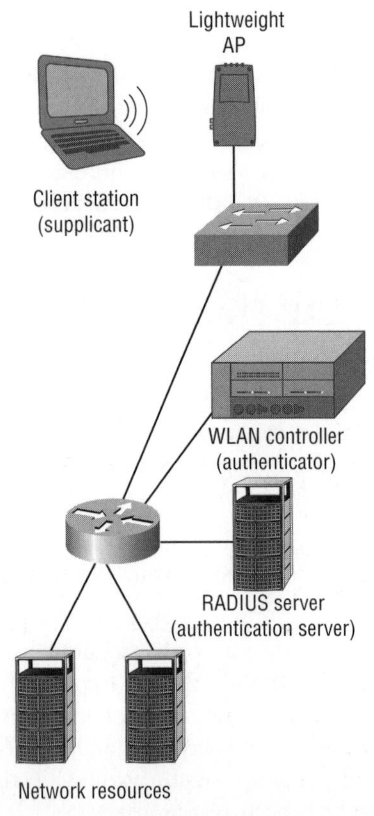

FIGURE 13.5 WLAN bridging and 802.1X

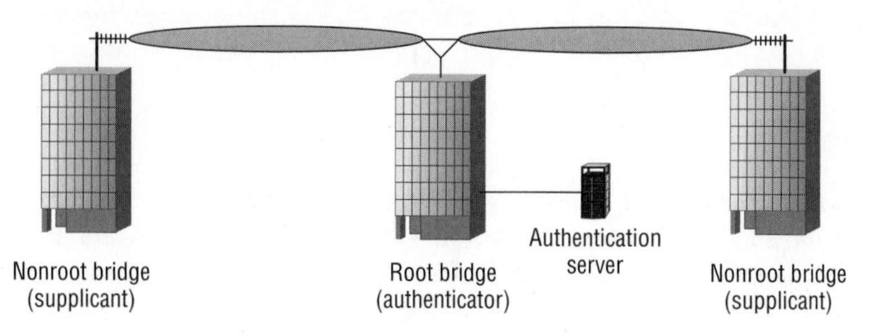

Although the *supplicant*, *authenticator*, and *authentication server* work together to provide the framework for 802.1X port-based access control, an authentication protocol is needed to actually perform the authentication process. *Extensible Authentication Protocol (EAP)* is used to provide user authentication. EAP is a flexible layer 2 authentication protocol that resides under Point-to-Point Protocol (PPP). The supplicant and the authentication server communicate with each other by using the EAP protocol. The authenticator allows the EAP traffic to pass through its virtual uncontrolled port. After the authentication server has verified the credentials of the supplicant, the server sends a message to the authenticator that the supplicant has been authenticated, and the authenticator is then authorized to open the virtual controlled port, allowing all other traffic to pass through. Figure 13.6 depicts the generic 802.1X/EAP frame exchanges.

FIGURE 13.6 802.1X/EAP authentication

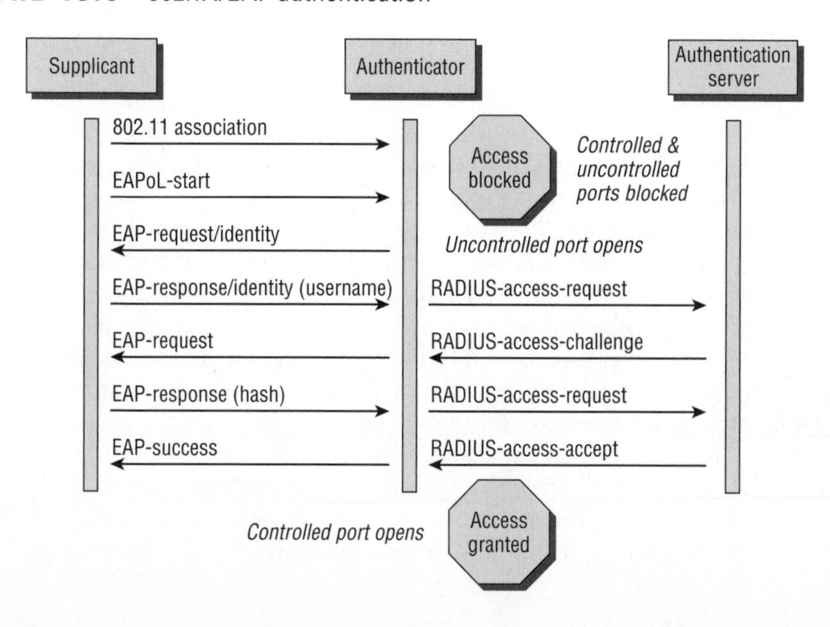

The 802.1X/EAP framework, when used with wireless networks, provides the necessary means of validating user identity as well as authorizing client stations onto the wired network infrastructure.

EAP Types

As noted earlier, *EAP* stands for *Extensible Authentication Protocol*. The key word in EAP is *extensible*. EAP is a layer 2 protocol that is very flexible, and many different flavors of EAP exist. Some, such as Cisco's Lightweight Extensible Authentication Protocol (LEAP), are proprietary, while others, such as Protected Extensible Authentication Protocol (PEAP), are considered standard-based. Some may provide for only one-way authentication, while others provide two-way authentication. Mutual authentication not only requires that the authentication server validate the client credentials, but the supplicant must also authenticate the validity of the authentication server. By validating the authentication server, the supplicant can ensure that the username and password are not inadvertently given to a rogue authentication server. Most types of EAP that require mutual authentication use a server-side digital certificate to validate the authentication server.

Table 13.2 provides a comparison chart of many of the various types of EAP. It is beyond the scope of this book to discuss in detail all the authentication mechanisms and differences between the different flavors of EAP. The CWSP exam will test you very heavily on the operations of the various types of EAP authentication. The CWNA exam will not test you on the specific EAP functions.

Dynamic Encryption-Key Generation

Although the 802.1X/EAP framework does not require encryption, it highly suggests the use of encryption. You have already learned that the purpose of 802.1X/EAP is authentication and authorization. However, a by-product of 802.1X/EAP is the generation and distribution of dynamic encryption keys. EAP protocols that utilize mutual authentication provide "seeding material" that can be used to generate encryption keys dynamically. Until now, you have learned about only static WEP keys. The use of static keys is typically an administrative nightmare, and when the same static key is shared among multiple users, the secret is easy to compromise via social engineering. The advantage of dynamic keys is that every user has a different and unique key that cannot be compromised by social engineering attacks.

After an EAP frame exchange where mutual authentication is required, both the AS and the supplicant know information about each other because of the exchange of credentials. This newfound information is used as seeding material or keying material to generate a matching dynamic encryption key for both the supplicant and the authentication server. These dynamic keys are generated *per session per user*, meaning that every time a client station authenticates, a new key is generated and every user has a unique and separate key.

Prior to 2004, many vendors implemented solutions that generated dynamic WEP encryption keys as a result of 802.1X/EAP authentication. Dynamic WEP was never standardized but was used by vendors until TKIP and CCMP became available to the marketplace.

TABLE 13.2 EAP Comparison Chart

	EAP-MD5	EAP-LEAP	EAP-TLS	EAP-TTLS	PEAPv0 (EAP-MSCHAPv2)	PEAPv0 (EAP-TLS)	PEAPv1 (EAP-GTC)	EAP-FAST
Security Solution	RFC-2284	Cisco proprietary	RFC-2716	IETF draft	IETF draft	IETF draft	IETF draft	IETF draft
Digital Certificates—Client	No	No	Yes	Optional	No	Yes	Optional	No
Digital Certificates—Server	No	No	Yes	Yes	Yes	Yes	Yes	No
Client Password Authentication	No	Yes	N/A	Yes	Yes	No	Yes	Yes
PACs—Client	No	No	No	No	No	No	No	Yes
PACs—Server	No	No	No	No	No	No	No	Yes
Credential Security	Weak	Weak (depends on password strength)	Strong	Strong	Strong	Strong	Strong	Strong (if Phase 0 is secure)
Encryption Key Management	No	Yes	Yes	Yes	Yes	Yes	Yes	Yes
Mutual Authentication	No	Yes	Yes	Yes	Yes	Yes	Yes	Yes
Tunneled Authentication	No	No	No	Yes	Yes	Yes	Yes	Yes
Wi-Fi Alliance supported	No	No	Yes	Yes	Yes	No	Yes	No
Man-in-the-Middle Protection	No	No	Yes	Yes	Yes	Yes	Yes	Yes
Dictionary Attack Resistance	No	No	Yes	Yes	Yes	N/A	Yes	Yes

Dynamic WEP was a short-lived encryption-key management solution that was often implemented prior to the release of WPA-certified WLAN products. The generation and distribution of dynamic WEP keys as a by-product of the EAP authentication process had many benefits and was preferable to the use of static WEP keys. Static keys were no longer used and did not have to be entered manually. Also, every user had a separate and independent key. If a user's dynamic WEP key was compromised, only that one user's traffic could be decrypted. However, a dynamic WEP key could still be cracked, and if compromised, it could indeed be used to decrypt data frames. Dynamic WEP still had risks. Please understand that a dynamic WEP key is not the same as TKIP or CCMP encryption keys that are also generated dynamically. WPA/WPA2 security defines the creation of stronger and safer dynamic TKIP/RC4 or CCMP/AES encryption keys that are also generated as a by-product of the EAP authentication process.

4-Way Handshake

As stated earlier, the 802.11-2007 standard defines what is known as a robust security network (RSN) and robust security network associations (RSNAs). Two stations (STAs) must establish a procedure to authenticate and associate with each other as well as create dynamic encryption keys through a process known as the 4-Way Handshake.

RSNAs utilize a dynamic encryption-key management method that actually involves the creation of five separate keys. It is beyond the scope of this book to fully explain this entire process, but a brief explanation is appropriate. Part of the RSNA process involves the creation of two master keys known as the Group Master Key (GMK) and the Pairwise Master Key (PMK). These keys are created as a result of 802.1X/EAP authentication. A PMK can also be created from a *preshared key (PSK)* authentication method instead of 802.1X/EAP authentication. These master keys are the seeding material used to create the final dynamic keys that are actually used for encryption and decryption. The final encryption keys are known as the Pairwise Transient Key (PTK) and the Group Temporal Key (GTK). The PTK is used to encrypt/decrypt unicast traffic, and the GTK is used to encrypt/decrypt broadcast and multicast traffic.

These final keys are created during a four-way EAP frame exchange that is known as the *4-Way Handshake*. The 4-Way Handshake will always be the final four frames exchanged during either 802.1X/EAP authentication or PSK authentication. Whenever TKIP/RC4 or CCMP/AES dynamic keys are created, the 4-Way Handshake must occur. Also, every time a client radio roams from one AP to another, a new 4-Way Handshake must occur so that new unique dynamic keys can be generated.

The CWNA exam currently does not test on the dynamic encryption-key creation process, which was originally defined by the 802.11i amendment. The process is heavily tested in the CWSP exam. Included on the CD of this book is a white paper titled "802.11i Authentication and Key Management (AKM)" authored by Devin Akin. This white paper is often referred to as the "chicken-and-egg" white paper and is recommended extra reading.

WPA/WPA2-Personal

Do you have a RADIUS server in your home or small business? The answer to that question will almost always be no. If you do not own a RADIUS server, 802.1X/EAP authentication will not be possible. WPA/WPA2-Enterprise solutions require 802.1X for mutual authentication using some form of EAP. Additionally, an authentication server will be needed. Because most of us do not have a RADIUS server in our basement, the 802.11-2007 standard offers a simpler method of authentication using a preshared key (PSK). This method involves manually typing matching passphrases on both the access point and all client stations that will need to be able to associate to the wireless network. A formula is run that converts the passphrase to a Pairwise Master Key (PMK) used with the 4-Way Handshake to create the final dynamic encryption keys.

This simple method of authentication and encryption key generation is known as WPA/WPA2-Personal. Other names include WPA/WPA2 Preshared Key and WPA/WPA2 PSK. Although this is certainly better than static WEP and Open System authentication, WPA/WPA2-Personal still requires significant administrative overhead and has potential social engineering issues in a corporate or enterprise environment. In Chapter 14, you will learn that WPA/WPA2-Personal is susceptible to offline dictionary attacks and should be avoided in an enterprise environment whenever possible. An 802.1X/EAP solution as defined by WPA/WPA2-Enterprise is the preferred method of security in a corporate and workplace environment.

TKIP Encryption

The optional encryption method defined for a robust security network is *Temporal Key Integrity Protocol (TKIP)*. This method uses the RC4 cipher just as WEP encryption does. As a matter of fact, TKIP is an enhancement of WEP encryption that addresses many of the known weaknesses of WEP.

TKIP starts with a 128-bit temporal key that is combined with a 48-bit Initialization Vector (IV) and source and destination MAC addresses in a complicated process known as per-packet key mixing. This key-mixing process mitigates the known IV collision and weak key attacks used against WEP. TKIP also uses a sequencing method to mitigate the reinjection attacks used against WEP. Additionally, TKIP uses a stronger data integrity check known as the *Message Integrity Check (MIC)* to mitigate known bit-flipping attacks against WEP. The MIC is sometimes referred to by the nickname *Michael*. All TKIP encryption keys are dynamically generated as a final result of the 4-Way Handshake.

WEP encryption adds an extra 8 bytes of overhead to the body of an 802.11 data frame. When TKIP is implemented, because of the extra overhead from the extended IV and the MIC, a total of 20 bytes of overhead is added to the body of an 802.11 data frame. Because TKIP uses the RC4 algorithm and is simply WEP that has been enhanced, most vendors released a WPA firmware upgrade that gave legacy WEP-only cards the capability of using TKIP encryption.

CCMP Encryption

The default encryption method defined under the 802.11i amendment is known as *Counter Mode with Cipher Block Chaining Message Authentication Code Protocol (CCMP)*. This method uses the Advanced Encryption Standard (AES) algorithm (Rijndael algorithm). CCMP/AES uses a 128-bit encryption-key size and encrypts in 128-bit fixed-length blocks. An 8-byte Message Integrity Check is used that is considered much stronger than the one used in TKIP. Also, because of the strength of the AES cipher, per-packet key mixing is unnecessary. All CCMP encryption keys are dynamically generated as a final result of the 4-Way Handshake.

CCMP/AES encryption will add an extra 16 bytes of overhead to the body of an 802.11 data frame. Because the AES cipher is processor intensive, older legacy 802.11 devices will not have the processing power necessary to perform AES calculations. Older 802.11 devices will not be firmware upgradeable, and a hardware upgrade is often needed to support WPA2. Because of the requirement to upgrade the hardware to implement AES, the initial transition to WPA2 was slow. For wireless security solutions, it is a recommended practice to choose hardware that handles the processing needs of CCMP/AES encryption. There are some vendors that still attempt to achieve this in software rather than through a hardware mechanism. Software solutions will always perform substantially slower. It is recommended that a device is selected with a CCMP/AES solution implemented on the card's chipset.

EXERCISE 13.2

802.1X/EAP and 4-Way Handshake Process

In this exercise, you will use a protocol analyzer to view the frame exchanges during an 802.1X/EAP authentication and to view a 4-Way Handshake used to create dynamic encryption keys. The following directions should assist you with the installation and use of the AirMagnet WiFi Analyzer product demo software. If you have already installed AirMagnet WiFi Analyzer, you can skip steps 1–5.

1. In your web browser, proceed to the following URL: `www.airmagnet.com/cwna/sybex`.
2. Fill out the product demo request forms. You will then be sent an email message with a private download URL. Install the proper driver for your Wi-Fi card.
3. Proceed to the private download URL. Verify that you have a supported Wi-Fi card. Review the system requirements and supported operating systems.
4. Download the Wi-Fi Analyzer demo to your desktop. This evaluation copy of Wi-Fi Analyzer will be licensed to work for seven days.
5. Double-click the file `AirMagnet_WiFi_DemoInstaller.exe` and follow the installation prompts.
6. From Windows, choose Start ➢ Programs ➢ AirMagnet and then click the demo icon. The Wi-Fi Analyzer application should appear.
7. In the bottom-left corner of the application, click the Decodes button.

EXERCISE 13.2 *(continued)*

8. From the AirMagnet Wi-Fi Analyzer main window, observe the EAP frame exchange from frame #12104 to frame #12123.

9. From the AirMagnet Wi-Fi Analyzer main window, observe the 4-Way Handshake from frame #5129 to frame #5136.

Real World Scenario

How Should Authentication and Encryption Be Deployed in the Enterprise?

As you have learned, the goal of authentication is to validate user credentials, whereas the goal of encryption is to ensure data privacy. However, you have also learned that the two processes are dependent on each other because dynamic TKIP or CCMP encryption keys are generated as a by-product of either 802.1X/EAP or PSK authentication. As depicted in the following graphic, when deploying a WPA/WPA2-Enterprise solution, the supplicant and the authentication server must support the same EAP protocol. Furthermore, the supplicant and the authenticator must support the same dynamic encryption method.

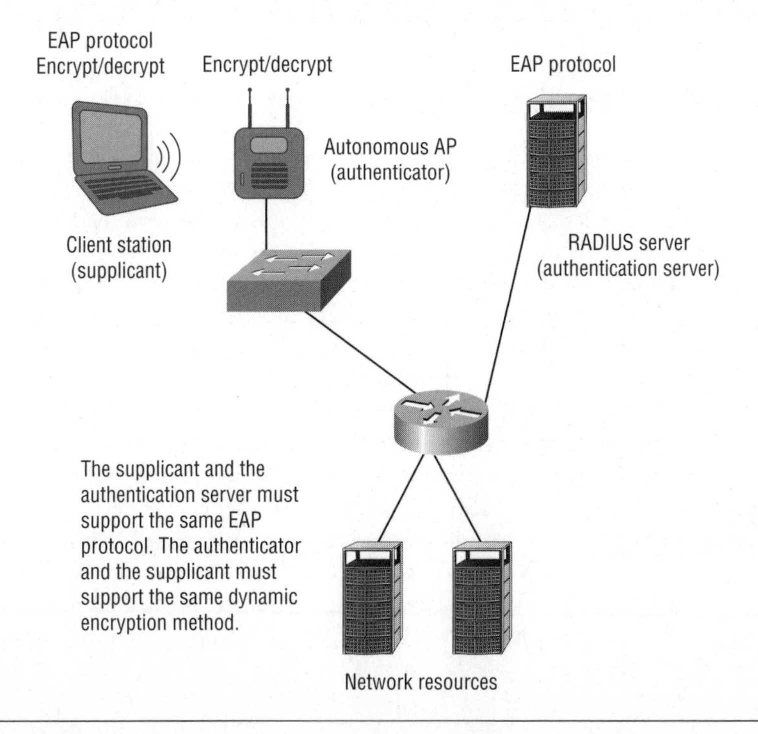

PSK authentication (WPA/WPA2Personal) should be avoided in the enterprise due to the risk of social engineering and offline dictionary attacks. One advantage of using preshared key (PSK) authentication is that it does not have the latency issues of 802.1X/EAP. Until fast, secure roaming mechanisms become more widely supported, PSK authentication is often used in the enterprise with VoWiFi phones that require authentication exchanges during roaming handoffs to occur in less than 150 milliseconds. Otherwise, 802.1X/EAP (WPA/WPA2-Enterprise) solutions should be used for authentication whenever possible.

The TKIP encryption has not been cracked. However, at the time this book was written, some new TKIP exploits based on integrity and checksums have been discovered. CCMP encryption uses the stronger AES cipher and is the preferred method of providing for data privacy.

Segmentation

As discussed earlier in this chapter, segmentation is a key part of a network design. Once authorized onto network resources, users can be further restricted as to what resources may be accessed and where they can go. Segmentation can be achieved through a variety of means, including firewalls, routers, VPNs, and VLANs. The most common wireless segmentation strategy used in 802.11 enterprise WLANs is layer 3 segmentation using virtual LANs (VLANs). Segmentation is also often intertwined with role-based access control (RBAC).

VLANs

Virtual local area networks (VLANs) are used to create separate broadcast domains in a layer 2 network and are often used to restrict access to network resources without regard to physical topology of the network. VLANs are used extensively in switched 802.3 networks for both security and segmentation purposes.

In a WLAN environment, individual SSIDs can be mapped to individual VLANs, and users can be segmented by the SSID/VLAN pair, all while communicating through a single access point. Each SSID can also be configured with separate security settings. Most vendors can have as many as 16 wireless VLANs with the capability of segmenting the users into separate layer 3 domains. A common strategy is to create a guest, voice, and data VLAN as pictured in Figure 13.7. The SSID mapped to the guest VLAN will have no security, and all users are restricted away from network resources and routed off to an Internet gateway. The voice VLAN SSID might be using a security solution such a WPA2 Passphrase, and the VoWiFi client phones are routed to a VoIP server that provides proprietary QoS services through the VLAN. The data VLAN SSID uses a stronger security solution such as WPA2-Enterprise, and the access control lists allow the data users to access full network resources once authenticated. In a WLAN controller environment, all VLAN, SSID, and security configurations are performed on the WLAN controller and then pushed or distributed to the thin access points. When using autonomous access points, the VLANs are created on a third-party managed switch and then the VLANs are mapped to SSID and security settings that are configured on the fat access points.

FIGURE 13.7 Wireless VLANs

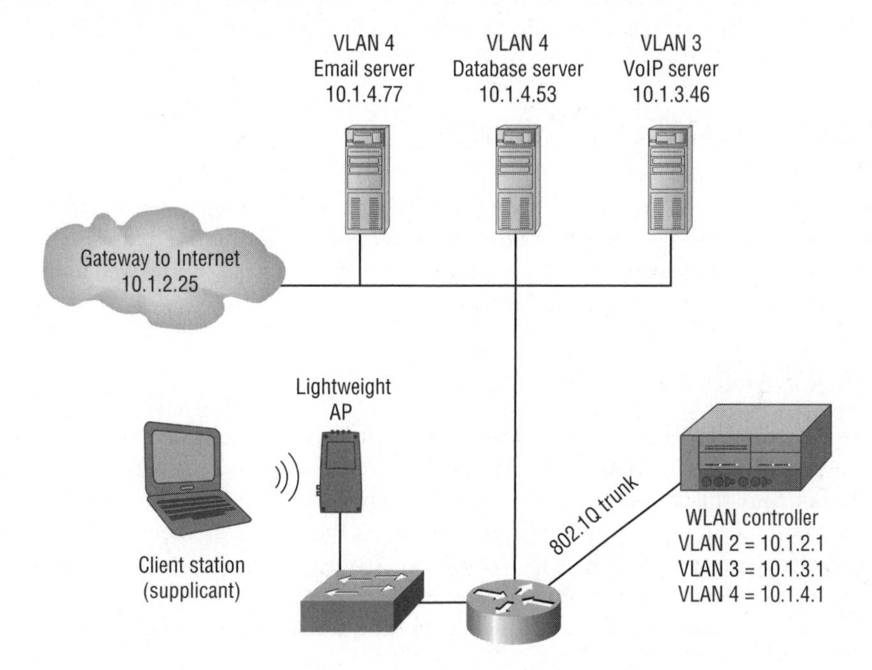

RBAC

Role-based access control (RBAC) is an approach to restricting system access to authorized users. The majority of WLAN controller solutions have RBAC capabilities. The three main components of an RBAC approach are users, roles, and permissions.

Separate roles can be created, such as the sales role or the marketing role. Individuals or groups of users are assigned to one of these roles. Permissions can be defined as layer 2 permissions (MAC filters), layer 3 permissions (access control lists), layers 4–7 permissions (stateful firewall rules), and bandwidth permissions. All of these permissions can also be time based. The permissions are mapped to the roles. When wireless users authenticate via the WLAN, they inherit the permissions of whatever roles they have been assigned. For example, users that associate with a "Guest" SSID are placed in a unique guest VLAN. The users then authenticate via a captive portal and are assigned a guest role. The guest role may have bandwidth permissions that restrict them to 100 kbps of bandwidth and allow them to use only ports 80 (HTTP), 25 (SMTP) and (110) POP during working hours. This scenario would restrict guest users who are accessing the Internet from hogging bandwidth and only allow them to view web pages and check email between 9 a.m. and 5 p.m. When used in a WLAN environment, role-based access control can provide granular wireless user management.

Infrastructure Security

An often-overlooked aspect of wireless security is protecting the infrastructure equipment. In addition to protecting Wi-Fi hardware from theft, you must also secure the management interfaces so that only authorized administrators have access. Protecting hardware and interfaces should never be ignored in an 802.11 enterprise.

Physical Security

Access points and other WLAN hardware can be quite expensive. Many enterprise access points can cost as much as $1,000 (in U.S. dollars). Although access points are usually mounted in or near the ceiling, theft can be a problem. Enclosure units with locks can be mounted in the ceiling or to the wall. Access points locked inside the enclosure units are safeguarded against theft. The enclosure units also prevent unwanted individuals from using a serial cable or console cable to try to gain access to the AP. Secure enclosure units may also meet aesthetic demands by keeping the access point out of plain sight.

Interface Security

All wireless infrastructure devices must be able to be accessed by administrators through a management interface. Enterprise equipment usually can be configured either through a command-line interface or a web interface or via Simple Network Management Protocol (SNMP). Any interface that is not used should be turned off. For example, if the administrator configures the access points only via a command-line interface, turn off the web interface capabilities on the access point. At a minimum, all the passwords for these configuration options should be changed from the factory defaults. It is important to note that some management interfaces have multiple default user levels. The default levels can include administrator, guest, and management. The passwords for all of these levels should be changed

Most infrastructure devices should also support some type of encrypted management capabilities. Newer Wi-Fi hardware should support either secure command-shell, HTTPS, or SNMPv3. Older legacy equipment may not support encrypted login capabilities. It is also a highly recommended practice to configure your infrastructure devices from only the wired side and never configure them wirelessly. If devices are configured from the wireless side, an intruder might capture your wireless packets and be able to watch what you are doing. There is also a very good chance that you will accidentally lock yourself out of the device while configuring Wi-Fi hardware wirelessly.

VPN Wireless Security

Although the 802.11-2007 standard clearly defines layer 2 security solutions, the use of upper-layer *virtual private network (VPN)* solutions can also be deployed with WLANs. VPNs are typically not recommended to provide wireless security in the enterprise due to the overhead and because faster, more-secure layer 2 solutions are now available. Although not usually a recommended practice, VPNs are often used for WLAN security because the VPN solution was already in place inside the wired infrastructure. VPNs do have their place in Wi-Fi security and should definitely be used for remote access. They are also often used in wireless bridging environments. The two major types of VPN topologies are router to router or client/server based.

Use of VPN technology is mandatory for remote access. Your end users will take their laptops off site and will most likely use public access Wi-Fi hot spots. Since there is no security at most hot spots, a VPN solution is needed. The VPN user will need to bring the security to the hot spot in order to provide a secure connection. It is imperative that users implement a VPN solution coupled with a personal firewall whenever accessing any public access Wi-Fi networks.

Layer 3 VPNs

VPNs have several major characteristics. They provide encryption, encapsulation, authentication, and data integrity. VPNs use secure tunneling, which is the process of encapsulating one IP packet within another IP packet. The first packet is encapsulated inside the second packet. The original destination and source IP address of the first packet is encrypted along with the data payload of the first packet. VPN tunneling therefore protects your original layer 3 addresses and also protects the data payload of the original packet. Layer 3 VPNs use layer 3 encryption; therefore, the payload that is being encrypted is the layer 4 to 7 information. The IP addresses of the second packet are seen in cleartext and are used for communications between the tunnel end points. The destination and source IP addresses of the second packet will point to the virtual IP address of the VPN server and VPN client software. Figure 13.8 depicts a layer 3 VPN in a wireless environment.

The two major types of layer 3 VPN technologies are *Point-to-Point Tunneling Protocol (PPTP)* and *Internet Protocol Security (IPsec)*. PPTP uses 128-bit *Microsoft Point-to-Point Encryption (MPPE)*, which uses the RC4 algorithm. PPTP encryption is considered adequate but not strong. PPTP uses MS-CHAP version 2 for user authentication. Unfortunately, the chosen authentication method can be compromised with offline dictionary attacks. VPNs using PPTP technology typically are used in smaller SOHO environments. IPsec VPNs use stronger encryption methods and more-secure methods of authentication. IPsec supports multiple ciphers including DES, 3DES, and AES. Device authentication is achieved by using either a server-side certificate or a preshared key. A full explanation of IPsec technology is beyond the scope of this book, but IPsec is usually the choice for VPN technology in the enterprise. VPN technologies do exist that operate at other layers of the OSI model, including layer 7 SSL tunneling and SSH2 VPNs.

FIGURE 13.8 Remote access VPN tunnel

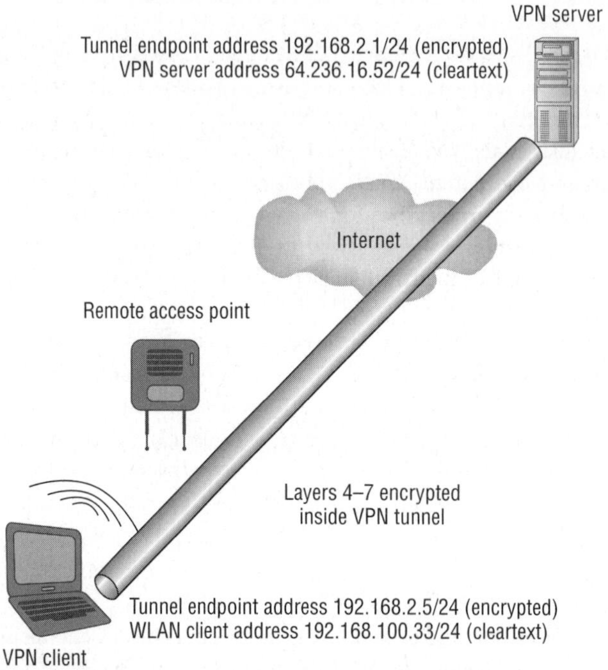

VPN server
Tunnel endpoint address 192.168.2.1/24 (encrypted)
VPN server address 64.236.16.52/24 (cleartext)
Internet
Remote access point
Layers 4–7 encrypted inside VPN tunnel
Tunnel endpoint address 192.168.2.5/24 (encrypted)
WLAN client address 192.168.100.33/24 (cleartext)
VPN client

Unlike 802.1X/EAP solutions, an IP address is needed before a VPN tunnel can be established. A downside to using a VPN solution is that access points are potentially open to attack because a potential attacker can get both a layer 2 and layer 3 connection before the VPN tunnel is established. 802.1X/EAP requires that all security credentials and transactions are completed before any layer 3 connectivity is even possible.

Summary

In this chapter, you learned that five major facets are needed for wireless security. A strong encryption solution is needed to protect the data frames. A mutual authentication solution is needed to ensure that only legitimate users are authorized to use network resources. A segmentation solution is necessary to further restrict users as to what resources they may access and where they can go. 802.11 wireless networks can be further protected with continuous monitoring and enforcement of WLAN security policy.

We discussed legacy 802.11 authentication and encryption solutions and why they are weak. We covered the stronger 802.1X/EAP authentication solutions and the benefits of dynamic encryption-key generation, as well as what is defined by the 802.11-2007 standard and the related WPA/WPA2 certifications. The 802.11-2007 standard defines a layer 2 robust security network using either 802.1X/EAP or PSK authentication and defines CCMP/AES or TKIP/RC4 dynamic encryption. Finally, we covered proper infrastructure and interface security as well as VPN technology in a WLAN environment.

It is important to understand the capabilities and limitations of the devices that will be deployed within your 802.11 wireless networks. Ideally, devices will be segmented into separate VLANs by using 802.1X/EAP authentication and CCMP/AES encryption. VoIP phones, mobile scanners, mobile printers, handheld devices, and so on are often not equipped with the ability to handle more-advanced security capabilities. Proper designs must take into account all of these components to ensure the most dynamic and secure network.

Exam Essentials

Define the concept of AAA. Be able to explain the differences between authentication, authorization, and accounting and why each is needed for a WLAN network.

Explain why data privacy and segmentation are needed. Be able to discuss why data frames must be protected with encryption. Know the differences between the various encryption ciphers. Understand how VLANs and RBAC mechanisms are used to further restrict network resources.

Understand legacy 802.11 security. Identify and understand Open System authentication and Shared Key authentication. Understand how WEP encryption works and all of its weaknesses.

Explain the 802.1X/EAP framework. Be able to explain all of the components of an 802.1X solution and the EAP authentication protocol. Understand that dynamic encryption-key generation is a by-product of mutual authentication.

Define the requirements of a robust security network (RSN). Understand what the 802.11-2007 standard specifically defines for robust security and be able to contrast what is defined by both the WPA and WPA2 certifications.

Understand TKIP/RC4 and CCMP/AES. Be able to explain the basics of both dynamic encryption types and why they are the end result of an RSN solution.

Explain VLANs and VPNs. Understand that VLANs are typically used for wireless segmentation solutions. Define the basics of VPN technology and when it might be used in a WLAN environment.

Key Terms

Before you take the exam, be certain you are familiar with the following terms:

4-Way Handshake
802.11i
802.1X
Advanced Encryption Standard (AES) algorithm
authentication server
authentication, authorization, and accounting (AAA)
authenticator
Counter Mode with Cipher Block Chaining Message Authentication Code Protocol (CCMP)
Extensible Authentication Protocol (EAP)
Initialization Vector (IV)
Internet Protocol Security (IPsec)
MAC Service Data Unit (MSDU)
Message Integrity Check (MIC)
Microsoft Point-to-Point Encryption (MPPE)
per session per user
Point-to-Point Tunneling Protocol (PPTP)
port-based access control
preshared key (PSK)
RC4 algorithm
robust security network (RSN)
robust security network associations (RSNAs)
role-based access control (RBAC)
RSN Information Element (IE)
supplicant
Temporal Key Integrity Protocol (TKIP)
transition security network (TSN)
virtual local area network (VLAN)
virtual private network (VPN)
Wi-Fi Protected Access (WPA)
Wired Equivalent Privacy (WEP)
WPA2

Review Questions

1. What WLAN security mechanism requires that each WLAN user has unique authentication credentials?
   - A. WPA-Personal
   - B. 802.1X/EAP
   - C. Open System
   - D. WPA2-Personal
   - E. WPA-PSK
2. Which wireless security standards and certifications call for the use of CCMP/AES encryption? (Choose all that apply.)
   - A. WPA
   - B. 802.11-2007
   - C. 802.1X
   - D. WPA2
   - E. 802.11 legacy
3. 128-bit WEP encryption uses a user-provided static key of what size?
   - A. 104 bytes
   - B. 64 bits
   - C. 124 bits
   - D. 128 bits
   - E. 104 bits
4. What three main components constitute an 802.1X/EAP framework? (Choose all that apply.)
   - A. Supplicant
   - B. Authorizer
   - C. Authentication server
   - D. Intentional radiator
   - E. Authenticator
5. The 802.11 legacy standard defines which wireless security solution?
   - A. Dynamic WEP
   - B. 802.1X/EAP
   - C. 64-bit static WEP
   - D. Temporal Key Integrity Protocol
   - E. CCMP/AES

6. Jimmy has been hired as a consultant to secure the Donahue Corporation's WLAN infrastructure. He has been asked to choose a solution that will both protect the company's equipment from theft and hopefully protect the access point's configuration interfaces from outside attackers. What recommendations would be appropriate? (Choose all that apply.)
   - **A.** Mounting all access points in lockable enclosure units
   - **B.** IPsec VPN
   - **C.** Configuring all access points via Telnet
   - **D.** Configuring access points from the wired side using HTTPS or Secure Command Shell
   - **E.** 802.1X/EAP
7. Which security solutions may be used to segment a wireless LAN? (Choose all that apply.)
   - **A.** VLAN
   - **B.** WEP
   - **C.** RBAC
   - **D.** CCMP/AES
   - **E.** TKIP/RC4
8. What wireless security solutions are defined by Wi-Fi Protected Access? (Choose all that apply.)
   - **A.** Passphrase authentication
   - **B.** LEAP
   - **C.** TKIP/RC4
   - **D.** Dynamic WEP
   - **E.** CCMP/AES
9. Name the three main components of a role-based access control solution.
   - **A.** EAP
   - **B.** Roles
   - **C.** Encryption
   - **D.** Permissions
   - **E.** Users
10. What does 802.1X/EAP provide when implemented for WLAN security? (Choose all that apply.)
    - **A.** Access to network resources
    - **B.** Verification of access point credentials
    - **C.** Dynamic authentication
    - **D.** Dynamic encryption-key generation
    - **E.** Verification of user credentials

11. Which technologies use the RC4 cipher? (Choose all that apply.)
   A. Static WEP
   B. Dynamic WEP
   C. CCMP
   D. TKIP
   E. MPPE

12. What must occur before dynamic TKIP/RC4 or CCMP/AES encryption keys are generated? (Choose all that apply.)
   A. Shared Key authentication and 4-Way Handshake
   B. 802.1X/EAP authentication and 4-Way Handshake
   C. Open System authentication and 4-Way Handshake
   D. PSK authentication and 4-Way Handshake

13. For an 802.1X/EAP solution to work properly, which two components must both support the same type of EAP? (Choose two.)
   A. Supplicant
   B. Authorizer
   C. Authenticator
   D. Authentication server

14. When using an 802.11 wireless controller solution, which device would be considered the authenticator?
   A. Access point
   B. RADIUS database
   C. LDAP
   D. WLAN controller
   E. VLAN

15. Identify some aspects of the Temporal Key Integrity Protocol. (Choose all that apply.)
   A. 128-bit temporal key
   B. 24-bit Initialization Vector
   C. Message Integrity Check
   D. 48-bit IV
   E. Diffe-Hellman Exchange

16. In a point-to-point bridge environment where 802.1X/EAP is used for bridge authentication, what device in the network acts as the 802.1X supplicant?
    A. Nonroot bridge
    B. Controller
    C. Root bridge
    D. RADIUS server
    E. Layer 3 core switch

17. CCMP encryption uses which AES key size?
    A. 192 bits
    B. 64 bits
    C. 256 bits
    D. 128 bits

18. Identify the security solutions that are defined by WPA2. (Choose all that apply.)
    A. 802.1X/EAP authentication
    B. Dynamic WEP encryption
    C. Optional CCMP/AES encryption
    D. Passphrase authentication
    E. DES encryption

19. The IEEE 802.11-2007 standard mandates __________ encryption for robust security network associations and optional use of __________ encryption.
    A. WEP, AES
    B. IPsec, AES
    C. MPPE, TKIP
    D. TKIP, WEP
    E. CCMP, TKIP

20. Which layer 2 protocol is used for authentication in an 802.1X framework?
    A. Extensible Authorization Protocol
    B. Extended Authentication Protocol
    C. Extensible Authentication Protocol
    D. CHAP/PPP
    E. Open System

Answers to Review Questions

1. B. As required by an 802.1X security solution, the supplicant is a WLAN client requesting authentication and access to network resources. Each supplicant has unique authentication credentials that are verified by the authentication server.

2. B, D. The 802.11-2007 standard defines CCMP/AES encryption as the default encryption method, while TKIP/RC4 is the optional encryption method. This was originally defined by the 802.11i amendment, which is now part of the 802.11-2007 standard. The Wi-Fi Alliance created the WPA2 security certification, which mirrors the robust security defined by the IEEE. WPA2 supports both CCMP/AES and TKIP/RC4 dynamic encryption-key management.

3. E. 128-bit WEP encryption uses a secret 104-bit static key that is provided by the user (26 hex characters) and combined with a 24-bit Initialization Vector for an effective key strength of 128 bits.

4. A, C, E. The supplicant, authenticator, and authentication server work together to provide the framework for an 802.1X/EAP solution. The supplicant requests access to network resources. The authentication server authenticates the identity of the supplicant, and the authenticator allows or denies access to network resources via virtual ports.

5. C. The original 802.11 standard ratified in 1997 defined the use of a 64-bit or 128-bit static encryption solution called Wired Equivalent Privacy (WEP). Dynamic WEP was never defined under any wireless security standard. The use of 802.1X/EAP, TKIP/RC4, and CCMP/AES are all defined under the current 802.11-2007 standard.

6. A, D, E. Access points may be mounted in lockable enclosure units to provide theft protection. All access points should be configured from the wired side and never wirelessly. Encrypted management interfaces such as HTTPS and Secure Command Shell should be used instead of HTTP or Telnet. An 802.1X/EAP solution guarantees that only authorized users will receive an IP address. Attackers can get an IP address prior to setting up an IPsec VPN tunnel and potentially attack the access points.

7. A, C. Virtual LANs are used to segment wireless users at layer 3. The most common wireless segmentation strategy often used in 802.11 enterprise WLANs is segmentation using VLANS combined with role-based access control (RBAC) mechanisms. CCMP/AES, TKIP/RC4, and WEP are encryption solutions.

8. A, C. The Wi-Fi Protected Access (WPA) certification was a snapshot of the not-yet-released 802.11i amendment, supporting only the TKIP/RC4 dynamic encryption-key generation. 802.1X/EAP authentication was required in the enterprise, and passphrase authentication was required in a SOHO or home environment. LEAP is Cisco proprietary and is not specifically defined by WPA. Neither dynamic WEP nor CCMP/AES were defined for encryption. CCMP/AES dynamic encryption is mandatory under the WPA2 certification.

9. B, D, E. Role-based access control (RBAC) is an approach to restricting system access to authorized users. The three main components of an RBAC approach are users, roles, and permissions.

10. A, D, E. The purpose of 802.1X/EAP is authentication of user credentials and authorization to network resources. Although the 802.1X/EAP framework does not require encryption, it highly suggests the use of encryption. A by-product of 802.1X/EAP is the generation and distribution of dynamic encryption keys.

11. A, B, D, E. All forms of WEP encryption use the Rivest Cipher 4 (RC4) algorithm. TKIP is WEP that has been enhanced and also uses the RC4 cipher. PPTP uses 128-bit Microsoft Point-to-Point Encryption (MPPE), which uses the RC4 algorithm. CCMP uses the AES cipher.

12. B, D. Open System and Shared Key authentication are legacy authentication methods that do not provide seeding material to generate dynamic encryption keys. A robust security network association requires a four-frame EAP exchange known as the 4-Way Handshake that is used to generate dynamic TKIP or CCMP keys. The handshake may occur either after an 802.1X/EAP exchange or as a result of PSK authentication.

13. A, D. An 802.1X/EAP solution requires that both the supplicant and the authentication server support the same type of EAP. The authenticator must be configured for 802.1X/EAP authentication but does not care which EAP type passes through. The authenticator and the supplicant must support the same type of encryption.

14. D. WLAN controllers use lightweight access points, which are dumb terminals with radio cards and antennas. The WLAN controller is the authenticator. When an 802.1X/EAP solution is deployed in a wireless controller environment, the virtual controlled and uncontrolled ports exist on the WLAN controller.

15. A, C, D. TKIP starts with a 128-bit temporal key that is combined with a 48-bit Initialization Vector (IV) and source and destination MAC addresses in a process known as per-packet key mixing. TKIP uses an additional data integrity check known as the Message Integrity Check (MIC).

16. A. The root bridge would be the authenticator, and the nonroot bridge would be the supplicant if 802.1X/EAP security is used in a WLAN bridged network.

17. D. The AES algorithm encrypts data in fixed data blocks with choices in encryption-key strength of 128, 192, or 256 bits. CCMP/AES uses a 128-bit encryption-key size and encrypts in 128-bit fixed-length blocks.

18. A, D. The WPA2 certification requires the use of an 802.1X/EAP authentication method in the enterprise and the use of a preshared key or a passphrase in a SOHO environment. The WPA2 certification also requires the use of stronger dynamic encryption-key generation methods. CCMP/AES encryption is the mandatory encryption method, and TKIP/RC4 is the optional encryption method.

19. E. The 802.11-2007 standard defines what is known as a robust security network (RSN) and robust security network associations (RSNAs). CCMP/AES encryption is the mandated encryption method, while TKIP/RC4 is an optional encryption method.

20. C. The supplicant, authenticator, and authentication server work together to provide the framework for 802.1X port-based access control, and an authentication protocol is needed to assist in the authentication process. The Extensible Authentication Protocol (EAP) is used to provide user authentication.

Chapter 14

Wireless Attacks, Intrusion Monitoring, and Policy

IN THIS CHAPTER, YOU WILL LEARN ABOUT THE FOLLOWING:

✓ **Wireless attacks**

- Rogue wireless devices
- Peer-to-peer attacks
- Eavesdropping
- Encryption cracking
- Authentication attacks
- MAC spoofing
- Management interface exploits
- Wireless hijacking
- Denial of service (DoS)
- Vendor-specific attacks
- Social engineering

✓ **Intrusion monitoring**

- Wireless intrusion detection system (WIDS)
- Wireless intrusion prevention system (WIPS)
- Mobile WIDS
- Spectrum analyzer

✓ **Wireless security policy**

- General security policy
- Functional security policy
- Legislative compliance
- 802.11 wireless policy recommendations

In Chapter 13, "802.11 Network Security Architecture," we discussed legacy 802.11 security solutions as well as the more robust security that is defined in the 802.11-2007 standard. In this chapter, we cover the wide variety of attacks that can be launched against 802.11 wireless networks. Some of these attacks can be mitigated by using the strong encryption and mutual authentication solutions that were discussed in Chapter 13. However, others cannot be prevented and can only be detected. Therefore, we also discuss the wireless intrusion detection systems that can be implemented to expose both layer 1 and layer 2 attacks. The most important component for a secure wireless network is a properly planned and implemented corporate security policy. This chapter also discusses some of the fundamental components of a wireless security policy that are needed to cement a foundation of Wi-Fi security.

Wireless Attacks

As you have learned throughout this book, the main function of an 802.11 WLAN is to provide a portal into a wired network infrastructure. The portal must be protected with strong authentication methods so that only legitimate users with the proper credentials will be authorized to have access to network resources. If the portal is not properly protected, unauthorized users can also gain access to these resources. The potential risks of exposing these resources are endless. An intruder could gain access to financial databases, corporate trade secrets, or personal health information. Network resources can also be damaged.

What would be the financial cost to an organization if an intruder used the wireless network as a portal to disrupt or shut down a SQL server or email server? If the Wi-Fi portal is not protected, any individual wishing to cause harm could upload data such as viruses, Trojan horse applications, keystroke loggers, or remote-control applications. Spammers have already figured out that they can use open wireless gateways to the Internet to commence spamming activities. Other illegal activities, such as software theft and remote hacking, may also occur through an unsecured gateway.

While an intruder can use the wireless network to attack wired resources, equally at risk are all of the wireless network resources. Any information that passes through the air can be captured and possibly compromised. If not properly secured, the management interfaces of Wi-Fi equipment can be accessed. Many wireless users are fully exposed for peer-to-peer attacks. Finally, the possibility of denial-of-service attacks against a wireless network always exists. With the proper tools, any individual with ill intent can temporarily disable a Wi-Fi network, thus denying legitimate users access to the network resources.

In the following sections, you will learn about all the potential attacks that can be launched against 802.11 wireless networks.

Rogue Wireless Devices

The big buzz-phrase in Wi-Fi security has always been the *rogue access point*. In Chapter 13, you learned about 802.1X/EAP authentication solutions that can be put in place to prevent unauthorized access. However, what is there to prevent an individual from installing their own wireless portal onto the network backbone? A rogue access point is any Wi-Fi device that is connected to the wired infrastructure but is not under the management of the proper network administrators. Any $50 SOHO Wi-Fi access point or router can be plugged into a live data port. The rogue device will just as easily act as a portal into the wired network infrastructure. Because the rogue device has no authorization and authentication security in place, any intruder could use this open portal to gain access to network resources.

It is not uncommon for a company to have a wireless network installed and not even know about its existence. The individuals most responsible for installing rogue access points are not hackers; they are employees not realizing the consequences of their actions. According to some statistics, well over 60 percent of home computer users have wireless access at home and have become accustomed to the convenience and mobility that Wi-Fi offers. As a result, employees often install their own wireless devices in the workplace because the company they work for has yet to deploy an enterprise wireless network. The problem is that, while these self-installed access points might provide the wireless access that the employees desire, they are rarely secured. Every rogue access point is a potential open and unsecured gateway straight into the wired infrastructure that the company wants to protect. Although only a single open portal is needed to expose network resources, many large companies have discovered literally dozens of rogue access points that have been installed by employees.

Ad hoc networks also have the potential of providing rogue access into the corporate network. Very often an employee will have a laptop or desktop plugged into the wired network via an Ethernet network card. On that same computer, the employee has a Wi-Fi radio and has set up an ad hoc Wi-Fi connection with another employee. This connection may be set up on purpose or may be accidental and occur as an unwitting result of the manufacturer's default configurations. Because the Ethernet connection and the Wi-Fi card can be bridged together, an intruder might also access the ad hoc wireless network and then route their way to the Ethernet connection and get onto the wired network. Many government agencies and corporations ban the use of ad hoc networks for this very reason. The ability to configure an ad hoc network can be disabled on most enterprise client devices.

As stated earlier, most rogue APs are installed by employees not realizing the consequences of their actions, but any malicious intruder can use these open portals to gain access. Furthermore, besides physical security, there is nothing to prevent an intruder from also connecting their own rogue access point via an Ethernet cable into any live data port provided in a wall plate. Later in this chapter, we discuss intrusion prevention systems that can both detect and disable rogue access points as well as ad hoc clients.

If an 802.1X solution is deployed for the wireless network, it can also be used to secure the network ports on the wired network. In that case, any new access points would need to authenticate to the network prior to being given access. This is a good way to not only utilize existing resources, but also provide better security for your wired network by protecting against rogue APs.

Peer-to-Peer Attacks

As mentioned earlier, wireless resources may also be attacked. A commonly overlooked risk is the *peer-to-peer attack*. As you learned in earlier chapters, an 802.11 client station can be configured in either Infrastructure mode or Ad Hoc mode. When configured in Ad Hoc mode, the wireless network is known as an independent basic service set (IBSS) and all communications are peer-to-peer without the need for an access point. Because an IBSS is by nature a peer-to-peer connection, any user who can connect wirelessly with another user can gain access to any resource available on either computer. A common use of ad hoc networks is to share files on-the-fly. If shared access is provided, files and other assets can accidentally be exposed. A personal firewall is often used to mitigate peer-to peer attacks. Some client devices can also disable this feature so that the device will connect to only certain networks and will not associate to a peer-to-peer without approval.

Users that are associated to the same access point are typically just as vulnerable to peer-to-peer attacks as IBSS users. Properly securing your wireless network often involves protecting authorized users from each other, because hacking at companies is often performed internally by employees. Any users associated to the same access point that are members of the same basic service set (BSS) and are in the same VLAN are susceptible to peer-to-peer attacks because they reside in the same layer 2 and layer 3 domains. In most WLAN deployments, Wi-Fi clients communicate only with devices on the wired network, such as email or web servers, and peer-to-peer communications are not needed. Therefore, most vendors provide some proprietary method of preventing users from inadvertently sharing files with other users. If connections are required to other wireless peers, the traffic is routed through a layer 3 switch or other network device prior to passing to the desired destination station.

Public Secure Packet Forwarding (PSPF) is a feature that can be enabled on WLAN access points or controllers to block wireless clients from communicating with other wireless clients on the same wireless VLAN. This isolates each user on the wireless network to ensure that a wireless station cannot be used to gain layer 2 or layer 3 access to another wireless station. With PSPF enabled, client devices cannot communicate directly with other client devices on the wireless network, as pictured in Figure 14.1. Although PSPF is a term most commonly used by Cisco, other vendors have similar capabilities under different names.

It should be noted that some applications require peer-to-peer connectivity. Many VoWiFi phones offer “push-to-talk” capabilities that use multicasting. VoWiFi phones are typically segmented in a separate wireless VLAN from the rest of wireless data clients. Peer-to-peer blocking should not be enabled in the VoWiFi VLAN if push-to-talk multicasting is required.

Eavesdropping

As you have learned throughout this book, 802.11 wireless networks operate in license-free frequency bands, and all data transmissions travel in the open air. Access to wireless transmissions is available to anyone within listening range, and therefore strong encryption is mandatory. Wireless communications can be monitored via two eavesdropping methods: casual eavesdropping and malicious eavesdropping.

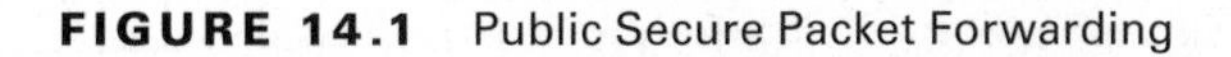

FIGURE 14.1 Public Secure Packet Forwarding

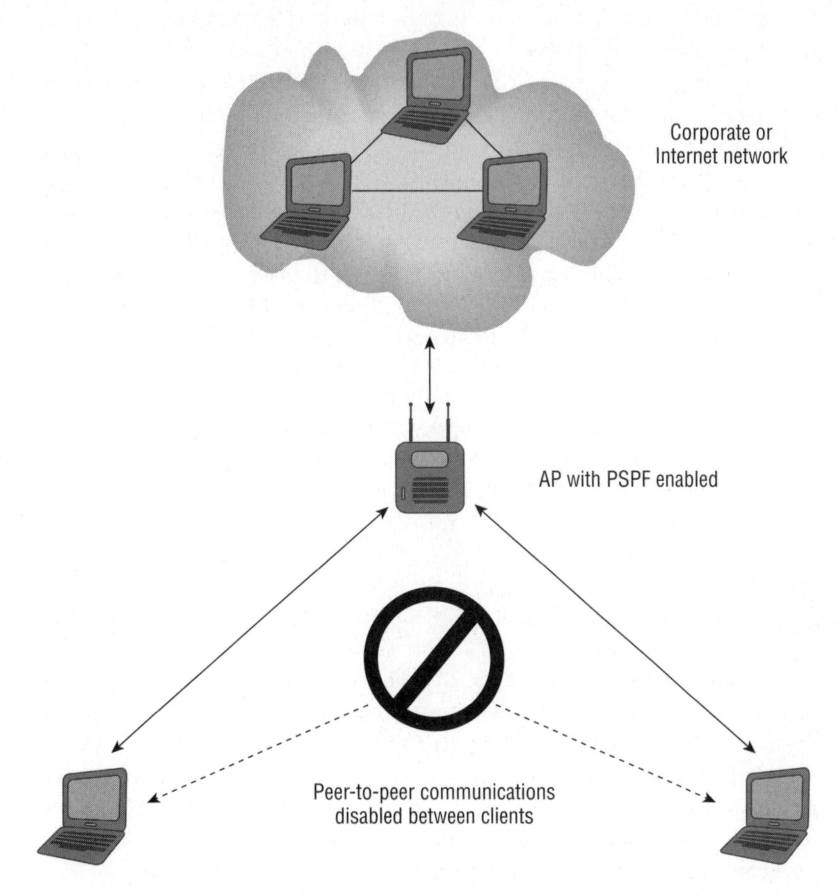

Casual eavesdropping is typically considered harmless and is also often referred to as *wardriving*. Software utilities known as WLAN discovery tools exist for the purpose of finding open WLAN networks. Wardriving is strictly the act of looking for wireless networks, usually while in a moving vehicle. The most common wardriving software tool is a freeware program called NetStumbler, pictured in Figure 14.2.

FIGURE 14.2 NetStumbler

Network Stumbler - [20060531200351]

File Edit View Device Window Help

Channels
SSIDs
Filters

MAC	SSID	Name	Chan	Speed	Vendor	Type	Enc...	SNR
000E38490580	AirSpy Networks		1	11 Mbps	Cisco	AP	WEP	66
000FB5ACC858	kentnet		6	54 Mbps		AP	WEP	15
0020A64D2070	SpectraLink Wi-Fi Phones		11	54 Mbps		AP	WEP	68
0020A64FD589	WonderPuppy Coffee Company		36	54 Mbps		AP		76

NetStumbler sends out null probe requests across all license-free 802.11 channels with the hope of receiving probe response frames containing wireless network information such as SSID, channel, encryption, and so on.

By technical design, the very nature of 802.11 passive and active scanning is to provide the identifying network information that is accessible to anyone with an 802.11 radio card. Because this is an inherent necessary function of 802.11, wardriving is not a crime. However, the goal of many wardrivers is to find open 802.11 wireless networks that can provide free gateway access to the Internet. Although the legality of using an open wireless gateway to the Internet remains unclear in most countries, the majority of wardrivers are not hackers intending harm but rather simply wireless users wanting temporary, free Internet access. The legality of using someone else's wireless network without permission is often unclear, but be warned that people have been arrested and prosecuted as a result of these actions.

We do not encourage or support the efforts of using wireless networks that you are not authorized to use. We recommend that you connect only to wireless networks that you are authorized to access.

What Tools Are Needed for Wardriving?

To get started wardriving, you will need an 802.11 client card, a software WLAN discovery application, and an automobile! Numerous freeware-based discovery tools exist, including NetStumbler for Windows, MiniStumbler for Windows CE, MacStumbler for Macintosh, and Kismet for Linux. A copy of NetStumbler is included on the CD that accompanies this book and can also be downloaded at `www.netstumbler.com`. Another, optional tool is a high-gain external antenna that can be connected to your wireless card via a pigtail connector. Many wardrivers also use Global Positioning System (GPS) devices in conjunction with NetStumbler to pinpoint longitude and latitude coordinates of the signal from access points that they discover. Wardriving capture files with GPS coordinates can be uploaded to large dynamic mapping databases on the Internet. One such database, called the Wireless Geographic Logging Engine (WIGLE), maintains a searchable database of more than 5 million access points. Go to `www.wigle.net` and type in your address to see whether any wireless access points have already been discovered in your neighborhood.

While casual eavesdropping is considered harmless, *malicious eavesdropping*, the unauthorized use of protocol analyzers to capture wireless communications, is typically considered illegal. Most countries have some type of wiretapping law that makes it a crime to listen in on someone else's phone conversation. Additionally, most countries have laws making it unlawful to listen in on any type of electromagnetic communications, including 802.11 wireless transmissions.

Many commercial and freeware 802.11 protocol analyzers exist that allow wireless network administrators to capture 802.11 traffic for the purpose of analyzing and troubleshooting their own wireless networks. Protocol analyzers are passive devices working in an RF monitoring mode that captures any transmissions within range. The problem is that anyone with malicious intent can also capture 802.11 traffic from any wireless network. Because protocol analyzers capture 802.11 frames passively, a wireless intrusion detection system (WIDS) cannot detect malicious eavesdropping. For this reason, a strong, dynamic encryption solution such as TKIP/RC4 or CCMP/AES is mandatory. Any cleartext communications such as email and Telnet passwords can be captured if no encryption is provided. Furthermore, any unencrypted 802.11 frame transmissions can be reassembled at the upper layers of the OSI model. Email messages can be reassembled and therefore read by an eavesdropper. Web pages and instant messages can also be reassembled. VoIP packets can be reassembled and saved as a WAV sound file. Malicious eavesdropping of this nature is highly illegal. Because of the passive and undetectable nature of this attack, encryption must always be implemented to provide data privacy.

It should be noted that the most common targets of malicious eavesdropping attacks are public-access hotspots. Public hotspots rarely offer security and usually transfer data without encryption, making hotspot users prime targets. As a result, it is imperative that a VPN-type solution be implemented for all mobile users who connect outside of your company's network.

Encryption Cracking

In Chapter 13, you learned that Wired Equivalent Privacy (WEP) encryption has been cracked. The current WEP-cracking tools that are freely available on the Internet can crack WEP encryption in as little as 5 minutes. There are several methods used to crack WEP encryption. However, an attacker usually needs only to capture several hundred thousand encrypted packets with a protocol analyzer and then run the captured data through a WEP-cracking software program, as pictured in Figure 14.3. The software utility will usually then be able to derive the secret 40-bit or 104-bit key in a matter of seconds. After the secret key has been revealed, the attacker can decrypt any and all encrypted traffic. In other words, an attacker can then eavesdrop on the WEP-encrypted network. Because the attacker can decrypt the traffic, they can reassemble the data and read it as if there was no encryption whatsoever.

Authentication Attacks

As you have already learned, authorization to network resources can be achieved by either an 802.1X/EAP authentication solution or the use of a preshared key for authentication. The 802.11-2007 standard does not define which type of EAP authentication method to use, and all flavors of EAP are not created equally. Some types of EAP authentication are more secure than others. As a matter of fact, Lightweight Extensible Authentication Protocol (LEAP), once one of the most commonly deployed 802.1X/EAP solutions, is susceptible to offline dictionary attacks. The hashed password response during the LEAP authentication process is crackable.

FIGURE 14.3 WEP-cracking utility

```
* Got  286716! unique IVs | fudge factor = 2
* Elapsed time [00:00:03] | tried 1 keys at 20 k/m

KB    depth   votes
 0    0/  1   DA(  60) 70(  23) 55(  15) A2(   5) CD(   5) 3E(   4)
 1    0/  2   BD(  57) 2A(  32) 29(  22) 1D(  13) F9(  13) 9F(  12)
 2    0/  1   8C(  51) 67(  23) 48(  15) DD(  15) D6(  13) FA(  12)
 3    0/  3   1D(  30) A5(  17) 07(  15) 7B(  12) 4B(  10) 63(  10)
 4    0/  1   43(  66) B1(  15) D2(   6) 1A(   5) 20(   5) 21(   5)
 5    0/  5   92(  27) 23(  25) 02(  18) 2F(  17) C1(  16) 36(  12)
 6    0/  1   C6(  51) 54(  17) 50(  15) 66(  15) 01(  13) 4A(  13)
 7    0/  2   84(  29) C0(  17) EE(  13) 80(  12) 49(  11) F6(  11)
 8    0/  1   81(1808) 09( 119) 99( 116) 32(  75) 49(  75) 9D(  65)
 9    0/  1   C4(1947) E1( 125) FC( 123) BD( 105) 8C(  98) 2F(  85)
10    0/  1   8A( 580) 41( 120) 18(  93) ED(  85) B0(  65) 97(  60)
11    0/  1   08(  97) FF(  29) 5D(  20) 1E(  17) 18(  15) 5E(  15)
12    0/  1   1B( 145) DD(  21) 46(  20) 1C(  15) 76(  15) 07(  13)

              KEY FOUND! [ DABD8C1D4392C68481C48A081B ]
```

An attacker merely has to capture a frame exchange when a LEAP user authenticates and then run the capture file through an offline dictionary attack tool, as shown in Figure 14.4. The password can be derived in a matter of seconds. The username is also seen in cleartext during the LEAP authentication process. After the attacker gets the username and password, they are free to impersonate the user by authenticating onto the WLAN and then accessing any network resources that are available to that user. Stronger EAP authentication protocols exist that are not susceptible to offline dictionary attacks.

FIGURE 14.4 Offline dictionary attack

```
<Finished> - /root/asleap - Konsole
Session Edit View Bookmarks Settings Help
        0025 0215 0025 1101 0018 b1b6 6613 94b9  .%...%......f...
        a076 15e7 07b3 5234 3033 0b55 4b30 f276  .v....R403.UK0.v
        12a4 7465 7374 32                        .. david

Captured LEAP auth success:

        0040 96a6 deca 0012 014d b400 888e 0100  .@.......M......
        0004 0315 0004 0000 0000 0000 0000 0000  ................
        0000 0000 0000 0000 0000 0000 0000 0000  ................
        0000 0000 0000 0000 0000 0000            ............

Captured LEAP exchange information:
       username:         david
       challenge:        373931a2d1888e58
       response:         b1b6661394b9a07615e707b3523430330b554b30f27612a4
       Attempting to recover last 2 of hash.
       hash bytes:       f2d8
       Starting dictionary lookups.
       NT hash:          f70da7fad38a37d803d9f737a286f2d8
       password:         123abc123abc
Reached EOF on pcapfile.
```

WPA/WPA2-Personal, using preshared keys, is also a weak authentication method that is vulnerable to offline dictionary attacks. Hacking utilities are available that can derive the WPA/WPA2 passphrase by using an offline dictionary attack. An attacker who obtains the passphrase can associate to the WPA/WPA2 access point. Even worse is that after obtaining the passphrase, the hacker can also begin to decrypt the dynamically generated TKIP/RC4 or CCMP/AES encryption key. In Chapter 13, you learned that an algorithm is run to convert the passphrase to a Pairwise Master Key (PMK), which is used with the 4-Way Handshake to create the final dynamic encryption keys. If a hacker has the passphrase and captures the 4-Way Handshake, they can re-create the dynamic encryption keys and decrypt traffic. WPA/WPA2-Personal is not considered a strong security solution for the enterprise because if the passphrase is compromised, the attacker can access network resources and decrypt traffic. In situations where there is no AAA server or the client devices do not support 802.1X authentication, a WPA/WPA2-Personal deployment may be necessary.

A policy mandating very strong passphrases of 20 characters or more should always be in place whenever a WPA/WPA2-Personal solution is deployed. Furthermore, because passphrases are static, they are susceptible to social engineering attacks. To prevent social engineering attacks, policy must dictate that only the administrator has knowledge of any static passphrases and that the passphrases are never shared with end users.

MAC Spoofing

All 802.11 wireless network cards have a physical address known as a *MAC address*. This address is a 12-digit hexadecimal number that is seen in cleartext in the layer 2 header of 802.11 frames. Wi-Fi vendors often provide MAC filtering capabilities on their access points. Usually, MAC filters are configured to apply restrictions that will allow traffic only from specific client stations to pass through. These restrictions are based on their unique MAC addresses. All other client stations whose MAC addresses are not on the allowed list will not be able to pass traffic through the virtual port of the access point and onto the distribution system medium.

Unfortunately, MAC addresses can be *spoofed*, or impersonated, and any amateur hacker can easily bypass any MAC filter by spoofing an allowed client station's address. MAC spoofing can often be achieved in the Windows operating system by simply editing the wireless card's MAC address in Device Manager or by performing a simple edit in the Registry. Third-party software utilities such as the one pictured in Figure 14.5 can also be used be assist in MAC spoofing.

Because of spoofing and because of all the administrative work that is involved with setting up MAC filters, MAC filtering is not considered a reliable means of security for wireless enterprise networks and should be implemented only as a last resort. In some cases, it is used as part of a layered security architecture to better secure client devices that are not capable of 802.1X/EAP protection.

It should be noted that some security options are not available on all devices or in all scenarios. For instance, mobile handheld scanners do not often support the stronger authentication and encryption techniques. In some cases, the stronger authentication and encryption may adversely affect performance. Careful consideration should be paid to selecting the best methods for your circumstances.

FIGURE 14.5 MAC spoofing software utility

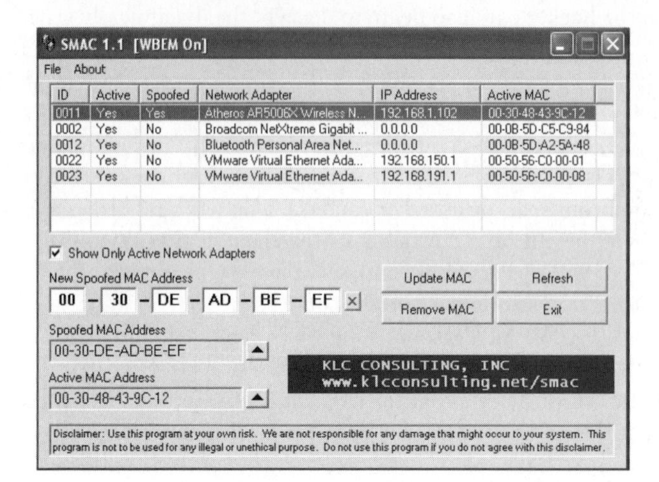

Management Interface Exploits

Wireless infrastructure hardware such as autonomous access points and WLAN controllers can be managed by administrators via a variety of interfaces, much like managing wired infrastructure hardware. Devices can typically be accessed via a web interface, a command-line interface, a serial port, a console connection, and/or Simple Network Management Protocol (SNMP).

As we discussed, it is imperative that these interfaces be protected. Interfaces that are not used should be disabled. Strong passwords should be used, and encrypted login capabilities such as Hypertext Transfer Protocol Secure (HTTPS) should be utilized if available. Lists of all the default settings of every major manufacturer's access points exist on the Internet and are often used for security exploits by hackers. It is not uncommon for intruders to use security holes left in management interfaces to reconfigure access points. Legitimate users and administrators can find themselves locked out of their own wireless networking equipment.

After gaining access via a management interface, an attacker might even be able to initiate a firmware upgrade of the wireless hardware and, while the upgrade is being performed, power off the equipment. This attack could likely render the hardware useless, requiring it to be returned to the manufacturer for repair.

Policy should dictate that all WLAN infrastructure devices be configured from only the wired side of the network. If an administrator attempts to configure a WLAN device while connected wirelessly, the administrator could lose connectivity due to configuration changes being made.

Wireless Hijacking

An attack that often generates a lot of press is *wireless hijacking*, also known as the *evil twin attack*. The attacker configures access point software on a laptop, effectively turning a Wi-Fi client card into an access point. The access point software is configured with the same SSID that is used by a public hotspot access point. The attacker then sends spoofed disassociation or deauthentication frames, forcing users associated with the hotspot access point to roam to the evil twin access point. At this point, the attacker has effectively hijacked wireless clients at layer 2 from the original access point.

The evil twin will typically be configured with a Dynamic Host Configuration Protocol (DHCP) server available to issue IP addresses to the clients. At this point, the attacker will have hijacked the users at layer 3 and now has a private wireless network and is free to perform peer-to-peer attacks on any of the hijacked clients.

The attacker may also be using a second wireless card with their laptop to execute what is known as a *man-in-the-middle attack*, as pictured in Figure 14.6. The second wireless card is associated to the hotspot access point as a client. In operating systems, networking cards can be bridged together to provide routing. The attacker has bridged together their second wireless card with the Wi-Fi card that is being used as the evil twin access point. After the attacker hijacks the users from the original AP, the traffic is then routed from the evil twin access point through the second Wi-Fi card, right back to the original access point from which the users have just been hijacked. The result is that the users remain hijacked; however, they still have a route back through the gateway to their original network, so they never know they have been hijacked. The attacker can therefore sit in the middle and execute peer-to-peer attacks indefinitely while remaining completely unnoticed.

These attacks can take another form in what is known as the *Wi-Fi phishing attack*. The attacker may also have web server software and captive portal software. After the users have been hijacked to the evil twin access point, they will be redirected to a login web page that looks exactly like the hotspot's login page. Then the attacker's fake login page may request a credit card number from the hijacked user. Phishing attacks are common on the Internet and are now appearing at your local hotspot. The only way to prevent a hijacking, man-in-the-middle, or Wi-Fi phishing attack is to use a mutual authentication solution. Mutual authentication solutions not only validate the user connecting to the network, they also validate the network to which the user is connecting. 802.1X/EAP authentication solutions require that mutual authentication credentials be exchanged before a user can be authorized. A user cannot get an IP address unless authorized; therefore, users cannot be hijacked.

Denial of Service (DoS)

The attack on wireless networks that seems to receive the least amount of attention is the *denial of service (DoS)*. With the proper tools, any individual with ill intent can temporarily disable a Wi-Fi network by preventing legitimate users from accessing network resources. The good news is that monitoring systems exist that can detect and identify DoS attacks immediately. The bad news is that usually nothing can be done to prevent denial-of-service attacks other than locating and removing the source of the attack.

FIGURE 14.6 Man-in-the-middle attack

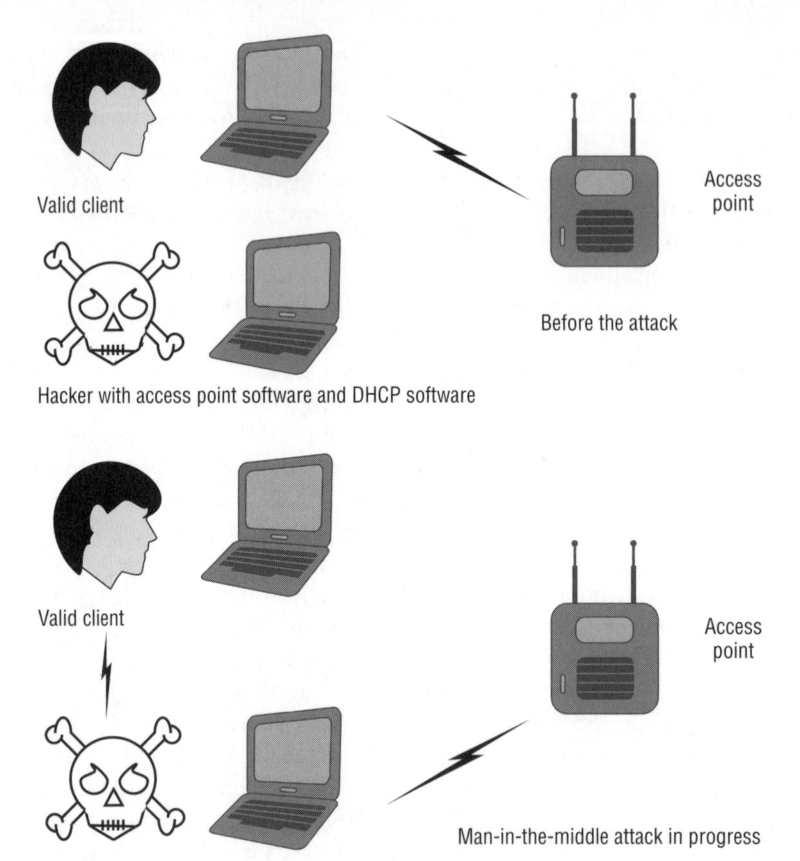

DoS attacks can occur at either layer 1 or layer 2 of the OSI model. Layer 1 attacks are known as *RF jamming attacks*. The two most common types of RF jamming attacks are intentional jamming and unintentional jamming. Intentional jamming attacks occur when an attacker uses some type of signal generator to cause interference in the unlicensed frequency space. Both narrowband and wideband jammers exist that will interfere with 802.11 transmissions, either causing all data to become corrupted or causing the 802.11 radio cards to continuously defer when performing a clear channel assessment (CCA).

While an intentional jamming attack is malicious, unintentional jamming is more common. Unintentional interference from microwave ovens, cordless phones, and other devices can also cause denial of service. Although unintentional jamming is not necessarily an attack, it can cause as much harm as an intentional jamming attack. The best tool to detect any type of layer 1 interference, whether intentional or unintentional, is a spectrum analyzer.

The more common type of denial-of-service attacks that originate from hackers are layer 2 DoS attacks. A wide variety of layer 2 DoS attacks exist that are a result of tampering with 802.11 frames. The most common involves spoofing disassociation or deauthentication frames. The attacker can edit the 802.11 header and spoof the MAC address of an access point or a client in either the destination address field or the source address field. The attacker then retransmits the spoofed disassociation or deauthentication frame repeatedly. Because these types of management frames are notification frames that cannot be ignored, the stations will constantly be denied service. Many more types of layer 2 DoS attacks exist, including association floods, authentication floods, PS-Poll floods, and virtual carrier attacks. Luckily, any good wireless intrusion detection system will be able to alert an administrator immediately to a layer 2 DoS attack. The 802.11w draft amendment is the proposed *management frame protection (MFP)* amendment with a goal of delivering management frames in a secure manner. The end result will hopefully prevent many of the layer 2 denial-of-service attacks that currently exist, but it is doubtful that all layer 2 DoS attacks will ever be circumvented.

A spectrum analyzer is your best tool to detect a layer 1 DoS attack, and a protocol analyzer or wireless IDS is your best tool to detect a layer 2 DoS attack. The best way to prevent any type of denial-of-service attack is physical security. The authors of this book recommend guard dogs and barbed wire. If that is not an option, there are several solutions that provide intrusion detection at layers 1 and 2.

Vendor-Specific Attacks

Hackers often find holes in the firmware code used by specific WLAN autonomous access points and WLAN controller vendors. Most of these vendor-specific exploits are in the form of buffer overflow attacks. When these vendor-specific attacks become known, the WLAN vendor usually makes a firmware fix available in a timely manner. These attacks can be best avoided by staying informed through your WLAN vendor's support services.

Social Engineering

Hackers do not compromise most wired or wireless networks with the use of hacking software or tools. The majority of breaches in computer security occur due to social engineering attacks. *Social engineering* is a technique used to manipulate people into divulging confidential information such as computer passwords. The best defense against social engineering attacks are strictly enforced policies to prevent confidential information from being shared.

Any information that is static is extremely susceptible to social engineering attacks. WEP encryption uses a static key, and WPA/WPA2-Personal requires the use of a static preshared key or passphrase. Both of these security methods should be avoided because of their static nature.

Intrusion Monitoring

When most people think of wireless, they think only in terms of access and not in terms of attacks or intrusions. However, it has become increasingly necessary to constantly monitor for the many types of attacks mentioned in this chapter because of the potential damage they can cause. Businesses of all sizes have begun to deploy 802.11 wireless networks for mobility and access and at the same time are running a wireless intrusion detection system (WIDS) to monitor for attacks. Many companies are very concerned about the potential damage that would result from rogue access points. It is not unusual for a company to actually deploy a WIDS before deploying the wireless network that is meant to provide access.

Wireless intrusion monitoring has evolved, and most current systems have methods to prevent and mitigate some of the known wireless attacks. While most systems are distributed for scalability across a large enterprise, single laptop versions of intrusion monitoring systems also exist. Most wireless intrusion monitoring exists at layer 2, but layer 1 wireless intrusion monitoring systems are now also available to scan for potential layer 1 attacks.

Wireless Intrusion Detection System (WIDS)

In today's world, a *wireless intrusion detection system (WIDS)* might be necessary even if there is no authorized 802.11 Wi-Fi network on site. Wireless can be an intrusive technology, and if wired data ports at a business are not controlled, any individual (including employees) can install a rogue access point. Because of this risk, many companies such as banks and other financial institutions as well as hospitals choose to install a WIDS prior to deploying a Wi-Fi network for employee access. After an 802.11 network is installed for access, it has become almost mandatory to also have a WIDS because of the other numerous attacks against Wi-Fi, such as denial of service, hijacking, and so on. The typical wireless intrusion detection system is a client/server model that consists of three components:

WIDS server A software or hardware server acting as a central point of management.

Management consoles Software-based management consoles that connect back to a WIDS server as clients. These consoles can be used for 24/7 monitoring of wireless networks.

Sensors Hardware or software-based sensors placed strategically to listen to and capture all 802.11 communications.

Figure 14.7 depicts the client/server model used by most wireless intrusion detection systems.

Sensors are basically radio devices that are in a constant listening mode as passive devices. The sensor devices are usually hardware based and resemble an access point. The sensors have some intelligence but also communicate with the centralized WIDS server. The centralized server can collect data from literally thousands of sensors from many remote locations, meeting the scalability needs of many large corporations. Management consoles can also be installed at remote locations, and while they talk back to the centralized server, they can also monitor all remote WLANs where sensors are installed. Figure 14.8 shows a WIDS management console and a hardware sensor.

Sensors do not provide access to WLAN clients because they are configured in a listen-only mode. The sensors will constantly scan all 14 channels in the 2.4 GHz ISM band as well as all 23 channels of the 5 GHz UNII bands. On rare occasions, the sensors can also be configured to listen on only one channel or a select group of channels.

FIGURE 14.7 Wireless intrusion detection system (WIDS)

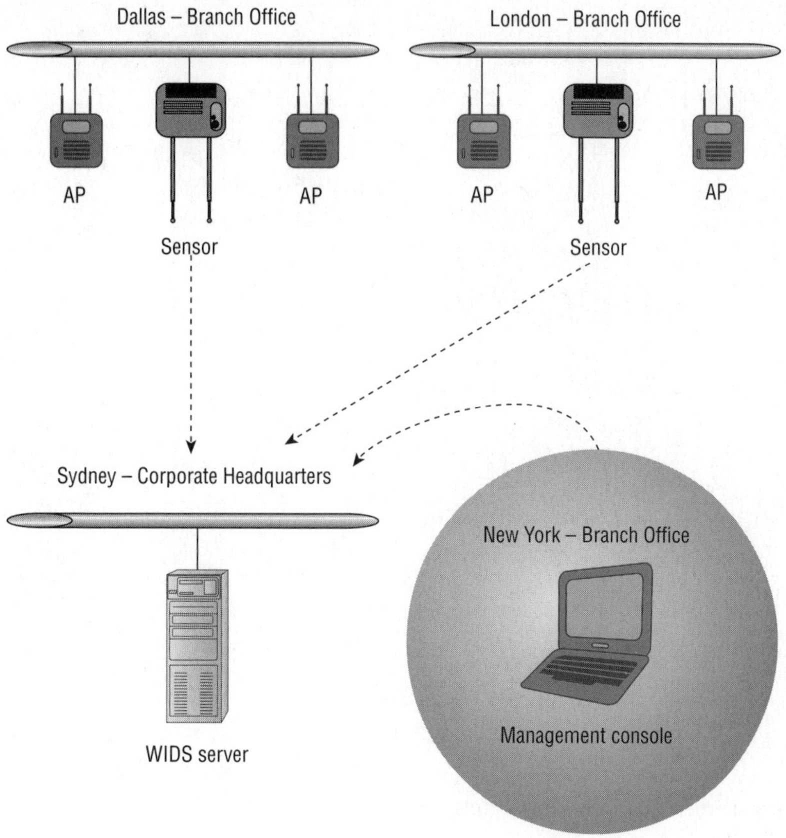

WIDS are best at monitoring layer 2 attacks such as MAC spoofing, disassociation attacks, and deauthentication attacks. Most WIDS will usually have alarms for as many as 60 potential security risks. An important part of deploying a WIDS is setting the policies and alarms. False positives are often a problem with intrusion detection systems, but they can be less of a problem if proper policies and thresholds are defined. Policies can be created to define the severity of various alerts as well as provide for alarm notifications. For example, an alert for broadcasting the SSID might not be considered severe and might even be disabled. However, a policy might be configured that classifies a deauthentication spoofing attack as severe, and an email message or pager notification might be sent automatically to the network administrator.

Although most of the scrutiny that is performed by a WIDS is for security purposes, many WIDS also have performance-monitoring capabilities. For example, performance alerts might be in the form of excessive bandwidth utilization or excessive reassociation and roaming of VoWiFi phones.

FIGURE 14.8 WIDS management console and hardware sensor

Currently, three WIDS design models exist:

Overlay The most common model is an overlay WIDS that is deployed on top of the existing wireless network. This model uses an independent vendor's WIDS and can be deployed to monitor any existing or planned WLAN. The overlay systems typically have more-extensive features, but they are usually more expensive. The overlay solution consists of a WIDS server and sensors that are not part of the WLAN solution that provides access to clients.

Integrated Many WLAN controller vendors have fully integrated WIDS capabilities. The wireless controller acts as the centralized IDS server. The lightweight access points can be configured in a full-time sensor-only mode or can act as part-time sensors when not transmitting as access points. In WLAN controller deployments, the lightweight access points use "off-channel scanning" procedures for dynamic RF spectrum management purposes. The lightweight access points are also effectively part-time sensors for the integrated IDS server when listening off channel. A recommended practice would be to also deploy some lightweight APs as full-time sensors. The integrated solution is a less-expensive solution but may not have all the capabilities that are offered in an overlay WIDS.

Integration enabled Wi-Fi vendors often integrate their access points and management systems with the major WIDS vendors. The Wi-Fi vendor's access points integrate software code that can be used to turn the APs into sensors that will communicate with the third-party WIDS server. Originally, integration-enabled solutions existed primarily for the purpose of converting autonomous access points into sensors that communicated with a third-party WIDS server. However, integration-enabled solutions now exist in WLAN controller-based deployments. Lightweight access points are converted into full-time sensors that communicate directly with a separate WIDS server and no longer communicate directly with the WLAN controller. At the same time, other lightweight access points still send WLAN traffic to the WLAN controller to provide access for WLAN clients. The integration-enabled WIDS server communicates directly with the WLAN controller to gather additional information from the lightweight access points that speak only with the controller.

Wireless Intrusion Prevention System (WIPS)

Most WIDS vendors prefer to call themselves a *wireless intrusion prevention system (WIPS)*. The reason that they refer to themselves as prevention systems is that they are all now capable of mitigating attacks from rogue access points and rogue clients. A WIPS characterizes access points and client radios in four or more classifications:

Infrastructure device This classification refers to any client station or access point that is an authorized member of the company's wireless network. A network administrator can manually label each radio as an infrastructure device after detection from the WIPS or can import a list of all the company's radio card MAC addresses into the system.

Unknown device The unknown device classification is assigned automatically to any new 802.11 radios that have been detected but not classified as rogues. Unknown devices are considered interfering devices and are usually investigated further to determine whether they are a neighbor's devices or a potential future threat.

Known device This classification refers to any client station or access point that is detected by the WIPS and whose identity is known. A known device is initially considered an interfering device. The known device label is typically manually assigned by an administrator to radio devices of neighboring businesses that are not considered a threat.

Rogue device The rogue classification refers to any client station or access point that is considered an interfering device and a potential threat. Most WIPS define rogue access points as devices that are actually plugged into the network backbone and are not known or managed by the organization. Most of the WIPS vendors use a variety of proprietary methods of determining whether a rogue access point is actually plugged into the wired infrastructure.

After a client station or access point has been classified as a rogue device, the WIPS can effectively mitigate an attack. WIPS vendors have several ways of accomplishing this, but the most common method is to use spoofed deauthentication frames. The WIPS will have the sensors go active and begin transmitting deauthentication frames that spoof the MAC addresses of the rogue access points and rogue clients. The WIPS uses a known layer 2 denial-of-service attack as a countermeasure. The effect is that communications between the rogue access point and clients are rendered useless. This countermeasure can be used to disable rogue access points, individual client stations, and rogue ad hoc networks.

Another method of rogue containment uses the Simple Network Management Protocol (SNMP). Most WIPS can determine that the rogue access point is connected to the wired infrastructure and may be able to use SNMP to disable the managed switch port that is connected to the rogue access point. If the switch port is closed, the attacker cannot access network resources that are behind the rogue AP.

The WIPS vendors have other proprietary methods of disabling rogue access points and client stations, and often their methods are not published. Currently, the main purpose of a wireless intrusion prevention system is to contain and disable rogue devices. In the future, other wireless attacks might be mitigated as well.

Real World Scenario

Will a WIPS Protect against All Known Rogue Devices?

The simple answer is no. Although the wireless intrusion prevention systems are outstanding products that can mitigate most rogue attacks, some rogue devices will go undetected. The radio cards inside the WIPS sensors typically monitor the 2.4 GHz ISM band and the 5 GHz UNII frequencies. Older legacy wireless networking equipment exists that transmits in the 900 MHz ISM band, and these devices will not be detected. The radio cards inside the WIPS sensors also use only direct sequencing spread spectrum (DSSS) and Orthogonal Frequency Division Multiplexing (OFDM) technologies. Wireless networking equipment exists that uses frequency hopping spread spectrum (FHSS) transmissions in the 2.4 GHz ISM band and will go undetected. The only tool that will 100 percent detect either a 900 MHz or frequency hopping rogue access point is a spectrum analyzer capable of operating in those frequencies. The WIPS should also monitor all the available channels and not just the ones permitted in your resident country. A common strategy used by hackers is to place rogue devices transmitting on 2.4 GHz channel 14, which is not permitted in many countries.

Mobile WIDS

Several of the wireless intrusion detection/prevention vendors also sell laptop versions of their distributed products. The software program is a protocol analyzer capable of decoding frames with some layer 1 analysis capabilities as well. The mobile WIDS software uses a standard Wi-Fi client radio as the sensor. However, the main purpose of the software is to provide a stand-alone mobile security and performance analysis tool. The mobile WIDS will have all the same policy, alarm, and detection capabilities as the vendor's distributed solution. Think of a mobile WIDS as a single sensor, server, and console all built into one unit. The mobile WIDS will be able to detect only attacks within its listening range, but the advantage is that the device is mobile. One useful feature of a mobile WIDS is that it can detect a rogue access point and client and then be used to track them down. The mobile WIDS locks onto the RF signal of the rogue device, and then an administrator can locate the transmitting rogue by using a directional antenna. Figure 14.9 pictures a location feature, common in a mobile WIDS.

FIGURE 14.9 Mobile WIDS locater tool

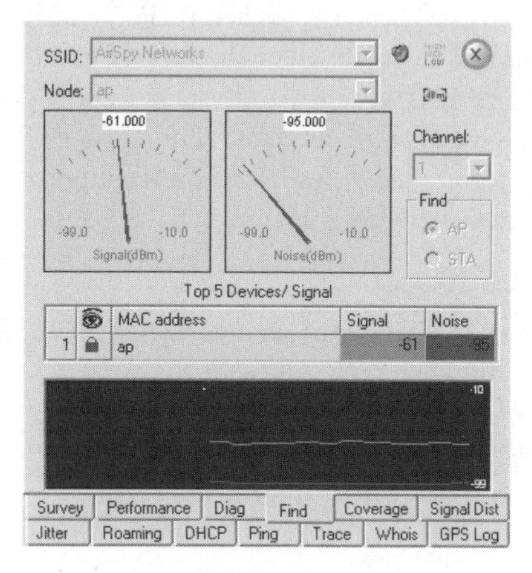

We highly recommend that you test-drive a mobile WIDS solution to gain hands-on experience with the capabilities that this type of product offers. One such solution is AirDefense Mobile. Download a fully working 30-day trial copy of AirDefense Mobile at `www.airdefense.net/products/admobile/trial.php`.

Spectrum Analyzer

In Chapter 15, "Radio Frequency Site Survey Fundamentals," and Chapter 16, "Site Survey Systems and Devices," we discuss the benefits of using a spectrum analyzer during a wireless site survey. WLAN administrators have also begun to realize the benefit of using spectrum analyzers for security purposes. The layer 2 WIDS vendors have long made claims that their products could detect layer 1 denial-of-service attacks, namely, RF jamming. The truth of the matter is that the WIDS vendors are excellent at detecting all of the numerous layer 2 attacks but have limited success with layer 1 detection because the sensor's radios are not spectrum analyzers.

A *spectrum analyzer* is a frequency domain tool that can detect any RF signal in the frequency range that is being scanned. A spectrum analyzer that monitors the 2.4 GHz ISM band will be able to detect both intentional jamming and unintentional jamming devices. Some spectrum analyzers can look at the RF signature of the interfering signal and classify the device. For example, the spectrum analyzer might identify the signal as a microwave oven, a Bluetooth transmitter, or an 802.11 FHSS radio. A spectrum analyzer might also be used to locate rogue 900 MHz or frequency hopping access points.

Two forms of spectrum analysis systems are available: mobile and distributed. Most spectrum analyzers are stand-alone mobile solutions; however, some WIPS vendors have begun to offer distributed spectrum analysis systems that use a centralized server and remote hardware spectrum analyzer cardbus sensors. Other WIPS vendors are providing distributed spectrum analysis by using the RF capabilities of the 802.11 radio cards already available in layer 2 WIDS sensors. A *distributed spectrum analysis system (DSAS)* is effectively a layer 1 wireless intrusion detection system that can detect and classify RF interference. The DSAS has the ability to categorize interference types based on frequency signatures. This can be useful to help classify and locate interfering devices.

Included on the CD that you received with this book is a white paper titled "Protecting Wi-Fi Networks from Hidden Layer 1 Security Threats," authored by David Coleman of AirSpy Training and Neil Diener of Cisco Systems. This white paper is recommended extra reading.

Wireless Security Policy

In Chapter 13, you learned about the various authentication, encryption, and RBAC methods that can be used to secure an 802.11 wireless network. In this chapter, you have learned about wireless intrusion detection systems that can be used to monitor for possible threats. Securing a wireless network and monitoring for threats are absolute necessities, but both are worthless unless proper security policies are in place. What good is an 802.1X/EAP solution if the end users share their passwords? Why purchase an intrusion detection system if a policy has not been established on how to deal with rogue access points?

More and more businesses have started to amend their network usage policies to include a wireless policy section. If you have not done so already, a WLAN section should absolutely be added to the corporate security policy. Two good resources for learning about best practices and computer security policies are the SANS Institute and the National Institute of Standards and Technology (NIST).

Security policy templates from the SANS Institute can be downloaded from `www.sans.org/resources/policies`. The NIST special publication document 800-48 regarding wireless security can be downloaded from `http://csrc.nist.gov/publications/nistpubs`.

General Security Policy

When establishing a wireless security policy, you must first define a *general policy*. A general wireless security policy establishes why a wireless security policy is needed for an organization. Even if a company has no plans for deploying a wireless network, there should be at a minimum a policy for how to deal with rogue wireless devices. A general wireless security policy will define the following items:

Statement of authority Defines who put the wireless policy in place and the executive management that backs the policy.

Applicable audience Defines the audience to whom the policy applies, such as employees, visitors, and contractors.

Violation reporting procedures Defines how the wireless security policy will be enforced, including what actions should be taken and who is in charge of enforcement.

Risk assessment and threat analysis Defines the potential wireless security risks and threats and what the financial impact will be on the company if a successful attack occurs.

Security auditing Defines internal auditing procedures as well as the need for independent outside audits.

Functional Security Policy

A *functional policy* is also needed to define the technical aspects of wireless security. The functional security policy establishes how to secure the wireless network in terms of what solutions and actions are needed. A functional wireless security policy will define the following items:

Policy essentials Defines basic security procedures such as password policies, training, and proper usage of the wireless network.

Baseline practices Defines minimum wireless security practices such as configuration checklists, staging and testing procedures, and so on.

Design and implementation Defines the actual authentication, encryption, and segmentation solutions that are to be put in place.

Monitoring and response Defines all wireless intrusion detection procedures and the appropriate response to alarms.

Legislative Compliance

In most countries, there are mandated regulations on how to protect and secure data communications within all government agencies. In the United States, NIST maintains the *Federal Information Processing Standards (FIPS)*. Of special interest to wireless security is the FIPS 140-2 standard, which defines security requirements for cryptography modules. The use of validated cryptographic modules is required by the U.S. government for all unclassified communications. Other countries also recognize the FIPS 140-2 standard or have similar regulations.

In the United States, other legislation exists for protecting information and communications in certain industries. These include the following:

HIPAA The Health Insurance Portability and Accountability Act (HIPAA) establishes national standards for electronic health care transactions and national standards for providers, health insurance plans, and employers. The goal is to protect patient information and maintain privacy.

Sarbanes-Oxley The Sarbanes-Oxley Act of 2002 defines more-stringent controls on corporate accounting and auditing procedures with a goal of corporate responsibility and enhanced financial disclosure.

GLBA The Gramm-Leach-Bliley Act (GLBA) requires banks and financial institutions to notify customers of policies and practices disclosing customer information. The goal is to protect personal information such as credit card numbers, social security numbers, names, addresses, and so forth.

Information about the FIPS regulations can be found at `http://csrc.nist.gov/publications/fips`. More information about HIPAA can be found at `www.hhs.gov/ocr/hipaa`. General information about Sarbanes-Oxley can be found via the Web at `www.sarbanes-oxley.com`. Further information about GLBA can be located at `www.ftc.gov`.

802.11 Wireless Policy Recommendations

Although a detailed and thorough policy document should be created, we highly recommend these five wireless security policies:

Remote-access WLAN policy End users will be taking their laptops and handheld devices off site and away from company grounds. Most users will likely use wireless networks at home and at wireless hotspots to access the Internet. By design, many of these remote wireless networks have absolutely no security in place, and it is imperative that a remote-access WLAN policy be strictly enforced. This policy should include the required use of an IPsec VPN solution to provide device authentication, user authentication, and strong encryption of all wireless data traffic. Hotspots are prime targets for malicious eavesdropping attacks. Personal firewalls should also be installed on all remote computers to prevent peer-to-peer attacks. Personal firewalls will not prevent hijacking attacks or peer-to-peer attacks, but will prevent attackers from accessing most critical information. Endpoint WLAN policy-enforcement software solutions exist that force end users to use VPN and firewall security when accessing any wireless network other than the corporate WLAN. The remote-access policy is mandatory because the most likely and vulnerable location for an attack to occur is at a public-access hotspot.

Rogue AP policy No end users should ever be permitted to install their own wireless devices on the corporate network. This includes access points, wireless routers, wireless hardware USB clients, and wireless cards. Any users installing their own wireless equipment could open unsecured portals into the main infrastructure network. This policy should be strictly enforced.

Ad hoc policy End users should not be permitted to set up ad hoc or peer-to-peer networks. Peer-to-peer networks rarely use encryption, are susceptible to peer attacks, and can serve as unsecured portals to the infrastructure network if the computer's Ethernet port is also in use.

Wireless LAN proper use policy A thorough policy should outline the proper use and implementation of the main corporate wireless network. This policy should include proper installation procedures, proper security implementations, and allowed application use on the wireless LAN.

IDS policy Policies should be written defining how to properly respond to alerts generated by the wireless intrusion detection system. An example would be how to deal with the discovery of rogue access points and all the necessary actions that should take place.

These five policies are simplistic but are a good starting point in writing a wireless security policy document. The authors of this book also recommend that the built-in Microsoft Windows XP Wi-Fi client utilities known as the Wireless Zero Configuration (WZC) service be disabled at all times because of numerous documented security risks. We recommend using one single vendor's software client or using third-party client utilities if multiple vendor cards must be supported.

Real World Scenario

What Type of Security Is Needed for Home Wireless Networks?

As you learned in Chapter 13, security for enterprise WLANs requires an 802.1X/EAP authentication solution and strong dynamic encryption such as TKIP/RC4 or CCMP/AES to provide data privacy. Usually SOHO Wi-Fi security is not as strong or complex as enterprise security because of cost considerations and available resources. However, many security steps can still be taken to offer a reasonably secure solution for a home wireless network.

Included with the CD of this book is a white paper from AirSpy Training titled "The Top 10 Security Checklist for SOHO Wireless LANs," which makes 10 commonsense suggestions toward small office and home Wi-Fi security. Although there is no such thing as perfect security, the implementation of these 10 suggestions will bring you well on your way toward being much more secure than the majority of your neighbors. Security in a SOHO environment is still dependent on the value of what you are trying to protect, and in some instances, enterprise-type security might be needed in a SOHO environment.

Several vendors now offer endpoint WLAN policy-enforcement software agents. These agents can protect mobile users at hotspots and other public Wi-Fi networks from wireless-specific risks that could expose private data and transactions. One example of a policy enforcement solution is AirDefense Personal. You can download a free copy of AirDefense Personal Lite at www.airdefense.net/products/adpersonal/index.php.

Summary

In this chapter, we discussed all the potential wireless attacks and threats. The rogue access point has always been the biggest concern in terms of wireless threats, followed immediately by social engineering. However, we discussed many other serious threats, such as peer-to-peer attacks and eavesdropping, that can have consequences that are just as serious. We also discussed denial-of-service attacks that cannot be mitigated and can only be monitored. We covered the various solutions that are available for intrusion monitoring. Most intrusion detection solutions use a distributed client/server model, and some offer rogue prevention capabilities. Finally, we discussed the need for sound wireless security policies that will act as a foundation for the wireless security solutions that you implement.

Exam Essentials

Understand the risk of the rogue access point. Be able to explain why the rogue AP provides a portal into network resources. Understand that employees are often the source of rogue APs.

Define peer-to-peer attacks. Understand that peer-to-peer attacks can happen via an access point or through an ad hoc network. Explain how to defend against this type of attack.

Know the risks of eavesdropping. Explain the difference between casual and malicious eavesdropping. Explain why encryption is needed for protection.

Define authentication and hijacking attacks. Explain the risks behind these types of attacks. Understand that a strong 802.1X/EAP solution is needed to mitigate them.

Explain wireless denial-of-service attacks. Know the difference between layer 1 and layer 2 DoS attacks. Explain why these attacks cannot be mitigated and can only be monitored.

Understand the types of wireless intrusion solutions. Explain the difference between a WIDS and a WIPS. Understand that most solutions are distributed client/server models. Know the various components of an intrusion monitoring solution as well as the various models. Understand which attacks can be monitored and which can be prevented.

Understand the need for a wireless security policy. Explain the difference between general and functional policies.

Key Terms

Before you take the exam, be certain you are familiar with the following terms:

casual eavesdropping
denial of service (DoS)
distributed spectrum analysis system (DSAS)
evil twin attack
Federal Information Processing Standards (FIPS)
functional policy
general policy
malicious eavesdropping
management frame protection (MFP)
man-in-the-middle attack
peer-to-peer attack
Public Secure Packet Forwarding (PSPF)
rogue access point
social engineering
spectrum analyzer
wardriving
Wi-Fi phishing attack
wireless hijacking
wireless intrusion detection system (WIDS)
wireless intrusion prevention system (WIPS)

Review Questions

1. Which of these attacks are considered denial-of-service attacks? (Choose all that apply.)
   - A. Man-in-the-middle
   - B. Jamming
   - C. Deauthentication spoofing
   - D. MAC spoofing
   - E. Peer-to-peer
2. Which of these attacks would be considered malicious eavesdropping? (Choose all that apply.)
   - A. NetStumbler
   - B. Peer-to-peer
   - C. Protocol analyzer capture
   - D. Packet reconstruction
   - E. PS polling attack
3. Which of these attacks will not be detected by a wireless intrusion detection system (WIDS)?
   - A. Deauthentication spoofing
   - B. MAC spoofing
   - C. Rogue access point
   - D. Protocol analyzer
   - E. Association flood
4. Which of these attacks can be mitigated with a mutual authentication solution? (Choose all that apply.)
   - A. Malicious eavesdropping
   - B. Deauthentication
   - C. Man-in-the-middle
   - D. Wireless hijacking
   - E. Authentication flood
5. Name two types of rogue devices that cannot be detected by a layer 2 wireless intrusion prevention system (WIPS).
   - A. 900 MHz radio
   - B. 802.11h-compliant device
   - C. FHSS radio
   - D. 802.11b routers
   - E. 802.11g mixed-mode device

6. When designing a wireless policy document, what two major areas of policy should be addressed?

   A. General policy

   B. Functional policy

   C. Rogue AP policy

   D. Authentication policy

   E. Physical security

7. What can happen when an intruder compromises the preshared key used during WPA/WPA2-Personal authentication? (Choose all that apply.)

   A. Decryption

   B. Hijacking

   C. Spoofing

   D. Encryption cracking

   E. Access to network resources

8. Which of these attacks are considered layer 2 denial-of-service attacks? (Choose all that apply.)

   A. Deauthentication spoofing

   B. Jamming

   C. Virtual carrier attacks

   D. PS-Poll floods

   E. Authentication floods

9. Which of these can cause unintentional RF jamming attacks against an 802.11 wireless network? (Choose all that apply.)

   A. Microwave oven

   B. Signal generator

   C. 2.4 GHz cordless phones

   D. 900 MHz cordless phones

   E. Deauthentication transmitter

10. Which of these tools will best detect frequency hopping rogue devices? (Choose all that apply.)

    A. Stand-alone spectrum analyzer

    B. Distributed spectrum analyzer

    C. Distributed layer 2 WIDS

    D. Mobile layer 2 WIDS

    E. Layer 2 WIPS

11. Name two solutions that can help mitigate peer-to-peer attacks from other clients associated to the same 802.11 access point?
    A. Personal firewall
    B. PSPF
    C. OSPF
    D. MAC filter
    E. Access control lists

12. What type of solution can be used to perform countermeasures against a rogue access point?
    A. WIDS
    B. 802.1X/EAP
    C. WIPS
    D. TKIP/RC4
    E. WINS

13. Name the four labels that a WIPS uses to classify an 802.11 device.
    A. Infrastructure
    B. Known
    C. Enabled
    D. Disabled
    E. Rogue
    F. Unknown

14. Scott is an administrator at the Williams Lumber Company, and his WIPS has detected a rogue access point. What actions should he take after discovering the rogue AP? (Choose the best two answers.)
    A. Enable the layer 2 rogue containment feature that his WIPS provides.
    B. Unplug the rogue AP from the electrical outlet upon discovery.
    C. Call the police.
    D. Call his mother.
    E. Unplug the rogue AP from the data port upon discovery.

15. Which of these attacks are wireless users susceptible to at a public-access hotspot? (Choose all that apply.)
    A. Wi-Fi phishing
    B. Happy AP attack
    C. Peer-to-peer attack
    D. Malicious eavesdropping
    E. 802.11 reverse ARP attack
    F. Man-in-the-middle
    G. Wireless hijacking

16. Name two components that should be mandatory in every remote-access wireless security policy.
    - A. IPsec VPN
    - B. 802.1X/EAP
    - C. Personal firewall
    - D. Captive portal
    - E. Wireless stun gun
17. MAC filters are typically considered useless in most cases because of what type of attack?
    - A. Spamming
    - B. Spoofing
    - C. Phishing
    - D. Cracking
    - E. Eavesdropping
18. WLAN controllers typically deploy which type of WIDS deployment model?
    - A. Integrated
    - B. Overlay
    - C. Access distribution
    - D. Edge distribution
    - E. Overlay enabled
19. Which of these encryption technologies have been cracked? (Choose all that apply.)
    - A. 64-bit WEP
    - B. TKIP/RC4
    - C. CCMP/AES
    - D. 128-bit WEP
    - E. Wired Equivalent Privacy
20. What is another name for a wireless hijacking attack?
    - A. Wi-Fi phishing
    - B. Man-in-the-middle
    - C. Fake AP
    - D. Evil twin
    - E. AirSpy

Answers to Review Questions

1. B, C. DoS attacks can occur at either layer 1 or layer 2 of the OSI model. Layer 1 attacks are known as RF jamming attacks. A wide variety of layer 2 DoS attacks exist that are a result of tampering with 802.11 frames, including the spoofing of deauthentication frames.

2. C, D. Malicious eavesdropping is achieved with the unauthorized use of protocol analyzers to capture wireless communications. Any unencrypted 802.11 frame transmission can be reassembled at the upper layers of the OSI model.

3. D. A protocol analyzer is a passive device that captures 802.11 traffic and can be used for malicious eavesdropping. A WIDS cannot detect a passive device. Strong encryption is the solution to prevent a malicious eavesdropping attack.

4. C, D. The only way to prevent a wireless hijacking, man-in-the-middle, and/or Wi-Fi phishing attack is to use a mutual authentication solution. 802.1X/EAP authentication solutions require that mutual authentication credentials be exchanged before a user can be authorized.

5. A, C. The radio cards inside the WIPS sensors monitor the 2.4 GHz ISM band and the 5 GHz UNII bands. Older legacy wireless networking equipment exists that transmits in the 900 MHz ISM band, and these devices will not be detected. The radio cards inside the WIPS sensors also use only DSSS and OFDM technologies. Wireless networking equipment exists that uses frequency hopping spread spectrum (FHSS) transmissions in the 2.4 GHz ISM band and will go undetected. The only tool that can detect either a 900 MHz or frequency hopping rogue access point is a spectrum analyzer.

6. A, B. The general wireless security policy establishes why a wireless security policy is needed for an organization. Even if a company has no plans for deploying a wireless network, there should be at a minimum a policy detailing how to deal with rogue wireless devices. The functional security policy establishes how to secure the wireless network in terms of what solutions and actions are needed.

7. A, E. After obtaining the passphrase, an attacker can also associate to the WPA/WPA2 access point and thereby access network resources. The encryption technology is not cracked, but the key can be re-created. If a hacker has the passphrase and captures the 4-Way Handshake, they can re-create the dynamic encryption keys and therefore decrypt traffic. WPA/WPA2-Personal is not considered a strong security solution for the enterprise because if the passphrase is compromised, the attacker can access network resources and decrypt traffic.

8. A, C, D, E. Numerous types of layer 2 DoS attacks exist, including association floods, deauthentication spoofing, disassociation spoofing, authentication floods, PS-Poll floods, and virtual carrier attacks. RF jamming is a layer 1 DoS attack.

9. A, C. Microwave ovens operate in the 2.4 GHz ISM band and are often a source of unintentional interference. 2.4 GHz cordless phones can also cause unintentional jamming. A signal generator is typically going to be used as a jamming device, which would be considered intentional jamming. 900 MHz cordless phones will not interfere with 802.11 equipment that operates in either the 2.4 GHz ISM band or the 5 GHz UNII bands. There is no such thing as a deauthentication transmitter.

10. A, B. The radio cards inside the WIPS/WIDS sensors currently use only DSSS and OFDM technologies. Wireless networking equipment exists that uses frequency hopping spread spectrum (FHSS) transmissions in the 2.4 GHz ISM and will go undetected by layer 2 WIPS/WIDS sensors. The only tool that can detect either a 900 MHz or a frequency hopping rogue access point is a spectrum analyzer. Some WIPS/WIDS vendors have begun to offer layer 1 distributed spectrum analysis system (DSAS) solutions.

11. A, B. Public Secure Packet Forwarding (PSPF) is a feature that can be enabled on WLAN access points or WLAN controllers to block wireless clients from communicating with other wireless clients on the same wireless segment. The use of a personal firewall can also be used to mitigate peer-to peer attacks.

12. C. A wireless intrusion prevention system (WIPS) is capable of mitigating attacks from rogue access points. A WIPS sensor can use layer 2 DoS attacks as a countermeasure against a rogue device. SNMP may be used to shut down ports that a rogue AP has been connected to. WIPS vendors also use unpublished methods for mitigating rogue attacks.

13. A, B, E, F. The WIPS solution labels 802.11 radios into four classifications. An infrastructure device refers to any client station or access point that is an authorized member of the company's wireless network. An unknown device is any new 802.11 radio that has been detected but not classified as a rogue. A known device refers to any client station or access point that is detected by the WIPS and has been identified as an interfering device but is not considered a threat. A rogue device refers to any client station or access point that is considered an interfering device and a potential threat.

14. A, E. Every company should have a policy forbidding installation of wireless devices by employees. Every company should also have a policy on how to respond to all wireless attacks, including the discovery of a rogue access point. If a WIPS discovers a rogue AP, temporarily implementing layer 2 rogue containment abilities is advisable until the rogue device can be physically located. After the device is found, immediately unplug it from the data port but not from the electrical outlet. It would be advisable to leave the rogue AP on so that the administrator can do some forensics and look at the association tables and log files to possibly determine who installed it.

15. A, C, D, F, G. Currently, there is no such thing as a Happy AP attack or an 802.11 reverse ARP attack. Wireless users are especially vulnerable to attacks at public-use hotspots because there is no security. Because no encryption is used, the wireless users are vulnerable to malicious eavesdropping. Because no mutual authentication solution is in place, they are vulnerable to hijacking, man-in-the-middle, and phishing attacks. The hotspot access point might also be allowing peer-to-peer communications, making the users vulnerable to peer-to-peer attacks. Every company should have a remote-access wireless security policy to protect their end users when they leave company grounds.

16. A, C. Public-access hotspots have absolutely no security in place, and it is imperative that a remote access WLAN policy be strictly enforced. This policy should include the required use of an IPsec VPN solution to provide device authentication, user authentication, and strong encryption of all wireless data traffic. Hotspots are prime targets for malicious eavesdropping attacks. Personal firewalls should also be installed on all remote computers to prevent peer-to-peer attacks.

17. B. MAC filters are configured to apply restrictions that will allow only traffic from specific client stations to pass through based on their unique MAC addresses. MAC addresses can be *spoofed*, or impersonated, and any amateur hacker can easily bypass any MAC filter by spoofing an allowed client station's address.

18. A. Many WLAN controller vendors have fully integrated WIDS capabilities. The wireless controller acts as the centralized server. Because the IDS capabilities are fully integrated, there is no need for an overlay solution.

19. A, D, E. Wired Equivalent Privacy (WEP) encryption has been cracked, and currently available tools may be able to derive the secret key within a matter of minutes. The size of the key makes no difference, and both 64-bit WEP and 128-bit WEP can be cracked. TKIP/RC4 and CCMP/AES encryption have not been cracked.

20. D. An attack that often generates a lot of press is wireless hijacking, also known as the evil twin attack. The attacker hijacks wireless clients at layer 2 and layer 3 by using an evil twin access point and a DHCP server. The hacker may take the attack several steps further and initiate a man-in-the-middle attack and/or a Wi-Fi phishing attack.

Radio Frequency Site Survey Fundamentals

IN THIS CHAPTER, YOU WILL LEARN ABOUT THE FOLLOWING:

✓ **WLAN site survey interview**

- Customer briefing
- Business requirements
- Capacity and coverage requirements
- Existing wireless network
- Infrastructure connectivity
- Security expectations
- Guest access

✓ **Documentation and reports**

- Forms and customer documentation
- Deliverables
- Additional reports

✓ **Vertical market considerations**

- Outdoor surveys
- Aesthetics
- Government
- Education
- Healthcare
- Hotspots
- Retail
- Warehouses
- Manufacturing
- Multitenant buildings

Chapter 16, "Site Survey Systems and Devices," discusses wireless site surveys from a technical perspective. You will learn about all the procedures and tools required for proper coverage, spectrum, and application analysis. In this chapter, however, we discuss the wireless site survey from an administrative perspective. Much preparation must take place before the actual WLAN site survey is conducted. The needs of the WLAN must be predetermined and the proper questions must be asked.

In this chapter, we cover all the necessary preparations for the site survey and the documentation that must be assembled prior to it. We also discuss all the final reports that are delivered upon completion of the WLAN site survey. Finally, we outline unique wireless site survey considerations that should be given to different vertical markets.

WLAN Site Survey Interview

Is a site survey even needed? The answer to that question is almost always a resounding yes. If an owner of a small retail flower shop desires a wireless network, the site survey that is conducted may be as simple as placing a residential wireless gateway in the middle of the shop, turning the transmit power to a lower setting, and making sure you have connectivity. Performing a site survey in a medium or large business entails much more physical work and time. Before the actual survey is conducted, a proper *site survey interview* should occur to both educate the customer and properly determine their needs.

Asking the correct questions during a site survey interview not only ensures that the proper tools are used during the survey, but also makes the survey more productive. Most important, the end result of a thorough interview and thorough survey will be a WLAN that meets all the intended mobility, coverage, and capacity needs. The following sections cover the questions that should be thoughtfully considered during the site survey interview.

Customer Briefing

Even though 802.11 technologies have been around since 1997, much misunderstanding and misinformation about wireless networking still exists. Because many businesses and individuals are familiar with Ethernet networks, a "just plug it in and turn it on" mentality is prevalent. If a wireless network is being planned for your company or for a prospective client, it is highly recommended that you sit management down, give them an overview of 802.11 wireless networking, and talk with them about how and why site surveys are conducted. You do not need to explain the inner workings of Orthogonal Frequency Division Multiplexing or Distributed Coordination Function; however, a conversation about the advantages of Wi-Fi as well as the limitations of a WLAN is a good idea.

For example, a brief explanation about the advantages of mobility would be an excellent start. Chances are that a wireless network is already being considered because the company's end users have requested mobility or because a specific application such as Voice over Wi-Fi (VoWiFi) is being contemplated.

Just as important is a discussion about the bandwidth and throughput limitations of current 802.11a/b/g/n technology. Enterprise users are accustomed to 100 Mbps full-duplex or better speeds on the wired network. Because of vendor hype, people often believe that a Wi-Fi network will provide them with similar bandwidth and throughput. Management will need to be educated that because of overhead, the aggregate throughput is usually one half or less of the advertised data rate.

As you learned in earlier chapters, the aggregate throughput of a 54 Mbps data rate is 20 Mbps or less. It should also be explained that the medium is a half-duplex shared medium and not full-duplex. Chances are that an 802.11b/g network is being considered, and it might be necessary to briefly explain the effect on throughput as a result of the 802.11g (ERP) protection mechanism. 802.11n WLAN equipment is starting to address greater throughput needs, thus making the bandwidth/throughput conversation less painful. However, with the demand for faster networks, in the future we are sure we will be explaining why 802.11n is so much slower than Gigabit Ethernet.

Another appropriate discussion is why a site survey is needed. A very brief explanation on how RF signals propagate and attenuate will provide management with a better understanding of why an RF site survey is needed to ensure the proper coverage and enhance performance. A discussion and comparison of a 2.4 GHz vs. a 5 GHz WLAN might also be necessary. If management is properly briefed on the basics of Wi-Fi as well as the importance of a site survey, the forthcoming technical questions will be answered in a more suitable fashion.

Business Requirements

The first question that should be proposed is, What is the purpose of the WLAN? If you have a complete understanding of the intended use of a wireless network, the result will be a better-designed WLAN. For example, a VoWiFi network has very different requirements than a heavily used data network. If the purpose of the WLAN is only to provide users a gateway to the Internet, security and segmentation recommendations will be different. A warehouse environment with 200 handheld scanners is very different from an office environment. A hospital's wireless network will have different business requirements than an airport's wireless network. Here are some of the business requirement questions that should be asked:

What applications will be used over the WLAN? This question could have both capacity and quality of service (QoS) implications. A wireless network for graphic designers moving huge graphics files across the WLAN would obviously need more bandwidth than a wireless network for nothing but wireless bar code scanners. If time-sensitive applications such as voice or video are required, proprietary QoS needs might have to be addressed, and standardized 802.11e/WMM solutions will need to be deployed to meet these QoS needs.

Who will be using the WLAN? Different types of users have different capacity and performance needs. Groups of users might be segmented into VLANs or even segmented by different frequencies. This is also an important consideration for security roles.

What types of devices will be connecting to the WLAN? Handheld devices may also be segmented into separate VLANs or by frequency. VoWiFi phones are always put in a separate VLAN than data users with laptops. Many handheld devices use older 802.11b (HR-DSSS) radios and can transmit in only the 2.4 GHz ISM band. The capabilities of the devices may also force decisions in security, frequency, technology, and data rates.

We discuss the varying business requirements of different vertical markets later in this chapter. Defining the purpose of the WLAN in advance will lead to a more productive site survey and is imperative to the eventual design of the WLAN.

Capacity and Coverage Requirements

After the purpose of the WLAN has been clearly defined, the next step is to begin asking all the necessary questions for planning the site survey and designing the wireless network. Although the final design of a WLAN is completed after the site survey is conducted, some preliminary design based on the *capacity* and *coverage* needs of the customer is recommended. You will need to sit down with a copy of the building's floor plan and ask the customer where they want RF coverage. The answer will almost always be everywhere.

If a VoWiFi deployment is planned, that answer is probably legitimate because VoWiFi phones will need mobility and connectivity throughout the building. If the WLAN is strictly a data network, the need for blanket coverage might not be necessary. Do laptop data users need access in a storage area? Do they need connectivity in the outdoor courtyard? Do handheld bar code scanners used in a warehouse area need access in the front office? The answer to these questions will often vary depending on the earlier questions that were asked regarding the purpose of the WLAN. However, if you can determine that certain areas of the facility do not require coverage, you will save the customer money and yourself time when conducting the physical survey. Depending on the layout and the materials used inside the building, some preplanning might need to be done as to what type of antennas to use in certain areas of the facility. A long hallway or corridor will most likely need an indoor semidirectional antenna for coverage as opposed to an omnidirectional antenna. When the survey is performed, this will be confirmed or adjusted accordingly.

The most often neglected aspect prior to the site survey is determining capacity needs of the WLAN. As mentioned in Chapter 11, "Network Design, Implementation, and Management," you must not just consider coverage; you must also plan for capacity. Cell sizing and/or colocation might be necessary to properly address your capacity requirements. In order for the wireless end user to experience acceptable performance, a ratio of average number of users per access point must be established. The answer to the capacity question depends on a host of variables, including answers to earlier questions about the purpose of the WLAN. Capacity will not be as big of a concern in a warehouse environment using mostly handheld data scanners. However, if the WLAN has average to heavy data requirements, capacity will absolutely be a concern. The following are among the many factors that need to be considered when planning for capacity:

Data applications The applications that are used will have a direct impact on the number of users who should be communicating on average through an access point. So the next question is, What is a good average number of data users per access point? Once again, it depends entirely on the purpose of the WLAN and the applications in use. However, in an average 802.11b/g network, 12 to 15 data users per access point is an often-quoted figure.

User density Three important questions need to be asked with regard to users. First, how many users currently need wireless access? Second, how many users may need wireless access in the future? These first two questions will help you to begin adequately planning for a good ratio of users per access point while allowing for future growth. The third question of great significance is, Where are the users? Sit down with network management and indicate on the floor plan of the building any areas of high user density. For example, one company might have offices with only 1 or 2 people per room, while another company might have 30 or more people in a common area separated by cubicle walls. Other examples of areas with high user density are call centers, classrooms, and lecture halls. Also plan to conduct the physical survey when the users are present and not during off-hours. A high concentration of human bodies can attenuate the RF signal because of absorption.

Peak on/off use Be sure to ask what the peak times are, that is, when access to the WLAN is heaviest. For example, a conference room might be used only once a day or once a month. Also, certain applications might be heavily accessed through the WLAN at specified times. Another peak period could be when one shift leaves and another arrives.

Existing transmitters This does not refer just to previously installed 802.11 networks. Rather, it refers to interfering devices such as microwaves, cordless headsets, cordless phones, wireless machinery mechanisms, and so on. Often this is severely overlooked. If a large open area will house the help desk after the wireless is installed, you may be thinking of capacity. However, if you don't know that the employees are using 2.4 GHz cordless headsets or Bluetooth keyboards and mice, you may be designing a network destined for failure.

Mobile vs. mobility There are two types of mobility. The first is related to being mobile and the other is true mobility. To help explain this, think of a marketing manager working on a presentation and saving it on a network share. He later wants to give that presentation in the boardroom. If he picks up his laptop, closes the lid, and walks to the conference room, where he opens the laptop, connects to the wireless network, and gives his presentation, that is being mobile. He may have disconnected in between points, and that is okay. However, having true mobility means that a user remains connected 100 percent of the time while traveling through the facility. This would be indicative of VoWiFi or warehouse scanning applications. Determining which type of connectivity is necessary can be key for not only troubleshooting an existing network but also designing a new one.

802.11g (ERP) protection mechanism It should be understood in advance that if there is any requirement for backward compatibility with 802.11b HR-DSSS clients, the protection mechanism for non-ERP clients will always adversely affect throughput. The majority of enterprise deployments will always require backward compatibility to provide access for older 802.11b (HR-DSSS) radios found in handhelds, VoWiFi phones, or older laptops.

Carefully planning coverage and capacity needs prior to the site survey will help you determine some of the design scenarios you may possibly need, including AP power settings, type of antennas, and cell sizes. The physical site survey will still have to be conducted to validate and further determine coverage and capacity requirements.

Real World Scenario

How Many Simultaneous VoWiFi Telephone Calls Can an Access Point Support?

Several factors come into play, including cell bandwidth, average use, and vendor specifics. One of the leading VoWiFi telephone vendors, Polycom, recommends a maximum of 12 calls per 11 Mbps cell. Because of bandwidth limitations, that number drops to a recommended maximum of 7 calls per 2 Mbps cell. A typical call requires 4.5 percent of AP bandwidth at 11 Mbps, and 12 percent of AP bandwidth at 2 Mbps. Different vendor-specific access point characteristics can also affect the number of concurrent calls, and extensive testing is recommended. Probability models also exist for predicting VoWiFi traffic. Not every Wi-Fi phone user will be making a call at the same time. Probabilistic traffic formulas use a telecommunications unit of measurement known as an *Erlang*. An Erlang is equal to 1 hour of telephone traffic in 1 hour of time. Some online VoWiFi Erlang traffic calculators can be found at `www.erlang.com`.

Existing Wireless Network

Quite often the reason you are conducting a WLAN site survey is that you have been called in as a consultant to fix an existing deployment. Professional site survey companies have reported that as much as 40 percent of their business is troubleshooting existing WLANs, which often requires conducting a second site survey or discovering that one was never conducted to begin with.

As more corporations and individuals become educated in 802.11 technologies, the percentage will obviously drop. Sadly, many untrained customers just install the access points wherever they can mount them and leave the default power and channel settings on every AP. Usually, site surveys must be conducted either because of performance problems or difficulty roaming. Performance problems are often caused by cochannel interference and multipath interference as well as other sources of interference. Roaming problems may also be interference related or caused by a lack of adequate coverage and/or by a lack of proper cell overlap. Here are some of the questions that should be asked prior to the reparative site survey:

What are the current problems with the existing WLAN? Ask the customer to clarify the problems. Are they throughput related? Are there frequent disconnects? Is there any difficulty roaming? In what part of the building do the problems occur most often? How often do they occur, and have there been any steps taken to duplicate the troubles?

Are there any known sources of RF interference? More than likely the customer will have no idea, but it does not hurt to ask. Are there any microwave ovens? Do they use cordless phones or headsets? Does anyone use Bluetooth for keyboards or mice? After asking these interference questions, you should always perform a spectrum analysis. This is the *only* way to determine whether there is any RF interference in the area that may inhibit future transmissions. Something like a new Wireless Internet Service Provider (WISP) in the area may simply be interfering with one of your channels.

Are there any known coverage dead zones? This is related to the roaming questions, and areas probably exist where proper coverage is not being provided. Remember, this could be too little or too much coverage. Both create roaming and connectivity problems.

Does prior site survey data exist? Chances are that an original site survey was not even conducted. However, if old site survey documentation exists, it may be helpful when troubleshooting existing problems. It is important to note that unless quantifiable data was collected that shows dBm strengths, the survey report should be viewed with extreme caution.

What equipment is currently installed? Ask what type of equipment is being used, such as 802.11a (5 GHz) or 802.11b/g (2.4 GHz) and which vendor has been used. Once again, the customer might have no idea, and it will be your job to determine what has been installed and why it is not working properly. Also check the configurations of the devices, including service set identifiers (SSIDs), WEP or WPA keys, channels, power levels, and firmware versions. Often issues can be as simple as all the access points are transmitting on the same channel or there is a buffer issue that is resolved with the latest firmware.

Depending on the level of troubleshooting that is required on the existing wireless network, a second site survey consisting of coverage and spectrum analysis will often be necessary. After the new site survey has been conducted, adjustments to the existing WLAN equipment typically is adequate. However, the worst-case scenario would involve a complete redesign of the WLAN. Keep in mind that whenever a second site survey is necessary, all the same questions that are asked as part of a survey for a new installation (Greenfield survey) should also be asked prior to the second site survey. If wireless usage requirements have changed, a redesign may be the best course of action.

Infrastructure Connectivity

You have already learned that the usual purposes of a WLAN are to provide client mobility and to provide access via an AP into a preexisting wired network infrastructure. Part of the interview process includes asking the correct questions so that the WLAN will integrate properly into the existing wired architecture. Asking for a copy of the wired network topology map is highly recommended.

For security reasons, the customer may not want to disclose the wired topology, and you may need to sign a nondisclosure agreement. It is a good idea to request that an agreement be signed to protect you legally as the integrator. Be sure that someone in your organization with the authority to sign finalizes the agreement.

Understanding the existing topology will also be of help when planning WLAN segmentation and security proposals and recommendations. With or without a topology map, the following topics are important to ensure the desired infrastructure connectivity:

Roaming Is roaming required? In most cases, the answer will be yes, because mobility is a key advantage of wireless networking. Any devices that run connection-oriented applications will need seamless roaming. Seamless roaming is mandatory if handheld devices and/or VoWiFi phones are deployed. Surprisingly, many customers do not require roaming capabilities and need coverage in only some areas of a building. In these cases, being mobile is sufficient, as mentioned previously. Some network administrators may want to be able to restrict certain areas where a user or a group of users can roam. For example, the sales team might be allowed to roam only between access points on floors 1 and 2 and not permitted to roam to APs on floors 3 and 4. The marketing team, however, could be allowed to roam between access points on all four floors. The role-based access control (RBAC) capabilities of a WLAN controller can deliver the granular control needed to segment and control roaming. The network may also have to be segmented with different SSIDs and VLANs. Another important roaming consideration is whether users will need to roam across layer 3 boundaries. A Mobile IP solution or a proprietary layer 3 roaming solution will be needed if client stations need to roam across subnets. Special consideration has to be given to roaming with VoWiFi devices because of the issues that can arise from network latency. With regard to the existing network, it is imperative that you determine whether the wired network infrastructure will support all the new wireless features. For instance, if you want to roll out five SSIDs with different VLANs but haven't checked to see if the customer's network switches can be configured with VLANs, you may have a serious problem.

Wiring closets Where are the wiring closets located? Will the locations that are being considered for AP installation be within a 100-meter (328-foot) cable drop from the wiring closets?

Antenna structure If an outdoor network or point-to-point bridging application is requested, some additional structure might have to be built to mount the antennas. Asking for building diagrams of the roof to locate structural beams and existing roof penetrations is a good idea. Depending on the weight of the installation, you may also need to consult a structural engineer.

Hubs/switches Will the access points be connected by category 5 (CAT5) cabling to hubs or managed switches? A layer 2 switch will be needed if VLANs are required. Connecting access points to hubs is not a recommended practice because of security and performance reasons. All traffic is broadcast to every port on a hub, and any traffic that traverses through an access point connected to a hub port can be heard on any of the other ports. Are there enough switch ports? Who will be responsible for programming the VLANs?

PoE How will the access points be powered? Because APs are often mounted in the ceiling, Power over Ethernet (PoE) will likely be required to remotely power the access points. Very often the customer will not yet have a PoE solution in place, and further investment will be needed. If the customer already does have a PoE solution installed, it must be determined whether the PoE solution is compliant with 802.3-2005 clause 33 (previously known as

802.3af) or is a proprietary PoE solution. Also, is the solution an endpoint or midspan solution? Regardless of what the customer has, it is important to make sure that it is compatible with the system you are proposing to install. If PoE injectors need to be installed, you will need to make sure there are sufficient power outlets. If not, who will be responsible for installing those? If you are installing 802.11n access points, they may require a proprietary PoE solution.

Segmentation How will the WLAN and/or users of the WLAN be segmented from the wired network? Will the entire wireless network be on a separate IP subnet? Will VLANs be used, and is a guest VLAN necessary? Will firewalls or VPNs be used for segmentation? Or will the wireless network be a natural extension to the wired network and follow the same wiring, numbering, and design schemes as the wired infrastructure? All these questions are also directly related to security expectations.

Naming convention Does the customer already have a naming convention for cabling and network infrastructure equipment, and will one need to be created for the WLAN?

User management Considerations regarding RBAC, bandwidth throttling, and load balancing should be discussed. Do they have an existing AAA server or does one need to be installed? Where will usernames and passwords be stored?

Infrastructure management How will the WLAN remote access points be managed? Is a central management solution a requirement? Will devices be managed using SSH2, SNMP, or HTTP/HTTPS? Do they have standard credentials that they would like to use to access these management interfaces?

A detailed site interview that provides detailed feedback about infrastructure connectivity requirements will result in a more thorough site survey and a well-designed wireless network. Seventy-five percent of the work for a good wireless network is in the pre-engineering. It creates the road map for all the other pieces.

Security Expectations

Network management should absolutely be interviewed about security expectations. All data privacy and encryption needs should be discussed. All authorization, authentication, and accounting (AAA) requirements must also be documented. It should also be determined whether the customer plans to implement a wireless intrusion detection or prevention system (WIDS or WIPS) for protection against rogue APs and the many other types of wireless attacks. Some security solutions, such as layer 3 virtual private networks (VPNs), may put extra overhead on the WLAN because of the type of encryption that is used. Overhead caused by encryption should be accounted for during the capacity planning stages. Special considerations will have to be given to VoWiFi devices because of the latency issues that might result from EAP authentication.

A comprehensive interview regarding security expectations will provide the necessary information to make competent security recommendations after the site survey has been conducted and prior to deployment. Industry-specific regulations such as the Health Insurance

Portability and Accountability Act (HIPAA), Gramm-Leach-Bliley, and Sarbanes-Oxley may have to be taken into consideration when making security recommendations. U.S. government installations may have to abide by the strict Federal Information Processing Standards (FIPS) 140-2 regulations, and all security solutions may need to be FIPS compliant.

All of these answers should also assist in determining whether the necessary hardware and software exists to perform these functions. If not, it will be your job to consider the requirements and recommendations that may be necessary.

Guest Access

Because of the widespread acceptance of Wi-Fi in business environments, most companies offer some sort of wireless guest access to the Internet. Guest users access the WLAN via the same access points and controllers. However, they usually connect via a guest SSID that redirects the guest users to a captive portal. The guest captive portal serves two purposes. First, the login screen forces guest users to read and abide by a legal disclaimer. Second, after logging in, the guest users are provided a gateway to the Internet. It should be noted that all users who connect with the guest SSID should be allowed to go only to the Internet gateway and should be properly segmented from all other network resources in a separate guest VLAN. Firewall restrictions and bandwidth throttling are also very common when deploying guest WLANs.

Documents and Reports

During the site survey interview (and prior to the site survey), proper documentation about the facility and network must be obtained. Additionally, site survey checklists should be created and adhered to during the physical survey. After the physical survey is performed, you will deliver to the customer a professional and comprehensive final report. Additional reports and customer recommendations may also be included with the final report. This report should provide detailed instructions on how to install and configure the proposed network so that anyone could read the report and understand your intent.

Forms and Customer Documentation

Prior to the site survey interview, you must obtain some critical documentation from the customer:

Blueprints You need a floor plan layout in order to discuss coverage and capacity needs with network administration personnel. As discussed earlier in this chapter, while reviewing floor plan layouts, keep in mind that capacity and coverage requirements will be preplanned. Photocopies of the floor plan will also need to be created and used to record the RF measurements that are taken during the physical site survey as well as to record the locations of hardware placement. Some software survey tools allow you to import floor plans, and the software will record the survey results on the floor plan for you. These are highly recommended and make the final report much easier to compile.

What if the customer does not have a set of blueprints? Blueprints can be located via a variety of sources. The original architect of the building will probably still have a copy of the blueprints. Many public and private buildings' floor plans might also be located at a public government resource such as city hall or the fire department. Businesses are usually required to post a fire escape plan. Many site surveys have been conducted using a simple fire escape plan that has been drawn to scale if blueprints cannot be located. In a worst-case scenario, you may have to use some graph paper and map out the floor plan manually. In Chapter 16, we discuss RF modeling software that can be used to create predictive capacity and coverage simulations. Predictive analysis tools require detailed information about building materials that may be found in blueprints. Blueprints may already be in a vector graphic format (with the extensions `.dwg` and `.dwf`) for importing into a predictive analysis application, or they may have to be scanned.

Topographic map If an outdoor site survey is planned, a topographic map, also called a *contour map*, will be needed. These contour maps display terrain information such as elevations, forest cover, and locations of streams and other bodies of water. Figure 15.1 depicts a typical topographic map. A topographic map will be a necessity when performing bridging calculations such as Fresnel zone.

FIGURE 15.1 Topographic map

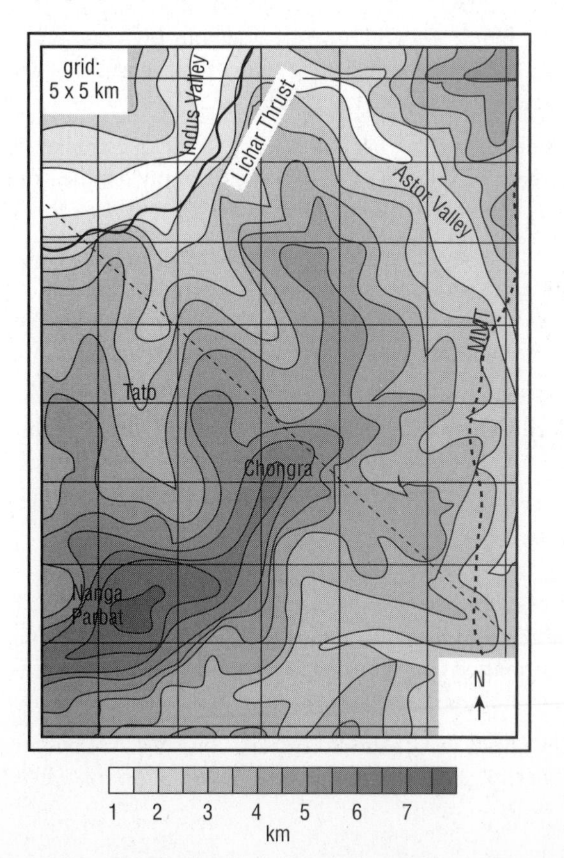

Network topology map Understanding the layout of the customer's current wired network infrastructure will speed up the site survey process and allow for better planning of the WLAN during the design phase. A computer network topology map will provide necessary information such as the location of the wiring closets and layer 3 boundaries. The WLAN topology will be integrated as seamlessly as possible into the wired infrastructure. VLANs will normally be used for segmentation and security for both the wired and wireless networks. Acquiring a network topology map from the customer is a highly recommended practice that will result in a well-designed and properly integrated WLAN. Some organizations may not wish to reveal their wired network topology, for security reasons. It may be necessary to obtain security clearance and/or sign nondisclosure agreements to gain access to these documents.

Security credentials You might need proper security authorization to access facilities when conducting the site survey. Hospitals, government facilities, and many businesses require badges, passes, and maybe even an escort for entrance into certain areas. A meeting with security personnel and/or the facilities manager will be necessary in order to meet all physical security requirements in advance of the survey. You do not want to show up at the customer site and be asked to return at another time because somebody forgot to schedule a security escort. Regardless of the security requirements, it is always a good idea to have the network administrator alert everyone that you will be in the area.

As a site survey professional, you will have created your own documentation or necessary checklists that will be used during the site survey interview as well as during the actual physical survey. There are several types of survey checklists:

Interview checklist A detailed checklist containing all the questions to be asked during the site survey interview should be created in advance. The many detailed interview questions discussed earlier in this chapter will all be outlined in the interview checklist.

Installation checklist Many site survey professionals prefer to record all installation details on the floor plan documents. An installation checklist detailing hardware placement and mounting for each individual access point is also an option. Information about AP location, antenna type, antenna orientation, mounting devices, and power sources may be logged.

Equipment checklist For organizational purposes, a checklist of all the hardware and software tools used during the survey might also be a good idea. All the necessary tools needed for both indoor and outdoor site surveys are covered in Chapter 16.

Deliverables

After the interview process has been completed and the actual survey has been conducted, a final report must be delivered to the customer. Information gathered during the site survey will be organized and formatted into a professional technical report for the customer's review. Compiled information contained in the *deliverables* will include the following:

Purpose statement The final report should begin with a WLAN purpose statement that stipulates the customer requirements and business justification for the WLAN.

Spectrum analysis Identifies potential sources of interference.

RF coverage analysis Defines RF cell boundaries.

Hardware placement and configuration Recommends AP placement, antenna orientation, channel reuse pattern, power settings, and any other AP-specific information such as installation techniques and cable routing.

Application analysis Includes results from application throughput testing, which is often an optional analysis report included with the final survey report.

Chapter 16 covers in detail the methods and tools used to compile all the necessary analytical information that belongs in the final report. A detailed site survey report may be hundreds of pages, depending on the size of the facility. Site survey reports often include pictures that were taken with a digital camera during the survey. Pictures can be used to record AP placement as well as identify problems such as interfering RF devices or potential installation problems such as a solid ceiling or concrete walls. Professional site survey software applications exist that also generate professional-quality reports using preformatted forms.

An example of a professional site survey report is included on the CD that comes with this book. The site survey report was provided by Netrepid, a professional wireless services company. This Adobe PDF file is called `sitesurvey.pdf`.

Additional Reports

Along with the site survey report, other recommendations will be made to the customer so that appropriate equipment and security are deployed. Usually, the individuals and/or company that performed the site survey are also hired for the installation of the wireless network. The customer, however, might use the information from the site survey report to conduct their own deployment. Regardless of who handles the installation work, other recommendations and reports will be provided along with the site survey report:

Vendor recommendations Many enterprise wireless vendors exist in the marketplace. It is a highly recommended practice to conduct the site survey using equipment from the same vendor who will supply the equipment that will later be deployed on site. Although the IEEE has set standards in place to ensure interoperability, every Wi-Fi vendor's equipment operates in some sort of proprietary fashion. You have already learned that many aspects of roaming are proprietary. The mere fact that every vendor's radio cards use proprietary RSSI thresholds is reason enough to stick with the same vendor during surveying and installation. Many site survey professionals have different vendor kits for the survey work. For example, a surveyor might own a kit that uses autonomous access points and might also own a WLAN controller survey kit with lightweight APs. It is not unheard of for a survey company to conduct two surveys with equipment from two different vendors and present the customer with two separate options. However, the interview process will usually determine in advance the vendor recommendations that will be made to the customer.

Implementation diagrams Based on information collected during the site survey, a final design diagram will be presented to the customer. The implementation diagram is basically a wireless topology map that illustrates where the access points will be installed and how the wireless network will be integrated into the existing wired infrastructure. AP placement, VLANs, and layer 3 boundaries will all be clearly defined.

Bill of materials Along with the implementation diagrams will be a detailed bill of materials (BOM) that itemizes every hardware and software component necessary for the final installation of the wireless network. The model number and quantity of each piece of equipment will be necessary. This includes access points, bridges, wireless switches, antennas, cabling, connectors, and lightning arrestors.

Project schedule and costs A detailed deployment schedule should be drafted that outlines all timelines, equipment costs, and labor costs. Particular attention should be paid to the schedule dependencies such as delivery times and licensing, if applicable.

Security solution recommendations As mentioned earlier in this chapter, security expectations should be discussed during the site survey interview. Based on these discussions, the surveying company will make comprehensive wireless security recommendations. All aspects of authentication, authorization, accounting, encryption, and segmentation should be included in the security recommendations documentation.

Wireless policy recommendations An extra addendum to the security recommendations might be corporate wireless policy recommendations. You might need to assist the customer in drafting a wireless network security policy if they do not already have one.

Training recommendations One of the most overlooked areas when deploying new solutions is proper training. It is highly recommended that wireless administration and security training sessions be scheduled with the customer's network personnel. Additionally, condensed training sessions should be scheduled with all end users.

Vertical Market Considerations

No two site surveys will ever be exactly alike. Every business has its own needs, issues, and considerations when conducting a survey. Some businesses may require an outdoor site survey instead of an indoor survey. A vertical market is a particular industry or group of businesses in which similar products or services are developed and marketed. The following sections outline the distinctive subjects that must be examined when a WLAN is being considered in specialized vertical markets.

Outdoor Surveys

Much of the focus of this book and the CWNA exam is on outdoor site surveys that are for establishing bridge links. Calculations necessary for outdoor bridging surveys are numerous, including the Fresnel zone, earth bulge, free space path loss, link budget, and fade margin.

However, outdoor site surveys for the purpose of providing general outdoor wireless access for users are becoming more commonplace. As the popularity of wireless mesh networking continues to grow, outdoor wireless access will become more commonplace. Outdoor site survey kits using mesh wireless routers will be needed.

Weather conditions such as lightning, snow and ice, heat, and wind must also be contemplated. Most important is the apparatus that the antennas will be mounted to. Unless the hardware is designed for outdoor use, the outdoor equipment must ultimately be protected from the weather elements by using NEMA-rated enclosure units (*NEMA* stands for National Electrical Manufacturers Association) like the one pictured in Figure 15.2. NEMA weatherproof enclosures are available with a wide range of options, including heating, cooling,.and PoE interfaces.

FIGURE 15.2 NEMA enclosure

PHOTO COURTESY OF NETREPID, INC.

Safety is also a big concern for outdoor deployments. Consideration should be given to hiring professional installers. Certified tower climbing courses and tower safety and rescue training courses are available.

Information about RF health and safety classes can be found at `www.sitesafe.com`. Also, tower climbing can be dangerous work. Information about tower climbing and safety training can be found at `www.comtrainusa.com`.

All RF power regulations as defined by the regulatory body of your country will need to be considered. If towers are to be used, several government agencies may need to be contacted. Local and state municipalities may have construction regulations, and a permit is almost always required. In the United States, if any tower exceeds a height of 200 feet above ground level (AGL) or is within a certain proximity to an airport, both the FCC and Federal Aviation Administration (FAA) must be contacted. If a roof mount is to be installed that is greater than 20 feet above the highest roof level, the FCC and FAA may need to be consulted as well. Other countries have similar height restrictions, and the proper RF regulatory authority and aviation authority must be contacted to find out the details.

Aesthetics

An important aspect of the installation of wireless equipment is the "pretty factor." The majority of businesses prefer that all wireless hardware remain completely out of sight. Aesthetics is extremely important in retail environments and in the hospitality industry (restaurants and hotels). Any business that is dealing with the public will require that the Wi-Fi hardware be hidden or at least secured. Many vendors are designing more aesthetic-looking access points and antennas. Some vendors have even camouflaged access points to resemble smoke detectors. Indoor enclosures, like the one pictured in Figure 15.3, can also be used to conceal access points from sight. It should also be noted that most enclosure units can be locked to help prevent theft of expensive Wi-Fi hardware.

FIGURE 15.3 Indoor enclosure

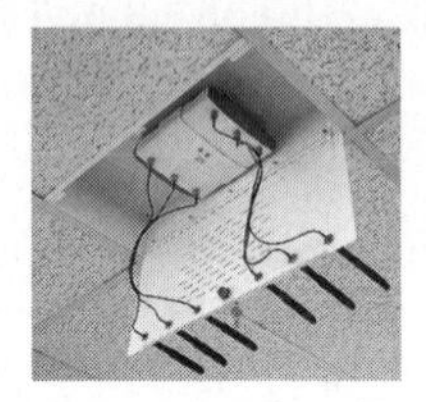

PHOTO COURTESY OF NETREPID, INC.

Government

The key concern during government wireless site surveys is security. When security expectations are addressed during the interview process, careful consideration should be given to all aspects of planned security. Many U.S. government agencies, including the military, require that all wireless solutions be FIPS 140-2 compliant. Other government agencies may require that the wireless network be completely shielded or shut off during certain times of the day. Be sure to check export restrictions before traveling to other countries with certain equipment. The United States forbids the export of AES encryption technology to some countries. Other countries have their own regulations and customs requirements.

Obtaining the proper security credentials will most likely be a requirement before conducting the government survey. An identification badge or pass often is required. In some government facilities, an escort is needed in certain sensitive areas.

Education

As with government facilities, obtaining the proper security credentials in an education environment usually is necessary. Properly securing access points in lockable enclosure units is also necessary to prevent theft. Apple Macintosh computers are used quite extensively in the education arena, so it is necessary to deploy client radio cards that support Macintosh drivers. Also, because of the high concentration of students, user density should be accounted for during capacity and coverage planning. In campus environments, wireless access is required in most buildings, and very often bridging solutions are needed between buildings across the campus. Some older educational facilities were constructed in such a manner as to serve as disaster shelters. That means that propagation in these areas is very limited.

Healthcare

One of the biggest concerns in a healthcare environment is sources of interference from the biomedical equipment that exists on site. Many biomedical devices operate in the ISM bands. For example, cauterizing devices in operating rooms have been known to cause problems with wireless networks. There is also a concern with 802.11 radios possibly interfering with biomedical equipment.

A meeting will be necessary with the biomedical department that maintains and services all biomedical equipment. Some hospitals have a person responsible for tracking and managing all RF devices in the facility.

A thorough spectrum analysis survey using a spectrum analyzer will be extremely important. It is recommended that several sweeps of these areas be conducted and compared to ensure the greatest probability of capturing all of the possible interferers. Because of the many potential sources of interference in the 2.4 GHz ISM band, it is likely that 5 GHz hardware will be deployed in many areas. Often, the dense environments require 5 GHz simply because you will need more channel options to prevent co-channel interference. Hospitals are usually large in scale, and a site survey may take many weeks. Long hallways, multiple floors, fire safety doors, reflective materials, concrete construction, lead-lined X-ray rooms, and wire mesh safety glass are some of the physical conditions that will be encountered during the survey.

The applications used in the medical environment should all be considered during the interview and the survey. Handheld PDAs are being used by doctors to transfer large files such as X-ray graphics. Medical carts use radios to transfer patient data back to the nursing stations. VoWiFi phone deployments are commonplace in hospitals because of the communication mobility that they provide to nurses. Wi-Fi real-time location systems (RTLS) using active 802.11 RFID tags are commonplace in hospitals for asset management tracking. Because of the presence of medical patients, proper security credentials and/or an escort will often be necessary. Many applications are connection oriented, and drops in connectivity can be detrimental to the operation of these applications.

Hotspots

Hotspots continue to grow in popularity, and many businesses are looking to provide wireless Internet access for their customers. Because most of the general public still uses 2.4 GHz radio cards, 5 GHz access is not often available at hotspots. Many hotspots are small, and care should be taken to limit the RF coverage area by using a single access point at a lower power setting. However, some large facilities such as airports and convention centers have begun offering wireless access, and obviously multiple access points and wider coverage will be needed. Security solutions at hotspots are usually limited to a captive portal solution for user authentication against a customer database.

Retail

A retail environment often has many potential sources of 2.4 GHz interference. Store demonstration models of cordless phones, baby monitors, and other ISM band devices can cause problems. The inventory storage racks and bins and the inventory itself are all potential sources of multipath problems. Heavy user density should also be considered, and a retail site survey should be done in the height of the shopping season as opposed to late January when the malls are empty.

Wireless applications that are used in retail stores include handheld scanners used for data collection and inventory control. Point-of-sale devices such as cash registers may also have radio cards. Often this equipment is older frequency hopping equipment that may cause all-band interference with an 802.11b/g (2.4 GHz) network. Steps may be necessary to upgrade the older equipment. Coverage is usually a greater concern than capacity because wireless data-collection devices require very little bandwidth, and the number used in a particular area is typically limited.

Warehouses

Some of the earliest deployments of 802.11 technology were in warehouses for the purpose of inventory control and data collection. A 2.4 GHz WLAN will likely be deployed because most handheld devices currently use 2.4 GHz radios. Coverage, not capacity, is the main objective when designing a wireless network in a warehouse. Warehouses are filled with metal racks and all sorts of inventory that can cause reflections and multipath. The use of directional antennas in a warehouse environment is almost a requirement. High ceilings often cause mounting problems as well as coverage issues. Indoor chain-link fences that are often used to secure certain areas will scatter and block a 2.4 GHz RF signal. Seamless roaming is also mandatory because the handheld devices will be mobile. Forklifts that can move swiftly through the warehouse may also have computing devices with radio cards. Many legacy deployments of 802.11 FHSS hardware and/or legacy 900 MHz radios still exist in many warehouse environments.

It is also important to keep stock levels in mind during the survey. Care should be taken to survey during peak times to create the worst-case scenario for coverage. It is also important to note that warehouses are probably the most dynamic RF environment. When stocks are low, the entire RF environment is different. It is recommended that the environment also be surveyed during low stocking levels to get a comparison. RF power levels will probably need to be adjusted as stock comes in and out. Some controller based WLANs are capable of monitoring the wireless network and adjusting the channels and power levels of each of the APs dynamically as the environment changes. However, the dynamic RF capabilities of WLAN controller solutions often can cause problems with VoWiFi.

Manufacturing

A manufacturing environment is often similar to a warehouse environment in terms of multipath interference and coverage design. However, a manufacturing plant presents many unique site survey challenges, including safety and the presence of employee unions. Heavy machinery and robotics may present safety concerns to the surveyor, and special care should be taken so as not to mount access points where they may be damaged by other machines. Many manufacturing plants also work with hazardous chemicals and materials. Proper protection may need to be worn, and ruggedized access points or enclosures may have to be installed.

Technology manufacturing plants often have clean rooms, and the surveyor will have to wear a clean suit and follow clean room procedures if they are even allowed in the room.

Many manufacturing plants are union shops with union employees. A meeting with the plant's union representative may be necessary to make sure that no union policies will be violated by the site surveyor team.

Multitenant Buildings

By far the biggest issue when conducting a survey in a multitenant building is the presence of other WLAN equipment used by nearby businesses. Office building environments are extremely cluttered with 802.11b/g wireless networks that operate at 2.4 GHz. Almost assuredly all of the other tenants' WLANs will be powered to full strength, and some equipment will be on nonstandard channels such as 2 and 8, which will likely interfere with your WLAN equipment. If at all possible, strong consideration should be given to deploying a WLAN using the 5 GHz UNII bands.

A recent survey conducted by Netrepid in Philadelphia, PA, revealed more than 50 2.4 GHz access points and not a single 5 GHz access point visible on one 8,000-square-foot floor. In this situation, using 2.4 GHz was not an option at all for the new installation.

Summary

In this chapter, you have learned about all the preparations and questions that must be asked prior to conducting a wireless site survey. The site survey interview is an important process necessary to both educate the customer and determine the customer's wireless needs. Defining the business purpose of the wireless network leads to a more productive survey. Capacity and coverage planning as well as planning for infrastructure connectivity is all part of the site survey interview. Prior to the site survey interview, you should obtain critical documentation such as blueprints or topographical maps from the customer. Interview and installation checklists are used during the site survey interview and during the actual physical survey. Different survey considerations are required for different vertical markets. After the site survey is completed, you will deliver to the customer a final site survey report as well as additional reports and recommendations.

Exam Essentials

Define the site survey interview. Be able to explain the importance of the interview process prior to the wireless site survey. Understand that the interview is for educating the customer and clearly defining all their wireless needs.

Identify the questions necessary to determine capacity and coverage needs. Understand the importance of proper capacity and coverage planning. Define all the numerous considerations when planning for RF cell coverage, bandwidth, and throughput.

Explain existing wireless network troubleshooting concerns. Be able to explain the questions necessary to troubleshoot an existing WLAN installation prior to conducting a secondary site survey.

Define infrastructure connectivity issues. Understand all the necessary questions that must be asked in order to guarantee proper integration of the WLAN into the existing wired infrastructure.

Identify site survey documentation and forms. Correctly identify all the documentation that must be assembled and created prior to the site survey. Be familiar with all the information and documentation that is needed in the final deliverables.

Explain vertical market considerations. Understand the business requirements of different vertical markets and how these requirements will alter the site survey and final deployment.

Key Terms

Before you take the exam, be certain you are familiar with the following terms:

capacity

coverage

deliverables

Erlang

site survey interview

Review Questions

1. You have been hired by the XYZ Company for a wireless site survey. Which statements best describe site survey best practices when choosing vendor equipment to be used during the survey? (Choose two answers.)
   - **A.** When conducting a wireless site survey with a WLAN switch, you should use both autonomous and lightweight access points.
   - **B.** When conducting a wireless site survey with autonomous access points, you should use different vendors' APs together.
   - **C.** When conducting a wireless site survey with a WLAN controller, you should use a WLAN controller and lightweight access points from the same vendor.
   - **D.** When conducting a wireless site survey with autonomous access points, you should use autonomous access points from the same vendor.
   - **E.** When a wireless site survey is conducted, proprietary security solutions are often implemented.

2. Name a unique consideration when deploying a wireless network in a hotel or other hospitality business. (Choose the best answer.)
   - **A.** Equipment theft
   - **B.** Aesthetics
   - **C.** Segmentation
   - **D.** Roaming
   - **E.** User management

3. Which of the following statements best describe security considerations during a wireless site survey? (Choose all that apply.)
   - **A.** Questions will be asked to define the customer's security expectations.
   - **B.** Wireless security recommendations will be made after the survey.
   - **C.** Recommendations about wireless security policies may also be made.
   - **D.** During the survey, both mutual authentication and encryption should be implemented.

4. The ACME Corporation has hired you to design a wireless network that will have data clients, VoWiFi phones, and access for guest users. The company wants the strongest security solution possible for the data clients and phones. Which design best fits the customer's requirements?

   A. Create one wireless VLAN. Segment the data clients, VoWiFi phones, and guest users from the wired network. Use an 802.1X/EAP authentication and CCMP/AES encryption for a wireless security.

   B. Create three separate VLANs. Segment the data clients, VoWiFi phones, and guest users into three distinct VLANs. Use an 802.1X/EAP authentication and TKIP encryption for security in the data VLAN. Use WPA2-Personal in the voice VLAN. The guest VLAN will have no security other than possibly a captive portal.

   C. Create three separate VLANs. Segment the data clients, VoWiFi phones, and guest users into three distinct VLANs. Use an 802.1X/EAP authentication with CCMP/AES encryption for security in the data VLAN. Use WPA2-Personal in the voice VLAN. The guest VLAN will have no security other than possibly a captive portal.

   D. Create two separate VLANs. The data and voice clients will share one VLAN while the guest users will reside in another. Use an 802.1X/EAP authentication and CCMP/AES encryption for security in the data/voice VLAN. The guest VLAN will have no security other than possibly a captive portal.

5. What are some additional recommendations that can be made along with the final site survey report? (Choose all that apply.)

   A. Training recommendations

   B. Security recommendations

   C. Coverage recommendations

   D. Capacity recommendations

   E. Roaming recommendations

6. What documents might be needed prior to performing an indoor site survey for a new wireless LAN? (Choose all that apply.)

   A. Blueprints

   B. Network topography map

   C. Network topology map

   D. Coverage map

   E. Frequency map

7. What roaming issues should be discussed during an interview for a future VoWiFi network? (Choose all that apply.)

   A. Layer 2 boundaries

   B. Layer 3 boundaries

   C. Layer 4 boundaries

   D. Latency

   E. Throughput

8. You have been hired by the Barry Corporation to conduct an indoor site survey. What information will be in the final site survey report that is delivered? (Choose all that apply.)
   - **A.** AP placement
   - **B.** Firewall settings
   - **C.** Router access control lists
   - **D.** Access point transmit power settings
   - **E.** Antenna orientation
9. The Kellum Corporation has hired you to troubleshoot an existing WLAN. The end users are reporting having difficulties when roaming. What are some of the possible causes? (Choose all that apply.)
   - **A.** The RF coverage cells have only 20 percent overlap. Fifty percent cell overlap is normally needed for seamless roaming.
   - **B.** The RF coverage cells have only 5 percent overlap. Fifteen to 20 percent cell overlap is normally needed for seamless roaming.
   - **C.** The RF coverage cells are colocated.
   - **D.** There is interference from the cellular network.
   - **E.** There is interference from 2.4 GHz portable phones.
10. After conducting a simple site survey in the office building where your company is located on the fifth floor, you have discovered that other businesses are also operating access points on nearby floors on channels 2 and 8. What is the best recommendation you will make to management about deploying a new WLAN for your company?
    - **A.** Install a 2.4 GHz access point on channel 6 and use the highest available transmit power setting to overpower the WLANs of the other businesses.
    - **B.** Speak with the other businesses. Suggest that they use channels 1 and 6 at lower power settings. Install a 2.4 GHz access point using channel 9.
    - **C.** Speak with the other businesses. Suggest that they use channels 1 and 11 at lower power settings. Install a 2.4 GHz access point using channel 6.
    - **D.** Recommend installing a 5 GHz access point.
    - **E.** Install a wireless intrusion prevention system (WIPS). Classify the other businesses' access points as interfering and implement de-authentication countermeasures.
11. The Harkins Corporation has hired you to make recommendations about a future wireless deployment that will require more than 300 access points to meet all coverage requirements. What is the most cost-efficient and practical recommendation in regard to providing electrical power to the access points?
    - **A.** Recommend that the customer replace older edge switches with new switches that have inline PoE.
    - **B.** Recommend that the customer replace the core switch with a new core switch that has inline PoE.
    - **C.** Recommend that the customer use single-port power injectors.
    - **D.** Recommend that the customer hire an electrician to install new electrical outlets.

12. The Chang Company has hired you to troubleshoot an existing 802.11b/g WLAN. The end users are reporting having difficulties with throughput performance. What are some of the possible causes of the difficulties? (Choose all that apply.)
    A. Multipath interference
    B. Co-channel interference
    C. Colocation interference
    D. Inadequate capacity planning
    E. Low client cards transmit power

13. What factors need to be considered when planning for capacity in a 5 GHz WLAN? (Choose all that apply.)
    A. Data applications
    B. User density
    C. Peak usage level
    D. ERP protection mechanism
    E. All of the above

14. During the interview process, which topics will be discussed so that the WLAN will integrate properly into the existing wired architecture? (Choose all that apply.)
    A. PoE
    B. Segmentation
    C. User management
    D. Infrastructure management
    E. All of the above

15. The Jackson County Regional Hospital has hired you for a wireless site survey. Prior to the site survey, employees from which departments at the hospital should be consulted? (Choose all that apply.)
    A. Network management
    B. Biomedical department
    C. Hospital security
    D. Custodial department
    E. Marketing department

16. Typically what are the biggest concerns when planning for a WLAN in a warehouse environment? (Choose all that apply.)
    A. Capacity
    B. Coverage
    C. Security
    D. Roaming

17. What type of hardware may be necessary when installing APs to be used for outdoor wireless coverage?
    - A. NEMA enclosure
    - B. Parabolic dish antennas
    - C. Patch antennas
    - D. Outdoor ruggedized core switch

18. What is a telecommunications unit of measurement of traffic equal to 1 hour of telephone traffic in 1 hour of time?
    - A. Ohm
    - B. dBm
    - C. Erlang
    - D. Call hour
    - E. Voltage Standing Wave Ratio

19. What additional documentation is usually provided along with the final site survey deliverable? (Choose all that apply.)
    - A. Bill of materials
    - B. Implementation diagrams
    - C. Network topology map
    - D. Project schedule and costs
    - E. Access point user manuals

20. The WonderPuppy Coffee Company has hired you to make recommendations about deploying wireless hotspots in 500 coffee shops across the country. What solutions might you recommend? (Choose all that apply.)
    - A. WPA2-Personal security solution
    - B. 802.11b/g access points at 100mW transmit power
    - C. 802.11b/g access points at 1 to 5mW transmit power
    - D. NEMA enclosures
    - E. Captive portal authentication
    - F. 802.1X/EAP security solution

Answers to Review Questions

1. C, D. It is a highly recommended practice to conduct the site survey by using equipment from the same vendor who will supply the equipment that will later be deployed on site. Mixing vendors during the survey is not recommended. Mixing an autonomous AP solution with a lightweight AP solution is also not recommended. Security is not implemented during the survey.

2. B. Although all the options are issues that may need addressing when deploying a WLAN in a hospitality environment, aesthetics is usually a top priority in the hospitality industry. The majority of customer service businesses prefer that all wireless hardware remain completely out of sight. It should also be noted that most enclosure units are lockable and help prevent theft of expensive Wi-Fi hardware. However, theft prevention is not unique to the hospitality business.

3. A, B, C. Although security in itself is not part of the WLAN site survey, network management should be interviewed about security expectations. The surveying company will make comprehensive wireless security recommendations. An extra addendum to the security recommendations might be corporate wireless policy recommendations. Authentication and encryption solutions are not usually implemented during the physical survey.

4. C. Segmentation, authentication, authorization, and encryption should all be considered during the site survey interview. In Chapter 13, you learned about the necessary components of wireless security. Segmenting three types of users into separate VLANs with separate security solutions is the best recommendation. The data users using 802.1X/EAP and CCMP/AES will have the strongest solution available. WPA-2 provides the voice users with CCMP/AES encryption as well but avoids using an 802.1X/EAP solution that will cause latency problems. The guest user VLAN requires minimal security for ease of use.

5. A, B. Training, security, and choice of vendor are extra recommendations that may also accompany the site survey report. The site survey report should already be addressing coverage, capacity, and roaming requirements.

6. A, C. Blueprints will be needed for the site survey interview to discuss coverage and capacity needs. A network topology map will be useful to assist in the design of integrating the wireless network into the current wired infrastructure.

7. B, D. Latency is an important consideration whenever any time-sensitive application such as voice or video is to be deployed. A Mobile IP solution or proprietary layer 3 roaming solution will be needed if layer 3 boundaries are crossed during roaming.

8. A, D, E. The final site survey report known as the deliverable will contain spectrum analysis information identifying potential sources of interference. Coverage analysis will also define RF cell boundaries. The final report also contains recommended access point placement, configuration settings, and antenna orientation. Application throughput testing is often an optional analysis report included in the final survey report. Firewall settings and router access control lists are not included in a site survey report.

9. B, E. Roaming problems may be interference related or caused by a lack of adequate coverage and/or cell overlap. In Chapter 12, "WLAN Troubleshooting," you learned that 15 to 20 percent cell overlap is typically needed for roaming. 2.4 GHz portable phones may be a source of interference. Cellular phones operate in a frequency space that will not interfere with the existing WLAN.

10. D. Although answer C is a possible solution, the best recommendation is to deploy hardware that operates at 5 GHz, and interference from the neighboring business's 2.4 GHz network will never be an issue.

11. A. The cheapest and most efficient solution will be to replace the older edge switches with newer switches that have inline power that can provide PoE to the access points. A core switch will not be used to provide PoE because of cabling distance limitations. Deploying single-port injectors is not practical, and hiring an electrician will be extremely expensive.

12. A, B, D. Multipath and co-channel interference are common causes of poor performance. Inadequate capacity planning can result in too many users per access point leading to throughput problems.

13. A, B, C. User density, data applications, peak usage levels, and the 802.11g (ERP) protection mechanism are all considerations when capacity planning for an 802.11b/g network. When planning for a 5 GHz WLAN, the ERP protection mechanism is not an issue.

14. E. Multiple questions are related to infrastructure integration. How will the access points be powered? How will the WLAN and/or users of the WLAN be segmented from the wired network? How will the WLAN remote access points be managed? Considerations such as role-based access control (RBAC), bandwidth throttling, and load balancing should also be discussed.

15. A, B, C. Network management will be consulted during most of the site survey and deployment process for proper integration of the WLAN. The biomedical department will be consulted about possible RF interference issues. Hospital security will be contacted in order to obtain proper security passes and possible escort.

16. B, C, D. Coverage, not capacity, is the main objective when designing a wireless network in a warehouse. Seamless roaming is also mandatory because handheld devices are typically deployed. Security is a major requirement for all WLAN enterprise installations.

17. A. Outdoor equipment must ultimately be protected from the weather elements by using either hardened APs or enclosure units rated by the National Electrical Manufacturers Association (NEMA). NEMA weatherproof enclosures are available with a wide range of options, including heating, cooling, and Power over Ethernet interfaces. Parabolic dishes and patch antennas are usually used with APs for outdoor bridge links.

18. C. Probabilistic traffic formulas use a telecommunications unit of measurement known as an Erlang. An Erlang is equal to 1 hour of telephone traffic in 1 hour of time.

19. A, B, D. Based on information collected during the site survey, a final design diagram will be presented to the customer. Along with the implementation diagrams will be a detailed bill of materials (BOM) that itemizes every hardware and software component necessary for the final installation of the wireless network. A detailed deployment schedule should be drafted that outlines all timelines, equipment costs, and labor costs.

20. C, E. Most of the general public still uses 2.4 GHz radio cards, so an 802.11b/g access point is recommended, although hotspots are also beginning to offer access at 5 GHz. Many hotspots are small, and care should be taken to limit the RF coverage area using a single access point at a lower power setting. Security solutions at hotspots are usually limited to a captive portal solution for user authentication against a customer database.

Site Survey Systems and Devices

IN THIS CHAPTER, YOU WILL LEARN ABOUT THE FOLLOWING:

✓ **Site survey defined**

- Mandatory spectrum analysis
- Mandatory coverage analysis
- AP placement and configuration
- Optional application analysis

✓ **Site survey tools**

- Indoor site survey tools
- Outdoor site survey tools

✓ **Coverage analysis**

- Manual
- Assisted
- Predictive
- Self-organizing wireless LANs

In Chapter 15, "Radio Frequency Site Survey Fundamentals," we discussed wireless site surveys from an administrative perspective. You learned what information to gather and what to plan for prior to the actual Wi-Fi site survey. In this chapter, we present the wireless site survey from a technical perspective. A proper site survey should include spectrum analysis as well as coverage analysis so that optimum 802.11 communications are realized. Determining the proper placement and configuration of the 802.11 equipment during the site survey is essential to reaching your expected performance goals for the wireless network. RF signal propagation studies are needed to determine existing and new RF coverage patterns. Many variables—such as walls, floors, doors, plumbing, windows, elevators, buildings, trees, and mountains—can have a direct effect on the coverage of an access point or wireless bridge.

In this chapter, we discuss how to perform a site survey, the types of site surveys, and the tools that can be used during a site survey. Site survey professionals often have their own unique technical approach for executing a site survey. We like to think of it as almost an art form, and in this chapter, we'll help you take the first steps in becoming a wireless site survey Picasso.

Site Survey Defined

When most individuals are asked to define a wireless site survey, the usual response is that a site survey is for determining RF coverage. Although that definition is absolutely correct, the site survey encompasses so much more, including looking for potential sources of interference as well as the proper placement, installation, and configuration of 802.11 hardware and related components. In the following sections, we cover the often overlooked, yet necessary, spectrum analysis requirement of the site survey and the often misunderstood coverage analysis requirement. During the coverage analysis process, a determination will be made for the proper placement of access points, the transmission power of the access point radio card, and the proper use of antennas.

Although not mandatory, capacity performance and application testing might also be an optional requirement of an 802.11 wireless survey. Depending on the purpose of the wireless network, different tools can be used, which is why the site survey interview and planning process is so important. Throughout the remainder of this chapter, we also cover the variety of tools that may be used as part of your site survey arsenal.

Mandatory Spectrum Analysis

Before conducting the coverage analysis survey, locating sources of potential interference is a must. Unfortunately, many site surveys completely ignore *spectrum analysis* because of the high cost generally associated with purchasing the necessary spectrum analyzer hardware. Spectrum analyzers are frequency domain measurement devices that can measure the amplitude and frequency space of electromagnetic signals. Spectrum analyzer hardware can cost upward of $40,000 (in U.S. dollars), thereby making them cost-prohibitive for many smaller and medium-size businesses. The good news is that several companies have solutions, both hardware and software based, that are designed specifically for 802.11 site survey spectrum analysis and are drastically less expensive. Figure 16.1 depicts a 2.4 GHz hardware spectrum analyzer.

FIGURE 16.1 2.4 GHz spectrum analyzer

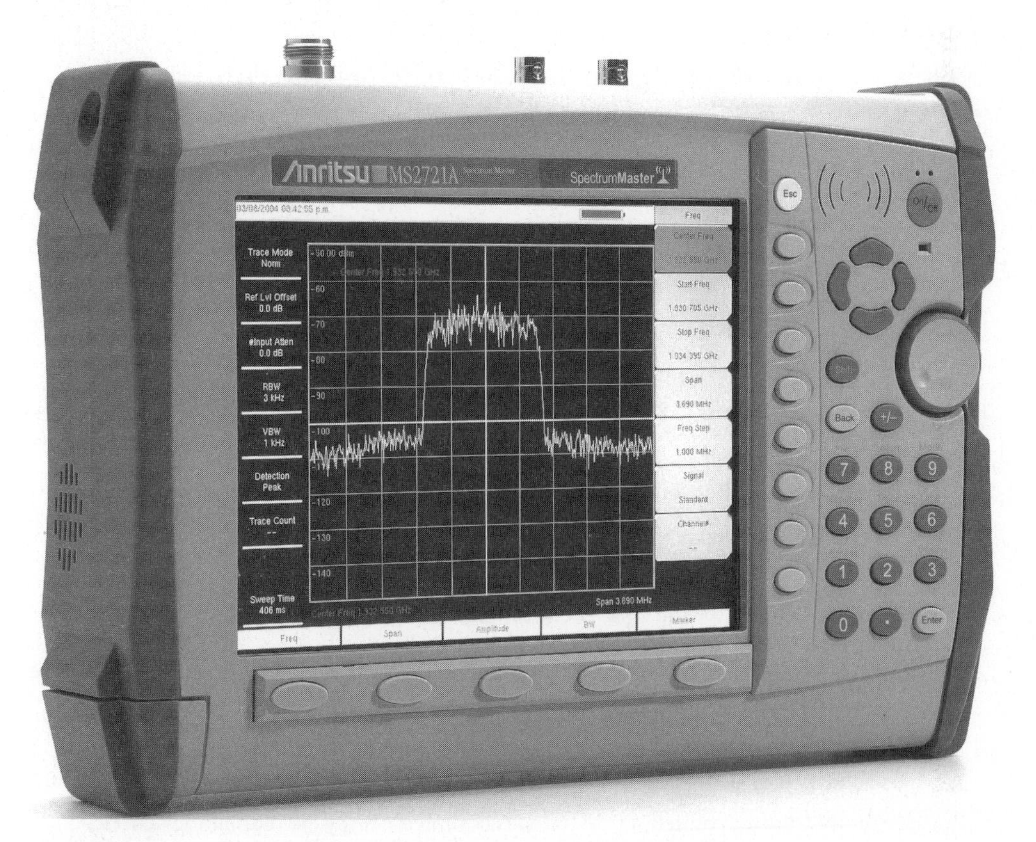

To conduct a proper 802.11 spectrum analysis survey, the *spectrum analyzer* will need to be capable of scanning both the 2.4 GHz ISM band and the 5 GHz UNII bands. Several companies now sell software-based solutions that work with special PCMCIA cards. These software-based spectrum analyzers are designed specifically for 802.11 site surveys and can correctly identify specific energy pulses such as those from a microwave oven or cordless phone. Be cautious in your research of these software tools. Some will measure only other 802.11 devices and are not true spectrum analyzers. A true spectrum analyzer picks up RF energy regardless of the source.

On the CD that accompanies this book is a live demo of the Fluke Networks AnalyzeAir Wi-Fi Spectrum Analyzer software. You will use this program to execute Exercise 16.1 by using playback files of spectrum captures.

EXERCISE 16.1

Spectrum Analysis

In this exercise, you will use a demo program of Fluke Networks AnalyzeAir Wi-Fi Spectrum Analyzer to view sources of interference and simulate spectrum analysis.

1. Install the AnalyzeAir program that's included on this book's CD. From the `AnalyzeAir` directory of the CD, copy the file `AnalyzeAir.exe` to your desktop. Click on the file and follow the default installation prompts.
2. Double-click the AnalyzeAir icon on your desktop. At the prompt, click the Screen Capture File button. Click the Browse button and browse to a file called `Sample_Spectrum_Capture.ccf`.
3. From the main screen, click the Spectrum (2) tab. Notice that the 2.4 GHz frequency space is very crowded.
4. From the main screen, under Control Panel/Frequency, change the band from 2.4–2.5 GHz to 5.15–5.35 GHz. Notice that the 5 GHz frequency space is not as crowded.
5. From the main screen, click the Devices tab. Notice all the Wi-Fi access points as well as interfering devices.
6. Continue to maneuver through the program and familiarize yourself with the spectrum analyzer's features and capabilities.

So why is spectrum analysis even necessary? If the background noise level exceeds –85 dBm in either the 2.4 GHz ISM band or 5 GHz UNII bands, the performance of the wireless network can be severely degraded. A noisy environment can cause the data in 802.11 transmissions to become corrupted. If the data is corrupted, the cyclic redundancy check (CRC) will fail and the receiving 802.11 radio will not send an ACK frame to the transmitting 802.11 radio. If an ACK frame is not received by the original transmitting radio, the unicast frame is not acknowledged and will have to be retransmitted. If an interfering device such as a microwave oven results in retransmissions above 10 percent, the performance or throughput of the wireless LAN will suffer significantly. Most data applications in a Wi-Fi network can handle a layer 2 retransmission rate of up to 10 percent without any noticeable degradation in performance. However, time-sensitive applications such as VoIP require that higher-layer IP packet loss be no greater than 2 percent. Therefore, Voice over Wi-Fi (VoWiFi) networks need to limit retransmissions at layer 2 to 5 percent or less to guarantee the timely delivery of VoIP packets.

Interfering devices may also prevent an 802.11 radio from transmitting. If another RF source is transmitting with strong amplitude, an 802.11 radio can sense the energy during the clear channel assessment (CCA) and defer transmission. If the source of the interference is a constant signal, an 802.11 radio will continuously defer transmissions until the medium is clear. In other words, a strong source of RF interference could actually prevent your 802.11 client stations and access point radios from transmitting at all.

It is a recommended practice to conduct spectrum analysis of all frequency ranges, especially in the 2.4 GHz ISM band. The 2.4 to 2.4835 GHz ISM band is an extremely crowded frequency space. The following are potential sources of interference in the 2.4 GHz ISM band:

- Microwave ovens
- 2.4 GHz cordless phones, DSSS and FHSS
- Fluorescent bulbs
- 2.4 GHz video cameras
- Elevator motors
- Cauterizing devices
- Plasma cutters
- Bluetooth radios
- Nearby 802.11, 802.11b, 802.11g, or 802.11n (2.4 GHz) WLANs
- Wireless Internet Service Providers (WISPs)

One of the first things you should determine during the site survey interview is the location of any microwave ovens. Microwave ovens typically operate at 800 to 1,000 watts. Although microwave ovens are shielded, they can become leaky over time. Commercial-grade microwave ovens will be shielded better than a discount microwave oven that you can buy at many retail outlets. A received signal of –40 dBm is about 1/10,000 of a milliwatt and is considered a very strong signal for 802.11 communications. If a 1,000 watt microwave oven is even 0.0000001 percent leaky, the oven will interfere with the 802.11 radio.

Figure 16.2 shows a spectrum view of a microwave oven. Note that this microwave operates dead center in the 2.4 GHz ISM band. Some microwave ovens can congest the entire frequency band. You should also check whether the call centers, receptionist, or other employees use a Bluetooth mouse, keyboard, or headset. These can also cause a great deal of interference.

FIGURE 16.2 Microwave oven spectrum use

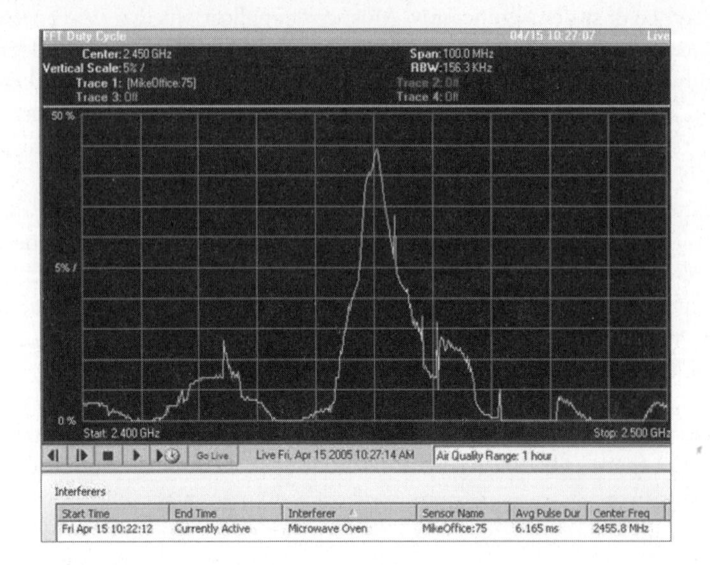

Because of the extreme crowding of the 2.4 GHz ISM band, many enterprise deployments are switching to 802.11a equipment that transmits in the 5 GHz UNII bands. Switching to a 5 GHz WLAN is often a wise choice in the enterprise because the 5 GHz UNII bands are currently not very crowded and there are more choices for channel reuse patterns. Not nearly as many interfering devices exist, and there are just not many neighboring 802.11a networks that can potentially cause interference. Although there is much less interference present at 5 GHz as compared to 2.4 GHz, this will change over time. Just as everyone moved from 900 MHz to 2.4 GHz to avoid interference, the "band jumping" effect will also catch up with 5 GHz. Current potential sources of interference in 5 GHz UNII bands include the following:

- 5 GHz cordless phones
- Radar
- Perimeter sensors
- Digital satellite
- Nearby 5 GHz WLANs
- Outdoor wireless 5 GHz bridges

The 802.11-2007 standard does define dynamic frequency selection (DFS) and transmit power control (TPC) mechanisms to satisfy regulatory requirements for operation in the 5 GHz band to avoid interference with 5 GHz radar systems. As you learned in earlier chapters, 802.11h-compliant radios are required to detect radar at 5 GHz and not transmit to avoid interfering with the radar systems. Using a 5 GHz spectrum analyzer during a site survey can help determine in advance whether radar transmissions exist in the area where the WLAN deployment is planned.

Although many devices can cause problems in both frequencies, one of the most common causes of interference is other wireless LANs. Strong signals from other nearby WLANs can be a huge problem, especially in a multitenant building environment. You may need to cooperate with neighboring businesses to ensure that their access points are not powered too high and that they are on channels that will not interfere with your access points. Once again, because of the proliferation of 2.4 GHz WLAN equipment, many businesses are now choosing to switch to 5 GHz WLAN solutions.

After locating the sources of interference, the best and simplest solution is to eliminate them entirely. If a microwave oven is causing problems, consider purchasing a more expensive commercial-grade oven that is less likely to be a nuisance. Other devices, such as 2.4 GHz cordless phones, should be removed and a policy should be strictly enforced that bans them. 5.8 GHz cordless phones operate in the 5.8 GHz ISM band, which overlaps with the upper UNII band (5.725 GHz to 5.825 GHz). Indoor use of 5.8 GHz phones will cause interference with 5 GHz radios transmitting in the upper UNII band.

If interfering devices cannot be eradicated in the 2.4 GHz bands, consider moving to the less crowded 5 GHz UNII bands. As stated earlier in this chapter, a VoWiFi network needs to limit layer 2 retransmissions to 5 percent or less, meaning that a very thorough spectrum analysis of the 2.4 GHz ISM band is a necessity. In the very recent past, VoWiFi phones operated using High-Rate DSSS (HR-DSSS) technology and the radios therefore transmitted only in the very crowded 2.4 GHz ISM band. WLAN vendors such as Polycom now manufacture OFDM-capable VoWiFi phones that can transmit in the less crowded 5 GHz UNII bands. A now common design strategy is to deploy the VoWiFi clients at 5 GHz and all other clients at 2.4 GHz in WLAN environments that deploy dual frequency access points. If your WLAN is being used for either data or voice or for both, a proper and thorough spectrum analysis is mandatory in an enterprise environment. It is important to make sure you know what your client devices are capable of before determining the spectrum to use. If all or some of your client devices are restricted to using 2.4 GHz, that may be your only option and you will need to be able to plan and engineer around the environment.

Mandatory Coverage Analysis

After conducting a spectrum analysis site survey, the next step is the all-important determination of proper 802.11 RF coverage inside your facility. During the site survey interview, capacity and coverage requirements are discussed and determined before the actual site survey is performed. In certain areas of your facility, smaller cells or colocation may be required because of a high density of users or heavy application bandwidth requirements.

After all of the capacity and coverage needs have been determined, RF measurements must be taken to guarantee that these needs are met and to determine the proper placement and configuration of the access points and antennas. Proper *coverage analysis* must be performed using some type of *received signal strength* measurement tool. This tool could be something as simple as the received signal strength meter in your wireless card's client utility, or it could be a more expensive and complex site survey software package. All of these measurement tools are discussed in more detail later in this chapter.

So how do you conduct proper coverage analysis? That question is often debated by industry professionals. Many site survey professionals have their own techniques; however, we will try to describe a basic procedure for coverage analysis. The first mistake that many people make during the site survey is leaving the access point radio at the default full-power setting. A 2.4 GHz 802.11b/g radio transmitting at 100 mW will often cause interference with other access point coverage cells simply because it is generating too much power. Also, many legacy client cards have a maximum transmit power of 30 mW. The RF signal of a 30 mW client might not be heard at the outer edge of an access point's 100 mW coverage cell. A good starting point for a 2.4 GHz access point is 30 mW transmit power. After the site survey is performed, the power can be increased if needed to meet unexpected coverage needs, or it can be decreased to meet capacity needs.

The hardest part of a coverage analysis site survey is often finding where to place the first access point and determining the boundaries of the first RF cell. The procedure outlined here is generally how this is achieved and is further illustrated in Figure 16.3:

1. Place an access point with a power setting of 30 mW in the corner of the building.
2. Walk diagonally away from the access point toward the center of the building until the received signal drops to –65 dBm. This is the location where you place your first access point.
3. Temporarily mount the access point in the first location and begin walking throughout the facility to find the –65 dBm end points, also known as *cell boundaries* or *cell edges*.
4. Depending on the shape and size of the first coverage cell, you may want to change the power settings and/or move the initial access point. A good portion of a proper coverage analysis involves starting over and trying again.

FIGURE 16.3 Starting coverage cell

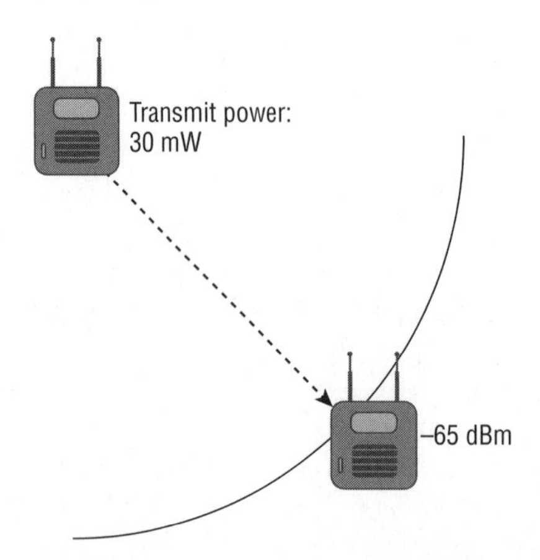

After the first coverage cell and boundaries have been determined, the next question is where to place the next access point. The placement of the next access point is performed by using a technique that's similar to the one you used to place the first access point:

1. Think of the cell boundary of the first access point, where the signal is –65 dBm, as the initial starting point, similar to the way you used the corner of the building as your initial starting point. From the first access point, walk parallel to the edge of the building, and place an access point at the location where the received signal is –65 dBm, as pictured in Figure 16.4.

FIGURE 16.4 Second coverage cell

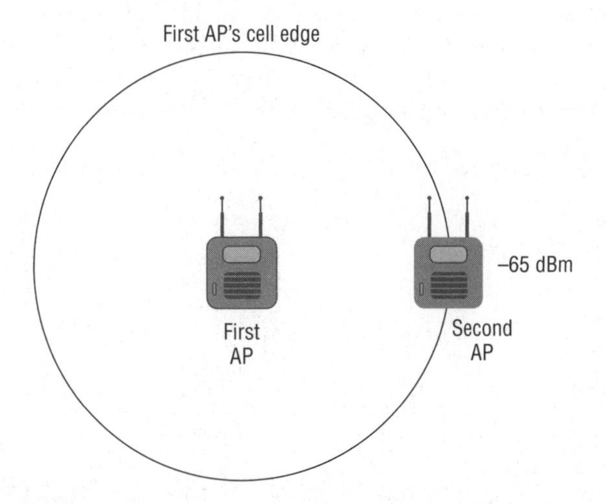

2. Now walk away from this access point, parallel to the edge of the building, until the received signal drops to –65 dBm.
3. This is the farthest point to place the access point if you do not want cell overlap.
4. Using the distance from the previous access point and this location, the placement of this next access point should be about 15 to 20 percent (depending on cell overlap requirements) closer to the previous access point.
5. Move to that location and temporarily mount the access point. Begin walking throughout the facility to find the –65 dBm end points, or cell boundaries.
6. Again, depending on the shape and size of the first coverage cell, you may want to change the power settings and/or move this access point.

It is important to avoid excessive overlap because it can cause frequent roaming and performance degradation. The shape and size of the building and the attenuation caused by the various materials of walls and obstacles will require you to change the distances between access points to ensure proper cell overlap. After finding the proper placement of the second access point and all of its cell boundaries, repeat the procedure all over again. The rest of the site survey is basically repeating this procedure over and over again, effectively daisy-chaining throughout the building until all coverage needs are determined.

The following cell edge measurements are taken during the site survey:

- Received signal strength (dBm), also known as received signal level (RSL)
- Noise level (dBm)
- Signal-to-noise ratio, or SNR (dB)

The received signal strength measurements that are recorded during a site survey typically depend on the intended use of the WLAN. If the intent of the WLAN is solely coverage and not capacity, a lower received signal of –85 dBm might be used as the boundary for your overlapping cells. If throughput and capacity are issues, using a stronger received signal of –65 dBm is recommended. The SNR is an important value because, if the background noise is too close to the received signal, data can get corrupted and retransmissions will increase. The SNR is simply the difference in decibels between the received signal and the background noise, as pictured in Figure 16.5. Many vendors recommended a minimum SNR of 18 dB for data networks and a minimum of 25 dB for voice networks.

FIGURE 16.5 Signal-to-noise ratio

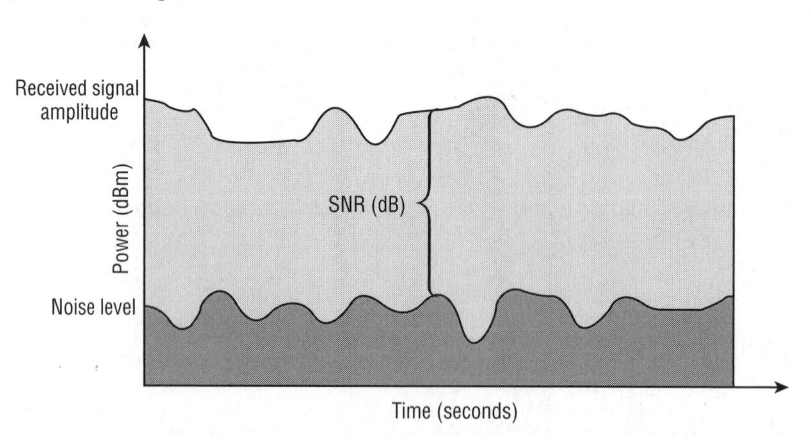

Some site survey professionals prefer to use data rate measurements as opposed to received signal strength measurements when determining their cell boundaries. The problem with using the data rate is that vendors have different received signal strength indicator (RSSI) thresholds, and different vendor cards will shift between data rates at different dBm levels. Cell design can be performed using one vendor's RSSI threshold values if the company deploying the WLAN intends to use just that one vendor's radios. If measurements are based on received signal levels (RSLs), the WLAN surveyor can always go back and map different client cards and data rates without having to resurvey. A site survey using just data rates or a proprietary signal strength measurement threshold does not allow for any flexibility between vendors. Table 16.1 depicts the recommended minimum received signal and minimum SNR for a WLAN data network using one vendor's highly sensitive radio card.

TABLE 16.1 WLAN Data Cell—Vendor Recommendations

Data Rate	Minimum Received Signal	Minimum Signal-to-Noise Ratio
54 Mbps	–71 dBm	25 dB
36 Mbps	–73 dBm	18 dB
24 Mbps	–77 dBm	12 dB
12/11 Mbps	–82 dBm	10 dB
6/5.5 Mbps	–89 dBm	8 dB
2 Mbps	–91 dBm	6 dB
1 Mbps	–94 dBm	4 dB

Most VoWiFi manufacturers require a minimum received signal of –65 dBm. Therefore, overlapping cells of –60 dBm is a good idea for VoWiFi wireless networks in order to provide a fade margin buffer. The recommended SNR ratio for a VoWiFi network is 25 dB or higher. Cell overlap of 15 to 20 percent will be needed, and the separation of same channel cells should be 20 dB or greater. Figure 16.6 depicts the recommended coverage for a VoWiFi network.

FIGURE 16.6 VoWiFi cell recommendations

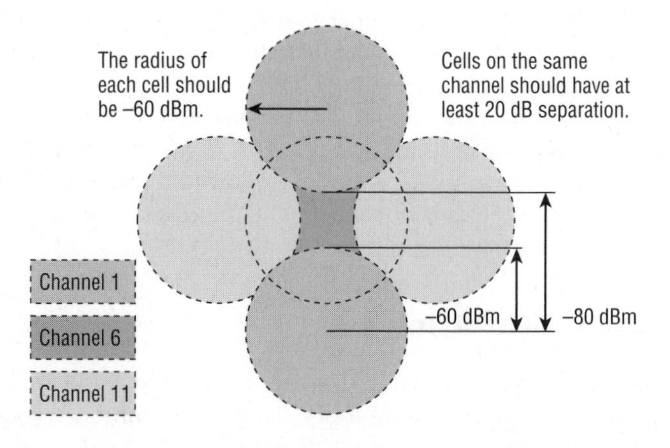

Although figures and drawings often depict the RF coverage as nice round symmetrical cells, the reality is that cell boundaries usually have an erratic shape that might resemble a starfish or elements in a Pablo Picasso painting.

AP Placement and Configuration

As you have just read, coverage analysis also determines the proper placement of access points and power settings. When the site survey is conducted, all the cell edge measurements will be recorded and written on a copy of the floor plan of the building. An entry with the exact location of each access point must also be recorded. Next to the entry of each access point should be the transmission power level of the AP's radio card when the survey was conducted. The location of all the wiring closets will also be noted on the floor plan, and care should be taken to ensure that the placement of any access point is within a 100-meter (328-foot) cable run back to the wiring closet because of CAT5 cabling distance limitations. Be sure to account for vertical cabling distances as well as horizontal runs.

Another often overlooked component in WLAN design during coverage analysis is the use of semidirectional antennas. Many deployments of WLANs use only the manufacturer's default low gain omnidirectional antenna, which typically has about 2.14 dBi of gain. Buildings come in many shapes and sizes and often have long corridors or hallways where using an indoor semidirectional antenna would be much more advantageous. Using a unidirectional antenna in areas where there are metal racks, file cabinets, and metal lockers can be advantageous because you can cut down on reflections. Using indoor semidirectional antennas to reduce reflections will cut down on the negative effects of multipath, namely the data corruption caused by the delay spread and intersymbol interference (ISI). If data corruption is reduced, so is the need for retransmissions, thus the performance of the WLAN is enhanced by the use of semidirectional antennas in the correct situations. Figure 16.7 depicts the use of semidirectional antennas in a warehouse with long corridors and metal racks that line the corridors.

A good site survey kit should have a variety of antennas, both omnidirectional and semidirectional. The best way to provide proper coverage in most buildings is to use a combination of both low-gain omnidirectional antennas and indoor semidirectional antennas together, as pictured in Figure 16.8.

When a semidirectional antenna is used, recording the received signal strength, SNR, and noise level measurements is still necessary to find the coverage edges. The coverage area should closely resemble the radiated pattern of the semidirectional antenna. Simply record the signal measurements along the directional path and the edges of the directional path where the antenna is providing coverage.

Optional Application Analysis

Whereas spectrum analysis and coverage analysis are considered mandatory during 802.11 wireless site surveys, *application analysis* is considered optional. Capacity planning is an important part of the site survey interview. This takes into account not only the user capacity but the bandwidth capacity as well. Although capacity testing is rare during a site survey, please understand that capacity planning is an important aspect of the survey. Cell sizing or colocation can be planned and surveyed during the coverage analysis portion of the survey. Capacity testing using application analysis and throughput verification is not normally part of a standard site survey. However, tools do exist that can perform application stress testing of a WLAN. These tools may be used at the tail end of a site survey, but they are more often used during the deployment stage of the WLAN network. Several companies offer 802.11a/b/g/n multistation emulation hardware that can simulate multiple concurrent virtual wireless client stations. The virtual client stations can have individual security settings. Roaming performance can also be tested. The 802.11a/b/g/n multistation emulator works in conjunction with another component that can emulate hundreds of protocols and generate traffic bidirectionally through the virtual client stations. A great use of such a device could be to test the performance of a simulated wireless data network along with simulated wireless VoIP traffic.

FIGURE 16.7 Proper use of semidirectional antennas

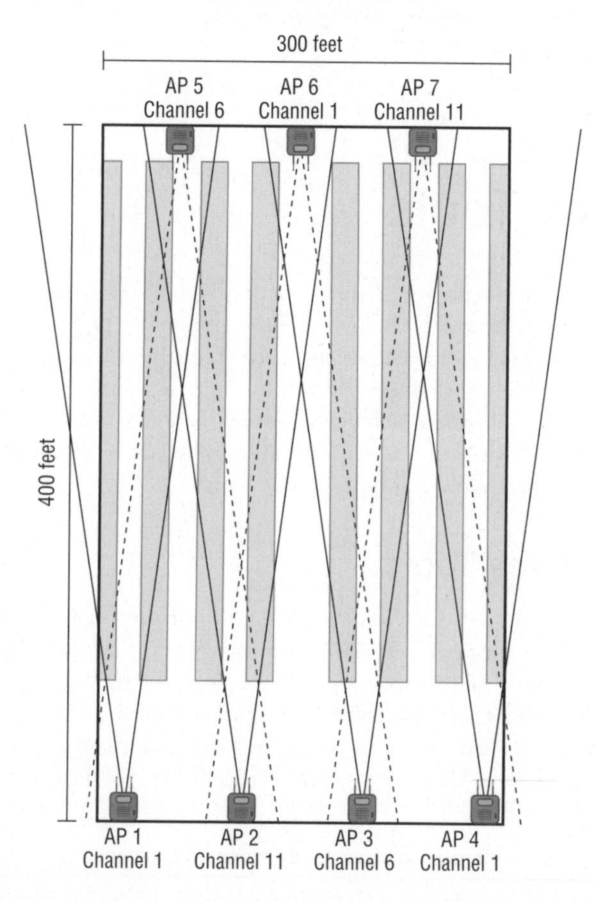

FIGURE 16.8 Omnidirectional and semidirectional antenna combination

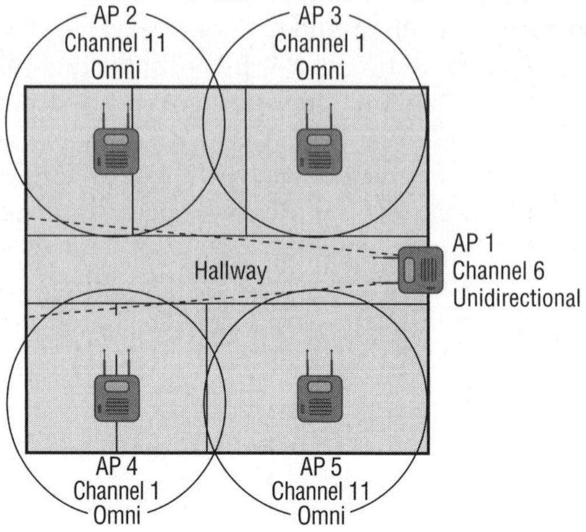

Site Survey Tools

Anyone who is serious about deploying wireless networks will put together a site survey toolbox with a multitude of products that can aid the site survey process. The main tool of course will be some sort of signal measurement software utility that interfaces with your wireless client card and is used for coverage analysis. Prepackaged site survey kits are often for sale on the Internet, but most site survey professionals prefer to put together their own kit. Indoor and outdoor site surveys are very different in nature, and in this section we discuss the different tools that are used in both types of surveys.

Indoor Site Survey Tools

As stated earlier, a spectrum analyzer will be needed for locating potential sources of interference. Your main weapon in your coverage analysis arsenal will be a received signal strength measurement tool. This tool could be something as simple as the received signal strength meter in your wireless card's client utility, or it could be a more expensive and complex site survey software package. However, there are many other tools that can assist you when you are conducting the physical site survey. These are some of the tools that you might use for an indoor site survey:

Spectrum analyzer This is needed for frequency spectrum analysis.

Blueprints Blueprints or floor plans of the facility are needed to map coverage and mark RF measurements. CAD software may be needed to view and edit digital copies of the blueprints.

Signal strength measurement software You'll need this for RF coverage analysis.

802.11 client card This is used with the signal measurement software. It is a recommended practice to use the vendor client card that is most likely to be deployed.

Access point At least one AP is needed, preferably two. Autonomous access points can be used as stand-alone devices during the site survey. Lightweight APs will require a controller.

WLAN controller The majority of WLAN deployments now use a centralized architecture that comprises a WLAN controller and lightweight access points. Depending on the vendor, lightweight APs often cannot transmit without also tunneling back to a WLAN controller. Most WLAN controller vendors manufacture small controllers that are designed for use in branch and remote offices. When performing a site survey, a small remote office WLAN controller that weighs 2 lbs will be easier to work with than a core WLAN controller that weighs 30 lbs.

Battery pack A battery pack is a necessity because the site survey engineer does not want to have to run electrical extension cords to power the access point while it is temporarily mounted for the site survey. Not only does the battery pack provide power to the access point, it also provides a safer environment because you don't have to run a loose power cord across the floor, and it makes it easier and quicker to move the access point to a new location.

Binoculars It may seem strange to have binoculars for an indoor sight survey, but they can be very useful in tall warehouses and convention centers. They can also be handy for looking at things in the plenum space above the ceiling.

Walkie-talkies or cellular phones When performing a site survey in an office environment, it is often necessary to be as quiet and unobtrusive as possible. Walkie-talkies or cellular phones are typically preferred over yelling across the room. You must also remember that RF is three-dimensional and it is common for one person to be on one floor with the access point while the other person is on another floor checking the received signal.

Antennas A wide variety of both indoor omnidirectional and indoor unidirectional antennas is a must in every indoor Wi-Fi site survey kit.

Temporary mounting gear During the site survey, you will be temporarily mounting the access point—often high up, just below the ceiling. Some sort of solution is needed to temporarily mount the AP. Bungee cords and plastic ties are often used as well as good old-fashioned duct tape. As shown in Figure 16.9, some professionals use a tripod and mount the access point on an extending mast. The tripod can then be moved within the building, thereby bypassing the need to temporarily mount the access point.

FIGURE 16.9 WLAN site survey tripod (WiFi Surveyor—courtesy of Caster Tray)

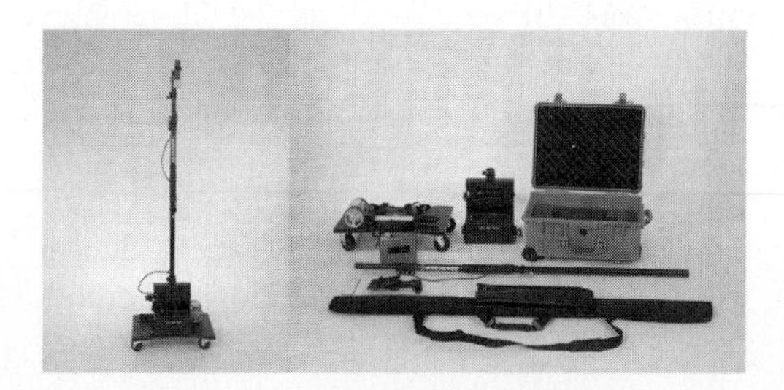

Digital camera A digital camera should be used to record the exact location of the access point's placement. Recording this information visually will assist whoever will do the final installation at a later date. Setting the date/time on the pictures may also come in handy when viewing the pictures later.

Measuring wheel or laser measuring meter A tool is needed to make sure the access point will in fact be close enough for a 100-meter cable run back to the wiring closet. Keep in mind that a 100-meter cable run includes running the CAT5 cabling through the plenum. A measuring wheel or a laser distance measuring tool could be used to measure the distance back to the wiring closet. A measuring wheel is usually the better tool because the laser devices can often yield bad results.

Colored electrical tape Everyone remembers the fable of Hansel and Gretel and how they used breadcrumbs to leave a trail to find their way home. The colored tape can be used to leave a trail back to where you want to mount the access points. Leave a small piece of colored electrical tape at the location where the access point was temporarily mounted during the site survey. This will assist whoever will do the final AP installation at a later date. A color scheme could even be used to track different channel frequencies: red for channel 1, green for channel 6, and blue for channel 11.

Ladder or forklift Ladders and/or forklifts may be needed to temporarily mount the access point to the ceiling.

When conducting a site survey, it is a highly recommended practice to use the same 802.11 access point hardware that you plan on deploying. Keep in mind that every vendor is different and implements RSSI differently. It is not advisable to conduct a coverage analysis survey using one vendor's access point and then deploy a completely different vendor's hardware. Many established site survey companies have put together different vendor site survey kits so that they can offer their customers several options.

Outdoor Site Survey Tools

As mentioned in Chapter 15, much of the focus of this book and the CWNA exam is on outdoor site surveys that are for establishing bridge links. However, outdoor site surveys for the purpose of providing general outdoor wireless access for users are becoming more commonplace. Outdoor site surveys are conducted using either outdoor access points or mesh routers, which are the devices typically used to provide access for client stations in an outdoor environment. These outdoor Wi-Fi surveys will use most of the same tools as an indoor site survey but may also use a Global Positioning System (GPS) device to record latitude and longitude coordinates. Although outdoor 802.11 deployments can be used to provide access, usually a discussion of outdoor site surveys is about wireless bridging. Wi-Fi bridging exists at the distribution layer and is used to provide a wireless link between two or more wired networks.

An entirely different set of tools is needed for an outdoor bridging site survey, and many more calculations are required to guarantee the stability of the bridge link. In earlier chapters, you learned that the calculations necessary when deploying outdoor bridge links are numerous, including the Fresnel zone, earth bulge, free space path loss, link budget, and fade margin. Other considerations may include the intentional radiator (IR) and equivalent isotropically radiated power (EIRP) limits as defined by the regulatory body of your country. Weather conditions are another major consideration in any outdoor site survey, and proper protection against lightning and wind will need to be deployed. An outdoor wireless bridging site survey usually requires the cooperative skills of two individuals. The following list includes some of the tools that you might use for an outdoor bridging site survey:

Topographic map Instead of a building floor plan, a topographic map that outlines elevations and positions will be needed.

Link analysis software Point-to-point link analysis software can be used with topographic maps to generate a bridge link profile and also perform many of the necessary calculations, such as Fresnel zone and EIRP. The bridge link analysis software is a predictive modeling tool.

Calculators Software calculators and spreadsheets can be used to provide necessary calculations for link budget, Fresnel zone, free space path loss, and fade margin. Other calculators can provide information about cable attenuation and voltage standing wave ratio (VSWR).

Maximum tree growth data Trees are a potential source of obstruction of the Fresnel zone, and unless a tree is fully mature, it will likely grow taller. A chainsaw is not always the answer, and planning antenna height based on potential tree growth might be necessary. The regional or local agricultural government agency should be able to provide you with the necessary information regarding the local foliage and what type of growth you can expect.

Binoculars Visual line of sight can be established with the aid of binoculars. However, please remember that determining RF line of sight means calculating and ensuring Fresnel zone clearance. For links longer than 5 miles or so, this will be almost impossible. A solid understanding of topography and earth bulge is necessary to plan a bridge link.

Walkie-talkies or cellular phones 802.11 bridge links can span many miles. Two site survey engineers working as a team will need some type of device for communicating during the survey.

Signal generator and wattmeter A signal generator is used together with a wattmeter, also known as a Bird meter, to test cabling, connectors, and accessories for signal loss and VSWR. This testing gear is necessary for testing cabling and connectors before deployment. The testing gear should also be used periodically after deployment to check that water and other environmental conditions have not damaged the cabling and connectors. Figure 16.10 depicts a signal generator and a wattmeter.

Variable-loss attenuator A variable-loss attenuator has a dial on it that enables you to adjust the amount of energy that is absorbed. These can be used during an outdoor site survey to simulate different cable lengths or cable losses.

Inclinometer This device is used to determine the height of obstructions. This is crucial when making sure that a link path is clear of obstructions.

GPS Recording the latitude and longitude of the transmit sites and any obstructions or points of interest along the path is important for planning. A GPS can easily provide this information.

FIGURE 16.10 Signal generator and wattmeter

Digital camera You will want to take pictures of mounting locations, cable paths, grounding locations, indoor mounting locations, obstructions, and so on.

Spectrum analyzer This should be used to test ambient RF levels at transmit sites.

High-power spotlight or sunlight reflector In the case of a wireless bridge, you will need to make sure you are surveying in the right direction. As the path gets farther away, the ability to make out a specific rooftop or tower becomes harder and harder. To aid in this task, the use of a high-power (3 million candle or greater) spotlight or a sunlight reflector may be used. Because light travels so well, it can be used to narrow in on the actual remote site and ensure that the survey is conducted in the right direction.

Antennas and access points are not typically used during the bridging site survey. Bridging hardware is rarely installed during the survey because most times a mast or some other type of structure has to be built. If all the bridging measurements and calculations are accurate, the bridge link will work. An outdoor site survey for a mesh network will require mesh APs and antennas.

EXERCISE 16.2

Cable Loss Calculations

On the CD that accompanies this book is a freeware software calculator courtesy of Times Microwave Systems.

In this exercise, you will use this attenuation calculator to see the dB loss per 100 feet for various grades of cabling.

1. From the enclosed CD, copy the file `losscalc.zip` to a temporary directory such as `C:\temp`. Unzip `losscalc.zip` to that same temporary directory.
2. From the temporary directory, double-click the file `TMSAPHC.exe`.
3. Under the Product text box, choose a grade of cable called LMR-1700-DB. In the Frequency text box, enter **2500**, and in the Run Length text box, enter **200** feet. Click the Calculate button. Note that this grade of cabling is rated for 1.7 dB loss per 100 feet.
4. Under the Product text box, choose a lower grade of cable called LMR-400. In the Frequency window, enter **2500**, and in the length window, enter **200** feet. Click the Calculate button.
5. Note that this grade of cabling is rated for 6.76 dB loss per 100 feet.

Coverage Analysis

We have already discussed the many considerations of coverage analysis in an earlier section of this chapter. In the following sections, we cover the three major types of coverage analysis site surveys: manual, assisted, and predictive. We also discuss the software tools that can be used to assist you with all three of these types of coverage analysis surveys. Finally, we discuss dynamic and adaptive WLAN technology.

Manual

The most common method of coverage analysis is the old-fashioned manual site survey. *Manual coverage analysis* involves the techniques described earlier, which are used to find the cell boundaries. There are two major types of manual coverage analysis surveys:

Passive During a *passive manual survey*, the radio card is collecting RF measurements, including received signal strength (dBm), noise level (dBm), and signal-to-noise ratio (dB). The client adapter, however, is not associated to the access point during the survey, and all information is received from radio signals that exist at layer 1.

Active During an *active manual survey*, the radio card is associated to the access point and has layer 2 connectivity, allowing for low-level frame transmissions. If layer 3 connectivity is also established, low-level data traffic such as Internet Control Message Protocol (ICMP) pings are sent in 802.11 data frame transmissions. Layer 1 RF measurements can also be recorded during the active survey. However, upper layer information such as packet loss and layer 2 retransmission percentages can be measured because the client card is associated to a single access point.

The main purpose of the active site survey is to look at the percentage of layer 2 retransmissions. You have learned throughout this book that both RF interference and multipath can cause layer 2 retransmissions. Why would you observe a very high rate of layer 2 retransmissions during the active coverage analysis if you have determined during the spectrum analysis portion of the site survey that there was no RF interference? More than likely the cause of the layer 2 retransmissions is multipath.

Most vendors recommend that both passive and active site surveys be conducted. The information from both manual surveys can then be compared, contrasted, and/or merged into one final coverage analysis report. So what measurement software tools can be used to collect the data required for both passive and active manual surveys? There are numerous freeware site survey utilities, including NetStumbler, which is a freeware utility that is included on the CD that accompanies this book. NetStumbler can be used for a passive coverage analysis survey. Most Wi-Fi vendors' client card utility software at the very least comes with a passive survey tool that can be used to measure received signal strength and SNR. Many vendors' software client utilities will also include active survey capabilities, such as the Cisco client software pictured in Figure 16.11.

FIGURE 16.11 Passive/active site survey utility

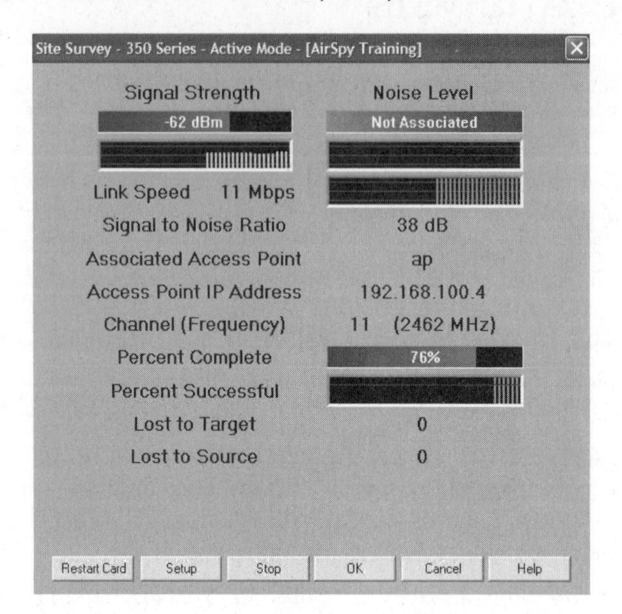

Some handheld devices such as VoWiFi phones or Wi-Fi bar code scanners may have site survey capabilities built into the internal software that runs on the handheld device. A common mistake is that a surveyor will hold the VoWiFi phone in a horizontal position when measuring RF signals during a manual site survey. The internal antenna of the VoWiFi phone is vertically polarized, and holding the phone in a horizontal position will result in misleading signal measurements. Therefore, the phone should be held in a vertical position to match the antenna polarization of the phones and the access points.

Many site survey professionals prefer working with vendors' client card site survey software tools as opposed to the more robust coverage analysis applications. However, commercial RF site survey applications like the one pictured in Figure 16.12 have gained wide acceptance.

Real World Scenario

Can I Perform a Site Survey Using Built-in OS Client Utilities?

The most commonly used software interface for wireless client adapters is the Microsoft Wireless Zero Configuration (WZC) service that is built into Windows XP. The WZC currently does not have received signal strength reporting capabilities, nor does it have any other site survey features. A recommended practice is not to use the wireless client software interface of any operating system (OS) and instead use the more professional site survey software tools.

FIGURE 16.12 Commercial coverage analysis application (AirMagnet Survey © courtesy of AirMagnet, Inc.)

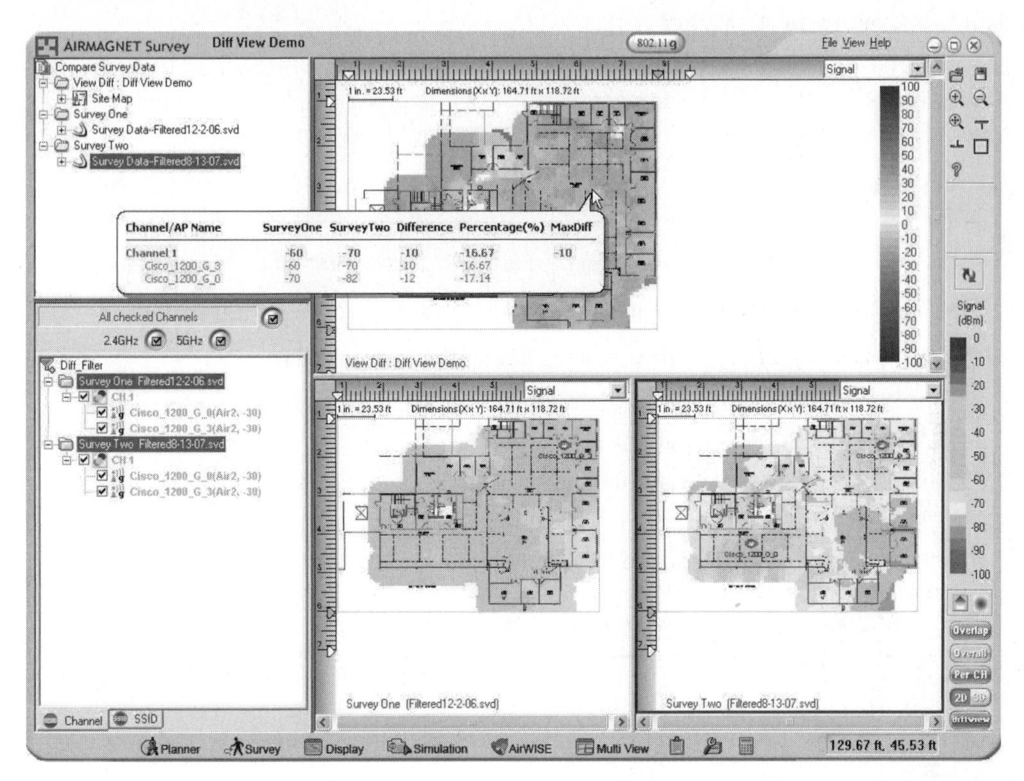

These commercial packages allow the site survey engineer to import a graphic of the building's floor plans into the application. A variety of graphic formats are usually supported, and the floor plan typically must be to scale. The commercial application works with an 802.11 client radio and takes measurements in either a passive manual mode or an active manual mode. The site survey engineer walks through the building capturing the RF information while also recording the location on the graphic of the floor plan that is displayed in the software. The information collected during both active and passive modes can then be merged, and a visual representation of the RF footprints or coverage cells is displayed over the graphic floor plan. These commercial packages can also retain the information, which can then be used for offline modeling so the WLAN design engineer can create some "what-if" scenarios by changing channel and power settings. Commercial site survey applications can also assist in capacity planning in regard to data rates per cell and per VLAN. Floor plans for multiple floors can be loaded into the applications, and 3D coverage analysis is often possible. For outdoor site surveys, GPS capabilities are included to log latitude and longitude coordinates.

EXERCISE 16.3

Manual Coverage Analysis

In this exercise, you will test-drive a commercial site survey software application called AirMagnet Survey to view manual coverage analysis. To properly execute this exercise, please download the software demo from `www.airmagnet.com/cwna/sybex`.

1. After downloading the AirMagnet Survey demo software, double-click on the executable file. Follow all the installation prompts. Install the proper driver for your Wi-Fi card.
2. From the Windows Start menu, under Programs, run the AirMagnet Survey demo.
3. When prompted to open a sample project, click the Yes button.
4. Click the Display button in the bottom menu of the program.
5. In the top-left window of the program, choose between the various surveys such as `PassiveSurvey1.svd`.
6. Observe the colored heat maps that give a graphical representation of RF coverage areas of a previous site survey.
7. Test-drive some of the other features on the bottom and right menu bars. The demo software will also work for 24 hours with your Wi-Fi card. Use the program to take some RF measurements in real time and observe the results.

Assisted

Some WLAN controllers and some centralized wireless network management system (WNMS) applications have the capability to conduct *assisted coverage analysis*. After the installation of access points, a centralized solution such as a wireless network management system (WNMS) or a WLAN controller scans the access point radio cards and collects the RF information, which is then used for visualization of coverage cells and for optimizing AP configurations such as channel and power settings. Most assisted solutions use the information gathered from the access point radio cards, but some solutions can also use a client radio to report information back to the centralized device during a client walk-through of the building.

Assisted site surveys typically are used as a starting point before final deployment and are often used as a calibration or planning stage tool with WLAN controllers. An assisted calibration process configures and reconfigures the access points based on analysis of all the collected RF data, as pictured in Figure 16.13.

Most WLAN controller solutions that have assisted site survey capabilities also go to the next level and offer dynamic radio frequency spectrum management (RFSM), which is discussed later in this chapter. Some system integrators bypass the site survey and install a grid pattern of lightweight access points. After the lightweight access points are installed, a wireless controller or wireless controllers working together use some form of RFSM technology to dynamically adjust power and channel settings. The Wi-Fi controller architecture constantly monitors the environment and makes adjustments as needed. Although assisted site survey features are an excellent starting point prior to deployment, most professionals still recommend a manual site survey for validation.

FIGURE 16.13 Assisted site survey

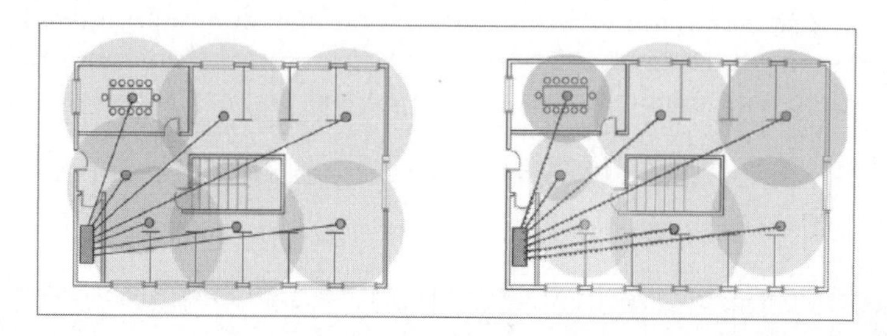

Predictive

The last method of RF coverage analysis uses applications that provide RF simulations and modeling design capabilities. *Predictive coverage analysis* is accomplished using an application that creates visual models of RF coverage cells, bypassing the need for actually capturing RF measurements. Projected cell coverage zones are created using modeling algorithms and attenuation values. One major controller vendor touts these modeling capabilities; however, most predictive coverage analysis tools are stand-alone software design applications.

Blueprints and floor plans often use vector graphic formats (`.dwg`, `.dwf`) and can contain layer information including the type of building materials that are used. Predictive analysis software supports both vector and raster graphics (`.bmp`, `.jpg`, `.tif`), allowing for the import of building floor plans. The WLAN design engineer will indicate in the software what materials are used in the floor plan. The predictive application already has attenuation values for various materials such as drywall, concrete, and glass programmed into the software. The software creates forecast models using the predictive algorithms and the attenuation information. The modeling forecast can include the following:

- Channel reuse patterns
- Coverage cell boundaries
- Access point placement
- Access point power settings
- Number of access points
- Data rates

Virtual access points are created and overlaid on the floor plan graphic. Multiple "what-if" scenarios can be created by changing the power settings, channel settings, or antenna type of the virtual access points, which can also be moved to any location on the floor plan. Predictive capacity analysis software vendors proclaim 85 percent accuracy in their modeling capabilities. Predictive applications are an excellent tool to use with blueprints of buildings that have yet to be built.

It should be noted that entering the initial data for a predictive survey can be very time-consuming. However, many site survey engineers have begun using predictive modeling software to cut down on the time needed for the actual site survey. A forecast model is first created with the predictive application and then the site survey engineer conducts a manual site survey to validate the projections. Predictive coverage analysis software can be a valuable tool, but a manual site survey is still necessary to confirm the coverage simulations.

EXERCISE 16.4

Predictive Coverage Analysis

In this exercise, you will test-drive a commercial site survey and predictive analysis software application from Fluke Networks called InterpretAir WLAN Survey to simulate predictive coverage analysis. A demo of this application is on the CD that accompanies this book.

1. From the `AnalyzeAir` directory of the CD, copy the file `InterpretAir_Demo.exe` to your desktop. Click on the file and follow the default installation prompts.
2. On your desktop, double-click the InterpretAir icon. Click the No button when asked to install a new license.
3. On the menu bar, click the New Map icon, which is the second icon from the left. Click the Select button, browse to the CD, and click on `floorplan.jpg` found in the `Fluke` directory on the CD. Click the Open button and then the OK button.
4. Click the AP icon and then click on the floor plan several times. This places several virtual access points.
5. Click the Visualization tab on the bottom of the screen. Under View, choose several different simulations, including data rate, access point placement, and so on.
6. Click the Wall icon. From the Wall Type option, choose different materials and place them on the floor plan. Notice the changes that take effect in the predictive analysis simulation.
7. Continue to maneuver through the program and familiarize yourself with the program's features and capabilities.

Self-Organizing Wireless LANs

Could the need for manual site surveys ever be eliminated? Most site survey professionals would argue that a manual coverage analysis of some type will always be needed. However, many WLAN vendors argue that dynamic RF technologies exist that eliminate or drastically reduce the need for manual coverage analysis.

Currently, software and hardware solutions already exist that provide *radio frequency spectrum management (RFSM)*, where a centralized device can dynamically change the configuration of thin or fat access points based on accumulated RF information gathered from the access points' radio cards. Based on the accumulated RF information, the centralized device controls the access points and adjusts their power and channel settings, dynamically changing the RF coverage cells. WLAN capacity needs can also be addressed with RFSM, which utilizes dynamic load balancing of clients between the access points. When implemented, RFSM provides automatic cell sizing, automatic monitoring, troubleshooting, and optimization of the RF environment, which can best be described as a self-organizing wireless LAN. Wireless network management systems (WNMSs) can offer RFSM capabilities with autonomous access points. However, this dynamic RF technology is much more commonly seen in a WLAN controller environment with lightweight access points. Some hardware vendors are also using the client radio cards as scouts to collect RF information to be used in a RFSM environment.

Real World Scenario

Is a Site Survey Even Needed If the WLAN Vendor Supports Dynamic RF Capabilities?

In some environments, a dynamic RF solution is not even a recommended or supported solution. RFSM technology has been known to cause problems with VoWiFi deployments as well as problems in high-multipath environments. A static channel reuse plan and static power settings might be a better solution in these environments. Most of the VoWiFi vendors currently will not support any type of dynamic RF deployment.

In recent years, RFSM technology has gained wide acceptance because almost all of the WLAN controller vendors offer some sort of dynamic RFSM solution. Many of the vendors' customers have had excellent success with dynamic RF deployments. Very often, sales representatives of the various WLAN controller vendors claim that a site survey is no longer necessary because of the dynamic and self-organizing nature of their RFSM solution. This is a big mistake. Although dynamic RFSM technology has come a long way in recent years, the authors of this book highly recommend that a site survey always take place prior to deployment. A site survey should always be considered mandatory.

In a WLAN controller environment, the lightweight access points will monitor their respective channels as well as use off-channel scanning capabilities to monitor other frequencies. Any RF information heard by the access points is reported back to the WLAN controller. Based on all the RF monitoring from multiple access points, the WLAN controller will make dynamic changes. Some lightweight access points may be told to change to a different channel while other APs may be told to change their transmit power settings.

As defined under the 802.11h amendment, *transmit power control (TPC)* and *dynamic frequency selection (DFS)* are examples of RFSM technology. Currently the dynamic RFSM technologies used by WLAN controller vendors are proprietary. However, the radio resource management (RRM) mechanisms defined by the recently ratified 802.11k amendment do define some aspects of 802.11 spectrum management. The 802.11v amendment also proposes load balancing standards.

Summary

In this chapter, you learned the mandatory and optional aspects of a wireless site survey. Spectrum and coverage analysis surveys are always mandatory, whereas application and throughput testing are considered optional. We discussed the importance of locating potential sources of interference by using a spectrum analyzer, and we defined all the steps necessary to conduct both a manual and passive coverage analysis site survey. This chapter also provided a discourse of all the tools necessary for either an indoor or outdoor site survey. We covered the three major types of coverage analysis as well as self-organizing WLAN technology. Conducting a well-defined and thorough wireless site survey will lay the foundation needed for proper WLAN design and WLAN management.

Exam Essentials

Define spectrum, coverage, and application analysis. Understand why both spectrum and coverage analysis are considered mandatory and application analysis is usually optional.

Identify sources of WLAN interference. Describe all of the various devices that are potential sources of interference in both the 2.4 GHz ISM and the 5 GHz UNII bands.

Explain RF measurements. Be able to explain the procedure used while conducting coverage analysis and the different types of RF measurements recorded, including received signal strength and signal-to-noise ratio.

Understand AP placement and configuration. Explain how AP placement, power, and channel settings are part of coverage analysis.

Identify all site survey tools. Understand the difference between an outdoor and indoor site survey, and identify all the necessary tools.

Explain the three major types of coverage analysis. Describe the differences between manual, assisted, and predictive site surveys, and explain self-organizing WLAN technology.

Key Terms

Before you take the exam, be certain you are familiar with the following terms:

active manual survey

application analysis

assisted coverage analysis

coverage analysis

dynamic frequency selection (DFS)

manual coverage analysis

passive manual survey

predictive coverage analysis

radio frequency spectrum management (RFSM)

received signal strength

spectrum analysis

spectrum analyzer

transmit power control (TPC)

Review Questions

1. The Crocker company has generated a visual model of RF coverage for their building by using predictive modeling site survey software. The next step requires validation with a manual site survey. What modeling parameters should be validated during the manual site survey? (Choose all that apply.)

   A. AP placement and power settings

   B. Throughput

   C. Coverage boundaries

   D. Encryption settings

   E. Roaming parameters

2. Which potential regional weather conditions can adversely affect an outdoor wireless bridge link and should be noted during an outdoor site survey? (Choose all that apply.)

   A. Lightning

   B. Dew point

   C. Wind

   D. Cloud cover

   E. Thunder

3. Name the major types of coverage analysis site surveys. (Choose all that apply.)

   A. Assisted

   B. Self-organizing

   C. Manual

   D. Capacity

   E. Predictive

4. ACME Hospital uses a connection-oriented telemetry monitoring system in the cardiac care unit. Management wants the application available over a WLAN. Uptime is very important because of the critical nature of the monitoring system. What should the site survey engineer be looking for that might cause a loss of communication over the WLAN? (Choose all that apply.)

   A. Medical equipment interference

   B. Safety glass

   C. Patients

   D. Bedpans

   E. Elevator shafts

5. Which type of coverage analysis requires a radio card to be associated to an access point?
   A. Associated
   B. Passive
   C. Predictive
   D. Assisted
   E. Active

6. Which of the following tools can be used in an indoor site survey? (Choose all that apply.)
   A. Measuring wheel
   B. GPS
   C. Ladder
   D. Battery pack
   E. Microwave oven

7. Which of the following tools might be used in an outdoor site survey used to provide outdoor coverage? (Choose all that apply.)
   A. Spectrum analyzer
   B. NEMA grade access point
   C. Outdoor blueprints or topography map
   D. Mesh routers
   E. GPS

8. Name potential sources of interference in the 5 GHz UNII band. (Choose all that apply.)
   A. Microwave oven
   B. Cordless phones
   C. FM radios
   D. Radar
   E. Nearby 802.11b/g WLAN

9. Which of these measurements are taken during a passive manual site survey? (Choose all that apply.)
   A. SNR
   B. dBi
   C. dBm
   D. dBd

10. Mr. Cherry is the site survey engineer for the ACME Corporation, which wants to deploy a VoWiFi network. Because of the latency requirements of VoIP, he is concerned about layer 2 retransmissions. A spectrum analysis site survey has determined that there are no sources of RF interference. During the active survey of manual coverage, he finds a layer 2 retry rate exceeding 15 percent in some areas of the facility but not others. What is the likely cause of the layer 2 retransmissions?

    A. Cellular phone system

    B. Free space path loss

    C. Brick walls

    D. Multipath

    E. High density of users

11. Name the necessary calculations for an outdoor bridging survey under 5 miles? (Choose all that apply.)

    A. Link budget

    B. Free space path loss

    C. Fresnel zone

    D. Fade margin

    E. Height of the antenna beamwidth

12. Name potential sources of interference that might be found during a 2.4 GHz site survey. (Choose all that apply.)

    A. Toaster oven

    B. Nearby 802.11 FHSS access point

    C. Plasma cutter

    D. Bluetooth headset

    E. 2.4 GHz video camera

13. Which of the following tools can be used in an indoor 802.11 site survey? (Choose all that apply.)

    A. Multiple antennas

    B. 902 to 928 MHz spectrum analyzer

    C. Client adapter

    D. Access point

    E. Floor plan map

14. Ms. Williams is a site survey engineer who is planning to deploy a wireless controller solution with dual-radio, dual-frequency lightweight access points. The employees will be assigned to the 5 GHz network, and the guest users will be assigned to the 2.4 GHz network. CCMP/AES encryption will be required for the employees, while the guest users will use only static WEP. Name the possible choices that Ms. Williams has for coverage analysis. (Choose the two best answers.)

A. Conduct a predictive site survey for the 5 GHz network and an assisted site survey for the 2.4 GHz network.

B. Install the thin access points in a grid and conduct an assisted site survey for both networks.

C. Conduct manual coverage analysis for the 2.4 GHz network first and then conduct manual coverage analysis for the 5 GHz network.

D. Conduct a predictive site survey for the 2.4 GHz network and an assisted site survey for the 5 GHz network.

E. Conduct manual coverage analysis for the 5 GHz network first and then conduct manual coverage analysis for the 2.4 GHz network based on the AP placement of the 5 GHz APs.

15. Bob has to perform a site survey for a WLAN by using a multiple channel architecture (MCA) system in a 20-story building with multiple tenants. What should Bob consider during the planning and implementation stages of the site survey? (Choose all that apply.)

A. Other tenants' WLANs.

B. Only WLAN controller solutions with lightweight APs should be deployed and not autonomous access points.

C. Access points should use high-gain omnidirectional antennas to provide coverage across multiple floors.

D. Access points should be at full transmit power to provide coverage across multiple floors.

E. The cell coverage of each access point should extend to only one floor above and one floor below to create a three-dimensional channel reuse pattern.

16. Which of the following tools may be found within an indoor site survey kit? (Choose all that apply.)

A. Digital camera

B. Colored electrical tape

C. Grid antenna

D. Access point enclosure unit

E. Temporary mounting gear

17. Ms. Turner is a site survey engineer who has to deploy eighty 2.4 GHz lightweight access points in a warehouse with long corridors and metal racks. A WLAN controller will be used to manage all the access points. Six hundred bar code scanners will be deployed throughout the warehouse using WPA2-Personal for security. Which is the most important site survey tool to ensure the best performance throughout the warehouse?

    A. 802.11a/b/g multistation emulator
    B. Directional antennas
    C. Predictive analysis software
    D. Security analysis software
    E. All of the above

18. What access point settings should be recorded during manual coverage analysis? (Choose all that apply.)

    A. Power settings
    B. Encryption settings
    C. Authentication settings
    D. Channel setting
    E. IP address

19. Which type of manual coverage analysis does not require a radio card to be associated to an access point?

    A. Associated
    B. Passive
    C. Predictive
    D. Assisted
    E. Active

20. Which type of site survey uses modeling algorithms and attenuation values to create visual models of RF coverage cells?

    A. Associated
    B. Passive
    C. Predictive
    D. Assisted
    E. Active

Answers to Review Questions

1. A, C. First a forecast model is created with the predictive software and then the site survey engineer conducts a manual site survey to validate the projections. Modeling forecasts that can be validated include channel reuse patterns, coverage cell boundaries, access point placement, access point power settings, number of access points, and data rates.

2. A, C. Lightning can cause damage to Wi-Fi bridging equipment and the network infrastructure equipment that resides behind the 802.11 bridges. Strong winds can cause instability between long-distance bridge links and a loss of RF line of sight. Potential weather conditions should be noted during the outdoor site survey. Proper protection against lightning, such as lightning arrestors and/or copper-fiber transceivers, must be recommended for deployment. In high-wind areas, consider the use of grid antennas. Dew point, cloud cover, and thunder have no effect on an 802.11 outdoor deployment and therefore need not be considered during a site survey.

3. A, C, E. Manual site surveys are usually conducted for coverage analysis using a signal strength measurement tool. After the installation of access points, a centralized solution such as a WLAN controller can be used for an assisted coverage survey. Predictive analysis tools can create a model of RF coverage cells.

4. A, B, E. Any type of RF interference could cause a denial of service to the WLAN. A spectrum analysis survey should be performed to determine if any of the hospital's medical equipment will cause interference in the 2.4 GHz ISM band or the 5 GHz UNII bands. Dead zones or loss of coverage can also disrupt WLAN communications. Most hospitals use metal mesh safety glass in many areas. The metal mesh will cause scattering and potentially create lost coverage on the opposite side of the glass. Elevator shafts are made of metal and often are dead zones if not properly covered with an RF signal.

5. E. During an active manual survey, the radio card is associated to the access point and has upper layer connectivity, allowing for low-level frame transmissions while RF measurements are also taken. The main purpose of the active site survey is to look at the percentage of layer 2 retransmissions.

6. A, C, D. A measuring wheel can be used to measure the distance from the wiring closet to the proposed access point location. A ladder or forklift might be needed when temporarily mounting an access point. Battery packs are used to power the access point. GPS devices are used outdoors and do not properly work indoors. Microwave ovens are sources of interference.

7. A, B, C, D, E. Outdoor site surveys are usually wireless bridge surveys; however, outdoor access points and mesh routers can also be deployed. Outdoor site surveys are conducted using either outdoor access points or mesh routers, which are the devices typically used to provide access for client stations in an outdoor environment. These outdoor Wi-Fi surveys will use most of the same tools as an indoor site survey but may also use a Global Positioning System (GPS) device to record latitude and longitude coordinates.

8. B, D. Cordless phones that operate in the same space as the 5GHz UNII bands may cause interference. Radar is also a potential source of interference at 5 GHz. Microwave ovens and 802.11b/g WLANs transmit in the 2.4 GHz ISM band. FM radios use narrowband transmissions in a lower-frequency licensed band.

9. A, C. During a passive manual survey, the radio card is collecting RF measurements, including received signal strength (dBm), noise level (dBm), and signal-to-noise ratio (dB). The SNR is a measurement of the difference in decibels (dB) between the received signal and the background noise. Received signal strength is an absolute measured in dBm. Antenna manufacturers predetermine gain using either dBi or dBd values.

10. D. If it has been determined during the spectrum analysis portion of the site survey that there was no RF interference, the most likely cause of the layer 2 retransmissions is multipath. Because VoWiFi is highly susceptible to layer 2 retransmissions, unidirectional antennas should be deployed wherever necessary to minimize reflections.

11. A, B, C, D. Outdoor bridging site surveys require many calculations that are not necessary during an indoor survey. Calculations for a link budget, FSPL, Fresnel zone clearance, and fade margin are all necessary for any bridge link.

12. B, C, D, E. Spectrum analysis for an 802.11b/g site survey should scan the 2.4 GHz ISM band. Bluetooth radios, plasma cutters, 2.4 GHz video cameras, and legacy 802.11 FHSS access points are all potential interfering devices.

13. A, C, D, E. Every indoor wireless site survey should use at least one access point and multiple antennas. A client radio card will be needed for coverage analysis as well as a floor plan to record measurements. A spectrum analyzer is needed that sweeps the 2.4 GHz ISM band and 5 GHz UNII bands.

14. B, E. Most wireless controllers have radio frequency spectrum management (RFSM) capabilities. Although not usually recommended, it is possible to bypass the manual site survey, deploy the access points in a grid, and perform an assisted survey, allowing the controller to automatically adjust the power and channel settings of the thin access points. If the survey was performed manually, the 5 GHz coverage analysis should be done first because of shorter range due to the smaller size 5 GHz wavelength. When performing a site survey for dual-radio access points, perform the initial site survey for the radios that provide the smallest coverage area, in this case the higher-frequency 5 GHz radios. The 2.4 GHz radios that provide the larger coverage area should be able to use the same access point location at a lower power setting to provide a similar coverage area as the 5 GHz radios. It may also be necessary to turn off some of the 2.4 GHz radios.

15. A, E. The number one source of RF interference in a multitenant environment is other WLANs. The odds are that most neighboring businesses will have deployed 2.4 GHz WLANs, and special consideration should be given to deploying a 5 GHz WLAN. Because RF propagates in all directions, it is necessary to always think three-dimensionally when designing a channel reuse pattern.

16. A, B, E. Temporary access point mounting gear is a necessity. A digital camera and colored electrical tape may also be used to record the locations of AP placement. Grid antennas are used outdoors for long-distance bridge links. An access point enclosure unit is used for permanent mounting.

17. B. Multipath is the biggest concern in the warehouse, and directional antennas will be needed for the survey. Reflections down the long corridors and metal racks will create multipath performance issues that can best be addressed by using a directional antenna.

18. A, D. Wherever an access point is placed during a site survey, the power and channel settings should be noted. Security settings and IP address are not necessary.

19. B. During a passive manual survey, the radio card is collecting RF measurements, including received signal strength (dBm), noise level (dBm), signal-to-noise ratio (dB), and bandwidth data rates. The client adapter, however, is not associated to the access point during a passive survey.

20. C. Predictive coverage analysis is accomplished using software that creates visual models of RF coverage cells, bypassing the need for actually capturing RF measurements. Projected cell coverage zones are created using modeling algorithms and attenuation values.

Power over Ethernet (PoE)

IN THIS CHAPTER, YOU WILL LEARN ABOUT THE FOLLOWING:

✓ **History of PoE**

- Nonstandard PoE
- IEEE 802.3af
- IEEE Std. 802.3-2005, Clause 33

✓ **PoE devices (overview)**

- Powered device (PD)
- Power-sourcing equipment (PSE)
- Endpoint PSE
- Midspan PSE
- Power-sourcing equipment pin assignments

✓ **Planning and deploying PoE**

- Power planning
- Redundancy

Before we begin this chapter, we need to explain what *Power over Ethernet (PoE)* is. Over the years, computer networking typically entailed connecting a stationary, electrically powered computer system to a wired network. The computers were typically anything from desktop PCs to servers and mainframes. As is typical with technology, larger computers gave way to smaller computers, and laptop and portable devices began to appear. Eventually, some of the networking devices became small enough, both physically and electronically, that it became possible and practical not only to use the Ethernet cable to transmit data to the device, but also to send the electricity necessary to power the device.

History of PoE

The concept of providing power from the network dates back to the birth of the telephone, which to this day still receives power from the telephone network. Computer networking devices that are often powered with PoE are desktop Voice over IP (VoIP) phones, cameras, and access points. Ethernet cables consist of four pairs of wires; typically two pairs are used for transmitting and receiving data and the other two pairs are unused. By providing power to devices via the same Ethernet cable that provides the data, a single low-voltage cable is all that is required to install a networked PoE device. The use of PoE devices alleviates the need to run electrical cables and outlets to every location that needs to be connected to the network. Not only does this greatly reduce the cost of installing network devices, it also increases flexibility in terms of where these devices can be installed and mounted. Moving devices is also easier, because all that is required at the new location is an Ethernet cable.

Nonstandard PoE

As with most new technologies, the initial PoE products were proprietary solutions created by individual companies that recognized the need for the technology. The IEEE process to create a PoE standard began in 1999; however, it would take about four years before the standard became a reality. In the meantime, vendor proprietary PoE continued to proliferate. Proprietary PoE solutions often used different voltages, and mixing proprietary solutions could result in damaged equipment.

IEEE 802.3af

The *IEEE 802.3af* Power over Ethernet committee created the PoE amendment to the 802.3 standard. It was officially referred to as IEEE 802.3 "Amendment: Data Terminal Equipment (DTE) Power via Media Dependent Interface." This amendment to the IEEE 802.3 standard was approved on June 12, 2003, and defined how to provide PoE to 10BaseT (Ethernet), 100BaseT (Fast Ethernet), and 1000BaseT (Gigabit Ethernet) devices.

IEEE Std. 802.3-2005, Clause 33

In June 2005, the IEEE revised the 802.3 standard, creating IEEE Std. 802.3-2005. 802.3af was one of four amendments that were incorporated into the revised standard. The four amendments were 802.3ae—10 Gigabit Ethernet, 802.3af—Power over Ethernet, 802.3ah—Ethernet in the First Mile, and 802.3ak—10GBase-CX4. In the revised 802.3 standard, Clause 33 is the section of the standard that specifically defines Power over Ethernet. Most people still refer to Power over Ethernet as 802.3af, but because this amendment has become part of the revised standard, it is now officially known as *IEEE Std. 802.3-2005, Clause 33*.

IEEE 802.3at

The IEEE 802.3at amendment had not yet been ratified at the time we were writing this book. 802.3at is also known as PoE+, since it is extending the capabilities of PoE. Two of the main objectives of the 802.3at Task Group are to be able to provide more power to powered devices and to maintain backward compatibility with Clause 33 devices. As access points become faster and incorporate newer technologies such as MIMO, they are requiring more power to operate. Switches and controllers that incorporate 802.3at technology should be able to provide power to legacy access points as well as newer access points that may require more power. Dual-frequency 802.11n access points with multiple radio chains will likely require greater than 15.4 watts of power. Some 802.11n enterprise vendors with 802.11n APs are currently using proprietary PoE methods, including some that provide more than 15.4 watts. The IEEE 802.3at amendment will hopefully soon standardize ways to provide as much as 30 watts of power to devices.

PoE Devices (Overview)

The PoE standard defines two types of PoE devices: powered devices and power-sourcing equipment. These devices communicate with each other, providing the PoE infrastructure.

Powered Device (PD)

The *powered device (PD)* either requests or draws power from the power-sourcing equipment. Powered devices must be capable of accepting up to 57 volts of power from either the data lines or the unused pairs of the Ethernet cable. The powered device must also be able to accept power with either polarity from the power supply in what is known as mode A or mode B, as seen in Table 17.1.

TABLE 17.1 PD Pinout

Conductor	Mode A	Mode B
1	Positive voltage, negative voltage	
2	Positive voltage, negative voltage	
3	Negative voltage, positive voltage	
4		Positive voltage, negative voltage
5		Positive voltage, negative voltage
6	Negative voltage, positive voltage	
7		Negative voltage, positive voltage
8		Negative voltage, positive voltage

The powered device must reply to the power-sourcing equipment with a *detection signature*, notifying the power-sourcing equipment whether it is in a state in which it will accept power or will not accept power. The detection signature is also used to indicate that the PD is compliant with 802.3-2005, Clause 33. If the device is determined not to be compliant, power to the device will be withheld. If the device is in a state in which it will accept power, it can optionally provide a *classification signature.* This classification signature lets the power-sourcing equipment know how much power the device will need. When a PD is first connected to a PSE, it presents itself as a nominal 25k-ohm (25,000-ohm) resistance. The PSE applies a 10.1-volt (V) current while measuring the data circuit to try to identify the detection signature. After the PD has been determined to be compliant, the PSE will increase the voltage up to 20.5 V and measure the resulting current to determine the class of the device.

Table 17.2 shows the current values used to identify the different classification signatures. If none of these current values are measured, the device is considered to be a Class 0 device. If the device is not identified, the PSE does not know how much power the device needs; therefore, it allocates the maximum power. If the device is classified, the PSE needs to allocate only the amount of power needed by the PD, thus providing better power management. Proper classification of the devices can lead to a reduction in power usage and can also enable you to connect more devices to a single PoE switch or controller.

TABLE 17.2 PD Classification Signature Measured Electrical Current Values

Parameter	Conditions	Minimum	Maximum	Unit
Class 0	14.5 V to 20.5 V	0	4	milliampere (mA)
Class 1	14.5 V to 20.5 V	9	12	mA
Class 2	14.5 V to 20.5 V	17	20	mA
Class 3	14.5 V to 20.5 V	26	30	mA
Class 4	14.5 V to 20.5 V	36	44	mA

Some vendors use proprietary techniques to perform classification. Although these techniques are good from the power-management and consumption perspective, they are proprietary and will not work with other manufacturers' products.

Table 17.3 shows the classes of PoE devices and the range of maximum power that they use.

TABLE 17.3 PD Power Classification and Usage

Class	Usage	Range of Maximum Power Used
0	Default	0.44 W to 12.95 W
1	Optional	0.44 W to 3.84 W
2	Optional	3.84 W to 6.49 W
3	Optional	6.49 W to 12.95 W
4	Not allowed	Reserved for future use

Power-Sourcing Equipment (PSE)

The *power-sourcing equipment (PSE)* provides power to the powered device. The power supplied is at a nominal 48 volts (44 to 57 volts). The power-sourcing equipment searches for powered devices by using a direct current (DC) detection signal. After a PoE-compliant device is identified, the power-sourcing equipment will provide power to that device. If a device does not respond to the detection signature, the PSE will withhold power. This prevents noncompliant PD equipment from becoming damaged.

As you can see in Table 17.4, the amount of power provided by the PSE is greater than what is used by the PD (Table 17.3). This is because the PSE needs to account for the worst-case scenario, in which there may be power loss due to the cables and connectors between the PSE and the PD. The maximum draw of any powered device is 12.95 watts. The power-sourcing equipment can also classify the powered device if the powered device provided a classification signature. Once connected, the power-sourcing equipment continuously checks the connection status of the powered device along with monitoring for other electrical conditions such as short circuits. When power is no longer required, the power-sourcing equipment will stop providing it. Power-sourcing equipment is divided into two types of equipment: endpoint and midspan.

TABLE 17.4 PSE Power

Class	Usage	Minimum Power from the PSE
0	Default	15.4 W
1	Optional	4.0 W
2	Optional	7.0 W
3	Optional	15.4 W
4	Reserved for future use	Treat as Class 0

Endpoint PSE

Endpoint power-sourcing equipment provides power and Ethernet data signals from the same device. Endpoint devices are typically PoE-enabled Ethernet switches or specialty devices such as WLAN controllers, as seen in Figure 17.1.

FIGURE 17.1 An Aruba 6000 wireless controller with PoE line card

Endpoint equipment can provide power by using two methods referred to as Alternative A and Alternative B. With *Alternative A*, the PSE places power on the data pair, as seen in Figure 17.2. With *Alternative B*, the PSE places power on the spare pair, as seen in

Figure 17.3. Endpoint power-sourcing equipment is compatible with 10BaseT (Ethernet), 100BaseTX (Fast Ethernet), and 1000BaseT (Gigabit Ethernet). 1000BaseT (Gigabit Ethernet) devices can receive PoE from only endpoint devices.

FIGURE 17.2 Endpoint PSE, Alternative A

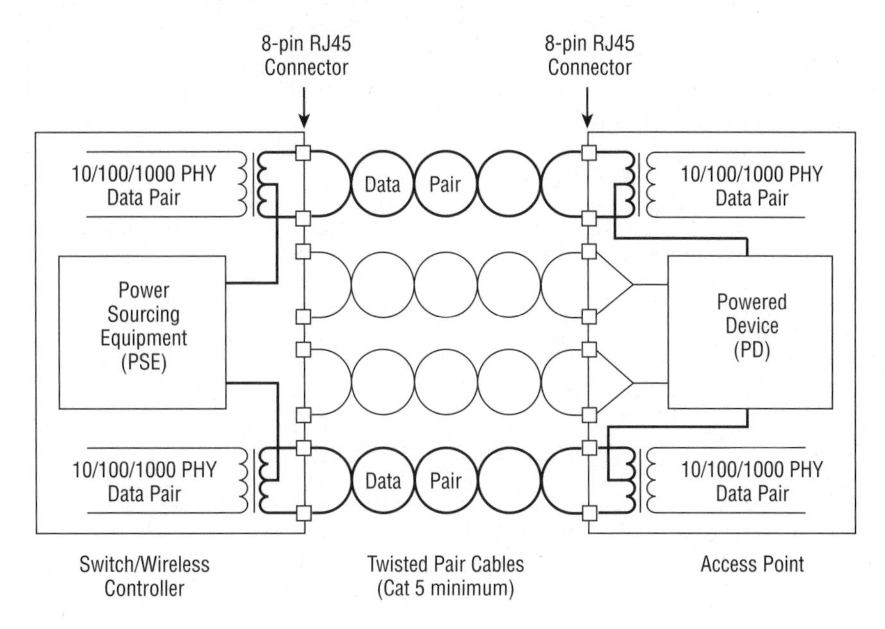

FIGURE 17.3 Endpoint PSE, Alternative B

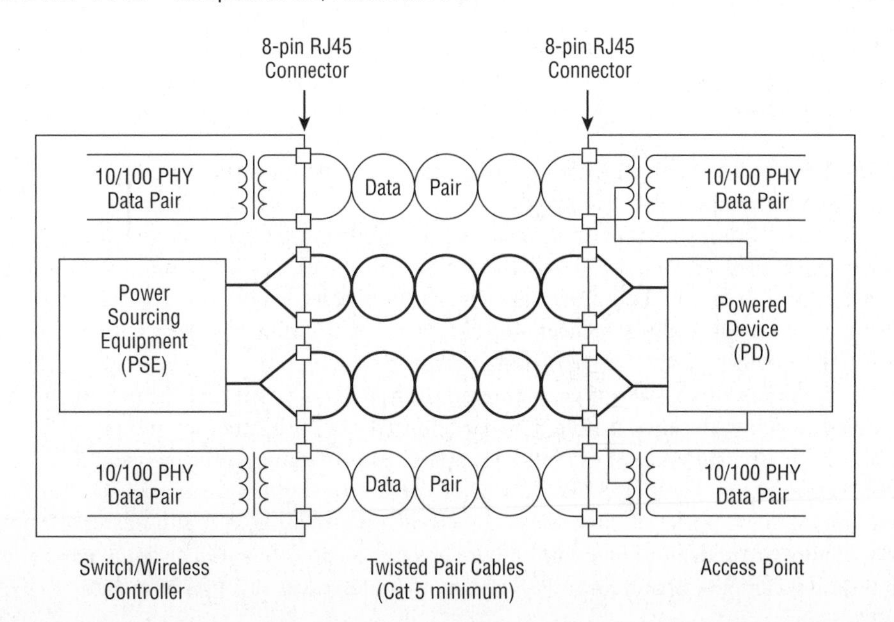

Midspan PSE

Midspan power-sourcing equipment acts as a pass-through device, adding power to an Ethernet segment. Midspan equipment enables you to provide PoE to existing networks without having to replace the existing Ethernet switches or WLAN controllers. Midspan power-sourcing equipment is placed between an Ethernet source (such as an Ethernet switch) and a powered device. The midspan power-sourcing equipment acts as an Ethernet repeater, while adding power to the spare pairs of the Ethernet cable, using Alternative B, as shown in Figure 17.4.

FIGURE 17.4 Midspan PSE, Alternative B

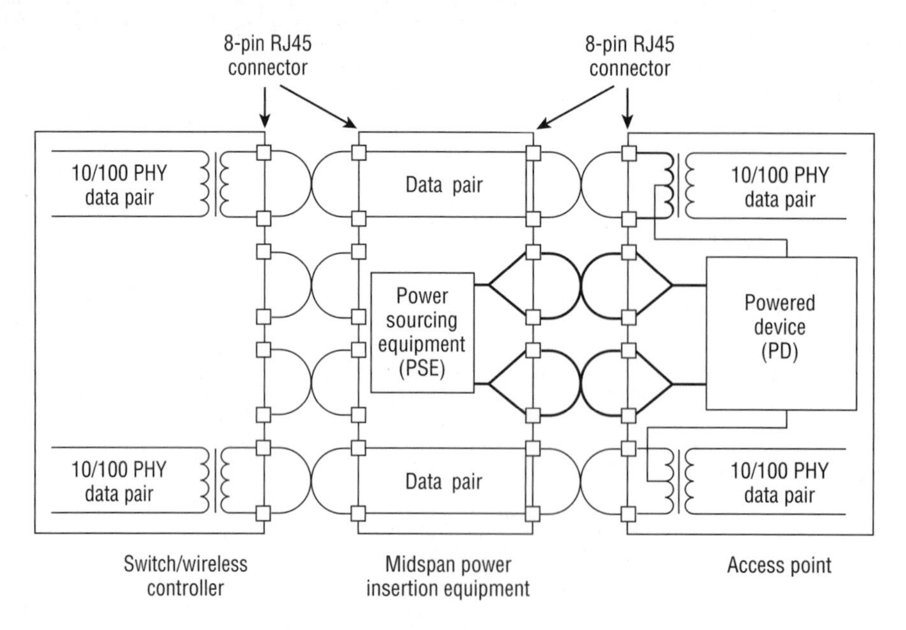

Figure 17.5 shows two single-port midspan devices sitting on top of two multiport devices. The midspan power-sourcing equipment is commonly known as a *power injector* (single-port device) or a PoE hub (multiport device). Midspan power-sourcing equipment only works with 10BaseT and 100BaseT Ethernet. 1000BaseT devices cannot be powered by midspan power-sourcing equipment because all four pairs (eight conductors) are used to transmit data, whereas midspan power-sourcing equipment will only work with spare pairs of cable, only providing power by using Alternative B.

Figure 17.6 shows three ways of providing power to a PD. Option 1 illustrates an endpoint PoE-enabled switch with inline power. This switch provides both Ethernet and power to the access point. The switch (endpoint) would be capable of providing power by using Alternative A or Alternative B. The switch would also be capable of supporting up to 1000BaseT devices. Option 2 and option 3 illustrate two methods of providing midspan power. Option 2 shows a multiport midspan PSE commonly referred to as an *inline power patch panel*, and option 3 shows a single-port midspan PSE commonly referred to as a *power injector.*

FIGURE 17.5 Midspan power-sourcing equipment

FIGURE 17.6 Three PSE solutions

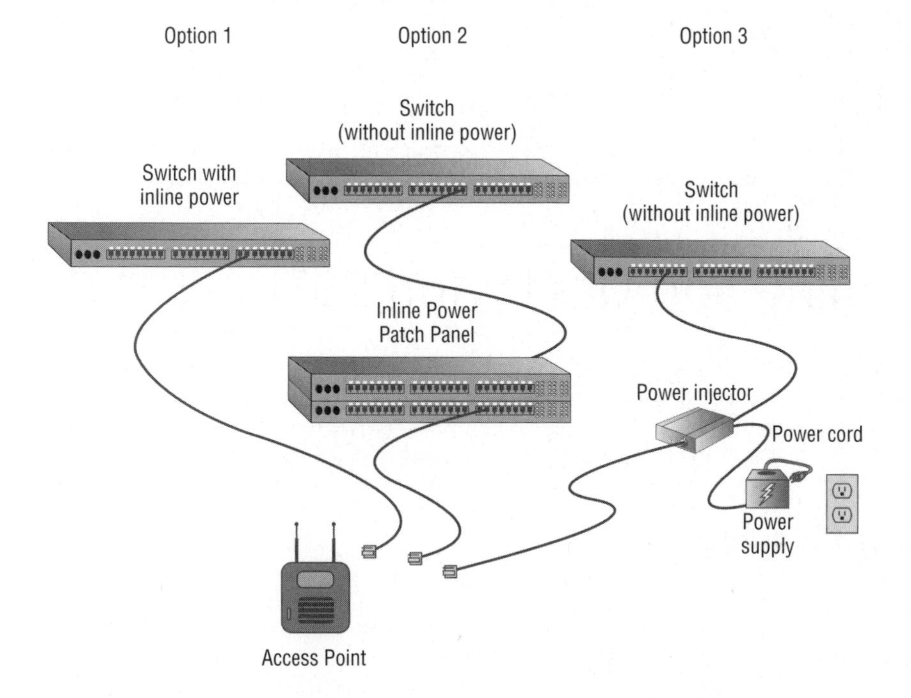

Power-Sourcing Equipment Pin Assignments

There are two valid four-wire pin connections that are used to provide Power over Ethernet. In each of these configurations, the two pairs of conductors carry the same nominal current in both magnitude and polarity. When powering a device by using Alternative A, the positive voltage is matched to the transmit pair of the PSE. If a PSE is configured to automatically configure MDI/MDI-X (also called Auto MDI-X, or automatic crossover), the port may choose either Alternative A polarity choice, as seen in Table 17.5.

TABLE 17.5 PSE Pinout Alternatives

Conductor	Alternative A (MDI)	Alternative A (MDI-X)	Alternative B (All)
1	Positive voltage	Negative voltage	
2	Positive voltage	Negative voltage	
3	Negative voltage	Positive voltage	
4			Positive voltage
5			Positive voltage
6	Negative voltage	Positive voltage	
7			Negative voltage
8			Negative voltage

Planning and Deploying PoE

In the past, when non-PoE desktop VoIP telephones and access points were connected to the network, each device needed to be individually plugged in to a power outlet. These outlets were spread around the building or campus, distributing the power needs. PoE now consolidates the power source to the wiring closet or data center.

Power Planning

Instead of distributing the power for hundreds or thousands of devices, the power for these devices is now being sourced from either a single or a limited number of locations. At maximum power for a powered device, the power-sourcing equipment must be capable of providing 15.4 watts of power to each PoE device. This means that a typical PoE-enabled 24-port Ethernet switch must be able to provide about 370 watts of power to provide PoE to all 24 ports (15.4 watts × 24 ports = 369.6 watts). This does not include the amount of power necessary for the switch to perform its networking duties. A simple way of determining whether the power supply of the switch is powerful enough is to determine the size of the power supply for the equivalent non-PoE switch and add 15.4 watts for each PoE device that you will be connecting to the switch.

The maximum power a 110-volt power supply is capable of providing is 3,300 watts (110 volts × 30 amperes). A typical U.S. wire closet is supplied with 1,650 W (110 V × 15 A). Enterprise-grade PoE-enabled switches often consist of multiple 48-port line cards housed in a chassis. The chassis itself may require 1,000 to 2,000 watts. If the 48-port line cards draw 15.4 watts per port, a total power draw of 740 watts would be required. Depending on the power requirements of the chassis, 3,300 watts would be able to power only the chassis and two to three fully populated 48-port line cards.

Because many devices such as 802.11 access points, video cameras, and desktop VoIP phones may require power, situations often arise where there simply is not enough available wattage to power all the PoE ports. Network engineers have begun to realize the need and importance for a *power budget*. Careful planning is now often needed to ensure that enough power is available for all the powered devices. Powered devices that are capable of classification greatly assist in conserving energy and subtracting less power from the power budget. A device that needs to draw 3 watts and is not capable of providing a classification signature would be classified as Class 0 by default and subtract 15.4 watts from the power budget. Effectively, 12 watts of power would be wasted. If that same device was capable of providing a classification signature and was classified as a Class 1 device, only 4 watts would be subtracted from the power budget. Classification of powered devices will grow in importance as the need for 802.11 PoE deployments grows.

Because of the demand for PoE-enabled devices, many switch manufacturers are starting to replace their 110 V / 30 A power supplies with 220 V / 20 A power supplies. The manufacturers are also putting larger power supplies in their switches to handle the additional requirements of PoE. Some PoE switches support power supplies as large as 9,000 watts. As the demand for PoE devices increases, the need to manage and troubleshoot PoE problems will also increase. Test equipment like that shown in Figure 17.7 can be placed between the PSE and the PD to troubleshoot PoE link issues.

The more PoE devices that you add to the network, the more you are concentrating the power requirements in the data center or wiring closet. As your power needs increase, electrical circuits supplying power to the PoE switches may need to be increased. Also, as the power increases, you increase the amount of heat that is generated in the wiring closets, often requiring more climate-control equipment. When you are using high-wattage power supplies, it is recommended that redundant power supplies be used.

Redundancy

As a child, it was always nice to know that even when there was an electrical failure, the telephone still worked and provided the ability to call someone. As an adult, it is reassuring that in an emergency the telephone still works, even during a power failure. This is a level of service that we have come to expect. As VoIP and VoWiFi telephones begin to replace traditional telephone systems, it is important to still provide this same level of continuous service. To achieve this, you should make sure that all of your PoE PSE equipment is connected to uninterruptible power sources. Additionally, it may be important enough to provide dual Ethernet connections to your PoE PD equipment.

FIGURE 17.7 Fluke NetTool Series II inline network tester

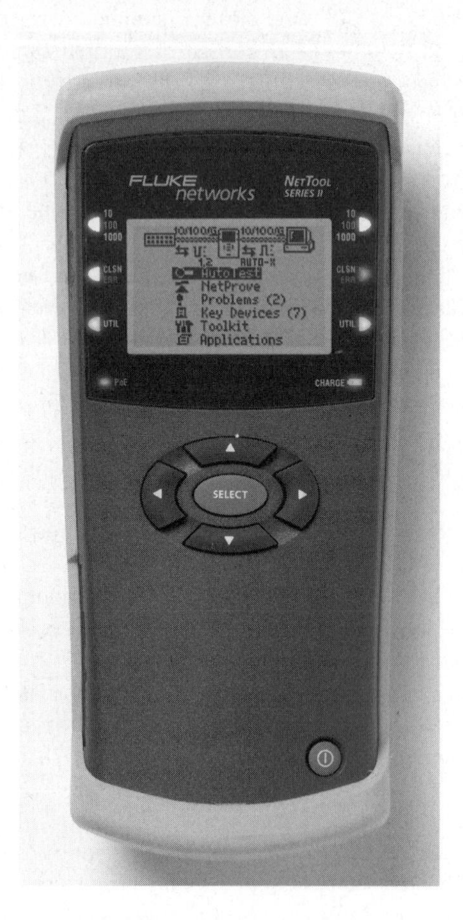

Summary

This chapter focused on Power over Ethernet and the equipment and techniques necessary to provide service to powered devices. Power over Ethernet can be provided in two general ways: through proprietary PoE or through standards-based PoE (802.3af, later referred to as IEEE Std. 802.3-2005, Clause 33).

Standards-based PoE consists of a few key components:

- Powered device (PD)
- Power-sourcing equipment (PSE)
- Endpoint PSE
- Midspan PSE

These components work together to provide a functioning PoE environment.

The final section of this chapter covered considerations that need to be made when planning and deploying PoE:

- Power planning
- Redundancy

Exam Essentials

Know the history of PoE. Make sure you know the history of PoE, the original 802.3af amendment, and current references to IEEE Std. 802.3-2005, Clause 33.

Know the different PoE devices and how they interoperate. Make sure you know the different PoE devices and their roles in providing PoE. Make sure you understand how the following devices work: powered device (PD), power-sourcing equipment (PSE), end-point PSE, and midspan PSE.

Know the different device classes and the classification process. Make sure you know the five device classes and how the classification process works to determine the class of a powered device. Know how much current each class of devices uses along with how much power the PSE generates for each class of devices.

Key Terms

Before you take the exam, be certain you are familiar with the following terms:

Alternative A
Alternative B
classification signature
detection signature
endpoint PSE
IEEE 802.3af
IEEE Std. 802.3-2005, Clause 33
midspan PSE
power budget
power injector
Power over Ethernet (PoE)
powered device (PD)
power-sourcing equipment (PSE)

Review Questions

1. The IEEE 802.3af amendment has been incorporated into the IEEE Std. 802.3-2005 revised standard and is defined in which clause?
   - A. Clause 15
   - B. Clause 17
   - C. Clause 19
   - D. Clause 33
   - E. Clause 43

2. If a classification signature is not provided, the device is considered to be in what class?
   - A. 0
   - B. 1
   - C. 2
   - D. 3
   - E. 4

3. Which types of PoE devices are defined by the standard? (Choose all that apply.)
   - A. PSE
   - B. PPE
   - C. PD
   - D. PT

4. A powered device (PD) must be capable of accepting up to how many volts from either the data lines or the unused pairs of the Ethernet cable?
   - A. 14.5 volts
   - B. 20.5 volts
   - C. 48 volts
   - D. 57 volts

5. To qualify as compliant with 802.3-2005, Clause 33, a powered device (PD) must do which of the following? (Choose all that apply.)
   - A. Be able to accept power in either of two ways: over the data pairs or over the unused pairs.
   - B. Reply to the PSE with a detection signature.
   - C. Accept power with either polarity from the PSE.
   - D. Reply to the PSE with a classification signature.

6. A VoIP telephone is connected to a 24-port PoE midspan PSE. If the telephone does not provide a classification signature, how much power will the PSE provide to the telephone?
   A. 0 watts (The telephone would not be recognized as a PD.)
   B. 4.0 watts
   C. 7.0 watts
   D. 15.4 watts

7. An endpoint PSE that provides power by using Alternative B is capable of providing power to devices by using which of the following Ethernet technologies? (Choose all that apply.)
   A. 10BaseT
   B. 100BaseT
   C. 1000BaseT
   D. 100BaseFX

8. Which of the following devices provide power over the spare pairs of the Ethernet cable? (Choose all that apply.)
   A. Single-port power injector
   B. Multiport power injector
   C. Midspan PSE
   D. Endpoint PSE, Alternative B

9. At maximum power requirements, a 24-port PoE Ethernet switch must be able to provide about how many total watts of power to PoE devices on all ports?
   A. 15.4 watts
   B. 370 watts
   C. 1,000 watts
   D. Not enough information is provided to answer the question.

10. If an 802.3-2005 Clause 33 access point is equipped with two radios and requires 7.5 watts of power, how much power will the PSE provide to it? (Choose all that apply.)
    A. 7.5 watts
    B. 10.1 watts
    C. 15 watts
    D. 15.4 watts

11. The PSE provides power within a range of __________ volts, with a nominal power of ______ volts.
    A. 14.5 to 20.5, 18
    B. 6.49 to 12.95, 10.1
    C. 12 to 19, 15.4
    D. 44 to 57, 48

12. You have installed an Ethernet switch that is compliant with 802.3-2005, Clause 33. You are having problems with your access points. Which of the following could be causing your problems?

 A. Many PoE VoIP telephones are connected to the same Ethernet switch.

 B. Most of the Ethernet cables running from the switch to the APs are 90 meters long.

 C. The Ethernet cables are only Cat 5e.

 D. The switch is capable of 1000BaseT, which is not compatible with VoIP telephones.

13. You are designing an 802.3-2005 Clause 33 network and are installing a 24-port Ethernet switch to support 10 Class-1 VoIP phones and 10 Class-0 access points. The switch requires 500 watts to perform its basic switching functions. How much total power will be needed?

 A. 500 watts

 B. 694 watts

 C. 808 watts

 D. 1,000 watts

14. You are designing an 802.3-2005 Clause 33 network and are installing a 24-port Ethernet switch to support 10 Class-2 cameras and 10 Class-3 access points. The switch requires 1,000 watts to perform its basic switching functions. How much total power will be needed?

 A. 1,080 watts

 B. 1,224 watts

 C. 1,308 watts

 D. 1,500 watts

15. When a powered device is first connected to a PSE, it presents itself as having a resistance level of what?

 A. 25-ohms

 B. 50-ohms

 C. 25k-ohms

 D. 50k-ohms

16. How much voltage does the PSE apply while trying to identify the detection signature?

 A. 5 volts

 B. 10.1 volts

 C. 12 volts

 D. 48 volts

17. What is the maximum power used by a PD Class 0 device?

 A. 3.84 W

 B. 6.49 W

 C. 12.95 W

 D. 15.4 W

18. The PSE will apply a voltage of between 14.5 and 20.5 and measure the resulting current to determine the class of the device. Which current range represents Class 2 devices?

A. 0 to 4 mA

B. 5 to 8 mA

C. 9 to 12 mA

D. 13 to 16 mA

E. 17 to 20 mA

19. A powered device must be capable of accepting power with either polarity from the power supply. In mode A, on which conductors/wires does the powered device accept power?

A. 1, 2, 3, 4

B. 5, 6, 7, 8

C. 1, 2, 3, 6

D. 4, 5, 7, 8

20. An 802.3-2005 Clause 33 midspan PSE can be used with which of the following networking devices? (Choose all that apply.)

A. 10BaseT repeater

B. 100BaseT layer 2 switch

C. 1000BaseT layer 2 switch

D. 1000BaseT layer 3 switch

Answers to Review Questions

1. D. Even when 802.3af was an amendment, PoE was defined in Clause 33. PoE is still defined in Clause 33. When an amendment is incorporated into a revised standard, the clause numbering remains the same.

2. A. Any device that does not provide a classification signature (which is optional) is automatically considered a Class 0 device, and the PSE will provide 15.4 watts of power to that device.

3. A, C. The PoE standard defines two types of devices: powered devices (PD) and power-sourcing equipment (PSE).

4. D. The power supplied to the PD is at a nominal 48 volts; however, the PD must be capable of accepting up to 57 volts.

5. A, B, C. The PD must be able to accept power over either the data pairs or the unused pairs. The PD must also reply to the PSE with a detection signature. The PD must accept power with either polarity. Replying to the PSE with a classification signature is optional.

6. D. Providing a classification signature is optional for the PD. If the PD does not provide a classification signature, the device is considered a Class 0 device, and the PSE will allocate the maximum power, or 15.4 watts.

7. A, B. Alternative B devices, either endpoint or midspan, provide power to the unused data pairs. Because 10BaseT and 100BaseT transmissions occur by using only two pairs of the Ethernet cable, they will work with Alternative B devices. 1000BaseT uses all four data pairs, so it is compatible with only endpoint PSE devices that support Alternative A. 100BaseFX uses fiber-optic cable and is not compatible with PoE.

8. A, B, C, D. All of these devices provide power over the spare pairs of Ethernet cable.

9. B. At maximum power, each PoE device will be provided with 15.4 watts of power from the PSE. If all 24 ports have powered devices connected to them, then a total of just under 370 watts (15.4 watts × 24 ports = 369.6 watts) is needed.

10. D. The power-sourcing equipment (PSE) provides three potential levels of power: Class 0 = 15.4 watts, Class 1 = 4.0 watts, Class 2 = 7.0 watts, and Class 3 = 15.4 watts. Because this device requires 7.5 watts of power, the PSE would be required to provide it with 15.4 watts.

11. D. The PSE provides power within a range of 44 to 57 volts, with a nominal power of 48 volts.

12. A. The maximum distance of 100 meters is an Ethernet limitation, not a PoE limitation. At 90 meters, this is not an issue. Although not specifically mentioned in the PoE standard, Category 5e cables support 1000BaseT communications and are therefore capable of also providing PoE. Because this is a 1000BaseT Ethernet switch that is also providing PoE, it would be using PSE Alternative A, providing PoE over the data pairs, which is compatible with 1000BaseT. The large number of PoE VoIP telephones could be requiring more power than the switch is capable of providing, thus causing problems with your access points.

13. B. The switch will provide the Class 0 devices with 15.4 W of power each, and the Class 1 devices with 4.0 W of power each. So the 10 VoIP phones will require 40 W of power, the 10 APs will require 154 W of power, and the switch will need 500 W—for a total of 694 W (40 W + 154 W + 500 W).

14. B. The switch will provide the Class 2 devices with 7.0 W of power each, and the Class 3 devices with 15.4 W of power each. So the 10 cameras will require 70 W of power, the 10 APs will require 154 W of power, and the switch will need 1,000 W—for a total of 1,224 W (70 W + 154 W + 1,000 W).

15. C . When a PD is first connected to a PSE, it presents itself as a nominal 25k-ohm resistance.

16. B. The PSE applies a 10.1 V current while measuring the data circuit to try to identify the detection signature.

17. C. The maximum power used by a Class 0 PD is 12.95 W. The PSE provides 15.4 W to account for a worst-case scenario, in which there may be power loss due to the cables and connectors between the PSE and the PD. The maximum power used by a Class 1 PD is 3.84 W, and the maximum power used by a Class 2 PD is 6.49 W.

18. E. The different class and range values are as follows:

    Class 0: 0 to 4 mA

    Class 1: 9 to 12 mA

    Class 2: 17 to 20 mA

    Class 3: 26 to 30 mA

    Class 4: 36 to 44 mA

19. C. Mode A accepts power with either polarity from the power supply on wires 1, 2, 3, and 6. With Mode B, the wires used are 4, 5, 7, and 8.

20. A, B. 10BaseT and 100BaseT transmissions occur by using only two pairs of the Ethernet cable. Midspan PSE provides power across the two unused cable pairs. 1000BaseT transmissions are performed using all four cable pairs. Midspan PoE therefore does not work because there are no unused cable pairs.

High Throughput (HT) and 802.11n

IN THIS CHAPTER, YOU WILL LEARN ABOUT THE FOLLOWING:

✓ **802.11n history**

- 802.11n draft amendment
- Wi-Fi Alliance certification

✓ **MIMO**

- Radio chains
- Spatial multiplexing (SM)
- MIMO diversity
- Transmit beamforming (TxBF)

✓ **HT channels**

- 20 MHz non-HT and HT channels
- 40 MHz channels
- Guard interval (GI)
- Modulation and coding scheme (MCS)

✓ **HT PHY**

- Non-HT legacy
- HT Mixed
- HT Greenfield

✓ **HT MAC**

- A-MSDU
- A-MPDU
- MTBA
- RIFS
- HT power management

✓ **HT operation**

- 20/40 channel operation
- HT protection modes (0–3)
- Dual-CTS protection
- Phased coexistence operation (PCO)

In this chapter, we discuss the most talked about new technology that is defined under the 802.11n draft amendment: *High Throughput (HT)*, which provides PHY and MAC enhancements to support throughput of 100 Mbps or greater. This and other technology defined by the 802.11n draft amendment is sometimes referred to as 802.11 *next generation* technology because of the promise of greater throughput as well as greater range. This chapter covers the history of the 802.11n technology and how the Wi-Fi industry is responding.

802.11n requires a whole new approach to the Physical layer, using a technology called MIMO that requires the use of multiple radios and antennas. As you learned in earlier chapters, multipath is an RF behavior that usually causes performance degradation in a WLAN. MIMO technology actually takes advantage of multipath to increase throughput as well as range.

Besides the use of MIMO technology, HT mechanisms defined by the 802.11n draft amendment provide for enhanced throughput using other methods. We discuss the use of 40 MHz channels that provide greater frequency bandwidth. Enhancements to the MAC sublayer also provide for greater throughput with the use of frame aggregation. The 802.11e amendment defined enhancements to power management, and the 802.11n amendment also provides for new power-management techniques.

Finally, we discuss the various modes of operation for an HT network and how HT radio transmissions can coexist in the same WLAN environment with radios that use the other technologies we have discussed throughout this book.

802.11n and HT technology are so complex that an entire book dedicated to the topic would probably not be able to fully cover every aspect of HT. However, in this chapter we cover all the key components of HT and the topics needed to properly prepare you for the CWNA exam.

802.11n History

Since 2004, the 802.11 Task Group n (TGn) has been working on improvements to the 802.11 standard to provide for greater throughput. Many of the past IEEE 802.11 amendments defined data bandwidth and data rates. 802.11n now defines data rates by using a modulation and coding scheme (MCS), which is discussed later in this chapter. The main objective of the 802.11n draft amendment is to increase the data rates and the throughput in both the 2.4 GHz and 5 GHz frequency bands. The 802.11n draft amendment defines a new operation known as High Throughput (HT), which provides PHY and MAC enhancements to provide for data rates potentially as high as 600 Mbps.

802.11n Draft Amendment

The 802.11n draft amendment defines High Throughput (HT) clause 20 radios that use multiple-input multiple-output (MIMO) technology in unison with Orthogonal Frequency Division Multiplexing (OFDM) technology. The beneficial consequences of using MIMO are increased throughput and even greater range. Enhancements to the MAC sublayer of the Data-Link layer are also defined in the 802.11n draft amendment to provide for greater throughput.

The mechanisms defined by the 802.11n draft amendment are vastly different from current 802.11 technologies. However, clause 20 radios (HT) are required to be backward compatible with older clause 18 radios (HR-DSSS), clause 17 radios (OFDM), and clause 19 radios (ERP). As you have learned, clause 18 HR-DSSS (802.11b) and clause 17 ERP (802.11g) radios can transmit in only the 2.4 GHz ISM band, while clause 17 OFDM (802.11a) radios transmit in the 5 GHz UNII bands. It should be noted that the technology defined for use by HT clause 20 radios is not frequency dependent. HT technology can be used in both the 2.4 GHz ISM band and the 5 GHz UNII bands.

At the time of writing this book, the 802.11n amendment had not yet been ratified. Current predictions are that final ratification will occur sometime in 2009. Very often, new 802.11 technologies do not find their way into the enterprise until a year or more after ratification of an 802.11 amendment. However, 802.11n technology is already being deployed in the enterprise prior to ratification of the 802.11n amendment. Most of the major Wi-Fi vendors debuted enterprise 802.11n solutions in 2007 and have already begun to direct their customers to HT technology.

Wi-Fi Alliance Certification

Because 802.11n technology is already being sold and deployed, the Wi-Fi Alliance currently has a vendor certification program called Wi-Fi CERTIFIED™ 802.11n draft 2.0. This certification program currently tests many of the HT capabilities that will eventually be seen in the final ratified 802.11n amendment. Most vendors claim that their Wi-Fi CERTIFIED™ 802.11n draft 2.0 products will be software-upgradeable and be compliant with the final ratified 802.11n amendment.

802.11n products are tested for both mandatory and optional baseline capabilities, as shown in Table 18.1. All certified products must also support both Wi-Fi Multimedia (WMM) quality-of-service mechanisms and WPA/WPA2 security mechanisms.

TABLE 18.1 Wi-Fi CERTIFIED™ 802.11n Draft 2.0 Baseline Requirements

Feature	Explanation	Type
Support for two spatial streams in transmit mode	Required for an AP device.	Mandatory
Support for two spatial streams in receive mode	Required for an AP and a client device, except for handheld devices.	Mandatory

TABLE 18.1 Wi-Fi CERTIFIED™ 802.11n Draft 2.0 Baseline Requirements *(continued)*

Feature	Explanation	Type
Support for A-MPDU and A-MSDU	Required for all devices.	Mandatory
Support for block ACK	Required for all devices.	Mandatory
2.4 GHz operation	Devices can be 2.4 GHz only, 5 GHz only, or dual-band. For this reason, both frequency bands are listed as optional.	Tested if implemented
5 GHz operation		Tested if implemented
40 MHz channels in the 5 GHz band	40 MHz operation is supported by the Wi-Fi Alliance in only the 5 GHz band. Operation of 40 MHz channels in the 2.4 GHz band is still being debated by the IEEE and may be supported by the Wi-Fi Alliance at a later time.	Tested if implemented
Greenfield preamble	Greenfield preamble cannot be interpreted by legacy stations. The Greenfield preamble improves efficiency of the 802.11n networks with no legacy devices.	Tested if implemented
Short guard interval (short GI), 20 and 40 MHz	Short GI is 400 nanoseconds vs. the traditional GI of 800 nanoseconds.	Tested if implemented
Concurrent operation in 2.4 and 5 GHz bands	This mode is tested for APs only.	Tested if implemented

It should be noted that prior to the Wi-Fi CERTIFIED 802.11n draft 2.0 certification program, many WLAN vendors offered *Pre-802.11n* products in the SOHO marketplace. The majority of these products were not interoperable with other vendors' products and are not compatible with certified Wi-Fi Alliance products.

A white paper from the Wi-Fi Alliance is also included on the CD of this book, called "Wi-Fi CERTIFIED™ 802.11n Draft 2.0: Longer-Range, Faster-Throughput, Multimedia-Grade Wi-Fi Networks."

MIMO

The heart and soul of the 802.11n amendment exists at the PHY layer with the use of a technology known as *multiple-input multiple-output (MIMO)*. MIMO requires the use of multiple radios and antennas, called radio chains. MIMO systems can also use multiple antennas to provide for better antenna diversity, which can increase range. Transmitting multiple streams of data with spatial multiplexing provides for greater throughput and takes advantage of the old enemy known as multipath. Transmit beamforming is an optional smart antenna technology that can be used in MIMO systems to "steer" beams and provide for greater range and throughput.

Radio Chains

Conventional 802.11 radios transmit and receive RF signals by using a *single-input single-output (SISO)* system. SISO systems use a single radio chain. A *radio chain* is defined as a single radio and all of its supporting architecture including mixers, amplifiers, and analog/digital converters.

A MIMO system consists of multiple radio chains, with each radio chain having its own antenna. A MIMO system is characterized by the number of transmitters and receivers used by the multiple radio chains. For example, a 2×3 MIMO system would consist of three radio chains with two transmitters and three receivers. A 3×3 MIMO system would use three radio chains with three transmitters and three receivers. In a MIMO system, the first number always references the transmitters (TX), and the second number references the receivers (RX).

Figure 18.1 illustrates both 2×3 and 3×3 MIMO systems. Please note that both systems utilize three radio chains; however, the 3×3 system has three transmitters, whereas the 2×3 system has only two transmitters.

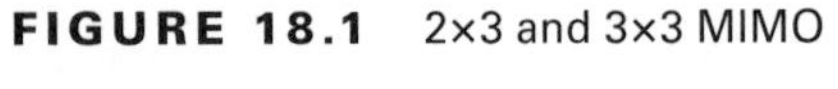

FIGURE 18.1 2×3 and 3×3 MIMO

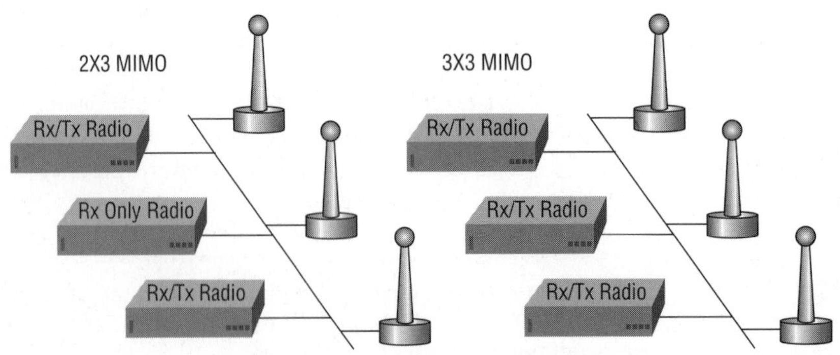

The use of multiple transmitters in a MIMO system provides for the transmission of more data via spatial multiplexing. The use of multiple receivers increases signal-to-noise ratio

(SNR) because of advanced MIMO antenna diversity. Both of these benefits are discussed in greater detail in the following paragraphs. The 802.11n standard allows for MIMO systems up to 4×4 using four radio chains. Each radio chain requires power. A 2×2 MIMO system would require much less of a power draw than a 4×4 MIMO system.

Spatial Multiplexing (SM)

In traditional 802.11 environments, the phenomenon of multipath has long caused problems. *Multipath* is a propagation phenomenon that results in two or more paths of the same signal arriving at a receiving antenna at the same time or within nanoseconds of each other. Due to the natural broadening of the waves, the propagation behaviors of reflection, scattering, diffraction, and refraction will occur. A signal may reflect off an object or may scatter, refract, or diffract. These propagation behaviors can each result in multiple paths of the same signal. As you learned in Chapter 2, "Radio Frequency Fundamentals," the negative effects of multipath can include loss of amplitude and data corruption. 802.11n MIMO systems, however, take advantage of multipath and, believe it or not, multipath then becomes our friend.

MIMO radios transmit multiple radio signals at the same time and take advantage of multipath. Each individual radio signal is transmitted by a unique radio and antenna of the MIMO system. Each independent signal is known as a *spatial stream*, and each unique stream can contain different data than the other streams transmitted by one or more of the other radios. Each stream will also travel a different path, because there is at least a half-wavelength of space between the multiple transmitting antennas. The fact that the multiple streams follow different paths to the receiver because of the space between the transmitting antennas is known as *spatial diversity*. Sending multiple independent streams of unique data using spatial diversity is often also referred to as *spatial multiplexing (SM)* or *spatial diversity multiplexing (SDM)*.

When using spatial multiplexing, both the transmitter and receiver must participate. In other words, both the transmitter and receiver must be MIMO systems. A simplistic description of spatial multiplexing would be to envision multiple unique data streams being transmitted via unidirectional antennas to multiple receiving unidirectional antennas. Spatial multiplexing can also be accomplished with omnidirectional antennas because of the advanced *digital signal processing (DSP)* techniques used by MIMO systems. The benefit of sending multiple unique data streams is that throughput is drastically increased. If a MIMO access point sends two unique data streams to a MIMO client station that receives both streams, the throughput is effectively doubled. If a MIMO access point sends three unique data streams to a MIMO client station that receives all three streams, the throughput is effectively tripled. Figure 18.2 depicts a 3×3 MIMO AP transmitting three independent streams of unique data to a 3× 3 MIMO client.

Currently, most 802.11n radios deploy 2×3 or 3×3 MIMO systems. The 802.11n amendment does allow for the use of up to a 4×4 MIMO system. As you have learned, transmitting multiple RF signals at the same time while carrying unique data takes advantage of multipath. As a matter of fact, if multiple signals sent by a MIMO transmitter all arrive simultaneously at the receiver, the signals will interfere with each other and the performance is basically the same as a non-MIMO system.

FIGURE 18.2 Multiple spatial streams

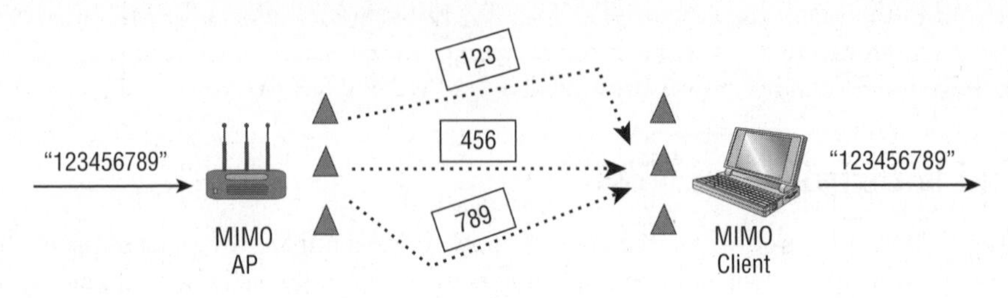

Multiple spatial streams can be sent with the same (equal) modulation or they can be sent using different (unequal) modulation. For example, a 3×3 MIMO radio can transmit three data streams using the same 64-QAM modulation technique. Another example is a 3×3 MIMO radio transmitting two streams by using 64-QAM modulation and the third stream using QPSK modulation because of a higher noise floor. A 3×3 MIMO system using *equal modulation* would accomplish greater throughput than a 3×3 MIMO system using *unequal modulation*. It remains to be seen how WLAN vendors will implement unequal modulation with 802.11n radios.

MIMO Diversity

If you cover one of your ears with your hand, will you hear better or worse with a single ear? Obviously, you will hear better with two ears. Do you think you would be able to hear more clearly if you had three or four ears instead of just two ears? Do you think you would be able to hear sounds from greater distances if you had three or four ears instead of just two ears? Yes, a human being would hear more clearly and with greater range if equipped with more than two ears. MIMO systems employ advanced antenna diversity capabilities that are analogous to having multiple ears.

Antenna diversity often is mistaken for the spatial multiplexing capabilities that are utilized by MIMO. Antenna diversity (both receive and transmit) is a method of using multiple antennas to survive the negative effects of multipath. As you just learned, MIMO takes advantage of multipath with spatial multiplexing to increase data capacity. *Antenna diversity* is a method of compensating for multipath as opposed to utilizing multipath. Multipath produces multiple copies of the same signal that arrive at the receiver with different amplitudes. In Chapter 4, "Radio Frequency Signal and Antenna Concepts," you learned about traditional antenna diversity, which consists of one radio with two antennas. Most pre-802.11n radios use *switched diversity*. When receiving RF signals, switched diversity systems listen with multiple antennas. Multiple copies of the same signal arrive at the receiver antennas with different amplitudes. The signal with the best amplitude is chosen, and the other signals are ignored. Switched diversity is also used when transmitting, but only one antenna is used. The transmitter will transmit out of the diversity antenna where the best amplitude signal was last heard.

As the distance between a transmitter and receiver increases, the received signal amplitude decreases to levels closer to the noise floor. As the signal-to-noise ratio (SNR) diminishes, the odds of data corruption grow. Listening with two antennas increases the odds of hearing at least one signal without corrupted data. Now imagine if you had three or four antennas listening for the best received signal by using switched diversity. The probabilistic odds of hearing signals with stronger amplitudes and uncorrupted data have increased even more. The increased probability of hearing at least one uncorrupted signal in a switched diversity system using three or four antennas often results in increased range.

When receive diversity is used, the signals may also be linearly combined by using a signal processing technique called *maximal ratio combining (MRC)*. MRC algorithms are used to combine multiple received signals by looking at each unique signal and optimally combining the signals in a method that is additive as opposed to destructive. MIMO systems using both switched diversity and MRC together will effectively raise the SNR level of the received signal. As shown in Figure 18.3, maximal ratio combining is most useful when a non-MIMO radio transmits to a MIMO receiver and multipath occurs. The MRC algorithm focuses on the signal with the least corruption; however, it may still combine information from the noisier signals. The end result is that less data corruption occurs because a better estimate of the original data has been reconstructed.

FIGURE 18.3 Maximal ratio combining (MRC)

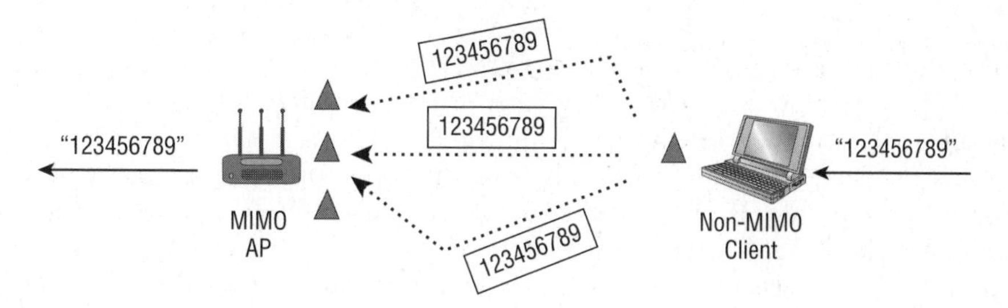

Transmit Beamforming (TxBF)

The 802.11n draft amendment also proposes an optional PHY capability called *transmit beamforming (TxBF)*. Beamforming technology, which has been used in radar systems for many years, is also known as a phased-array antenna system and is often referred to as *smart antenna* technology.

The two major types of smart antenna array systems include a switched array and an adaptive array. As pictured in Figure 18.4, a *switched antenna array* uses a number of fixed beam patterns while an *adaptive antenna array* maneuvers the beam in the direction of a targeted receiver.

Transmit beamforming is a method that allows a MIMO transmitter using multiple antennas to "focus" the transmissions in a coordinated method much like an adaptive antenna array. The focused transmissions are sent in the best direction of a receiver (RX). When multiple copies of the same signal are sent to a receiver, the signals will usually arrive out of phase with each other. If the transmitter (TX) knows about the receiver's location, the phase of the multiple signals sent by a MIMO transmitter can be adjusted. When the multiple signals arrive at the receiver, they are in-phase, resulting in constructive multipath instead of the destructive multipath caused by out-of-phase signals. Carefully controlling the phase of the signals transmitted from multiple antennas has the effect of emulating a high-gain unidirectional antenna or "steering" the beams.

FIGURE 18.4 Antenna arrays and beamforming

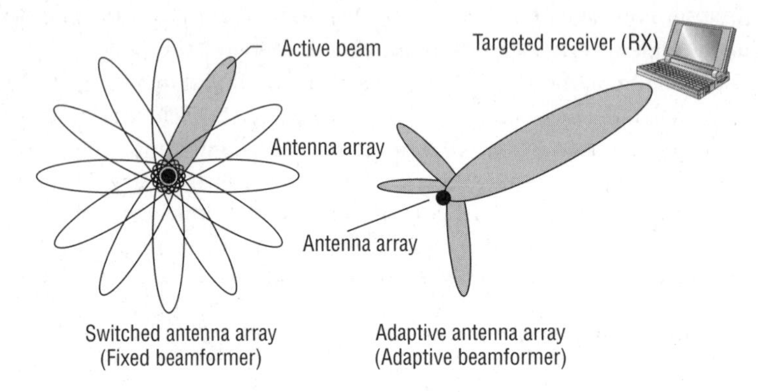

Because transmit beamforming results in constructive multipath communication, the result is a higher signal-to-noise ratio and greater received amplitude. Therefore, transmit beamforming will result in greater range for individual clients communicating with an access point. Transmit beamforming will also result in higher throughput because of the higher SNR that allows for the use of more-complex modulation methods that can encode more data bits. The higher SNR also results in fewer layer 2 retransmissions.

Transmit beamforming could be used together with spatial diversity multiplexing (SDM); however, the number of spatial streams is constrained by the number of receiving antennas. For example, a 4×4 MIMO radio might be transmitting to a 2×2 MIMO radio which can only receive only two spatial streams. The 4×4 MIMO radio will send only two spatial streams but might also use the other antennas to form beams that are more focused to the receiving 2×2 MIMO receiver. In practice, transmit beamforming will probably be used when SDM is not the best option. As pictured in Figure 18.5, when utilizing transmit beamforming, the transmitter will not be sending multiple unique spatial streams but will instead be sending multiple streams of the same data with the phase adjusted for each RF signal.

Transmit beamforming relies on *implicit feedback* and *explicit feedback* from both the transmitter and receiver. A good analogy for explicit feedback is sonar. Sonar is a method in which submarines use sound propagation underwater to detect other vessels. A submarine sends out a sound wave, and, based on the characteristics of the returning sound wave, the crew can determine the type of vessel that might be in the path of the submarine. 802.11n transmitters that use beamforming will try to adjust the phase of the signals based on feedback from the receiver by using *sounding frames*. The transmitter is considered the *beamformer*, while the receiver is considered the *beamformee*. The beamformer and the beamformee work together to educate each other about the characteristics of the MIMO channel. The beamformer will send a sounding request frame and will make phase adjustments based on the information that is returned in a sounding response frame from the beamformee. Any frame can be used as a sounding frame. Null function data frames can be used if another frame is not used. When using implicit feedback, the beamformer receives long training symbols transmitted by the beamformee, which allow the MIMO channel between the beamformee and beamformer to be estimated. Much more information is exchanged between two HT radios that are capable of explicit feedback. The beamformee makes a direct estimate of the channel from training symbols sent to the beamformee by the beamformer. The beamformee then takes that information and sends additional feedback back to the beamformer. The beamformer then transmits based on the feedback from the beamformee.

FIGURE 18.5 Transmit beamforming data

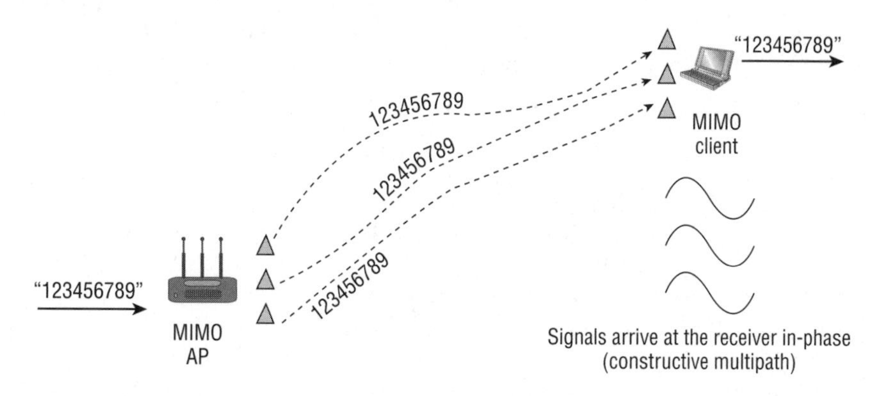

Transmit beamforming will work best between 802.11n clients that have similar capabilities, because the technology depends heavily on feedback from the receiving radio. 802.11 a/b/g radios cannot provide the implicit/explicit feedback used by the 802.11n HT radios that are capable of transmit beamforming. Access points using beamforming can target multiple clients, but only for unicast transmissions. Transmit beamforming is not used for broadcast or multicast transmissions. Currently, only a few 802.11n chipset manufacturers have incorporated transmit beamforming, and the Wi-Fi Alliance does not yet test the technology. After the 802.11n amendment is ratified, WLAN enterprise vendors will probably begin to deploy the technology.

HT Channels

In previous chapters, you learned that the 802.11a amendment defined the use of clause 17 radios using *Orthogonal Frequency Division Multiplexing (OFDM)* technology in the 5 GHz UNII bands. 802.11g defined the use of clause 19 radios using ERP-OFDM, which is effectively the same technology except that transmissions occur in the 2.4 GHz ISM band. The 802.11n draft amendment also defines the use of OFDM channels. However, key differences exist for HT clause 20 radios. As mentioned earlier in this chapter, HT clause 20 radios can operate in either frequency.

You have already learned that 802.11n radios use spatial multiplexing to send multiple independent streams of unique data. Spatial multiplexing is one method of increasing the throughput. The OFDM channels used by 802.11n radios are larger in size and bandwidth. The greater frequency bandwidth provided by the OFDM channels used by HT clause 20 radios also provides for greater eventual throughput.

20 MHz Non-HT and HT Channels

As you learned in Chapter 6, "Wireless Networks and Spread Spectrum Technologies," 802.11a and 802.11g radios use 20 MHz OFDM channels. As pictured in Figure 18.6, each channel consists of 52 subcarriers. Forty-eight of the subcarriers transmit data, while four of the subcarriers are used as pilot tones for dynamic calibration between the transmitter and receiver. OFDM technology also employs the use of convolutional coding and forward error correction.

FIGURE 18.6 20 MHz non-HT (802.11a/g) channel

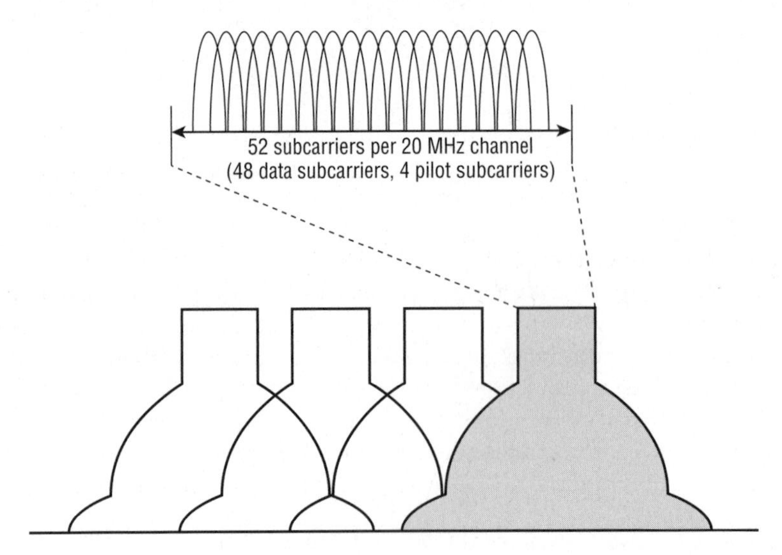

HT clause 20 radios also use the same OFDM technology and have the capability of using either 20 MHz channels or 40 MHz channels. The 20 MHz channels used by HT radios have four extra subcarriers and can carry a little more data than a non-HT OFDM channel. As a result, the HT 20 MHz channel can provide greater aggregate throughput for the same frequency space. As pictured in Figure 18.7, an HT 20 MHz OFDM channel has 56 subcarriers. Fifty-two of the subcarriers transmit data, while four of the subcarriers are used as pilot tones for dynamic calibration between the transmitter and receiver.

FIGURE 18.7 20 MHz HT (802.11n) channel

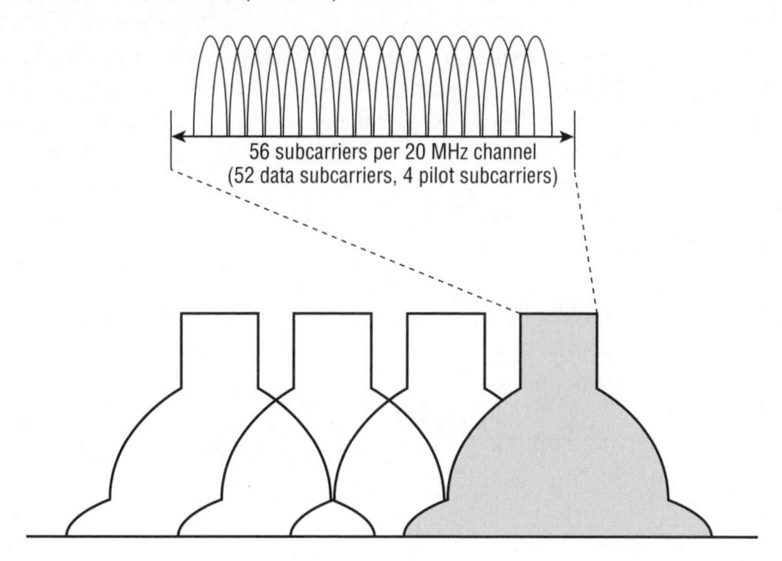

40 MHz Channels

HT clause 20 radios also have the capability of using 40 MHz OFDM channels. As pictured in Figure 18.8, the 40 MHz HT channels use 114 OFDM subcarriers. One hundred and eight of the subcarriers transmit data, while six of the subcarriers are used as pilot tones for dynamic calibration between the transmitter and receiver. A 40 MHz channel effectively doubles the frequency bandwidth available for data transmissions.

FIGURE 18.8 40 MHz HT (802.11n) channel

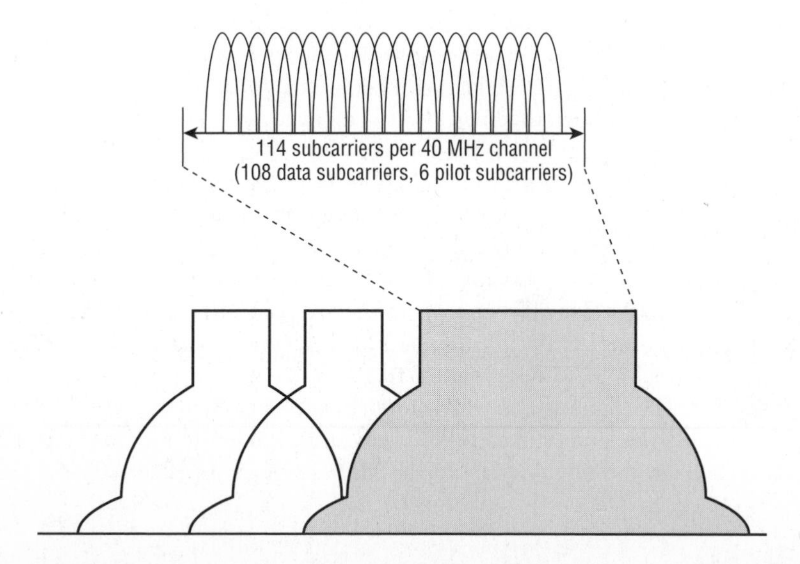

The 40 MHz channels used by HT radios are essentially two 20 MHz OFDM channels that are bonded together. Each 40 MHz channel consists of a primary and secondary 20 MHz channel. The primary and secondary 20 MHz channels must be adjacent 20 MHz channels in the frequency in which they operate. As pictured in Figure 18.9, the two 20 MHz channels used to form a 40 MHz channel are designated as primary and secondary and are indicated by two fields in the body of certain 802.11 management frames. The primary field indicates the number of the primary channel. A positive or negative offset indicates whether the secondary channel is one channel above or one channel below the primary channel.

FIGURE 18.9 Channel bonding

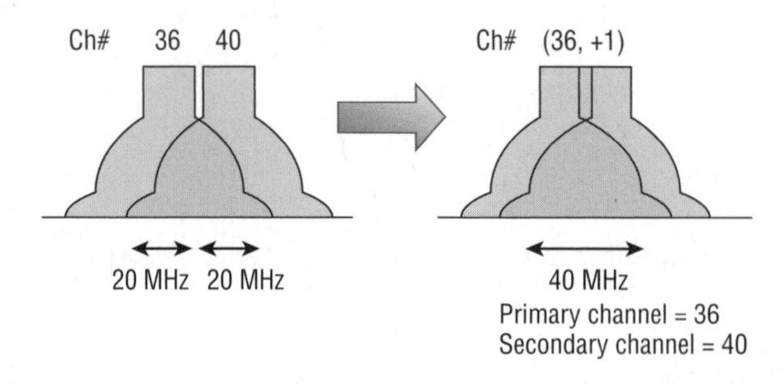

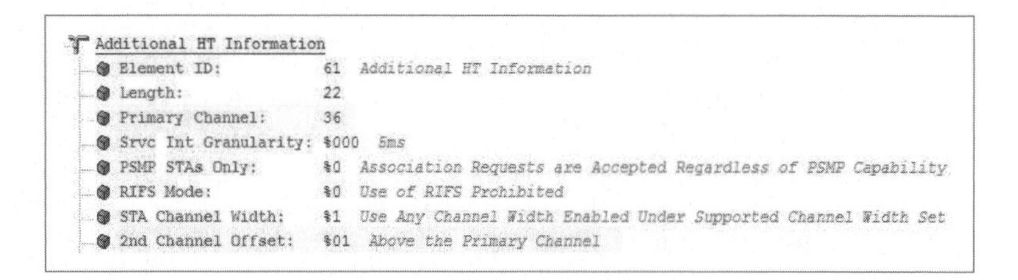

A standard 20 MHz HT channel reserves some frequency bandwidth at the top and bottom of the channel to avoid interference with adjacent 20 MHz HT channels. When two 20 MHz HT channels are bonded together, there is no need to reserve this bandwidth at the bottom of the higher channel and at the top end of the lower channel. Therefore, an HT (802.11n) 40 MHz channel uses a total of 114 subcarriers instead of 112 subcarriers.

As you learned in Chapter 12, "WLAN Troubleshooting," channel reuse patterns are needed in multiple channel architecture (MCA). Channel reuse patterns using 40 MHz channels at 5 GHz are still feasible because of all the possible combinations within the UNII bands. The use of 40 MHz HT channels in the 5 GHz UNII bands makes perfect sense because there are a total of twenty-three 20 MHz channels that can be bonded together in various pairs, as pictured in Figure 18.10.

Deploying 40 MHz HT channels at 2.4 GHz unfortunately does not scale well in multiple channel architecture. As you learned in earlier chapters, although fourteen channels are available at 2.4 GHz, there are only three nonoverlapping 20 MHz channels available in the 2.4 GHz ISM band. When the smaller channels are bonded together to form 40 MHz channels in the 2.4 GHz ISM band, any two 40 MHz channels will overlap, as pictured in Figure 18.11. In other words, only one 40 MHz channel can be used at 2.4 GHz, and the possibility of a channel reuse pattern is essentially impossible. Currently, 40 MHz channels are supported by the Wi-Fi Alliance in only the 5 GHz UNII bands.

FIGURE 18.10 Channel bonding—5 GHz UNII bands

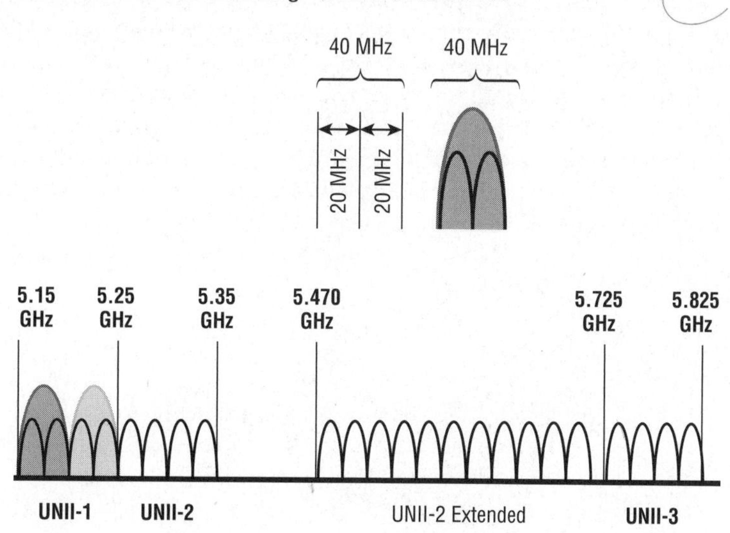

FIGURE 18.11 Channel bonding—2.4 GHz ISM band

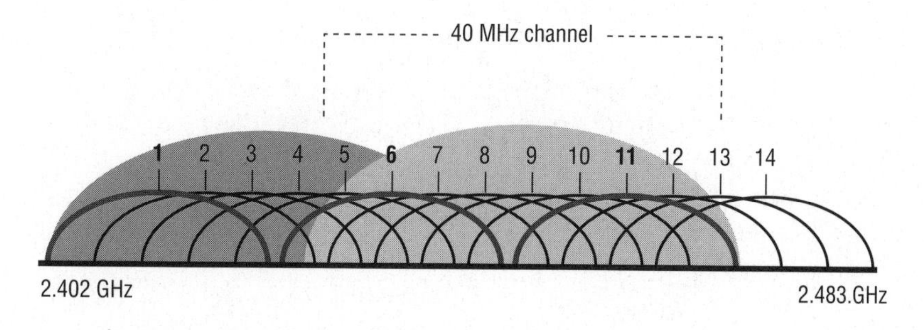

Guard Interval (GI)

For digital signals, data is modulated onto the carrier signal in bits or collections of bits called *symbols*. When 802.11a/g radios transmit at 54 Mbps, each OFDM symbol contains 288 bits; 216 of these bits are data, and 72 of the bits are error-correction bits. All the data bits of an OFDM symbol are transmitted across the 48 data subcarriers of a 20 MHz non-HT channel.

802.11a/g radios use an 800-nanosecond *guard interval (GI)* between OFDM symbols. The guard interval is a period of time between symbols that accommodates for the late arrival of symbols over long paths. In a multipath environment, symbols travel different paths, and therefore some symbols arrive later. A "new" symbol may arrive at a receiver before a "late" symbol has been completely received. This is known as *intersymbol interference (ISI)* and often results in data corruption.

In earlier chapters, we discussed ISI and delay spread. The delay spread is the time differential between multiple paths of the same signal. Normal delay spread is 50–100 nanoseconds, and a maximum delay spread is about 200 nanoseconds. The guard interval should be two to four times the length of the delay spread. Think of the guard interval as a buffer for the delay spread. The normal guard interval is an 800-nanosecond buffer between symbol transmissions. As pictured in Figure 18.12, a guard interval will compensate for the delay spread and help prevent intersymbol interference. If the guard interval is too short, intersymbol interference can still occur.

FIGURE 18.12 Guard interval

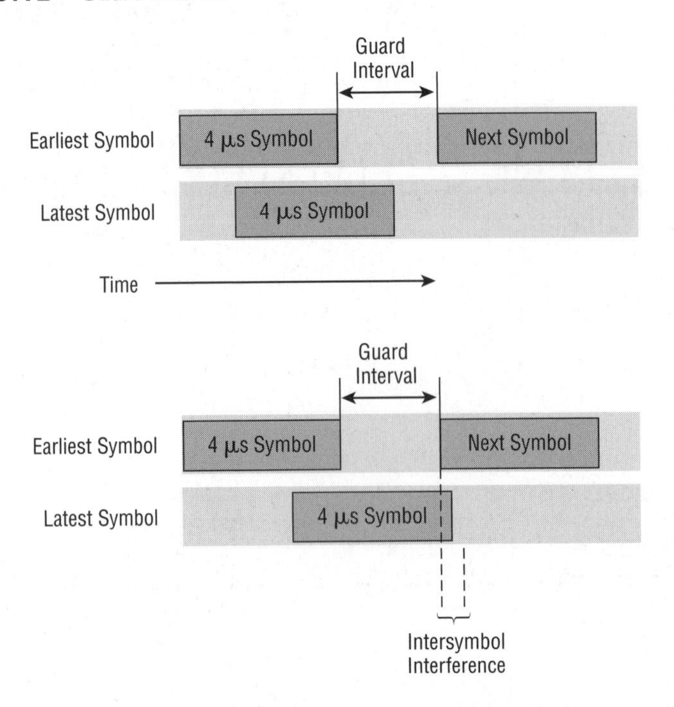

802.11n also uses an 800-nanosecond guard interval; however, a shorter 400-nanosecond guard interval is optional. A shorter guard interval results in a shorter symbol time, which has the effect of increasing data rates by about 10 percent. If the optional, shorter 400-nanosecond guard interval is used with an 802.11n radio, throughput will increase, however, the odds of an intersymbol interference occurrence increases. If intersymbol interference does indeed occur because of the shorter GI, the result is data corruption. If data corruption occurs, layer 2 retransmissions will increase and the throughput will be adversely affected. Therefore, a 400-nanosecond guard interval should be used in only good RF environments. If throughput goes down because of a shorter GI setting, the default guard interval setting of 800 nanoseconds should be used instead.

Modulation and Coding Scheme (MCS)

As defined by the 802.11n draft amendment, data rates are now defined with a *modulation and coding scheme (MCS)*. Non-HT radios that used OFDM technology (802.11a/g) defined data rates of 6 Mbps to 54 Mbps based on the modulation that was used. HT radios, however, define data rates based on numerous factors including modulation, the number of spatial streams, channel size, and guard interval. Each modulation coding scheme is a variation of these multiple factors. Seventy-seven modulation coding schemes exist for both 20 MHz HT channels and 40 MHz HT channels. There are eight mandatory modulation and coding schemes for 20 MHz HT channels, as shown in Table 18.2. The eight mandatory MCSs for 20 MHz channels are comparable to basic (required) rates.

TABLE 18.2 Mandatory Modulation and Coding Schemes—20 MHz Channel

MCS Index	Modulation	Spatial Streams	Data Rates 800 ns GI	400 ns GI
0	BPSK	1	6.5 Mbps	7.2 Mbps
1	QPSK	1	13.0 Mbps	14.4 Mbps
2	QPSK	1	19.5 Mbps	21.7 Mbps
3	16-QAM	1	26.0 Mbps	28.9 Mbps
4	16-QAM	1	39.0 Mbps	43.3 Mbps
5	64-QAM	1	52.0 Mbps	57.8 Mbps
6	64-QAM	1	58.5 Mbps	65.0 Mbps
7	64-QAM	1	65.0 Mbps	72.2 Mbps

As you can see from Table 18.2, the modulation type, the guard interval, and the number of spatial streams all determine the eventual data rate. Table 18.3 depicts the modulation and coding schemes for a 20 MHz channel using four spatial streams.

TABLE 18.3 MCS—20 MHz Channel, Four Spatial Streams

MCS Index	Modulation	Spatial Streams	Data Rates 800 ns GI	Data Rates 400 ns GI
24	BPSK	4	26.0 Mbps	28.9 Mbps
25	QPSK	4	52.0 Mbps	57.8 Mbps
26	QPSK	4	78.0 Mbps	86.7 Mbps
27	16-QAM	4	104.0 Mbps	115.6 Mbps
28	16-QAM	4	156.0 Mbps	173.3 Mbps
29	64-QAM	4	208.0 Mbps	231.1 Mbps
30	64-QAM	4	234.0 Mbps	260.0 Mbps
31	64-QAM	4	260.0 Mbps	288.9 Mbps

Table 18.4 depicts the modulation and coding schemes for a 40 MHz channel using one spatial stream.

TABLE 18.4 MCS—40 MHz Channel, One Spatial Stream

MCS Index	Modulation	Spatial Streams	Data Rates 800 ns GI	Data Rates 400 ns GI
0	BPSK	1	13.5 Mbps	15.0 Mbps
1	QPSK	1	27.0 Mbps	30.0 Mbps
2	QPSK	1	40.5 Mbps	45.0 Mbps
3	16-QAM	1	54.0 Mbps	60.0 Mbps
4	16-QAM	1	81.0 Mbps	90.0 Mbps
5	64-QAM	1	108.0 Mbps	120.0 Mbps
6	64-QAM	1	121.5 Mbps	135.0 Mbps
7	64-QAM	1	135.0 Mbps	150.0 Mbps

Table 18.5 depicts the modulation and coding schemes for a 40 MHz channel using four spatial streams.

TABLE 18.5 MCS—40 MHz Channel, Four Spatial Streams

MCS Index	Modulation	Spatial Streams	Data Rates 800 ns GI	400 ns GI
24	BPSK	4	54.0 Mbps	60.0 Mbps
25	QPSK	4	108.0 Mbps	120.0 Mbps
26	QPSK	4	162.0 Mbps	180.0 Mbps
27	16-QAM	4	216.0 Mbps	240.0 Mbps
28	16-QAM	4	324.0 Mbps	360.0 Mbps
29	64-QAM	4	432.0 Mbps	480.0 Mbps
30	64-QAM	4	486.0 Mbps	540.0 Mbps
31	64-QAM	4	540.0 Mbps	600.0 Mbps

Other factors such as the use of unequal modulation can also determine the final data rate. As depicted in Table 18.6, different spatial streams might use different modulation methods.

TABLE 18.6 MCS—40 MHz Channel, Four Spatial Streams, Unequal Modulation

	Modulation				Data Rates	
MCS Index	Stream 1	Stream 2	Stream 3	Stream 4	800 ns GI	400 ns GI
67	16-QAM	16-QAM	16-QAM	QPSK	283.4 Mbps	315.0 Mbps
68	64-QAM	QPSK	QPSK	QPSK	243.0 Mbps	270.0 Mbps

HT PHY

In earlier chapters, you learned that a MAC Service Data Unit (MSDU) is the layer 3–7 payload of an 802.11 data frame. You also learned that a MAC Protocol Data Unit (MPDU) is a technical name for an entire 802.11 frame. An MPDU consists of a layer 2 header, body, and trailer.

When an MPDU (802.11 frame) is sent down from layer 2 to the Physical layer, a preamble and PHY header are added to the MPDU. This creates what is called a *PLCP Protocol Data Unit (PPDU)*. Describing all the details of the PHY preamble and header is well beyond the scope of the CWNA exam. The main purpose of the preamble is to use bits to synchronize transmissions at the Physical layer between two 802.11 radios. The main purpose of the PHY header is to use a signal field to indicate how long it will take to transmit the 802.11 frame (MPDU). The 802.11 draft amendment defines the use of three PPDU structures that use three different preambles. One of the preambles is a legacy format, and two are newly defined HT preamble formats.

Non-HT Legacy

The first PPDU format is called *non-HT* and is often also referred to as a legacy format because it was originally defined by clause 17 of the 802.11-2007 standard for OFDM transmissions. As pictured in Figure 18.13, the non-HT PPDU consists of a preamble that uses short and long training symbols, which are used for synchronization. An OFDM symbol consists of 12 bits. The header contains the signal field, which indicates the time needed to transmit the payload of the non-HT PPDU, which of course is the MPDU (802.11 frame).

FIGURE 18.13 802.11n PPDU formats

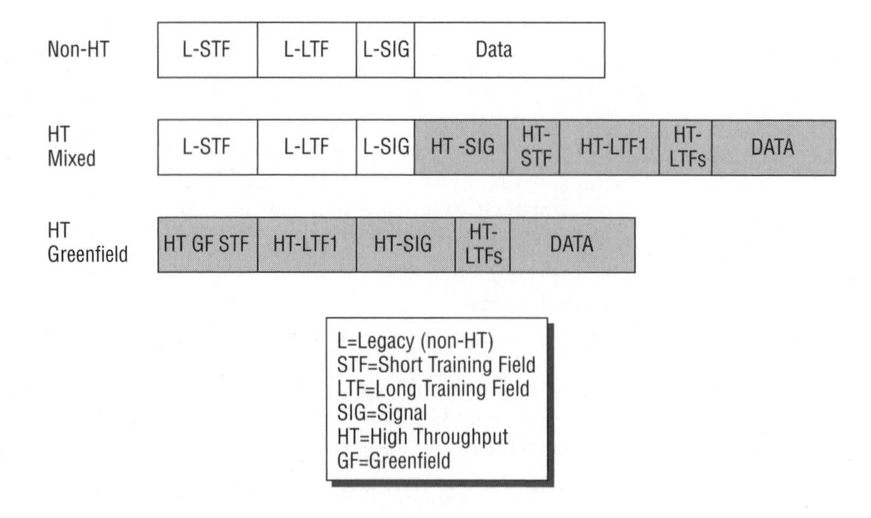

Support for the non-HT legacy format is mandatory for 802.11n radios, and transmissions can occur in only 20 MHz channels. The non-HT format effectively is the same format used by legacy 802.11a and 802.11g radios.

HT Mixed

The first of the two new PPDU formats defined in the 802.11n amendment is the *HT Mixed* format. As seen in Figure 18.13, the beginning of the preamble contains the non-HT short and long training symbols that can be decoded by legacy 802.11a and 802.11g radios. The rest of the HT Mixed preamble and header cannot be decoded by legacy 802.11a/g devices. The HT Mixed format will likely be the most commonly used format because it supports both HT and legacy 802.11a/g OFDM radios. The HT Mixed format is also considered mandatory, and transmissions can occur in both 20 MHz and 40 MHz channels. When a 40 MHz channel is used, all broadcast traffic must be sent on a legacy 20 MHz channel so as to maintain interoperability with the 802.11a/g non-HT clients. Also, any transmissions to and from the non-HT clients will have to use a legacy 20 MHz channel.

HT Greenfield

The second of the two new PPDU formats defined by the 802.11n amendment is the *HT Greenfield* format. As pictured in Figure 18.13, the preamble is not compatible with legacy 802.11a/g radios, and only HT radios can communicate when using the HT Greenfield format. Support for the HT Greenfield format is optional, and the HT radios can transmit by using both 20 MHz and 40 MHz channels.

HT MAC

So far, we have discussed all the enhancements to the Physical layer that 802.11n radios use to achieve greater bandwidth and throughput. The 802.1n amendment also addresses new enhancements to the MAC sublayer of the Data-Link layer to increase throughput and improve power management. Medium contention overhead is addressed by using two new methods of frame aggregation. New methods are also addressed using interframe spacing and block acknowledgments to limit the amount of fixed MAC overhead. Finally, two new methods of power management are defined for HT clause 20 radios.

A-MSDU

As pictured in Figure 18.14, every time a unicast 802.11 frame is transmitted, a certain amount of fixed overhead exists as a result of the PHY header, MAC header, MAC trailer, interframe spacing, and acknowledgment frame. Medium contention overhead also exists because of the time required when each frame must contend for the medium.

FIGURE 18.14 802.11 unicast frame overhead

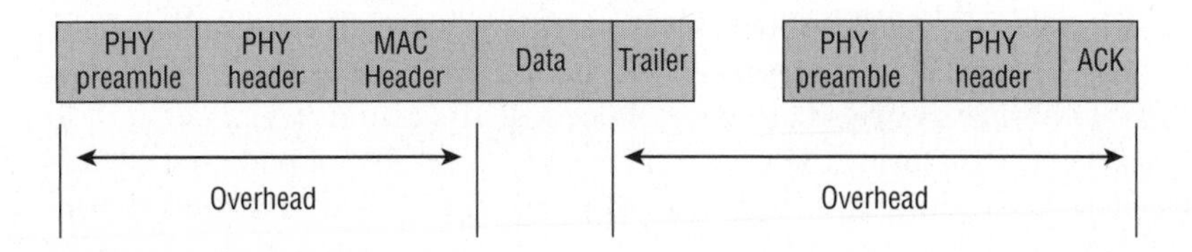

The 802.11n amendment introduces two new methods of frame aggregation to help reduce the overhead. *Frame aggregation* is a method of combining multiple frames into a single frame transmission. The fixed MAC layer overhead is reduced, the odds of collision are lowered, and overhead caused by the random back-off timer during medium contention is also minimized.

The first method of frame aggregation is known as *Aggregate MAC Service Data Unit (A-MSDU)*. As you learned in earlier chapters, the MSDU is the layer 3–7 payload of a data frame. As pictured in Figure 18.15, multiple MSDUs can be aggregated into a single frame transmission.

FIGURE 18.15 A-MSDU

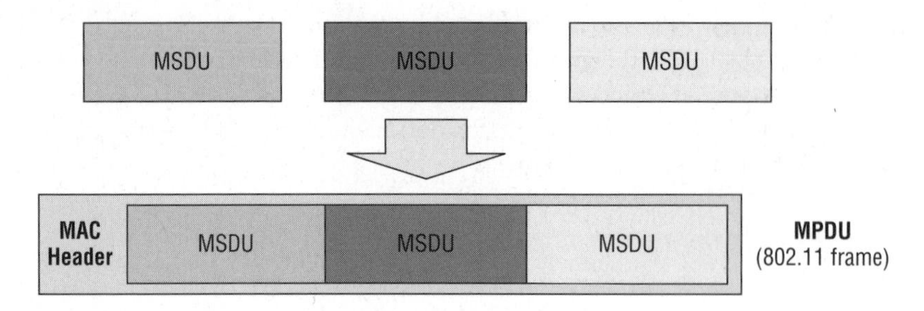

An 802.11n access point using A-MSDU aggregation would receive multiple 802.3 frames, remove the 802.3 headers and trailers, and then wrap the multiple MSDU payloads into a single 802.11 frame for transmission. The aggregated MSDUs will have a single destination when wrapped together in a single frame.

The entire aggregated frame can be encrypted by using either TKIP or CCMP. It should be noted, however, that the individual MSDUs must all be of the same 802.11e quality-of-service access category. Voice MSDUs cannot be mixed with Best Effort or Video MSDUs inside the same aggregated frame.

A-MPDU

The second method of frame aggregation is known as *Aggregate MAC Protocol Data Unit (A-MPDU)*. As you learned in earlier chapters, the MPDU is an entire 802.11 frame including the MAC header, body, and trailer. As pictured in Figure 18.16, multiple MPDUs can be aggregated into a single frame transmission.

The individual MPDUs within an A-MPDU must all have the same receiver address. Also, the data payload of each MPDU is encrypted separately by using either TKIP or CCMP. Much like MSDU aggregation, individual MPDUs must all be of the same 802.11e quality-of-service access category. Voice MPDUs cannot be mixed with Best Effort or Video MPDUs inside the same aggregated frame. Please note that MPDU aggregation has more overhead than MSDU aggregation because each MPDU has an individual MAC header and trailer.

FIGURE 18.16 A-MPDU

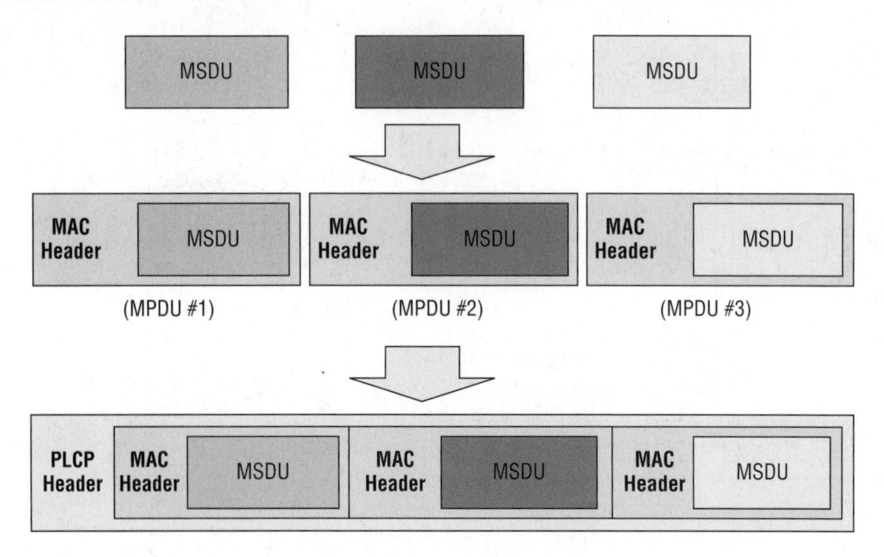

MTBA

As you learned in earlier chapters, all 802.11 unicast frames must be followed by an ACK frame for delivery verification purposes. Multicast and broadcast frames are not acknowledged. An A-MSDU contains multiple MSDUs all wrapped in a single frame with one MAC header and one destination. Therefore, only normal acknowledgments are required when using MSDU aggregation. However, an A-MPDU contains multiple MPDUs, each with their own unique MAC header. Each of the individual MPDUs must be acknowledged; this is accomplished by using a *multiple traffic ID block acknowledgment (MTBA)* frame. Block ACKs were first introduced by the 802.11e amendment as a method of acknowledging multiple individual 802.11 frames during a *frame burst*. As pictured in Figure 18.17, block acknowledgments are also needed to cover the multiple MPDUs that are aggregated inside a single A-MPDU transmission.

RIFS

The 802.11e QoS amendment introduced the capability for a transmitting radio to send a burst of frames during a transmit opportunity (TXOP). During the frame burst, a short interframe space (SIFS) was used between each frame to ensure that no other radios transmitted during the frame burst. The 802.11n amendment defines a new interframe space that is even shorter in time, called a *reduced interframe space (RIFS)*. A SIFS interval is 16 μ, whereas a RIFS interval is only 2 μ. A RIFS interval can be used in place of a SIFS interval, resulting is less overhead during a frame burst. It should be noted that RIFS intervals can be used only when a Greenfield HT network is in place. RIFS can only be used between HT radios, and no legacy devices can belong to the basic service set.

FIGURE 18.17 Block acknowledgments

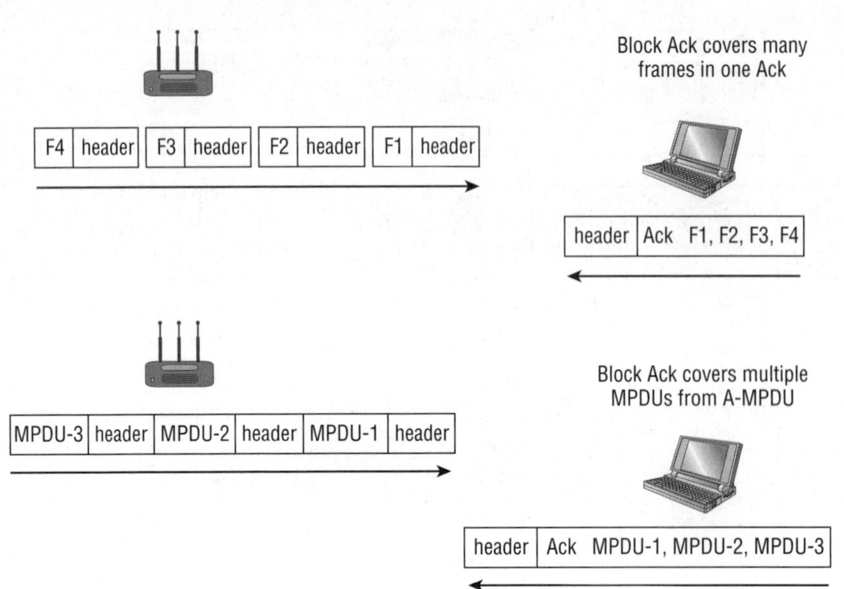

HT Power Management

As the 802.11 standard is amended, power-management capabilities continue to be enhanced. The 802.11e QoS amendment introduced unscheduled automatic power save delivery (U-APSD), which is the mechanism used by WMM Power Save. The 802.11n amendment introduces two new power-management mechanisms that can be used by HT clause 20 radios.

802.11n radios still support a "basic" Power Save mode, which is based on the original 802.11 power-management mechanisms. Access points buffer frames for stations in basic Power Save mode. The stations wake up when delivery traffic indication message (DTIM) beacons are broadcast and the stations download their buffered frames.

The first new power-management method is called *spatial multiplexing power save (SM power save)*. The purpose of SM power save is to allow a MIMO 802.11n device to power down all but one of its radios. For example, a 4×4 MIMO device with four radio chains would power down three of the four radios, thus conserving power. SM power save defines two methods of operation: static and dynamic.

When static SM power save is utilized, a MIMO client station powers down all the client's radios except for one single radio. Effectively, the client station is now the equivalent of an 802.11a/b/g radio that is capable of sending and receiving only one spatial stream. The client uses an SM power save action frame to inform the access point that the MIMO client is using only one radio and is capable of receiving only one spatial stream from the AP. The SM power save action frame is also used to tell the AP that the client station has powered up all of its radios and now is capable of transmitting and receiving multiple spatial streams once again.

When dynamic SM power save is utilized, the MIMO client can also power down all but one of the client's radios, but can power up the radios again much more rapidly. The client station disables all but one of the radios after a frame exchange. An access point can trigger the client to wake up the sleeping radios by sending a request-to-send (RTS) frame. The client station receives the RTS frame, powers up the sleeping radios, and sends a clear-to-send (CTS) frame back to the access point. The client can now once again transmit and receive multiple spatial streams. The client uses an SM power save action frame to inform the AP of the client's dynamic power save state.

The second new power-management method, *Power Save Multi Poll (PSMP)*, has also been defined for use by HT clause 20 radios. PSMP is an extension of automatic power save delivery (APSD) that was defined by the 802.11e amendment. Unscheduled PSMP is similar to U-APSD and uses the same delivery-enabled and trigger-enabled mechanisms. Scheduled PSMP is also very similar to S-APSD and is an effective method for streaming data and other scheduled transmissions.

HT Operation

802.11n access points can operate in several modes of channel operation. An access point could be manually configured to only transmit on legacy 20 MHz channels, although most 802.11n APs are configured to operate as a 20/40 basic service set. A 20/40 BSS allows 20 MHz 802.11a/g client stations and 20/40 MHz–capable 802.11n stations to operate within the same cell at the same time. In earlier chapters, you learned about the protection mechanisms used in an ERP (802.11g) network. RTS/CTS and CTS-to-Self mechanisms are used to ensure that 802.11b HR-DSSS clients do not transmit when ERP-OFDM transmissions are occurring. The 802.11n amendment requires backward compatibility with 802.11a and 802.1b/g radios. Therefore, the 802.11n amendment defines *HT protection modes* that enable HT clause 20 radios to be backward compatible with older clause 18 radios (HR-DSSS), clause 17 radios (OFDM), and clause 19 radios (ERP).

Much like the RTS/CTS mechanisms used by 802.11a/b/g networks, an HT network can use a Dual-CTS protection method to prevent hidden nodes as well as prevent interference with nearby legacy cells. The 802.11n amendment defines an optional mode of operation called phased coexistence operation (PCO) that divides time and alternates between 20 MHz and 40 MHz transmissions.

20/40 Channel Operation

20 MHz 802.11a/g stations and 20/40 MHz–capable 802.11n stations can operate within the same cell at the same time when they are associated to an HT access point. Older legacy 802.11a/g stations will obviously use 20 MHz transmissions. HT radios that are 20/40 capable can use 40 MHz transmissions when communicating with each other; however, they would need to use 20 MHz transmissions when communicating with the legacy stations. Several rules

apply for the operation of 20 MHz and 40 MHz stations within the same HT 20/40 basic service set. These rules include the following:

- The HT access point must declare 20 or 20/40 support in the beacon management frame.
- Client stations must declare 20 or 20/40 in the association or reassociation frames.
- Client stations must reassociate when switching between 20 and 20/40 modes.
- If 20/40-capable stations transmit by using a single 20 MHz channel, they must transmit on the primary channel and not the secondary channel.

HT Protection Modes (0–3)

To ensure backward compatibility with older 802.11 a/b/g radios, HT access points may signal to HT clause 20 radios when to use one of four HT protection modes. A field in the beacon frame called the HT Protection field has four possible settings of 0–3. Much like an ERP access point, the protection modes may change dynamically depending on devices that are nearby or associated to the HT access point. The protection mechanisms that are used are either RTS/CTS, CTS-to-Self, Dual-CTS, or other protection methods. The four modes are as follows:

Mode 0—Greenfield mode This mode is referred to as *Greenfield* because only HT radios are in use. All the HT client stations must also have the same operational capabilities. If the HT basic service set is a 20 MHz BSS, all the stations must be 20 MHz capable. If the HT basic service set is a 20/40 MHz BSS, all the stations must be 20/40 capable. If these conditions are met, there is no need for protection.

Mode 1—HT nonmember protection mode In this mode, all the stations in the BSS must be HT stations. Protection mechanisms kick in when a non-HT client station or non-HT access point is heard that is not a member of the BSS. For example, an HT AP and stations may be transmitting on a 40 MHz HT channel. A non-HT 802.11a access point or client station is detected to be transmitting in a 20 MHz space that interferes with either the primary or secondary channel of the 40 MHz HT channel.

Mode 2—HT 20 MHz protection mode In this mode, all the stations in the BSS must be HT stations and are associated to a 20/40 MHz access point. If a 20 MHz-only HT station associates to the 20/40 MHz AP, protection must be used. In other words, the 20/40 capable HT stations must use protection when transmitting on a 40 MHz channel in order to prevent the 20 MHz-only HT stations from transmitting at the same time.

Mode 3—HT Mixed mode This protection mode is used when one or more non-HT stations are associated to the HT access point. The HT basic service set can be either 20 MHz or 20/40 MHz capable. If any clause 18 radios (HR-DSSS), clause 17 radios (OFDM), or clause 19 radios (ERP) associate to the BSS, protection will be used. Mode 3 will probably be the most commonly used protection mode because most basic service sets will most likely have legacy devices as members.

Dual-CTS Protection

The 802.11n amendment defines an operational mechanism that accounts for coexistence between 802.11n HT coverage cells and nearby legacy 802.11a/b/g coverage cells. As pictured in Figure 18.18, when either an HT or non-HT station transmits a frame, the station first sends a request-to-send (RTS) frame to the HT access point. The access point will then reply with two clear-to-send (CTS) frames. One CTS frame is in the legacy non-HT format, while the other frame is in the HT format. The RTS/CTS process resets the NAV timer for all stations within the HT cell as well as any neighboring clients or APs that hear the RTS/CTS transmissions.

The HT access point protects its transmissions by sending both a legacy CTS-to-Self frame and an HT CTS-to-Self frame. *Dual-CTS protection* has two benefits. First, using the legacy RTS/CTS and legacy CTS-to-Self frames to reset NAV timers, interference with any nearby legacy 802.11a/b/g cells is prevented. Second, the hidden node problem is resolved within the HT cell. It should be noted that the use of Dual-CTS protection will adversely affect the throughput of the cell because of the extra overhead from the RTS/CTS and CTS-to-Self frames.

FIGURE 18.18 Dual-CTS protection

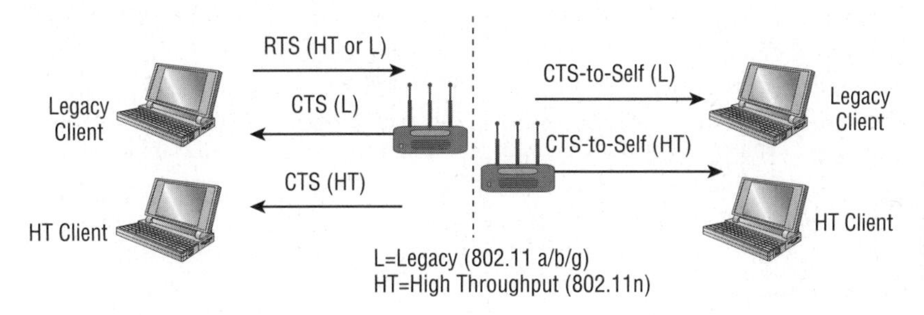

Phased Coexistence Operation (PCO)

Another operational mechanism that accounts for coexistence between 802.11n HT coverage cells and nearby legacy 802.11a/b/g coverage cells is *phased coexistence operation (PCO)*. This is an optional mode of operation that divides time and alternates between 20 MHz and 40 MHz transmissions. The HT access point designates time slices for 20 MHz operations in both primary and secondary 20 MHz channels and designates time slices for 40 MHz transmissions. As pictured in Figure 18.19, the HT access point uses CTS-to-Self frames to reset the NAV timers of the appropriate clients when switching back and forth between 20 MHz and 40 MHz channels. The main advantage of PCO is that no protection mechanisms are needed during the 40 MHz operational phase. PCO might improve throughput in some situations. However, switching back and forth between channels could increase jitter, and therefore PCO mode would not be recommended when VoWiFi phones are deployed.

FIGURE 18.19 Phased coexistence operation

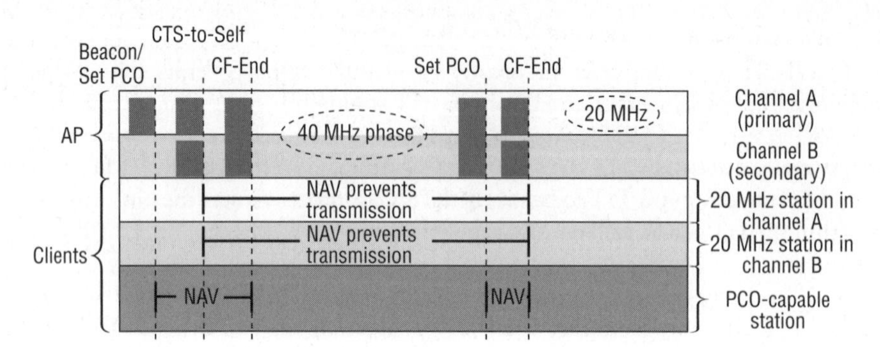

Summary

In this chapter, you learned the history of the 802.11n amendment and how the Wi-Fi Alliance has already begun to certify 802.11n equipment for interoperability. We also discussed all the methods used by clause 20 HT radios to increase throughput and range at the Physical layer. In addition to PHY enhancements, HT radios also utilize MAC layer mechanisms to enhance throughput and power management. Finally, we discussed HT modes of operation that are used for protection mechanisms and co-existence with older legacy 802.11a/b/g technologies. 802.11n technology is "next generation" technology because of the promise of greater throughput as well as greater range. Although 802.11n is still in its infancy in regards to enterprise acceptance and deployment, the future for this technology is remarkable.

Exam Essentials

Define the differences between MIMO and SISO. Understand that SISO devices use only one radio chain, whereas MIMO systems use multiple radio chains.

Understand spatial multiplexing. Describe how SM takes advantage of multipath and sends multiple spatial streams resulting in increased throughput.

Explain MIMO diversity. Be able to explain the differences between simple switched diversity and the advanced diversity used by MIMO antenna systems. Explain the use of maximal ratio combining with MIMO diversity.

Understand transmit beamforming. Explain how optional transmit beamforming can be used to steer beams in an optimal path toward a receiving radio and the benefits of the beamforming process.

Understand 20 MHz and 40 MHz channels. Understand legacy 20 MHz channels, 20 MHz HT channels, and 40 MHz channels and how they use OFDM. Explain why 40 MHz channels work best in the 5 GHz UNII bands. Explain primary and secondary channels.

Explain the guard interval. Describe how the guard interval compensates for intersymbol interference. Discuss the use of both 800- and 400-nanosecond GIs.

Understand modulation coding schemes. Explain how modulation coding schemes are used to define data rates and all the variables that can affect the data rates.

Explain the three HT PPDU formats. Describe the differences between non-HT legacy, HT Mixed, and HT Greenfield.

Understand HT MAC enhancements. Explain how the use of A-MSDU, A-MPDU, block ACKs, and RIFS are used to increase throughput at the MAC sublayer. Define the two new power-management methods used by HT radios.

Explain the HT protection modes. Describe the differences between protection modes 0–3. Explain the use of Dual-CTS.

Key Terms

Before you take the exam, be certain you are familiar with the following terms:

adaptive antenna array
Aggregate MAC Protocol Data Unit (A-MPDU)
Aggregate MAC Service Data Unit (A-MSDU)
antenna diversity
beamformee
beamformer
digital signal processing (DSP)
Dual-CTS protection
equal modulation
explicit feedback
frame aggregation
guard interval (GI)
High Throughput (HT)
HT Greenfield
HT Mixed
HT protection modes
implicit feedback
intersymbol interference (ISI)
maximal ratio combining (MRC)
modulation and coding scheme (MCS)
multipath
multiple traffic ID block acknowledgment (MTBA)
multiple-input multiple-output (MIMO)
non-HT
Orthogonal Frequency Division Multiplexing (OFDM)
phased coexistence operation (PCO)
PLCP Protocol Data Unit (PPDU)
Power Save Multi Poll (PSMP)
radio chain
reduced interframe space (RIFS)
single-input single-output (SISO)
sounding frames
spatial diversity
spatial diversity multiplexing (SDM)
spatial multiplexing (SM)
spatial multiplexing power save (SM power save)
spatial stream
switched antenna array
switched diversity
symbols
transmit beamforming (TxBF)
unequal modulation

Review Questions

1. A WLAN consultant has been asked to plan an 802.11n Greenfield deployment using dual-frequency 4×4 MIMO access points. What are the options available to provide power for the 802.1n access points? (Choose all that apply.)
   - A. 802.3-2005 clause 33 PoE
   - B. 802.3AT PoE
   - C. Proprietary PoE solution
   - D. Battery pack
   - E. Electrical outlet power
2. How can a MIMO system increase throughput at the Physical layer? (Choose all that apply.)
   - A. Spatial multiplexing
   - B. A-MPDU
   - C. Transmit beamforming
   - D. 40 MHz channels
   - E. Dual-CTS protection
3. Which new power-management method defined by the 802.11n amendment conserves power by powering down all but one radio?
   - A. A-MPDU
   - B. Power Save protection
   - C. PSMP
   - D. SM power save
   - E. PS mode
4. The guard interval is used as a buffer to compensate for what type of interference?
   - A. Co-channel interference
   - B. Adjacent cell interference
   - C. RF interference
   - D. HT interference
   - E. Intersymbol interference
5. Name some of the factors that a modulation and coding scheme (MCS) uses to define data rates for an HT radio. (Choose all that apply.)
   - A. Modulation method
   - B. Equal/unequal modulation
   - C. Number of spatial streams
   - D. GI
   - E. Channel size

6. How can an HT radio increase throughput at the MAC sublayer of the Data-Link layer? (Choose all that apply.)
   - A. A-MSDU
   - B. RIFS
   - C. A-MPDU
   - D. Guard interval
   - E. MTBA
7. Transmit beamforming uses what type of frames to analyze the MIMO channel before transmitting directed paths of data?
   - A. Action frames
   - B. Trigger frames
   - C. Beaming frames
   - D. Sounding frames
   - E. SM power save action frames
8. A 3×4 MIMO radio transmits how many spatial streams of unique data?
   - A. Three.
   - B. Four.
   - C. Three equal and four unequal streams.
   - D. None—the streams are not unique data.
9. Name a capability not defined for A-MPDU.
   - A. Multiple QoS access categories
   - B. Independent data payload encryption
   - C. Individual MPDUs have the same receiver address
   - D. MPDU aggregation
10. Which HT protection modes allow only for the association of HT stations in the HT basic service set? (Choose all that apply.)
    - A. Mode 0—Greenfield mode
    - B. Mode 1—HT nonmember protection mode
    - C. Mode 2—HT 20 MHz protection mode
    - D. Mode 3—HT Mixed mode
11. Which of these capabilities are considered mandatory for the Wi-Fi Alliance's vendor certification program called Wi-Fi CERTIFIED™ 802.11n draft 2.0? (Choose all that apply.)
    - A. Three spatial streams in receive mode
    - B. WPA/WPA2
    - C. WMM
    - D. Two spatial streams in transmit mode
    - E. 2.4 GHz–40 MHz channels

12. Which HT operational mode uses time slices to alternate between 20 MHz channel transmissions and 40 MHz transmissions?

A. Non-HT legacy

B. PCO mode

C. HT Mixed mode

D. HT Phased mode

13. HT clause 20 radios are backward compatible with which of the following type of 802.11 radios? (Choose all that apply.)

A. Clause 18 radios (HR-DSSS)

B. Clause 17 radios (OFDM)

C. Clause 14 radios (FHSS)

D. Clause 19 radios (ERP)

14. Which HT operational mode is not recommended when VoWiFi phones are deployed?

A. Non-HT legacy

B. PCO mode

C. HT Mixed mode

D. HT Greenfield mode

15. Which HT PPDU formats support both 20 MHz and 40 MHz channels? (Choose all that apply.)

A. Non-HT legacy format

B. PCO mode

C. HT Mixed format

D. HT Greenfield format

16. A WLAN consultant has recommend that a new 802.11n HT network be deployed by using channels in the 5 GHz UNII bands. Why would he recommend 5 GHz over 2.4 GHz?

A. HT radios do not require DFS and TPC in the 5 GHz bands.

B. HT radios get better range using TxBF in the 5 GHz bands.

C. 40 MHz channels do not scale in the 2.4 GHz ISM band.

D. 5 GHz HT radios are less expensive than 2.4 GHz HT radios.

17. What are the benefits of using Dual-CTS protection in an HT basic service set? (Choose all that apply.)

A. Resolves hidden node problems

B. Prevents intersymbol interference

C. Prevents interference with nearby 802.11a/b/g basic service sets

D. Allows for 20 MHz and 40 MHz channel operation

18. What frequencies are defined for clause 20 HT radio transmissions? (Choose all that apply.)

A. 902–928 MHz

B. 2.4–2.4835 GHz

C. 5.15–5.25 GHz

D. 5.25–5.35 MHz

19. What PHY layer mechanism might be used to increase throughput for an HT radio in a clean RF environment with minimal reflections and low multipath?

A. Maximum ratio combining

B. 400-nanosecond guard interval

C. Switched diversity

D. Spatial multiplexing

E. Spatial diversity

20. What PHY layer mechanisms might be used to increase the range for a 802.11n radio using a MIMO system? (Choose all that apply.)

A. Switched diversity

B. Guard interval

C. Transmit beamforming

D. Spatial multiplexing

Answers to Review Questions

1. C, E. The 802.11n standard allows for MIMO systems to have up to four transmit and receive radios, using four radio chains. Each radio chain requires power. A 2×2 MIMO system would require much less power draw than a 4×4 MIMO system. The 802.3-2005 clause 33 PoE standard allows for a maximum of 15.4 watts. 802.11n dual-frequency access points with 3×3 and 4×4 MIMO systems will require more than 15.4 watts. The 802.3AT draft amendment may provide for more power in the future. In the meantime, the best options for powering many 802.11n access points are proprietary PoE solutions or an electrical outlet.

2. A, C, D. Spatial multiplexing transmits multiple streams of unique data at the same time. If a MIMO access point sends two unique data streams to a MIMO client who receives both streams, the throughput is effectively doubled. If a MIMO access point sends three unique data streams to a MIMO client who receives all three streams, the throughput is effectively tripled. Because transmit beamforming results in constructive multipath communication, the result is a higher signal-to-noise ratio and greater received amplitude. Transmit beamforming will result in higher throughput because of the higher SNR that allows for the use of more-complex modulation methods that can encode more data bits. 40 MHz HT channels effectively double the frequency bandwidth which results in greater throughput. A-MPDU and Dual-CTS protection are MAC layer mechanisms.

3. D. Spatial multiplexing power save (SM power save) allows a MIMO 802.11n device to power down all but one of its radios. For example, a 4×4 MIMO device with four radio chains would power down three of the four radios, thus conserving power. SM power save defines two methods of operation: static and dynamic.

4. E. The guard interval acts as a buffer for the delay spread, and the normal guard interval is an 800-nanosecond buffer between symbol transmissions. The guard interval will compensate for the delay spread and help prevent intersymbol interference. If the guard interval is too short, intersymbol interference will still occur. HT radios also have the capability of using a shorter 400-nanosecond GI.

5. A, B, C, D, E. HT radios use modulation and coding schemes to define data rates based on numerous factors including modulation type, the number of spatial streams, channel size, guard interval, equal/unequal modulation, and other factors. Each modulation and coding scheme (MCS) is a variation of these multiple factors. A total of 77 modulation and coding schemes exist for both 20 MHz HT channels and 40 MHz HT channels.

6. A, B, C, E. The 802.11n amendment introduces two new methods of frame aggregation to help reduce overhead and increase throughput. Frame aggregation is a method of combining multiple frames into a single frame transmission. The two types of frame aggregation are A-MSDU and A-MPDU. Multiple traffic ID block acknowledgment (MTBA) frames are used to acknowledge A-MPDUs. Block ACKs result in less overhead. RIFS is a 2-microsecond interframe space that can be used in an HT Greenfield network during frame bursts. The 2-microsecond interframe space is less overhead than the more commonly used SIFS. Guard intervals are used at the Physical layer.

7. D. An 802.11n transmitter that uses beamforming will try to adjust the phase of the signals based on feedback from the receiver using sounding frames. The transmitter is considered the beamformer, while the receiver is considered the beamformee. The beamformer and the beamformee work together to educate each other about the characteristics of the MIMO channel.

8. A. MIMO radios transmit multiple radio signals at the same time and take advantage of multipath. Each individual radio signal is transmitted by a unique radio and antenna of the MIMO system. Each independent signal is known as a spatial stream, and each stream can contain different data than the other streams transmitted by one or more of the other radios. A 3×4 MIMO system can transmit three unique data streams. The 3 is the number of transmitters, and the 4 is the number of receivers.

9. A. Multiple MPDUs can be aggregated into one frame. The individual MPDUs within an A-MPDU must all have the same receiver address. However, individual MPDUs must all be of the same 802.11e quality-of-service access category.

10. A, B, C. Modes 0, 1, and 2 all define protection to be used in various situations where only HT stations are allowed to associate to an HT access point. Mode 3—HT Mixed mode—defines the use of protection when both HT and non-HT radios are associated to an HT access point.

11. B, C, D. Some of the mandatory baseline requirements of Wi-Fi CERTIFIED™ 802.11n draft 2.0 include two spatial streams in both transmit and receive mode, WPA/WPA2 certification, WMM certification, and support for 40 MHz channels in the 5 GHz UNII bands. 40 MHz channels in 2.4 GHz are not required.

12. B. Phased coexistence operation (PCO) is an optional mode of operation that divides time and alternates between 20 MHz and 40 MHz transmissions. The HT access point designates time slices for 20 MHz operations in both the primary and secondary 20 MHz channels and designates time slices for 40 MHz transmissions.

13. A, B, D. HT clause 20 radios are backward compatible with older clause 18 radios (HR-DSSS), clause 17 radios (OFDM), and clause 19 radios (ERP). In other words, 802.11n radios are backward compatible with 802.11b, 802.11a, and 802.11g radios. HT radios are not backward compatible with legacy frequency hopping radios.

14. B. Phased coexistence operation (PCO) is an optional mode of operation that divides time and alternates between 20 MHz and 40 MHz transmissions. The HT access point designates time slices for 20 MHz operations in both primary and secondary 20 MHz channels and designates time slices for 40 MHz transmissions. PCO might improve throughput in some cases but would potentially increase jitter; therefore, PCO mode would not be recommended when VoWiFi phones are deployed.

15. C, D. The HT Mixed format is considered mandatory and transmissions can occur in both 20 MHz and 40 MHz channels. Support for the HT Greenfield format is optional, and the HT radios can transmit by using both 20 MHz and 40 MHz channels. Support for the non-HT legacy format is mandatory for 802.11n radios, and transmissions can occur in only 20 MHz channels. PCO is not a PPDU format.

16. C. Deploying 40 MHz HT channels at 2.4 GHz does not scale properly in multiple channel architecture. Although fourteen channels are available at 2.4 GHz, there are only three non-overlapping 20 MHz channels available in the 2.4 GHz ISM band. When the smaller channels are bonded together to form 40 MHz channels in the 2.4 GHz ISM band, any two 40 MHz channels will overlap. Channel reuse patterns are not possible with 40 MHz channels in the 2.4 GHz ISM band.

17. A, C. Dual-CTS protection has two benefits. First, using the legacy RTS/CTS and legacy CTS-to-Self frames to reset NAV timers prevents interference with any nearby legacy 802.11a/b/g cells. Second, the hidden node problem is resolved within the HT cell.

18. B, C. Other 802.11 technologies are frequency dependent on a single RF band. For example, clause 14 radios (FHSS), clause 18 radios (HR-DSSS), and clause 19 radios (ERP) can transmit in only the 2.4 GHz ISM band. Clause 17 radios (OFDM) are restricted to the 5 GHz UNII bands. Clause 20 radios (HT) are not locked to a single frequency band and can transmit on both the 2.4 GHz ISM band and the 5 GHz UNII bands.

19. B. 802.11n also uses an 800-nanosecond guard interval; however, a shorter 400-nanosecond guard interval is optional. A shorter guard interval results in a shorter symbol time, which has the effect of increasing data rates by about 10 percent. If the optional shorter 400-nanosecond guard interval is used with an 802.11n radio, throughput should increase. However, if inter-symbol interference occurs because of multipath, the result is data corruption. If data corruption occurs, layer 2 retransmissions will increase and the throughput will be adversely affected. Therefore, a 400-nanosecond guard interval should be used in only good RF environments. If throughput goes down because of a shorter GI setting, the default guard interval setting of 800 nanoseconds should be used instead.

20. A, C. As the distance between a transmitter and receiver increases, the received signal amplitude decreases to levels closer to the noise floor. Using switched diversity to listen with three or four antennas for the best received signal increases the probabilistic odds of hearing signals with stronger amplitude and better SNR. Switched diversity using three or four antennas often results in increased range. Because transmit beamforming results in constructive multipath communication, the result is a higher signal-to-noise ratio and greater received amplitude. Therefore, transmit beamforming will result in greater range for individual clients communicating with an access point.

About the Companion CD

IN THIS APPENDIX:

- ✓ What you'll find on the CD
- ✓ System requirements
- ✓ Using the CD
- ✓ Troubleshooting

What You'll Find on the CD

The following sections are arranged by category and summarize the software and other goodies you'll find on the CD. If you need help installing the items provided on the CD, refer to the installation instructions in the "Using the CD" section of this appendix.

Some programs on the CD might fall into one of these categories:

Shareware programs are fully functional, free, trial versions of copyrighted programs. If you like particular programs, register with their authors for a nominal fee and receive licenses, enhanced versions, and technical support.

Freeware programs are free, copyrighted games, applications, and utilities. You can copy them to as many computers as you like—for free—but they offer no technical support.

GNU software is governed by its own license, which is included inside the folder of the GNU software. There are no restrictions on distribution of GNU software. See the GNU license at the root of the CD for more details.

Trial, demo, or evaluation versions of software are usually limited either by time or by functionality (such as not letting you save a project after you create it).

White papers serve as additional reference material.

Sybex Test Engine

For Windows

The CD contains the Sybex Test Engine, which includes all of the assessment test and chapter review questions in electronic format, as well as three bonus exams located only on the CD.

Electronic Flashcards

For PC, Pocket PC, and Palm

These handy electronic flashcards are just what they sound like. One side contains a question or fill-in-the-blank statement, and the other side shows the answer.

System Requirements

Make sure your computer meets the minimum system requirements shown in the following list. If your computer doesn't match up to most of these requirements, you may have problems using the software and files on the companion CD. For the latest and greatest information, please refer to the readme file located at the root of the CD-ROM.

- A PC running Microsoft Windows 98, Windows 2000, Windows NT4 (with SP4 or later), Windows Me, Windows XP, or Windows Vista
- An Internet connection
- A CD-ROM drive

Using the CD

To install the items from the CD to your hard drive, follow these steps:

1. Insert the CD into your computer's CD-ROM drive. The license agreement appears.

Windows users: The interface won't launch if you have autorun disabled. In that case, click Start ➢ Run (for Windows Vista, Start ➢ All Programs ➢ Accessories ➢ Run). In the dialog box that appears, type **D:\Start.exe**. (Replace *D* with the proper letter if your CD drive uses a different letter. If you don't know the letter, see how your CD drive is listed under My Computer.) Click OK.

2. Read the license agreement, and then click the Accept button if you want to use the CD.

The CD interface appears. The interface enables you to access the content with just one or two clicks.

Troubleshooting

Wiley has attempted to provide programs that work on most computers with the minimum system requirements. Alas, your computer may differ, and some programs may not work properly for some reason.

The two likeliest problems are that you don't have enough memory (RAM) for the programs you want to use or you have other programs running that are affecting installation or running of a program. If you get an error message such as "Not enough memory" or

"Setup cannot continue," try one or more of the following suggestions and then try using the software again:

Turn off any antivirus software running on your computer. Installation programs sometimes mimic virus activity and may make your computer incorrectly believe that it's being infected by a virus.

Close all running programs. The more programs you have running, the less memory is available to other programs. Installation programs typically update files and programs; so if you keep other programs running, installation may not work properly.

Have your local computer store add more RAM to your computer. This is, admittedly, a drastic and somewhat expensive step. However, adding more memory can really help the speed of your computer and allow more programs to run at the same time.

Customer Care

If you have trouble with the book's companion CD-ROM, please call the Wiley Product Technical Support phone number at (877) 762-2974. Outside the United States, call +1(317) 572-3994. You can also contact Wiley Product Technical Support at `http://sybex.custhelp.com`. John Wiley & Sons will provide technical support only for installation and other general quality-control items. For technical support on the applications themselves, consult the program's vendor or author.

To place additional orders or to request information about other Wiley products, please call (877) 762-2974.

Appendix B

Abbreviations, Acronyms, and Regulations

Certifications

CWNA Certified Wireless Network Administrator

CWNE Certified Wireless Network Expert

CWNP Certified Wireless Network Professional

CWNT Certified Wireless Network Trainer

CWSP Certified Wireless Security Professional

CWTS Certified Wireless Technology Specialist

Organizations and Regulations

ACMA Australian Communications and Media Authority

ARIB Association of Radio Industries and Businesses (Japan)

ATU African Telecommunications Union

CEPT European Conference of Postal and Telecommunications Administrations

CITEL Inter-American Telecommunication Commission

CTIA Cellular Telecommunications and Internet Association

ERC European Radiocommunications Committee

EWC Enhanced Wireless Consortium

FCC Federal Communications Commission

FIPS Federal Information Processing Standards

GLBA Gramm-Leach-Bliley Act

HIPAA Health Insurance Portability and Accountability Act

IEEE Institute of Electrical and Electronics Engineers

IETF Internet Engineering Task Force

ISO International Organization for Standardization

NEMA National Electrical Manufacturers Association

NIST National Institute of Standards and Technology

RCC Regional Commonwealth in the field of Communications

SEEMesh Simple, Efficient, and Extensible Mesh

TGn Sync Task Group n Sync

WECA Wireless Ethernet Compatibility Alliance

WIEN Wireless InterWorking with External Networks

Wi-Fi Alliance Wi-Fi Alliance

WiMA Wi-Mesh Alliance

WNN Wi-Fi Net News

WWiSE World-Wide Spectrum Efficiency

Measurements

dB decibel

dBd decibel referenced to a dipole antenna

dBi decibel referenced to an isotropic radiator

dBm decibel referenced to 1 milliwatt

GHz gigahertz

Hz hertz

KHz kilohertz

mA milliampere

MHz megahertz

mW milliwatt

SNR signal-to-noise ratio

V volt

VDC voltage direct current

W watt

Technical Terms

AAA authorization, authentication, and accounting

AC access category

AC alternating current

ACK acknowledgment

AES Advanced Encryption Standard

AGL above ground level

AID association identifier

AIFS arbitration interframe space

AKM Authentication and Key Management

AM amplitude modulation

A-MPDU Aggregate MAC Protocol Data Unit

A-MSDU Aggregate MAC Service Data Unit

AP access point

APSD automatic power save delivery

ARS adaptive rate selection

ARS automatic rate selection

AS authentication server

ASK Amplitude Shift Keying

ATF airtime fairness

ATIM announcement traffic indication message

BA Block Acknowledgment

BER bit error rate

BPSK Binary Phase Shift Keying

BSA basic service area

BSS basic service set

BSSID basic service set identifier

BT Bluetooth

BVI Bridged Virtual Interface

CAD computer-aided design

CAM content addressable memory

CAM Continuous Aware mode

CAPWAP Control and Provisioning of Wireless Access Points

CCA clear channel assessment

CC-AP cooperative control access point

CCI co-channel interference

CCK Complementary Code Keying

CCMP Counter Mode with Cipher Block Chaining Message Authentication Code Protocol

CCX Cisco Compatible Extensions

CF CompactFlash

CF contention free

CFP contention-free period

CLI command-line interface

CP contention period

CRC cyclic redundancy check

CSMA/CA Carrier Sense Multiple Access with Collision Avoidance

CSMA/CD Carrier Sense Multiple Access with Collision Detection

CTS clear to send

CW contention window

CWG-RF Converged Wireless Group–RF Profile

DA destination address

DBPSK Differential Binary Phase Shift Keying

DC direct current

DCF Distributed Coordination Function

DDF distributed data forwarding

DFS dynamic frequency selection

DHCP Dynamic Host Configuration Protocol

DIFS Distributed Coordination Function interframe space

DoS denial of service

DQPSK Differential Quadrature Phase Shift Keying

DRS dynamic rate switching

DS distribution system

DSAS distributed spectrum analysis system

DSCP differentiated services code point

DSM distribution system medium

DSP digital signal processing

DSRC Dedicated Short Range Communications

DSS distribution system services

DSSS direct sequencing spread spectrum

DTIM delivery traffic indication message

EAP Extensible Authentication Protocol

EDCA Enhanced Distributed Channel Access

EEG enterprise encryption gateway

EIFS extended interframe space

EIRP equivalent isotropically radiated power

EM electromagnetic

EQM equal modulation

ERP Extended Rate Physical

ERP-CCK Extended Rate Physical–Complementary Code Keying

ERP-DSSS Extended Rate Physical–Direct Sequencing Spread Spectrum

ERP-OFDM Extended Rate Physical–Orthogonal Frequency Division Multiplexing

ERP-PBCC Extended Rate Physical–Packet Binary Convolutional Coding

ESA extended service area

ESS extended service set

ESSID extended service set identifier

EUI extended unique identifier

EWG enterprise wireless gateway

FAST Flexible Authentication via Secure Tunnel

FCS frame check sequence

FEC forward error correction

FHSS frequency hopping spread spectrum

FM frequency modulation

FMC fixed mobile convergence

FSK Frequency Shift Keying

FSPL free space path loss

FSR fast secure roaming

FT fast BSS transition

FZ Fresnel zone

GFSK Gaussian Frequency Shift Keying

GI guard interval

GMK Group Master Key

GPS Global Positioning System

GRE Generic Routing Encapsulation

GSM Global System for Mobile Communications

GTC Generic Token Card

GTK Group Temporal Key

GUI graphical user interface

HC hybrid coordinator

HCCA Hybrid Coordination Function Controlled Channel Access

HCF Hybrid Coordination Function

HR-DSSS High-Rate Direct Sequencing Spread Spectrum

HSRP Hot Standby Router Protocol

HT High Throughput

HT-GF-STF high-throughput Greenfield short training field

HT-LTF high-throughput long training field

HT-SIG high-throughput SIGNAL field

HT-STF high-throughput short training field

HTTPS Hypertext Transfer Protocol Secure

HWMP Hybrid Wireless Mesh Protocol

IAPP Inter-Access Point Protocol

IBSS independent basic service set

ICMP Internet Control Message Protocol

ICV Integrity Check Value

IDS intrusion detection system

IE Information Element

IFS interframe space

IP Internet Protocol

IPsec Internet Protocol Security

IR infrared

IR intentional radiator

IS integration service

ISI intersymbol interference

ISM Industrial, Scientific, and Medical

ITS Intelligent Transportation Systems

IV Initialization Vector

L2TP Layer 2 Tunneling Protocol

LAN local area network

LEAP Lightweight Extensible Authentication Protocol

LLC Logical Link Control

L-LTF legacy (non-HT) long training field

LOS line of sight

L-SIG legacy (non-HT) signal

L-STF legacy (non-HT) short training field

LWAPP Lightweight Access Point Protocol

MAC media access control

MAHO Mobile Assisted Hand-Over

MAN metropolitan area network

MAP mesh access point

MCA multiple channel architecture

MCS modulation and coding schemes

MD5 Message Digest 5

MDI media dependent interface

MFP management frame protection

MIB Management Information Base

MIC Message Integrity Check

MIMO multiple-input multiple-output

MMPDU Management MAC Protocol Data Unit

MPDU MAC Protocol Data Unit

MP mesh point

MPP mesh point collocated with a mesh portal

MPPE Microsoft Point-to-Point Encryption

MRC maximal ratio combining

MSDU MAC Service Data Unit

MTBA multiple traffic ID block acknowledgment

MTU maximum transmission unit

NAT Network Address Translation

NAV network allocation vector

nQSTA Non–Quality of Service Station

OFDM Orthogonal Frequency Division Multiplexing

OS operating system

OSI model Open Systems Interconnection model

OUI Organizationally Unique Identifier

PAN personal area network

PAT Port Address Translation

PBCC Packet Binary Convolutional Coding

PBX private branch exchange

PC point coordinator

PCI Peripheral Component Interconnect

PCF Point Coordination Function

PCO phased coexistence operation

PCMCIA Personal Computer Memory Card International Association (PC Card)

PD powered device

PEAP Protected Extensible Authentication Protocol

PHY physical layer

PIFS Point Coordination Function interframe space

PLCP Physical Layer Convergence Procedure

PMD Physical Medium Dependent

PMK Pairwise Master Key

PN pseudo-random number

PoE Power over Ethernet

POP Post Office Protocol

PPDU PLCP Protocol Data Unit

PPP Point-to-Point Protocol

PPTP Point-to-Point Tunneling Protocol

PSE power-sourcing equipment

PSK Phase Shift Keying

PSK preshared key

PSMP Power Save Multi Poll

PSPF Public Secure Packet Forwarding

PS-Poll power save poll

PSTN public switched telephone network

PTK Pairwise Transient Key

PtMP point-to-multipoint

PtP point-to-point

QAM quadrature amplitude modulation

QAP quality-of-service access point

QBSS quality-of-service basic service set

QoS quality of service

QSTA quality-of-service station

QPSK Quadrature Phase Shift Keying

RA receiver address

RADIUS Remote Authentication Dial-In User Service

RBAC role-based access control

RF radio frequency

RFC request for comment

RFSM radio frequency spectrum management

RIFS reduced interframe space

RRM radio resource measurement

RSL received signal level

RSN robust security network

RSNA robust security network association

RSSI received signal strength indicator

RTLS real-time location system

RTS request to send

RTS/CTS request to send/clear to send

RWG residential wireless gateway

RX receive or receiver

SA source address

S-APSD scheduled automatic power save delivery

SCA single channel architecture

SD Secure Digital

SDR software defined radio

SIFS short interframe space

SISO single-input single-output

SM spatial multiplexing

SMTP Simple Mail Transfer Protocol

SNMP Simple Network Management Protocol

SNR signal-to-noise ratio

SOHO small office/home office

SOM system operating margin

SQ signal quality

SSH Secure Shell

SSID service set identifier

SSL Secure Sockets Layer

STA station

STC Space Time Coding

STP Spanning Tree Protocol

TA transmitter address

TBTT target beacon transmission time

TCP/IP Transmission Control Protocol/Internet Protocol

TIM traffic indication map

TKIP Temporal Key Integrity Protocol

TLS Transport Layer Security

TPC transmit power control

TSN transition security network

TTLS Tunneled Transport Layer Security

TX transmit or transmitter

TxBF transmit beamforming

TXOP transmit opportunity

U-APSD unscheduled automatic power save delivery

UEQM unequal modulation

UNII Unlicensed National Information Infrastructure

UP user priority

USB Universal Serial Bus

VLAN virtual local area network

VoIP Voice over IP

VoWiFi Voice over Wi-Fi

VoWIP Voice over Wireless IP

VPN virtual private network

VRRP Virtual Router Redundancy Protocol

VSWR voltage standing wave ratio

WAN wide area network

WAVE Wireless Access in Vehicular Environments

WDS wireless distribution system

WEP Wired Equivalent Privacy

WGB workgroup bridge

WIDS wireless instruction detection system

Wi-Fi Sometimes said to be an acronym for *wireless fidelity*, a term that has no formal definition; Wi-Fi is a general marketing term used to define 802.11 technologies.

WIGLE Wireless Geographic Logging Engine

WiMAX Worldwide Interoperability for Microwave Access

WIPS wireless intrusion prevention system

WISP Wireless Internet Service Provider

WLAN wireless local area network

WLSE Wireless LAN Solution Engine

WM wireless medium

WMAN wireless metropolitan area network

WMM Wi-Fi Multimedia

WMM-PS Wi-Fi Multimedia Power Save

WMM-SA Wi-Fi Multimedia Scheduled Access

WNMS wireless network management system

WPA Wi-Fi Protected Access

WPAN wireless personal area network

WPP Wireless Performance Prediction

WWAN wireless wide area network

WZC Wireless Zero Configuration

XOR exclusive or

Power Regulations

The Federal Communications Commission (FCC) regulates communications to and from the United States. The FCC and the respective controlling agencies in other countries regulate the amount of power at the intentional radiator (IR) and the amount of power radiated from the antenna (EIRP) for 802.11 radios. Power output regulations are typically created to minimize interference within the band and to minimize interference to adjacent or nearby bands.

The rules regarding the amount of power that is permitted are typically divided into two categories: point-to-multipoint communications (PtMP) and point-to-point communications (PtP). The regulations for PtMP communications are generally more restrictive than the regulations for PtP communications. The reasoning is fairly straightforward. PtMP signals are generated in all directions, covering a broad area, and thus are more likely to interfere with other devices. PtP signals are focused using high-gain antennas, making the area of potential interference very small. The following sections review the FCC power regulations.

2.4 GHz ISM Point-to-Multipoint (PtMP) Communications

PtMP communications consist of a central communications device communicating to multiple other devices. If the central device is connected to an omnidirectional antenna, the FCC automatically classifies the communications as PtMP. The central PtMP device does not have to be connected to an omnidirectional antenna, as is the case with many access points that are connected to semidirectional patch antennas.

The FCC limits the maximum power at the intentional radiator (IR) to 1 watt (+30 dBm) and the maximum radiated power from the antenna (EIRP) to 4 watts (+36 dBm). This means that if the IR is at the maximum power of 1 watt, or 30 dBm, the maximum gain antenna that can be used is 6 dBm, which creates a total EIRP of 36 dBm, or 4 watts. Remember that IR + antenna gain = EIRP.

No matter what you want to do, the EIRP cannot be greater than 36 dBm, or 4 watts. This means that if you want to use a higher-gain antenna, you must subtract the antenna gain from the EIRP to calculate the maximum IR that you can have. As an example, if you wanted to use a 9 dBi patch antenna, the maximum IR would be 27 dBm, or 500 mW (36 dBm – 9 dBi = 27 dBm). For every dBi increase in the antenna above 6 dBi, the IR must decrease by the same amount. This is often known as the one-to-one, or 1:1, rule.

5 GHz UNII Point-to-Multipoint (PtMP) Communications

The FCC PtMP rules for the 5 GHz UNII bands follow the same basic rules of the 2.4 GHz ISM PtMP communications. A 6 dBi antenna can be connected to the PtMP device without affecting the maximum EIRP. Any additional increase in antenna gain requires an equal decrease in IR. Figures B.1 and B.2 show the maximum IR and EIRP values for the UNII bands in both the United States (FCC) and Europe (ERC).

FIGURE B.1 5 GHz PtMP—intentional radiator power regulations

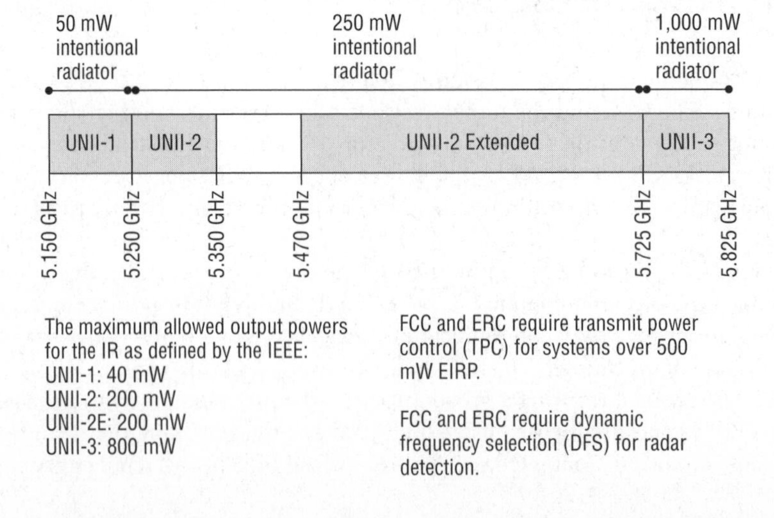

FIGURE B.2 5 GHz PtMP—Equivalent isotropically radiated power (EIRP) regulations

Equivalent isotropically radiated power (EIRP)

UNII-1: FCC: 50 mW intentional radiator + up to 6 dBi antenna gain = EIRP (IEEE defines MAX IR 40 mW)

UNII-2 & UNII-2E: FCC: 250 mW intentional radiator + up to 6 dBi antenna gain = EIRP (IEEE defines MAX IR 200 mW)

UNII-3: FCC: 1000 mW intentional radiator + up to 6 dBi antenna gain = EIRP (IEEE defines MAX IR 800 mW)

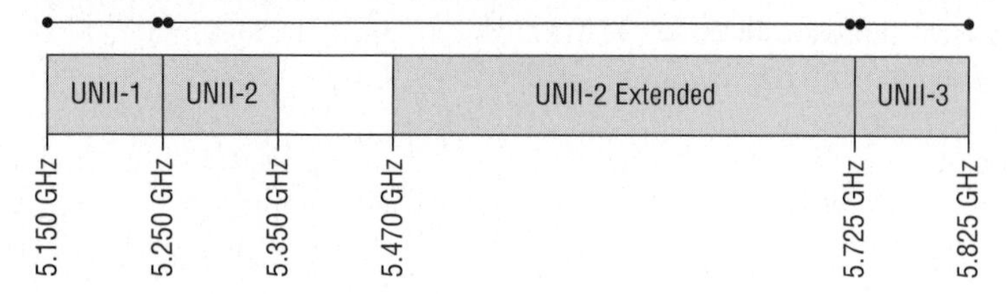

2.4 GHz ISM Point-to-Point (PtP) Communications

Point-to-point communication consists of two devices communicating to each other by using directional antennas. The FCC PtP rules for the 2.4 GHz ISM band start with the same initial values as the PtMP rules: 1-watt IR, 6 dBi antenna, 4-watt EIRP. The maximum allowed IR is still 1 watt; however, because the antenna is directional and communicating with only one other device, the FCC allows the antenna gain and the EIRP to be increased. For every 3 dB additional increase of the antenna (above the initial 6 dBi value), the IR must be decreased by 1 dB. This is often known as the three-to-one, or 3:1, rule.

Remember that IR + antenna gain = EIRP, so if the antenna is increased by 3 dB and the IR is decreased by 1 dB, the EIRP is increased by 2 dB. To help understand the rule, just remember 3-2-1. A 3 dB increase in the antenna creates a 2 dB increase in the EIRP because it requires a 1 dB decrease in the IR.

IR	Antenna Gain	Maximum EIRP
+30 dBm (1 watt)	6 dBi	+36 dBm (4 watts)
+29 dBm	9 dBi	+38 dBm (6.3 watts)
+28 dBm	12 dBi	+40 dBm (10 watts)
+27 dBm	15 dBi	+42 dBm (16 watts)
+26 dBm	18 dBi	+44 dBm (25 watts)
+25 dBm	21 dBi	+46 dBm (39.8 watts)
+24 dBm	24 dBi	+48 dBm (63 watts)
+23 dBm	27 dBi	+50 dBm (100 watts)
+22 dBm	30 dBi	+52 dBm (158 watts)

5 GHz UNII Point-to-Point (PtP) Communications

The FCC PtP rules for the 5 GHz UNII-1 and UNII-2 bands are identical to the PtMP rules for these bands. For the UNII-3 band, the FCC has a separate set of rules because the UNII-3 band is often used for long-distance point-to-point communications. A fixed PtP transmitter with a maximum IR of +30 dBm (1 watt) is allowed to be connected to a directional antenna with a gain of up to 23 dBi without making any change to the IR. The maximum allowed EIRP is therefore +53 dBm (200 watts). For every dBi increase in the antenna above 23 dBi, the IR must decrease by the same amount. So if you have any gain above the 23 dBi, you must adhere to the 1:1 rule.

Glossary

Numbers

4-Way Handshake Under the 802.11i amendment, two stations (STAs) must establish a procedure to authenticate and associate with each other as well as create dynamic encryption keys through a process known as the 4-Way Handshake.

6 dB rule Doubling the distance between a transmitter and receiver will decrease the received signal by 6 dB. Halving the distance between a transmitter and receiver will increase the received signal by 6 dB.

802.11-2007 standard On March 8, 2007, a new iteration of the standard was approved, IEEE Std. 802.11-2007. This new standard is an update of the IEEE Std. 802.11-1999 revision. The following documents have been rolled into this latest revision, providing users a single document with all of the amendments that had been published to date. This new standard includes the following:

- IEEE Std 802.11-1999 (R2003)
- IEEE Std 802.11a-1999
- IEEE Std 802.11b-1999
- IEEE Std 802.11d-2001
- IEEE Std 802.11g-2003
- IEEE Std 802.11h-2003
- IEEE Std 802.11i-2004
- IEEE Std 802.11j-2004
- IEEE Std 802.11e-2005

802.1X The 802.1X standard is a port-based access control standard. 802.1X provides an authorization framework that allows or disallows traffic to pass through a port and thereby access network resources. An 802.1X framework may be implemented in either a wireless or wired environment. The three main components of an 802.1X framework are the supplicant, the authenticator, and the authentication server.

A

absorption The most common RF behavior is absorption. If a signal does not bounce off an object, pass through it, or move around it, the signal will be absorbed to some extent by the object. Most materials will absorb some amount of an RF signal to varying degrees.

access layer The access layer of the network is responsible for delivery of the traffic directly to the end user or end node. The access layer ensures the final delivery of packets to the end user. It is connected to the distribution layer, which is connected to the core.

access point The CWNP definition is a half-duplex wireless device with switchlike intelligence. In reality, an access point is simply a hub with a radio card and an antenna.

Access points must contend for the half-duplex medium in the same fashion that the client station radio cards do.

acknowledgment (ACK) The ACK frame is one of the six control frames and one of the key components of the 802.11 CSMA/CA media access control method. Because 802.11 is a wireless medium that cannot guarantee successful data transmission, the only way for a station to know that a frame it transmitted was properly received is for the receiving station to notify the transmitting station. This notification is performed using an ACK. The ACK frame is a very simple frame consisting of 14 octets of information.

active gain Active gain is usually the increase of a signal's amplitude caused by the use of an amplifier on the wire that connects the transceiver to the antenna. The amplifier is usually bidirectional, meaning that it increases the AC voltage both inbound and outbound. Active gain devices require the use of an external power source. Active gain devices must also be certified by the FCC for the system to be legal to use in the United States.

active manual survey An active manual survey is a site survey method in which the radio card is associated to the access point and has layer 2 connectivity allowing for low-level frame transmissions. Layer 3 connectivity can also be established by generating ping traffic. RF measurements can be recorded along with packet loss and retransmission percentages.

Active mode Active mode is the default power-management mode for most 802.11 stations. When a station is set for Active mode, the wireless station is always ready to transmit or receive data. Active mode is sometimes referred to as Continuous Aware mode, and it provides no battery conservation. In the MAC header of an 802.11 frame, the Power Management field is 1 bit in length and is used to indicate the power-management mode of the station. A value of 0 indicates that the station is in Active mode. Stations running in Active mode will achieve higher throughput than stations running in Power Save mode, but the battery life will typically be much shorter.

active scanning In order for a station to be able to connect to an access point, it needs to first discover an access point. Active scanning is one of the methods that stations use to discover access points. The station and access point will exchange probe requests and probe responses to establish the capabilities of the basic service set.

ad hoc The 802.11 standard defines three topologies known as service sets. One topology known as an independent basic service set (IBSS) involves direct communications between 802.11 client stations without the use of an access point. An 802.11 IBSS network is also known as a peer-to-peer network or an ad hoc network.

Ad Hoc mode A common term used to refer to a station that is configured to connect to an independent basic service set.

adjacent-cell interference Degradation of performance caused by layer 2 retransmissions resulting from overlapping frequency space that occurs because of an improper channel reuse design.

adjacent channel The next or previous numbered channel.

Advanced Encryption Standard (AES) The AES algorithm, originally named the Rijndael algorithm, is a block cipher that offers much stronger protection than the RC4 streaming cipher. AES is used to encrypt 802.11 wireless data by using an encryption method known as Counter mode with Cipher Block Chaining–Message Authentication Code (CCMP). The AES algorithm encrypts data in fixed data blocks with choices in encryption key strength of 128, 192, or 256 bits.

Aggregate MAC Protocol Data Unit (A-MPDU) A frame aggregation technique that combines multiple frames into a single frame transmission. All of the 802.11 frames (MPDUs) do not have to have the same destination address. Also, the data payload of each MPDU is encrypted separately by using the multiple dynamic encryption keys that are unique between the access point and each individual client.

Aggregate MAC Service Data Unit (A-MSDU) A frame aggregation technique that combines multiple frames into a single frame transmission. The aggregated MSDUs will have a single destination when wrapped together in a single frame. Multiple MSDUs are encrypted by using the same dynamic encryption key.

airtime fairness Mechanisms on a WLAN controller that prioritizes transmissions from stations with higher data rates over stations using lower data rates.

all-band interference All-band interference is RF interference that occurs across the entire frequency range that is being used. The term *all-band interference* is typically associated with frequency hopping spread spectrum (FHSS) communications that disrupt HR-DSSS and/or ERP-OFDM channel communications.

ALOHAnet In 1970, the University of Hawaii developed the first wireless network, called ALOHAnet, to wirelessly communicate data between the Hawaiian Islands.

alternating current (AC) An electrical current with a magnitude and direction that varies cyclically—as opposed to direct current, the direction of which stays in a constant form. The shape and form of the AC signal—defined as the *waveform*—is known as a sine wave.

amplification The increase of a signal's amplitude by the use of an external device.

amplifier An RF amplifier takes the signal that is generated by the transceiver, increases it, and sends it to the antenna. An amplifier provides an overall increase in power by adding electrical energy to the signal, which is referred to as *active gain*. Unidirectional amplifiers perform the amplification in only one direction, either when transmitting or when receiving. Bidirectional amplifiers perform the amplification in both directions.

amplitude The height, force, or power of a wave. Often referred to as *signal strength*.

Amplitude Shift Keying (ASK) Amplitude Shift Keying (ASK) varies the amplitude, or height, of a signal to represent the binary data. ASK is a current state technique, where one level of amplitude can represent a 0 bit and another level of amplitude can represent a 1 bit.

announcement traffic indication message (ATIM) A unicast frame that is used in an IBSS network when Power Save mode is enabled. If a station has buffered data for another station, it will send an ATIM frame to the other station, informing it that it must stay awake until the next ATIM window so that it can receive the buffered data. Any station that either has buffered data for another station or has received an ATIM will stay awake so that the buffered data can be exchanged.

antenna An antenna provides two functions in a communication system. When connected to the transmitter, it collects the AC signal that it receives from the transmitter and directs, or radiates, the RF waves away from the antenna in a pattern specific to the antenna type. When connected to the receiver, it takes the RF waves that it receives through the air and directs the AC signal to the receiver.

antenna diversity Antenna diversity occurs when an access point has two antennas and receivers functioning together to minimize the negative effects of multipath. When the access point senses an RF signal, it compares the signal that it is receiving on both antennas and uses the antenna with the higher signal strength to receive the frame of data. This sampling is performed on a frame-by-frame basis, choosing whichever antenna has the higher signal strength.

antenna polarization Antennas radiate RF signals with the amplitude of the waves fluctuating either vertically or horizontally. The orientation of the antenna is referred to as either vertically or horizontally polarized.

antenna reciprocity The concept that antennas amplify received signals just as they amplify transmitted signals.

application analysis Testing procedures that are used to determine how an application performs on the wireless network and how the application affects the wireless network. Tools exist that can simulate multiple concurrent virtual wireless client stations.

assisted coverage analysis An RF coverage analysis typically performed by a centralized wireless network management system (WNMS) or a WLAN controller. The centralized device scans the access point radio cards and collects the RF information, which is then used for visualization of coverage cells and for optimizing AP configurations such as channel and power settings. Most assisted solutions use the information gathered from the access point radios; however, some solutions can also use a client radio to report information back to the centralized device during a client walk-through of the building.

association After a station has authenticated with the access point, the next step is for it to associate with the access point. When a client station associates, it becomes a member of a basic service set (BSS). Association means that the client station can send data through the access point and on to the distribution system medium.

association identifier (AID) Any time a station associates to an access point, the station receives an association identifier (AID). The access point uses this AID to keep track of the stations that are associated and the members of the BSS.

attenuation The decrease of amplitude or signal strength. Also known as signal loss.

attenuator Attenuators are small devices about the size of a C-cell battery, with cable connectors on both sides. They absorb energy, decreasing the signal as it travels through. A variable-loss attenuator has a dial or switches on it that enable you to adjust the amount of energy that is absorbed. Fixed-loss attenuators provide a set amount of loss. Attenuators are based on frequency.

authentication Authentication is the verification of user identity and credentials. Users must identify themselves and present credentials such as usernames and passwords or digital certificates. More-secure authentication systems exist that require multifactor authentication, where at least two sets of different credentials must be presented.

authentication server (AS) When an 802.1X/EAP solution is deployed, an authentication server validates the credentials of the supplicant that is requesting access and notifies the authenticator that the supplicant has been authorized. The authentication server will maintain a user database or may proxy with an external user database to authenticate user credentials.

authenticator When an 802.1X/EAP solution is deployed, a device that blocks or allows traffic to pass through its port entity is known as the authenticator. Authentication traffic is normally allowed to pass through the authenticator while all other traffic is blocked until the identity of the supplicant has been verified.

authorization, authentication, and accounting (AAA) AAA is a security concept. Authorization involves granting access to network resources and services. Before authorization to network resources can be granted, proper authentication must occur. Authentication is the verification of user identity and credentials. Accounting is tracking the use of network resources by users. It is an important aspect of network security, used to keep a paper trail of who used what resource and when and where.

automatic power save delivery An enhanced power-management method introduced by the 802.11e amendment.

autonomous AP A term for the traditional access point. An autonomous access point contains at least two physical interfaces, usually an RF radio card and a 10/100BaseT port. All configuration settings exist in the autonomous access point itself, and, therefore, management and configuration occurs at the access layer. All encryption and decryption mechanisms and MAC layer mechanisms also operate within the autonomous AP. The distribution system service (DSS) and integration service (IS) function within an autonomous AP.

azimuth chart The azimuth chart, labeled H-plane, shows the top-down view of the radiation pattern of the antenna.

B

back lobe Unintended RF signal that radiates from the back side of an antenna.

bandwidth Wireless communication is typically performed within a constrained set of frequencies known as a frequency band. This frequency band is the bandwidth. The term bandwidth is sometimes also used to refer to 802.11 *data rates* which are the speed of data transfer. The proper term for the changes in speed due to modulation and coding is data rates; however they are also often referred to as *data bandwidth*.

Barker code A spreading/coding technique used by 802.11 cards.

basic rates The set of data rates that a client station must be capable of communicating with in order to successfully associate with an access point. Basic rates are required rates with a basic service set (BSS).

basic service area (BSA) The physical area of coverage provided by an access point in a BSS is known as the basic service area (BSA). Client stations may move throughout the coverage area and maintain communications with the AP as long as the received signal between the radios remains above RSSI thresholds. Client stations may also shift between concentric zones of variable data rates that exist within the BSA.

basic service set (BSS) The 802.11 standard defines three topologies known as service sets. One topology, known as the basic service set (BSS), involves communications between a single access point and client stations that are associated to the access point.

basic service set identifier (BSSID) The BSSID address is a 48-bit (6-octet) MAC address used as a unique identifier of a basic service set. In either a BSS or ESS topology, the BSSID address is simply the MAC address of a single access point. In an IBSS topology, the BSSID address is a virtual address.

beacon management frame One of the most important 802.11 frame types. Commonly referred to as the beacon. Beacons are essentially the heartbeat of the wireless network. They are sent only by the access point of a basic service set. Client stations transmit beacons only when participating in an IBSS also known as Ad Hoc mode. Each beacon contains a time stamp, which client stations use to keep their clocks synchronized with the access point. Because so much of successful wireless communications is timing based, it is imperative that all stations are in synch with each other.

beamwidth The measurement of how broad or narrow the focus of an antenna is. Beamwidth is measured both horizontally and vertically. It is the measurement from the center, or strongest point, of the antenna signal to each of the points along the horizontal and vertical axes where the signal decreases by half power (–3 dB). These –3 dB points are often referred to as half-power points. The distance between the two half-power points on the horizontal axis is measured in degrees, giving the horizontal beamwidth measurement. The distance between the two half-power points on the vertical axis is also measured in degrees, giving the vertical beamwidth measurement. Beamwidth of the antenna does not affect the wavelength or Fresnel zone requirements of the signal.

Bluetooth A short-distance RF technology defined by the 802.15 standard. Bluetooth operates using FHSS and hops across the 2.4 GHz ISM band at 1,600 hops per second. Older Bluetooth devices were known to cause all-band interference. Newer Bluetooth devices utilize adaptive mechanisms to avoid interfering with 802.11 WLANs.

Bridged Virtual Interface (BVI) Autonomous access points contain at least two physical interfaces, usually an RF radio card and a 10/100BaseT port. The majority of the time, these physical interfaces are bridged together by a virtual interface known as a Bridged Virtual Interface (BVI). The BVI is assigned an IP address that is shared by the two physical interfaces.

broadcast key When an 802.1X/EAP solution is used with dynamic WEP encryption, a static key known as the broadcast key exists on the access point. The broadcast key is used to encrypt and decrypt all broadcast and multicast 802.11 data frames.

C

capacity Proper network design entails providing the necessary coverage while trying to limit the number of devices connected to any single access point at the same time. This design process ensures the highest level of throughput to the individual stations by limiting contention. This is what is meant by capacity.

carrier frequency The nominal frequency of a carrier wave.

Carrier Sense Multiple Access with Collision Avoidance (CSMA/CA) Media access control method used by 802.11 networks.

Carrier Sense Multiple Access with Collision Detection (CSMA/CD) Media access control method used by 802.3 networks. Four mechanisms are used together to ensure that only one station is transmitting at any given time on the half-duplex RF medium. The four mechanisms are physical carrier-sense, virtual carrier-sense, interframe spaces, and the random back-off algorithm.

carrier signal If a signal fluctuates or is altered, even slightly, data can be properly sent and received. This modulated signal is now capable of distinguishing between 0s and 1s and is referred to as a carrier signal.

casual eavesdropping Casual eavesdropping is not considered malicious and is also often referred to as wardriving. Software utilities known as WLAN discovery tools exist for the purpose of finding open WLAN networks. Wardriving is strictly the act of looking for wireless networks, usually while in a moving vehicle. The most common wardriving software tool is a freeware program called NetStumbler.

cell-sizing Determining how big the cell size needs to be to provide the desired coverage, and adjusting the power level of the access point in order to create a cell of the desired size. Cell-sizing is almost always the preferable method for meeting capacity needs in an MCA environment.

channel blankets In a single channel architecture, each layer of multiple APs on a single channel and using the same virtual BSSID is known as a channel blanket.

channel reuse In order to avoid co-channel interference, a channel reuse pattern is necessary. Overlapping RF coverage cells are needed for roaming, but overlap frequencies must be avoided. The only three channels that meet these criteria in the 2.4 GHz ISM band are channels 1, 6, and 11. Overlapping coverage cells therefore should be placed in a channel reuse pattern that minimizes co-channel interference.

Channel reuse patterns should also be used in the 5 GHz UNII bands. Because of the frequency overlap of channel sidebands, there should always be at least two cells between access points on the same channel. Also, it is a recommend practice that any adjacent cells use a frequency that is at least two channels apart and not use an adjacent frequency.

channel spans See *channel blankets.*

channel stacking Colocation design in a single channel architecture.

chip A series of bits that represent a single bit of data. To prevent confusion, the data is referred to as a bit, and the series of bits are referred to as chips instead of bits.

chipping The process of converting a single data bit into a sequence of bits known as chips is often called *spreading* or *chipping.*

chipset A group of integrated circuits designed to work together. Many 802.11 chipset manufacturers exist and sell their chipset technology to the various radio card manufacturers.

classification signature In PoE, the classification signature lets the power-sourcing equipment know how much power the device will need.

clear channel assessment (CCA) A layer 1 process that determines whether the RF medium is busy. This is performed by 802.11 radios prior to transmitting data.

client station A radio card that is not used in an access point is typically referred to as a client station. Client station radio cards are typically used in laptops, PDAs, scanners, phones, and many other mobile devices.

client utilities Software used to configure a wireless client card. The software interface will usually have the ability to create multiple connection profiles. Configuration settings of a client utility typically include the SSID, transmit power, security settings, 802.11e/QoS capabilities, and power management.

co-channel interference (CCI) Unnecessary medium contention overhead that occurs when access points with overlapping coverage cells are configured to transmit on the same frequency channel. Because of the CSMA/CA, all nearby access points and clients on the same channel will defer transmissions. The unnecessary medium contention overhead caused by co-channel interference is a result of improper channel reuse design in multiple channel architecture (MCA).

colocation Placing multiple access points in the same physical space to provide for greater capacity. In a multiple channel architecture (MCA) three access points operating in the 2.4 GHz ISM band could be colocated in the same physical area.

CompactFlash (CF) A peripheral expansion slot often found on handheld PDAs, laptops, and other mobile devices.

Complementary Code Keying (CCK) A spreading/coding technique used by 802.11b cards to provide higher data rates (HR-DSSS).

contention-free period (CFP) Occurs when the access point is functioning in PCF mode. During the contention-free period, the access point polls only clients in PCF mode about their intention to send data. This is a method of prioritizing clients.

contention period (CP) Occurs when the access point is functioning in DCF mode.

contention window After a station has waited while performing both virtual and physical carrier-senses, the station may contend for the medium during a window of time known as the *contention window*.

Control and Provisioning of Wireless Access Points (CAPWAP) A set of standards proposed by the Internet Engineering Task Force (IETF) for WLAN controller protocols.

control frames Control frames help with the delivery of the data frames. Control frames must be able to be heard by all stations; therefore, they must be transmitted at one of the basic rates. Control frames are also used to clear the channel, acquire the channel, and provide unicast frame acknowledgments. They contain only layer 2 header information.

cooperative control A wireless architecture that consists of groups of autonomous access points with WLAN controller intelligence and capabilities.

cooperative control access points (CC-AP) An access point that combines an autonomous access point with a suite of cooperative control protocols, but without requiring a WLAN controller.

Converged Wireless Group–RF Profile Converged Wireless Group–RF Profile (CWG-RF) was developed jointly by the Wi-Fi Alliance and the Cellular Telecommunications and Internet Association (CTIA). CWG-RF defines performance metrics for Wi-Fi and cellular radios in a converged handset to help ensure that both technologies perform well in the presence of the other.

convolutional coding A form of error correction. Convolutional coding is not part of OFDM but rather part of 802.11a and 802.11g. It is a forward error correction (FEC) that allows the receiving system to detect and repair corrupted bits. There are many levels of convolutional coding.

core The high-speed backbone of the network. The goal of the core is to carry large amounts of information between key data centers or distribution areas. The core layer does not route traffic nor manipulate packets but rather performs high-speed switching. Redundant solutions are usually designed at the core layer to ensure the fast and reliable delivery of packets.

Counter Mode with Cipher Block Chaining Message Authentication Code (CCMP) The default encryption method defined under the 802.11i amendment. This method uses the Advanced Encryption Standard (AES) cipher. CCMP/AES uses a 128-bit encryption key size and encrypts in 128-bit fixed-length blocks. An 8-byte Message Integrity Check is used that is considered much stronger than the one used in TKIP. CCMP/AES is the default encryption method defined by WPA2.

coverage A wireless network design in which access points are configured with the power set to the maximum level to provide the largest coverage area possible. Coverage also defines the physical area where a usable signal can be received by the station.

coverage analysis Determining the proper placement of access points, the transmission power of the access point radio card, and the proper use of antennas.

CTS-to-Self A protection mechanism for mixed-mode environments. One of the benefits of using CTS-to-Self over RTS/CTS as a protection mechanism is that the throughput will be higher, because there are fewer frames being sent.

cyclic redundancy check (CRC) An error-detecting code.

D

data frames Data frames carry the actual data that is passed down from the higher-layer protocols.

data privacy One of the key components of a wireless security solution. Data privacy is achieved by using encryption.

data rates Data rates are the transmission rates specified by the 802.11 standard and amendments, not actual throughput. Because of medium access methods, aggregate throughput is typically half or less of the available data rate bandwidth.

dBd The increase in gain of an antenna, compared to the signal of a dipole antenna. Another way of phrasing this is *decibel gain relative to a dipole antenna*.

dBi The gain, or increase of power, from an antenna when compared to what an isotropic radiator would generate is known as decibels isotropic (dBi). Another way of phrasing this is *decibel gain relative to an isotropic radiator*.

dBm Compares a signal to 1 milliwatt of power. dBm means *decibels relative to 1 milliwatt*. Because dBm is a measurement that is compared to a known value, 1 milliwatt, then dBm is actually a measure of power.

deauthentication A notification frame used to terminate an authentication. Because authentication is a prerequisite for association, disassociation will also occur. Deauthentication cannot be refused by either party.

decibel (dB) Decibel is derived from the term *bel*. It is a measurement of the ratio between two powers: decibels = $10 \times \log^{10}(P_1/P_2)$

decibels dipole (dBd) The increase in gain of an antenna compared to the signal of a dipole antenna.

decibels isotropic (dBi) The increase in gain of an antenna compared to what an isotropic radiator would generate.

delay spread The delay between the reception of the main signal and the reflected signal.

deliverables The site survey information contained in the final report delivered to the customer. This can include spectrum analysis, RF coverage analysis, hardware placement and configuration, and application analysis.

delivery traffic indication message (DTIM) A special type of TIM that is used to ensure that all stations are awake when multicast or broadcast traffic is sent.

denial of service (DoS) Any individual with ill intent can temporarily disable a Wi-Fi network by preventing legitimate wireless users from accessing network resources. Layer 1 and layer 2 attacks exist that can deny 802.11 wireless services to legitimate authorized users. 802.11 DoS attacks cannot be prevented, but they can be detected with the proper intrusion detection tools.

detection signature A signal that a PoE powered device uses to notify the power-sourcing equipment whether it is in a state where it will accept power.

Differential Binary Phase Shift Keying (DBPSK) A modulation technique used to transmit 802.11 DSSS data at 1 Mbps.

Differential Quadrature Phase Shift Keying (DQPSK) A modulation technique used to transmit 802.11 DSSS data at 2 Mbps.

diffraction The bending of an RF signal around an object.

digital signal processing (DSP) Techniques used to shape and direct wireless transmissions.

dipole antenna An antenna that consists of two elements. A half-wave dipole antenna consists of two elements, each one-quarter of the wavelength long.

direct link setup (DLS) In most WLAN environments, all frame exchanges between client stations that are associated to the same access point must pass through the access point. DLS allows for client stations to bypass the access point and communicate with direct frame exchanges.

direct sequence spread spectrum (DSSS) A spread spectrum technology originally specified in the 802.11 standard. Provides 1 and 2 Mbps RF communications using the 2.4 GHz ISM band. DSSS 802.11 radio cards are often known as clause 15 devices.

direct sequence spread spectrum-OFDM (DSSS-OFDM) An optional PHY defined by the 802.11g ratified amendment.

directed probe request Probe request with a specific SSID.

disassociation A notification frame used to terminate an association. A polite way of terminating the association. Disassociation cannot be refused by either party.

Distributed Coordination Function (DCF) CSMA/CA is provided by DCF, which is the mandatory access method of the 802.11 standard.

distributed data forwarding (DDF) The use of multiple wireless controllers as data distribution gateways onto the wired network.

distributed spectrum analysis system (DSAS) A centralized server that uses remote hardware spectrum analyzer sensors.

Distribution layer The distribution layer of the network routes or directs traffic toward the smaller clusters of the network's nodes. The distribution layer routes traffic between VLANs and subnets.

distribution system (DS) The DS is a system used to interconnect a set of basic service sets (BSSs) and integrated local area networks (LANs) to create an extended service set (ESS). The DS consists of a medium used for transport of traffic as well as services used for transport of traffic.

distribution system medium (DSM) The DSM is a logical physical medium used to connect access points. Normally, the DSM is an 802.3 Ethernet backbone; however, the medium can also be wireless or some other type of medium.

distribution system service (DSS) A system service built inside an autonomous access point or WLAN controller usually in the form of software. The distribution system service is used to transport 802.11 traffic.

downfade Decreased signal strength caused by multiple RF signal paths arriving at the receiver at the same time, and the signals are out of phase with the primary wave.

Dual-CTS protection The 802.11n amendment defines an operational mechanism that accounts for coexistence between 802.11n HT coverage cells and nearby legacy 802.11a/b/g coverage cells. When either an HT or non-HT station transmits a frame, the station first sends a request-to-send (RTS) frame to the HT access point. The access point will then reply with two clear-to-send (CTS) frames. One CTS frame is in the legacy non-HT format, while the other frame is in the HT format.

Duration/ID A field in an 802.11 frame header that is typically used to set the NAV timer in other stations. Used with virtual carrier-sense.

dwell time A defined amount of time that the FHSS system transmits on a specific frequency before it switches to the next frequency in the hop set. The local regulatory body typically limits the amount of dwell time.

dynamic frequency selection (DFS) Used for spectrum management of 5 GHz channels for 802.11a radio cards. The European Radiocommunications Committee (ERC) originally mandated that radio cards operating in the 5 GHz band implement a mechanism to avoid interference with radar systems as well as provide equable use of the channels. The DFS service is used to meet the ERC regulatory requirements. This requirement has since become a requirement of other regulatory bodies, such as the FCC in the United States.

dynamic rate switching (DRS) Also known as dynamic rate shifting, adaptive rate selection, or automatic rate selection. A process that client stations use to shift to lower-bandwidth capabilities as they move away from an access point and to higher-bandwidth capabilities as they move toward an access point. The objective of DRS is upshifting and downshifting for rate optimization and improved performance.

dynamic RF An environment in which a WLAN controller is a centralized device that can dynamically change the configuration of the lightweight access points based on accumulated RF information gathered from the access points' radio cards.

E

earth bulge The curvature of the earth, which must be considered when installing long-distance point-to-point RF communications.

electromagnetic (EM) spectrum The range of all possible electromagnetic radiation. This radiation exists as self-propagating electromagnetic waves that can move through matter or space.

elevation chart The elevation chart, labeled E-plane, shows the side view of the radiation pattern of the antenna.

endpoint Defined under the PoE standard as a switch with integrated power-supplying equipment, or more specifically a switch with PSE (power-sourcing equipment).

Enhanced Distributed Channel Access (EDCA) As defined by the 802.11e amendment, Enhanced Distributed Channel Access (EDCA) is an extension to DCF. The EDCA medium access method provides for the prioritization of traffic via the use of 802.1d priority tags.

enterprise encryption gateway (EEG) A specialty 802.11 device that provides for segmentation and encryption. The EEG typically sits behind several fat access points and segments the wireless network from the protected wired network infrastructure. Proprietary encryption technology using the AES algorithm at layer 2 is provided by the enterprise encryption gateway.

enterprise wireless gateway (EWG) A specialty 802.11 device used to segment autonomous access points from the protected wired network infrastructure. An EWG can segment the unprotected wireless network from the protected wired network by acting either as a router or a VPN endpoint and/or as a firewall.

equal modulation In a MIMO system, multiple spatial streams are sent with the same (equal) modulation.

equivalent isotropically radiated power (EIRP) The highest RF signal strength that is transmitted from a particular antenna.

Erlang Probabilistic traffic formulas use a telecommunications unit of measurement known as an Erlang. An Erlang is equal to 1 hour of telephone traffic in 1 hour of time.

evil twin attack The evil twin attack, also known as wireless hijacking, occurs when a hacker disrupts communications between client stations and a legitimate AP. Client stations lose their connection to the legitimate AP and reconnect to the evil twin access point. The evil twin hijacks the client stations at layer 1 and layer 2, allowing the hacker to proceed with peer-to-peer attacks.

explicit feedback This occurs in 802.11n when a transmitter and receiver work together to educate each other about the characteristics of the MIMO channel. The transmitter will make phase adjustments based on the information that is returned from the receiver. When using explicit feedback, the beamformee makes a direct estimate of the channel from training symbols sent to the beamformee by the beamformer. The beamformee takes that information and sends additional feedback back to the beamformer. The beamformer then transmits based on the feedback from the beamformee.

ExpressCard A hardware standard that is replacing PCMCIA cards.

Extended Rate Physical (ERP) A physical layer specification (PHY) defined for clause 19 radios. This PHY operates in the 2.4 GHz ISM band and uses ERP-OFDM to support data rates of 6–54 Mbps. ERP/DSSS/CCK technology is used to maintain backward compatibility with HR-DSSS (clause 18) radios and DSSS (clause 15) radios.

Extended Rate Physical DSSS/CCK 802.11g clause 19 radios must maintain backward compatibility with 802.11 (DSSS only) and 802.11b (HR-DSSS) radios. A Physical layer (PHY) technology called Extended Rate Physical DSSS (ERP-DSSS/CCK) is used for backward compatibility and support for the data rates of 1, 2, 5.5, and 11 Mbps. This PHY operates in the 2.4 GHz ISM band.

Extended Rate Physical OFDM (ERP-OFDM) A Physical layer (PHY) technology used by 802.11g clause 19 radios to achieve greater bandwidth. Uses OFDM as defined in the 802.11a amendment. Therefore, data rates of 6, 9, 12, 18, 24, 36, 48, and 54 Mbps are possible using OFDM technology. This PHY operates in the 2.4 GHz ISM band.

Extended Rate Physical PBCC (ERP-PBCC) An optional PHY defined by the 802.11g ratified amendment for clause 19 radios.

extended service set (ESS) The 802.11 standard defines three topologies known as service sets. One topology, known as the extended service set (ESS), involves communications between multiple access points that share a network infrastructure. An ESS is one or more basic service sets that share a distribution system medium.

Extensible Authentication Protocol (EAP) Extensible Authentication Protocol (EAP) is used to provide user authentication for an 802.1X port-based access control solution. EAP is a flexible layer 2 authentication protocol that resides under Point-to-Point Protocol (PPP).

F

fade margin A level of desired signal above the minimum required signal for successful communications.

fast basic service set transition (FT) More commonly referred to as fast secure roaming, because it defines faster handoffs when roaming occurs between cells in a wireless LAN using the strong security defined in a robust security network (RSN). Fast and secure 802.11 roaming is needed to meet latency requirements for time-sensitive applications in a WLAN.

fast secure roaming (FSR) See *fast basic service set transition (FT)*.

Federal Communications Commission (FCC) The FCC is an independent U.S. government agency, directly responsible to the U.S. Congress. It was established by the Communications Act of 1934 and is responsible for regulating interstate and international communications by radio, television, wire, satellite, and cable. The FCC's jurisdiction covers all of the 50 states, the District of Columbia, and U.S. possessions.

Federal Information Processing Standards (FIPS) In the United States, the National Institute of Standards and Technology (NIST) maintains the Federal Information Processing Standards (FIPS). The FIPS 140-2 standard defines security requirements for cryptography modules. The use of validated cryptographic modules is required by the U.S. government for all unclassified communications. Other countries also recognize the FIPS 140-2 standard or have similar regulations.

fixed mobile convergence (FMC) An environment in which a single device, with a single telephone number that is capable of switching between networks, always uses the lowest-cost network. FMC devices typically are capable of communicating via either a cellular telephone network or a VoWiFi network.

forward error correction (FEC) A technology that allows a receiving system to detect and repair corrupted bits.

frame A unit of data at the Data-Link layer.

frame aggregation Combining multiple frames into a single frame transmission.

frame check sequence (FCS) The extra characters added to a frame and used for error detection and correction.

free space path loss (FSPL) The loss of signal strength caused by the natural broadening of the waves, often referred to as beam divergence. RF signal energy spreads over larger areas as the signal travels farther away from an antenna, and as a result, the strength of the signal attenuates.

frequency A term describing a behavior of waves. How fast the waves travel, or more specifically, how many waves are generated over a 1-second period of time is known as *frequency*.

frequency hopping spread spectrum (FHSS) A spread spectrum technology that was first patented during World War II. FHSS was used in the original 802.11 standard and provided 1 and 2 Mbps RF communications using the 2.4 GHz ISM band. FHSS works by using a small frequency carrier space to transmit data, and then hopping to another small frequency carrier space and transmitting data, and then to another frequency, and so on.

Frequency Shift Keying (FSK) Frequency Shift Keying (FSK) varies the frequency of the signal to represent the binary data. FSK is a current state technique, where one frequency can represent a 0 bit and another frequency can represent a 1 bit.

Fresnel zone An imaginary football-shaped area (American football) that surrounds the path of the visual LOS between two point-to-point antennas. Theoretically, there are an infinite number of Fresnel zones, or concentric ellipsoids (the football shape), that surround the visual LOS. The closest ellipsoid is known as the first Fresnel zone, the next one is the second Fresnel zone, and so on. If the first Fresnel zone becomes even partly obstructed, the obstruction will negatively influence the integrity of the RF communication.

functional policy A functional security policy defines the technical aspects of wireless security. The functional security policy establishes how to secure the wireless network in terms of what solutions and actions are needed. A functional policy defines essentials, baseline practices, design, implementation, and monitoring procedures.

G

gain Also known as amplification. Gain is the increase of amplitude or signal strength. The two types of gain are active gain and passive gain.

general policy A general wireless security policy establishes why a wireless security policy is needed for an organization. The general wireless security policy defines a statement of authority, applicable audience, violating policy procedures, risk assessment, threat analysis, and auditing.

Generic Routing Encapsulation (GRE) A process in which frames such as 802.11 frames are encapsulated in a packet, transmitted between two devices on a network, and then removed from the packet and forwarded.

grid antenna A highly directional antenna that resembles the rectangular grill of a barbecue, with the edges slightly curved inward. The spacing of the wires on a grid antenna is determined by the wavelength of the frequencies that the antenna is designed for.

guard interval (GI) A period of time between OFDM symbols that accommodates for the late arrival of symbols over long paths.

H

hertz (Hz) A standard measurement of frequency, which was named after the German physicist Heinrich Rudolf Hertz. An event that occurs once in 1 second is equal to 1 Hz. An event that occurs 325 times in 1 second is measured as 325 Hz.

hidden node Hidden node occurs when one client station's transmissions are unheard by other client stations in the basic service set (BSS). Every time the hidden node transmits, there is a risk another station is also transmitting and a collision can occur.

High Throughput (HT) High Throughput (HT) provides PHY and MAC enhancements to support wireless throughput of 100 Mbps and greater. HT is defined by the 802.11n draft amendment for clause 20 radios.

High Throughput Greenfield An 802.11n mode that is not compatible with legacy 802.11a/b/g radios; only HT radios can communicate when using the HT Greenfield.

High Throughput Mixed An 802.11n protection mode that is used when one or more non-HT stations are associated to the HT access point.

High Throughput protection modes Four protection modes used by 802.11n to ensure backward compatibility with older 802.11 a/b/g radios.

highly directional antenna Strictly used for point-to-point communications, typically to provide network bridging between two buildings. Highly directional antennas provide the most focused, narrow beamwidth of any of the antenna types. There are two types of highly directional antennas: parabolic dish and grid antennas.

High-Rate DSSS (HR-DSSS) The 802.11b 5.5 and 11 Mbps speeds are known as High-Rate DSSS, or HR-DSSS.

hop time In a frequency hopping spread spectrum network, the amount of time it takes for the transmitter to change from one frequency to another.

hopping sequence A predefined hopping pattern or set used in frequency hopping spread spectrum. The hopping sequence comprises a series of small carrier frequencies, or *hops*. Instead of transmitting on one set channel or finite frequency space, an FHSS radio card transmits on a sequence of subchannels called hops. Each time the hop sequence is completed, it is repeated.

hotspot A free or pay-for-use wireless network that is normally used to provide guest access to the Internet.

Hybrid Coordination Function (HCF) The 802.11e amendment defines enhanced medium access methods to support QoS requirements. Hybrid Coordination Function (HCF) is an additional coordination function that is applied in an 802.11e QoS wireless network. HCF has two access mechanisms to provide QoS: Enhanced Distributed Channel Access (EDCA) and Hybrid Coordination Function Controlled Channel Access (HCCA).

Hybrid Coordination Function Controlled Channel Access (HCCA) As defined by the 802.11e amendment, Hybrid Coordination Function Controlled Channel Access (HCCA) is similar to PCF. HCCA gives the access point the ability to provide for prioritization of stations via a polling mechanism. Certain client stations are given a chance to transmit before others.

hybrid coordinator (HC) A QoS-aware centralized coordinator that works within HCCA. The HC is built into the access point and has a higher priority of access to the wireless medium. Using this higher priority level, it can allocate TXOPs to itself and other stations to provide a limited-duration controlled access phase (CAP), providing contention-free transfer of QoS data.

Hybrid Wireless Mesh Protocol (HWMP) As defined by the 802.11s draft amendment, HWMP is a mandatory mesh routing protocol that uses a default path selection metric. Vendors may also use proprietary mesh routing protocols and metrics.

I

implicit feedback Occurs in 802.11n when a transmitter and receiver work together to educate each other about the characteristics of the MIMO channel. The transmitter will make phase adjustments based on the information that is returned from the receiver. When using implicit feedback, the beamformer receives long training symbols transmitted by the beamformee, which allow the MIMO channel between the beamformee and beamformer to be estimated.

independent basic service set (IBSS) The 802.11 standard defines three topologies known as service sets. One topology, known as an independent basic service set (IBSS), involves direct communications between 802.11 client stations without the use of an access point. An 802.11 IBSS network is also known as a peer-to-peer network or an ad hoc network.

Industrial, Scientific, and Medical (ISM) The ISM bands are defined by the ITU-T in S5.138 and S5.150 of the Radio Regulations. Although the FCC ISM bands are the same as defined by the ITU-T, the usage of these bands in other countries may be different because of local regulations. The 900 MHz band is known as the Industrial band, the 2.4 GHz band is known as Scientific band, and the 5.8 GHz band is known as the Medical band. It should be noted that all three of these bands are license-free bands, and there are no restrictions on what types of equipment can be used in any of the three ISM bands.

The ISM bands are as follows:
902–928 MHz (26 MHz wide)
2.4000–2.4835 GHz (83.5 MHz wide)
5.725–5.875 GHz (150 MHz wide)

information elements Variable-length fields that are optional in the body of a management frame.

information fields Fixed-length mandatory fields in the body of a management frame.

infrared (IR) A communication technology that uses a light-based medium. Information about modern implementations of infrared can be found at the Infrared Data Association's website at `www.irda.org`.

infrastructure mode A common term used to refer to a client station that is configured to connect to a basic service set or an extended service set.

Initialization Vector (IV) The IV is utilized by the RC4 streaming cipher that WEP encryption uses. The IV is a block of 24 bits that is combined with a static key. It is sent in cleartext and is different on every frame. The effective key strength of combining the IV with the 40-bit static key is 64-bit encryption. TKIP uses an extended IV.

Institute of Electrical and Electronics Engineers (IEEE) The IEEE is a global professional society with more than 350,000 members. The IEEE's mission is to "promote the engineering process of creating, developing, integrating, sharing, and applying knowledge about electro and information technologies and sciences for the benefit of humanity and the profession."

integration service (IS) Enables delivery of MSDUs between the distribution system (DS) and a non-IEEE-802.11 local area network (LAN), via a portal.

intentional radiator (IR) A device that intentionally generates and emits radio frequency energy by radiation or induction.

Inter-Access Point Protocol (IAPP) A protocol used for announcement and handover processes that result in how autonomous APs inform other autonomous APs about roamed clients and that define a method of delivery for buffered packets.

interframe space (IFS) A period of time that exists between transmissions of wireless frames.

International Organization for Standardization (ISO) The ISO is a global, nongovernmental organization that identifies business, government, and society needs and develops standards in partnership with the sectors that will put them to use. The ISO is responsible for the creation of the Open Systems Interconnection (OSI) model, which has been a standard reference for data communications between computers since the late 1970s.

The OSI model is the cornerstone to data communications, and understanding it is one of the most important and fundamental tasks a person in the networking industry can undertake.

The layers of the OSI model are as follows:

Layer 1—Physical
Layer 2—Data-Link
Layer 3—Network
Layer 4—Transport
Layer 5—Session
Layer 6—Presentation
Layer 7—Application

International Telecommunications Union Radiocommunication Sector (ITU-R) The United Nations has tasked the ITU-R with global spectrum management. The ITU-R maintains a database of worldwide frequency assignments and coordinates spectrum management through five administrative regions.

The five regions are broken down as Region A (North and South America), Region B (Western Europe), Region C (Eastern Europe and Northern Asia), Region D (Africa), and Region E (Asia and Australasia).

Internet Protocol Security (IPsec) IPsec is a layer 3 VPN technology. IPsec can use RC4, DES, 3DES, and AES ciphers for encryption. It provides for encryption, encapsulation, data integrity, and device authentication.

intersymbol interference (ISI) Data corruption cause by the delay spread in a multipath environment. The difference in time between the primary signal and the reflected signals causes problems for the receiver when demodulating the RF signal's information. The delay spread time differential can cause bits to overlap with each other, and the end result is corrupted data.

isotropic radiator A point source that radiates signal equally in all directions. The sun is probably one of the best examples of an isotropic radiator. It generates equal amounts of energy in all directions.

K

keying method When data is sent, a signal is transmitted from the transceiver. In order for the data to be transmitted, the signal must be manipulated so that the receiving station has a way of distinguishing 0s and 1s. This method of manipulating a signal so that it can represent multiple pieces of data is known as a keying method. A keying method is what changes a signal into a carrier signal. It provides the signal with the ability to encode data so that it can be communicated or transported.

L

last-mile The term *last-mile* is often used by the telephone and cable companies to refer to the last segment of their service that connects the home subscriber to their network.

layer 3 roaming Any roaming technology that allows mobile-device users to move from one layer 3 network to another while maintaining their original IP address.

lightning arrestor A device that redirects (shunts) transient currents caused by nearby lightning strikes or ambient static away from your electronic equipment and into the ground. Lightning arrestors are used to protect electronic equipment from the sudden surge of power that a nearby lightning strike or static buildup can cause. The lightning arrestor is installed between the transceiver and the antenna.

lightweight access point Lightweight access points are used in a centralized WLAN architecture together with WLAN controllers. A lightweight AP has minimal intelligence and is functionally just a radio card and an antenna. All the intelligence resides in the centralized WLAN controller, and all of the AP configuration settings such as channel and power are distributed to the lightweight APs from the WLAN controller and stored in the RAM of the lightweight AP. The encryption and decryption capabilities might reside in the centralized WLAN controller or may still be handled by the lightweight APs, depending on the vendor. Lightweight access points tunnel 802.11 wireless traffic to the WLAN controller which is typically deployed at either the distribution or core layer.

line of sight (LOS) When light travels from one point to another, it travels across what is perceived to be an unobstructed straight line, known as visual line of sight.

link budget The calculation of the amount of RF signal that is received minus the amount of signal required by the receiver.

lobes Even though the majority of the RF signal that is generated from an antenna is focused within the beamwidth of the antenna, there is still a significant amount of signal that radiates from outside of the beamwidth and from what are known as the antenna's side or rear lobes.

Logical Link Control (LLC) The upper portion of the Data-Link layer is the IEEE 802.2 Logical Link Control (LLC) sublayer, which is identical for all 802-based networks, although not used by all IEEE 802 networks.

loss Also known as attenuation. Loss is the decrease of amplitude or signal strength.

M

MAC Protocol Data Unit (MPDU) An 802.11 frame. The components include a MAC header, an MSDU (data payload) and a trailer.

MAC Service Data Unit (MSDU) The MSDU contains data from the LLC and layers 3–7. A simple definition of the MSDU is the data payload that contains the IP packet plus some LLC data.

malicious eavesdropping The unauthorized use of protocol analyzers to capture wireless communications is known as malicious eavesdropping and is typically considered illegal. Most countries have laws making it unlawful to listen in on any type of electromagnetic communications such as phone conversations. Unauthorized monitoring of 802.11 wireless transmissions is considered malicious and normally illegal. The most common target of malicious eavesdropping attacks is public access hotspots.

management frame protection (MFP) Techniques used to deliver management frames in a secure manner, with the hope of preventing many layer 2 denial-of-service attacks.

management frames A majority of the frame types in an 802.11 network. Used by wireless stations to join and leave the network. Another name for an 802.11 management frame is a Management MAC Protocol Data Unit (MMPDU). Management frames do not carry any upper-layer information. There is no MSDU encapsulated in the MMPDU frame body, which carries only layer 2 information fields and information elements.

man-in-the-middle attack After successfully completing wireless hijacking, an attacker may use a second wireless card with their laptop to execute what is known as a man-in-the-middle attack. The second wireless card is associated with the original legitimate AP. The hacker bridges the second wireless card to the evil twin access point radio and routes all hijacked traffic right back to the gateway of the original network. The attacker can therefore sit in the middle and execute peer-to-peer attacks indefinitely while remaining completely unnoticed.

manual coverage analysis Determining the proper placement of access points, the transmission power of the access point radio card, and the proper use of antennas by performing a manual site survey. There are two major types of manual coverage analysis surveys: passive and active.

maximal ratio combining (MRC) A signal-processing technique used to combine multiple received signals, which looks at each unique signal and optimally combines them in a method that is additive as opposed to destructive.

maximum transmission unit (MTU) The largest-size packet or frame that can be transmitted across the network. The size varies depending on the protocol.

Media Access Control (MAC) The bottom portion of the Data-Link layer is the Media Access Control (MAC) sublayer, which is identical for all 802.11-based networks.

mesh access point (MAP) Any 802.11 device that provides both mesh functionality and AP functionality simultaneously.

mesh networking A network environment in which wireless mesh access points communicate with each other using proprietary layer 2 routing protocols, creating a self-forming and self-healing wireless infrastructure (a mesh) over which edge devices can communicate.

mesh point (MP) Any 802.11 device capable of using a mandatory mesh routing protocol called Hybrid Wireless Mesh Protocol (HWMP).

mesh point portal (MPP) Any 802.11 device that provides both mesh functionality and acts as a gateway to one or more external networks such as an 802.3 wired backbone.

Message Integrity Check (MIC) TKIP uses a data integrity check known as the Message Integrity Check (MIC) to mitigate known bit-flipping attacks against WEP. The MIC is sometimes referred to by the nickname *Michael.*

Microsoft Point-to-Point Encryption (MPPE) MPPE is a 128-bit encryption method that uses the RC4 algorithm. MPPE is used with Point-to-Point Tunneling Protocol (PPTP) VPN technology.

midspan Defined under the PoE standard as a pass-through device with integrated power-supplying equipment. The midspan device does not regenerate the Ethernet signal and must not disrupt the Ethernet signal. Midspan devices can send power over only the unused twisted pairs on the Ethernet cable. A midspan solution will work with 10BaseT and 100BaseTX but not Gigabit Ethernet.

milliwatt (mW) A unit of power equal to 1/1000 of a watt.

Mini PCI A small form factor PCI expansion card. The Mini PCI is a variation of the Peripheral Component Interconnect (PCI) bus technology and was designed for use mainly in laptops. A Mini PCI radio is often used inside access points and is also the main type of radio used by manufacturers as the internal 802.11 wireless adapter inside laptops.

Mixed mode The default operational mode of most 802.11g access points. Support for both DSSS/HR-DSSS and ERP is enabled; therefore, both 802.11b and 802.11g clients can communicate with the access point. See *protection mechanism*.

mobile assisted handover (MAHO) A technique used by digital phones and cellular systems working together to provide better handover between cells.

Mobile IP An Internet Engineering Task Force (IETF) standard protocol that allows mobile-device users to move from one layer 3 network to another while maintaining their original IP address. Mobile IP is defined in IETF Request for Comments (RFC) 3344.

modulation The manipulation of a signal so that the receiving station has a way of distinguishing 0s and 1s.

modulation coding schemes (MCS) As mandated by the 802.11n draft amendment, data rates for clause 20 HT radios are defined by multiple variables known as *modulation coding schemes (MCS)*. Non-HT radios that used OFDM technology (802.11a/g) defined data rates of 6 Mbps–54 Mbps based on the modulation that was used. HT radios, however, define data rates based on numerous factors including modulation, the number of spatial streams, channel size, and guard interval.

multipath A propagation phenomenon that results in two or more paths of a signal arriving at a receiving antenna at the same time or within nanoseconds of each other.

multiple channel architecture (MCA) A WLAN channel reuse pattern with overlapping coverage cells that utilizes three channels at 2.4 GHz or numerous channels at 5 GHz.

multiple traffic ID block acknowledgment (MTBA) A block acknowledgment technique used to acknowledge each of the individual MPDUs when using aggregate MPDU.

multiple-input multiple-output (MIMO) Any RF communications system that has multiple antennas at both ends of the communications link and being used concurrently.

N

near/far A low-powered client station that is a great distance from the access point could become an unheard client if other high-powered stations are very close to the access point. The transmissions of the high-powered stations can raise the noise floor to a higher level at which the lower-powered station cannot be heard. This scenario is referred to as the near/far problem.

Network Allocation Vector (NAV) A timer mechanism that maintains a prediction of future traffic on the medium based on duration value information seen in a previous frame transmission. When an 802.11 radio is not transmitting, it is listening. When the listening radio hears a frame transmission from another station, it looks at the header of the frame and determines whether the Duration/ID field contains a Duration value or an ID value. If the field contains a Duration value, the listening station will set its NAV timer to this value. The listening station will then use the NAV as a countdown timer, knowing that the RF medium should be busy until the countdown reaches 0.

Newton's inverse square law This law states that the change in power is equal to 1 divided by the square of the change in distance.

noise floor A measurable level of background noise. This is often compared to received signal amplitudes. See *signal-to-noise ratio (SNR)*.

nonroot bridge Wireless bridges support two major configuration settings: root and non-root. Bridges work in a parent/child type of relationship, so think of the root bridge as the parent and the non-root bridge as the child.

null probe request Probe request management frame with no SSID information.

O

omnidirectional antenna A type of antenna that radiates RF signal in all directions. The small rubber dipole antenna, often referred to as a *rubber duck* antenna, is the classic example of an omnidirectional antenna and is the default antenna of most access points. A perfect omnidirectional antenna would radiate RF signal like a theoretical isotropic radiator. Because of manufacturing limitations, a perfect omnidirectional antenna cannot be made.

Open System authentication Open System authentication is the simpler of the two 802.11 authentication methods. It provides authentication without performing any type of client verification. It is essentially an exchange of hellos between the client and the access point.

Orthogonal Frequency Division Multiplexing (OFDM) Orthogonal Frequency Division Multiplexing is one of the most popular communications technologies, used in both wired and wireless communications. As part of 802.11 technologies, OFDM is specified in the 802.11a and 802.11g amendments and can transmit at speeds of up to 54 Mbps. OFDM technology is also used by 802.11n HT radios. OFDM transmits across separate, closely and precisely spaced frequencies, often referred to as *subcarriers*.

oscillation A single change from up to down to up, or a single change from positive to negative to positive. Also known as a *cycle*.

oscilloscope A time domain tool that can be used to measure how a signal's amplitude changes over time.

P

packet A unit of data at the Network layer.

Packet Binary Convolutional Coding (PBCC) A modulation technique that supports data rates of 5.5, 11, 22, and 33 Mbps. Both the transmitter and receiver must support the technology to achieve the higher speeds. PBCC was developed by Alantro Communications, which was purchased by Texas Instruments. During the development of the 802.11b amendment, PBCC was adopted as an optional modulation technique.

panel antenna A type of semidirectional planar antenna designed to direct a signal in a specific direction. Used for short- to medium-distance communications.

parabolic dish antenna A highly directional parabolic dish antenna that is similar to the small digital satellite TV antennas that can be seen on the roofs of many houses. Parabolic dish antennas are normally used for long distance point-to-point bridge links.

passive gain Passive gain is accomplished by focusing the RF signal with the use of an antenna. Antennas are passive devices that do not require an external power source. The internal workings of an antenna can focus the signal more powerfully in one direction than another.

passive manual survey When performing coverage analysis with a passive manual survey, the radio card is collecting RF measurements, including received signal strength (dBm), noise level (dBm), signal-to-noise ratio (dB), and bandwidth data rates. The client adapter, however, is not associated to the access point during the survey.

passive scanning In order for a station to be able to connect to an access point, it needs to first discover an access point. Passive scanning involves the client station listening for the beacon frames that are continuously being sent by the access points.

patch antenna A type of semidirectional planar antenna designed to direct a signal in a specific direction. Used for short- to medium-distance communications.

PC Card The PC Card Standard specifies three types of PC Cards. The three card types are the same length and width and use the same 68-pin connector. The thickness of the cards are as follows: Type I = 3.3 mm, Type II = 5.0 mm, and Type III = 10.5 mm.

PCMCIA See *Personal Computer Memory Card International Association.*

peer-to-peer See *independent basic service set (IBSS).*

peer-to-peer attack A wireless client station can attack the resources of any peer 802.11 client station in the same 802.11 service set. Peer-to-peer attacks can occur in any ad hoc network or through any access point or Wi-Fi switch where client stations share an association. Wireless peer-to-peer attacks can be mitigated with a personal firewall on the client side or through the use of PSPF on the access point or Wi-Fi switch.

per session per user After an EAP frame exchange where mutual authentication is required, both the AS and the supplicant know information about each other because of the exchanging of credentials. This newfound information is used as seeding material or keying material to generate a matching dynamic encryption key for both the supplicant and the authentication server. These dynamic keys are generated per session per user, meaning that every time a client station authenticates, a new key is generated and every user has a unique and separate key.

Personal Computer Memory Card International Association (PCMCIA) PCMCIA is an international standards body and trade association. The PCMCIA has more than 100 member companies and was founded in 1989 to establish standards for peripheral cards and to promote interchangeability with mobile computers. A PCMCIA adapter is also known as a PC Card. A radio card can be used in any laptop or handheld device that has a PC Card slot. Most PC Cards have integrated antennas. Some cards have only external antenna connectors, while others have external antennas and external connectors.

phase The relationship between two waves with the same frequency. To determine phase, a wavelength is divided into 360 pieces referred to as *degrees*. If you think of these degrees as starting times, then if one wave begins at the 0-degree point and another wave begins at the 90-degree point, these waves are considered to be 90 degrees out of phase.

Phase Shift Keying (PSK) Phase Shift Keying varies the phase of the signal to represent the binary data. PSK is a state transition technique, where one phase can represent a 0 bit and another phase can represent a 1 bit. This shifting of phase determines the data that is being transmitted.

phased array antenna An antenna system made up of multiple antennas that are connected to a signal processor. The processor feeds the individual antennas with signals of different relative phases, creating a directed beam of RF signal aimed at the client device. They are capable of creating narrow beams and transmitting multiple beams to multiple users simultaneously. Because of this unique capability, they are often regulated differently by the local RF regulatory agency.

phased coexistence operation (PCO) An optional 802.11n mode of operation that divides time and alternates between 20 MHz and 40 MHz transmissions.

physical carrier-sense Performed constantly by all stations that are not transmitting or receiving data. Determines whether a frame transmission is inbound for a station to receive or whether the medium is busy before transmitting. This is known as the clear channel assessment (CCA).

Physical Layer Convergence Procedure (PLCP) The upper portion of the Physical layer. PLCP prepares the frame for transmission by taking the frame from the MAC sublayer and creating the PLCP Protocol Data Unit (PPDU).

Physical Medium Dependent (PMD) The lower portion of the Physical layer. The PMD sublayer modulates and transmits the data as bits.

PLCP Protocol Data Unit (PPDU) When the PLCP receives the PSDU, it prepares it to be transmitted and creates the *PLCP Protocol Data Unit (PPDU)*. The PLCP adds a preamble and PHY header to the PSDU.

PLCP Service Data Unit (PSDU) Equivalent to the MPDU. The MAC layer refers to the frame as the MPDU, while the Physical layer refers to this same exact frame as the PSDU.

Point Coordination Function (PCF) An optional 802.11 medium access method that uses a form of polling. Although defined by the standard, the medium access method has not been implemented.

point coordinator The polling device in an 802.11 PCF network.

point source A point that radiates signal equally in all directions. The sun is one of the best examples of this.

point-to-multipoint (PtMP) A wireless network configuration that has a central communications device such as a bridge or an access point providing connectivity to multiple devices such as other bridges or clients.

point-to-point (PtP) A wireless network configuration that connects only two devices together. This is typically a wireless bridge link.

Point-to-Point Tunneling Protocol (PPTP) PPTP is a layer 3 VPN technology. It uses 128-bit Microsoft Point-to-Point Encryption (MPPE), which uses the RC4 algorithm. MPPE encryption is considered adequate but not strong. PPTP also uses MS-CHAP version 2 for user authentication, which is susceptible to offline dictionary attacks.

polarization Wave polarization is defined as the position and direction of the electric field (E-field) as referenced to the surface of the earth. If an antenna element is positioned vertically, the E-field is also vertical. Vertical polarization occurs when the E-field is perpendicular to the earth. If an antenna element is positioned horizontally, the electric field is also horizontal. Horizontal polarization occurs when the E-field is parallel to the earth.

port-based access control The 802.1X standard defines port-based access control. 802.1X provides an authorization framework that allows or disallows traffic to pass through a port and thereby access network resources. 802.1X defines two virtual ports: an uncontrolled port and a controlled port. The uncontrolled port allows EAP authentication traffic to pass through, while the controlled port blocks all other traffic until the supplicant has been authenticated.

power budget Planning and calculations necessary to ensure that enough power is available for all PoE powered devices.

power injectors Midspan power-sourcing equipment. Powers equipment via Ethernet cable.

power management bit A bit in the 802.11 MAC header that is used by a client station to notify the AP that the station is going into Power Save mode.

Power over Ethernet (PoE) A solution that can be used to power remote network devices over the same Ethernet cabling that carries data to the remote device. Using PoE to provide power to 802.11 access points is often a simpler and more cost-effective solution than hiring an electrician to install new electrical drops and outlets for every AP.

Power Save mode An optional mode for 802.11 stations. A wireless station can shut down some of the transceiver components for a period of time to conserve power. The station indicates that it is using Power Save mode by changing the value of the Power Management field to 1.

Power Save Multi Poll (PSMP) Power management method defined for use by HT radios. PSMP is an extension of automatic power save delivery (APSD) that was defined by the 802.11e amendment.

power-sourcing equipment (PSE) One of the two main components in a POE solution. Power is sent from the PSE to the PD over Ethernet cable. The power-sourcing equipment might be housed inside an inline switch or an injector. The PSE provides power for the connected device.

powered device (PD) One of the two main components in a POE solution. Power is sent from the PSE to the PD over Ethernet cable. An example of a PD is an access point. The powered device (PD) requires power from the PSE. The PD must be able to accept power through either the data pairs or unused pairs.

predicted coverage analysis An RF coverage analysis method that uses an application that provides RF simulation and modeling design capabilities. Predicted coverage analysis is accomplished by using an application that creates visual models of RF coverage cells, bypassing the need for actually capturing RF measurements. Projected cell coverage zones are created by using modeling algorithms and attenuation values. Virtual access points are created and overlaid on the floor plan graphic. Multiple "what if" scenarios can be created by changing the power settings, channel settings, or antenna type of the virtual access points, which can also be moved to any location on the floor plan.

preshared keys (PSKs) A method of distributing encryption passphrases or keys by manually typing the matching passphrases or keys on both the access point and all client stations that will need to be able to associate to the wireless network. This information is shared ahead of time (preshared) by using a manual distribution method such as telephone, email, or face-to-face conversation.

private branch exchange (PBX) A telephone exchange that serves a particular business or office.

probe request An 802.11 management frame that is transmitted during active scanning. A client station that is looking for an SSID sends a probe request. Access points that hear the probe request will send a probe response, notifying the client of the access points' presence. If a client station receives probe responses from multiple access points, signal strength and quality characteristics are typically used by the client station to determine which access point has the best signal and thus which access point to connect to.

probe response An 802.11 management frame that is transmitted during active scanning. After a client station sends a probe request, access points that hear the probe request will send a probe response, notifying the client of the access points' presence. The information that is contained inside the body of a probe response frame is the exact same information that can be found in a beacon frame with the exception of the traffic indication map (TIM).

processing gain The task of adding additional, redundant information to data. In this day and age of data compression, it seems strange that we would use a technology that adds data to our transmission, but by doing so, the communication is more resistant to data corruption. The system converts 1 bit of data into a series of bits that are referred to as *chips*.

propagation The movement or motion of the RF waves through the air.

protection mechanism In order for the legacy 802.11 DSSS stations, 802.11b HR-DSSS stations and 802.11g ERP stations to coexist, the ERP stations enable a protection mechanism, also known as protected mode. RTS/CTS or CTS-to-Self is used by the ERP stations to avoid interfering with the DSSS and HR-DSSS stations.

Public Secure Packet Forwarding (PSPF) PSPF is a feature that can be enabled on WLAN access points or switches to block wireless clients from communicating with other wireless clients on the same wireless segment. With PSPF enabled, client devices cannot communicate with other client devices on the wireless network. *PSPF* is a term most commonly used by Cisco; other vendors have similar capabilities under different names. PSPF is useful in preventing peer-to-peer attacks through an access point.

public switched telephone network (PSTN) The public network of telephone systems and providers.

Q

quadrature amplitude modulation (QAM) A modulation technique that is a hybrid of phase and amplitude modulation. Used for transmission of OFDM 24 Mbps, 36 Mbps, 48 Mbps, and 54 Mbps data by 802.11a and 802.11g radios.

quality of service (QoS) The attempt to prioritize and provide certain levels of predictable throughput along a shared access medium.

quality-of-service basic service set (QBSS) An 802.11 basic service set that provides quality of service (QoS). An infrastructure QBSS contains an 802.11e-compliant access point.

R

radio chain A single radio and all of its supporting architecture including mixers, amplifiers, and analog/digital converters.

radio frequency spectrum management (RFSM) Software and hardware solutions that can dynamically change the configuration of lightweight or autonomous access points based on accumulated RF information gathered from the access points' radio cards. Based on the accumulated RF information, the centralized device controls the access points and adjusts their power and channel settings, dynamically changing the RF coverage cells.

radio resource measurement (RRM) A mechanism in which client station resource data is gathered and processed by an access point or WLAN controller.

range The area or distance that an RF signal can provide effective usable coverage.

Rayleigh fading Because of the differences in phase of the multiple paths, a combined signal will often attenuate, amplify, or become corrupted. These effects are sometimes called Rayleigh fading, named after British physicist Lord Rayleigh.

RC4 The RC4 algorithm is a streaming cipher used in technologies that are often used to protect Internet traffic, such as Secure Sockets Layer (SSL). The RC4 algorithm is used to protect 802.11 wireless data and is incorporated into two encryption methods known as WEP and TKIP.

real-time location system (RTLS) Software and hardware solutions that can track the location of any 802.11 radio device as well as active Wi-Fi RFID tags.

reassociation When a client station decides to roam to a new access point, it will send a reassociation request frame to the new access point. It is called a *reassociation* not because it is reassociating to the access point, but because it is reassociating to the SSID of the wireless network.

receive diversity A method of listening for the best received signal from multiple antennas.

receive sensitivity The amount of signal a wireless station must receive in order to distinguish between data and noise.

received amplitude The received signal strength is most often referred to as *received amplitude*. RF signal strength measurements taken during a site survey is an example of received amplitude.

received signal strength A measurement of the amount of signal received.

received signal strength indicator (RSSI) An optional 802.11 parameter with a value from 0 to 255. It is designed to be used by the hardware manufacturer as a relative measurement of the RF power that is received. The RSSI is one of the indicators that is used by a wireless device to determine whether another device is transmitting, also known as a *clear channel assessment*.

receiver The receiver is the final component in the wireless medium. The receiver takes the carrier signal that is received from the antenna and translates the modulated signals into 1s and 0s. It then takes this data and passes it to the computer to be processed.

reduced interframe space (RIFS) A new interframe space that is used in 802.11n and is even shorter in time than a SIFS. A RIFS interval can be used in place of a SIFS interval, resulting is less overhead during a frame burst.

reflection One of the most important RF propagation behaviors to be aware of is reflection. When a wave hits a smooth object that is larger than the wave itself, depending on the media, the wave may bounce in another direction. This behavior is categorized as reflection.

refraction If certain conditions exist, an RF signal can be bent in a behavior known as refraction. Refraction is the bending of an RF signal as it passes through a medium with a different density, thus causing the direction of the wave to change. RF refraction most commonly occurs as a result of atmospheric conditions.

regulatory domain authority Local regulatory domain authorities of individual countries or regions define the spectrum policies and transmit power rules.

request to send/clear to send (RTS/CTS) A mechanism that performs a NAV distribution and helps to prevent collisions from occurring. This NAV distribution reserves the medium prior to the transmission of the data frame. RTS/CTS can be used to discover hidden node problems. RTS/CTS is one of the two protection mechanisms used in mixed-mode environments.

residential wireless gateway (RWG) A fancy term for a home wireless router. The main function of a residential wireless gateway is to provide shared wireless access to a SOHO Internet connection while providing a level of security from the Internet. These SOHO Wi-Fi routers are generally inexpensive yet surprisingly full featured.

RF line of sight (RF LOS) An area around the visual LOS, also known as the Fresnel zone.

RF shadow The area directly behind an RF obstruction. Depending on the change in direction and velocity of the diffracted signals, the area of the RF shadow can become a dead zone of coverage or still possibly receive degraded signals.

roaming The ability for the client stations to transition from one access point and basic service set (BSS) to another while maintaining network connectivity for the upper-layer applications.

robust security network (RSN) A robust security network (RSN) is a network that only allows for the creation of robust security network associations (RSNAs). An RSN utilizes CCMP/AES encryption as well as 802.1X/EAP authentication.

robust security network associations (RSNAs) As defined by the 802.11i security amendment, two stations (STAs) must establish a procedure to authenticate and associate with each other as well as create dynamic encryption keys through a process known as the

4-Way Handshake. This association between two stations is referred to as a *robust security network association (RSNA)*.

rogue access point A rogue access point is any wireless device that is connected to the wired infrastructure but is not under the management of the proper network administrators. The rogue device acts as a portal into the wired network infrastructure. Because the rogue device has no authorization or authentication security in place, any intruder can use the open portal to gain access to network resources.

role-based access control (RBAC) Role-based access control (RBAC) is an approach to restricting system access to authorized users. The three main components of an RBAC approach are users, roles, and permissions. Separate roles can be created such as the sales role or the marketing role. Individuals or groups of users are assigned to one of these roles. Permissions can be defined as firewall permissions, layer 2 permissions, layer 3 permissions, and bandwidth permissions and can be time based. The permissions are then mapped to the roles. When wireless users authenticate via the WLAN, they inherit the permissions of whatever roles they have been assigned to.

root bridge Wireless bridges support two major configuration settings: root and nonroot. Bridges work in a parent/child type of relationship, so think of the root bridge as the parent and the nonroot bridge as the child.

rule of 10s and 3s Provides approximate values when performing RF math calculations. All calculations are based on the following four rules:

- For every 3 dB gain, double the absolute power (mW).
- For every 3 dB loss, halve the absolute power (mW).
- For every 10 dB gain, multiply the absolute power (mW) by a factor of 10.
- For every 10 dB loss, divide the absolute power (mW) by a factor of 10.

S

scattering Scattering can most easily be described as multiple reflections. Scattering can happen in two ways. The first type of scatter is on a smaller level and has a lesser effect on the signal quality and strength. This type of scatter may manifest itself when the RF signal moves through a substance and the individual electromagnetic waves are reflected off the minute particles within the medium. Smog in our atmosphere and sandstorms in the desert can cause this type of scattering.

The second type of scattering occurs when an RF signal encounters some type of uneven surface and is reflected into multiple directions. Chain-link fences, tree foliage, and rocky terrain commonly cause this type of scattering. When striking the uneven surface, the main signal dissipates into multiple reflected signals, which can cause substantial signal downgrade and may even cause a loss of the received signal.

scheduled automatic power save delivery (S-APSD) An enhanced power management method introduced by the IEEE 802.11e amendment.

sector antenna A special type of high-gain, semidirectional antenna that provides a pie-shaped coverage pattern. These antennas are typically installed in the middle of the area where RF coverage is desired and placed back to back with other sector antennas. Individually, each antenna services its own piece of the pie—but as a group, all of the pie pieces fit together and provide omnidirectional coverage for the entire area. A sector antenna generates very little RF signal behind the antenna (back lobe) and therefore does not interfere with the other sector antennas that it is working with.

sectorized array Multiple sector antennas combined together to provide 360 degrees of horizontal coverage. This can be done with a single transmitter or multiple transmitters.

Secure Digital (SD) A peripheral expansion slot often found on handheld PDAs.

semidirectional antenna A type of antenna that is designed to direct a signal in a specific direction. Semidirectional antennas are used for short- to medium-distance communications, with long-distance communications being served by highly directional antennas.

service set identifier (SSID) The SSID is a network name used to identify an 802.11 wireless network. The SSID wireless network name is the logical name of the WLAN. The SSID can be made up of as many as 32 characters and is case sensitive.

service sets Three separate 802.11 topologies that describe how radio cards may be used to communicate with each other. The three 802.11 topologies are known as a basic service set (BSS), extended service set (ESS), and independent basic service set (IBSS).

Shared Key authentication The more complex of the two 802.11 authentication methods. Shared Key authentication uses WEP to authenticate client stations and requires that a static WEP key be configured on both the station and the access point. In addition to WEP being mandatory, authentication will not work if the static WEP keys do not match. The authentication process is similar to Open System authentication but includes a challenge and response between the AP and client station.

short interframe space (SIFS) A short gap or period of time that is used during the transmission of data.

signal quality (SQ) The 802.11-2007 standard defines a metric called signal quality (SQ), which is a measure of pseudonoise (PN) code correlation quality received by a radio. In simpler terms, the signal quality could be a measurement of what might affect coding techniques such as the Barker code or Complementary Code Keying (CCK). Anything that might increase the bit error rate (BER) such as a low SNR or multipath might trigger SQ metrics.

signal-to-noise ratio (SNR) The SNR is the difference in decibels between a received signal and the background noise. The SNR is an important value because, if the background noise is too close to the received signal, data can get corrupted and retransmissions will increase.

simple data frame An 802.11 data frame whose subtype is *data*. Simple data frames carry MSDU payloads.

single channel architecture (SCA) A WLAN architecture in which all access points in the network can be deployed on one channel in either the 2.4 GHz or 5 GHz frequency bands. Uplink and downlink transmissions are coordinated by a WLAN controller on a single 802.11 channel in such a manner that the effects of co-channel and adjacent-channel interference are minimized.

single-input single-output (SISO) A system that makes use of a single radio chain.

site survey A process performed to determine RF coverage, potential sources of interference, and the proper placement, installation, and configuration of 802.11 hardware.

slot time A period of time that differs between the different spread spectrum technologies. It is a large enough time to allow for receive-to-transmit radio turnaround, MAC processing, and clear channel assessment (CCA).

small office/home office (SOHO) A common reference to the office environment of self-employed people, satellite employees, or an environment in which someone is likely to bring work home with them.

smart antenna An antenna system made up of multiple antennas that are connected to a signal processor. The processor feeds the individual antennas with signals of different relative phases, creating a directed beam of RF signal aimed at the client device.

social engineering A technique used to manipulate people into divulging confidential information such as computer passwords.

software defined radio (SDR) A future technology that will be able to dynamically switch across a wide range of frequency bands, transmission techniques, and modulation schemes so that a single radio could replace multiple products.

sounding frames Frames that are sent by 802.11n HT radios with transmit beamforming capabilities. Sounding frames are used to exchange implicit and explicit feedback. The transmitter is considered the beamformer, while the receiver is considered the beamformee. The beamformer and the beamformee work together to educate each other about the characteristics of the MIMO channel. The beamformer will send a sounding request frame and will make phase adjustments based on the information that is returned in a sounding response frame from the beamformee. Any frame can be used as a sounding frame.

spatial diversity Antenna diversity that results in multiple RF streams following different paths to the receiver because of the space between the transmitting antennas. Each stream travels a different path, because there is at least a half-wavelength of space between the multiple transmitting antennas.

spatial multiplexing power save (SM power save) Power-saving mechanism used to allow a MIMO 802.11n device to power down all but one of its radios.

spatial stream MIMO radios transmit multiple radio signals at the same time and take advantage of multipath. Each individual radio signal is transmitted by a unique radio and antenna of the MIMO system. Each independent signal is known as a *spatial stream*, and each stream can contain different data than the other streams transmitted by one or more of the other radios. Each stream will also travel a different path because there is at least a half-wavelength of space between the multiple transmitting antennas.

spectrum analysis Locating sources of interference in the 2.4 GHz ISM and 5 GHz UNII bands is considered mandatory when performing an 802.11 wireless site survey. Using a spectrum analyzer to determine the state of the RF environment within a certain frequency range is known as *spectrum analysis.*

spectrum analyzer Spectrum analyzers are frequency domain measurement devices that can measure the amplitude and frequency space of electromagnetic signals. A spectrum analyzer is a tool that should always be used to locate sources of interference during an 802.11 wireless site survey. Spectrum analyzers are also used for security purposes to locate layer 1 denial-of-service attacks. Most spectrum analyzers are stand-alone devices, but distributed solutions exist that can be used as layer 1 intrusion detection systems.

split MAC architecture With this type of WLAN architecture, some of the MAC services are handled by the WLAN controller and some are handled by the lightweight access point. For example, integration service (IS) and distribution system service (DSS) are handled by the controller. WMM QoS methods are usually handled by the controller. Depending on the vendor, encryption and decryption of 802.11 data frames might be handled by the controller or by the AP. Some 802.11 management frames such as beacons and ACKs might originate at the AP instead of the WLAN controller.

splitter A splitter takes an RF signal and divides it into two or more separate signals. Only under an unusually special or unique situation would you need to use an RF splitter.

spread spectrum Spread spectrum transmission uses more bandwidth than is necessary to carry its data. Spread spectrum technology takes the data that is to be transmitted and spreads it across the frequencies that it is using.

station (STA) The main component of an 802.11 wireless network is the radio card, which is referred to by the 802.11 standard as a station (STA). The radio card can reside inside an access point or be used as a client station.

subcarrier OFDM transmits across separate, closely and precisely spaced frequencies referred to as *subcarriers.*

supplicant When an 802.1X/EAP solution is deployed, a host with software that is requesting authentication and access to network resources is known as the *supplicant.*

supported rates The set of data rates that the access point will use when communicating with an associated station.

swarm logic Swarm logic occurs when 802.11 radio cards establish collective intelligence and decentralize any decision-making in a BSS. Client stations behave collectively with a higher-level intelligence to dynamically manage the RF environment. A client station can sense the other client stations' RF transmissions and dynamically make adjustments. All client radios and access point radios work together in a collective RF domain.

switched diversity When receiving, multiple copies of the same signal arrive at the receiver antennas with different amplitudes. The signal with the best amplitude is chosen, and the other signals are ignored. Also used when transmitting, but only one antenna is used. The transmitter will transmit out of the diversity antenna where the best amplitude signal was last heard.

system operating margin (SOM) The calculation of the amount of RF signal that is received minus the amount of signal required by the receiver.

T

task group Various 802.11 task groups are in charge of revising and amending the original standard that was developed by the MAC Task Group (MAC) and the PHY Task Group (PHY). Each group is assigned a letter from the alphabet, and it is common to hear the term *802.11 alphabet soup* when referring to all the amendments created by the multiple 802.11 task groups. Quite a few of the 802.11 task group projects have been completed, and amendments to the original standard have been ratified. Other 802.11 task group projects still remain active and exist as draft amendments.

Temporal Key Integrity Protocol (TKIP) TKIP is an enhancement of WEP encryption that addresses many of the known weaknesses of WEP. TKIP starts with a 128-bit temporal key that is combined with a 48-bit Initialization Vector (IV) and source and destination MAC addresses in a complicated process known as per-packet key mixing. TKIP also uses sequencing and uses a stronger data integrity check known as the Message Integrity Check (MIC). TKIP is the mandatory encryption method under WPA and is optional under WPA2.

throughput A measurement of the amount of user data that successfully traverses the network over a period of time.

topology The physical and/or logical layout of nodes in a computer network.

traffic indication map (TIM) The traffic indication map (TIM) is used when stations have enabled Power Save mode. The TIM is a list of all stations that have undelivered data buffered on the access point waiting to be delivered. Every beacon will include the AID of the station until the data is delivered.

transceiver A radio that is capable of both transmitting and receiving.

transition security network (TSN) An 802.11 wireless network that allows for the creation of pre-robust security network associations (pre-RSNAs) as well as RSNAs is known as a *transition security network*. A TSN supports 802.11i-defined security as well as legacy security such as WEP within the same BSS.

transmit amplitude The amount of initial amplitude that leaves the radio transmitter.

transmit beamforming (TxBF) Multiple antennas that are connected to a signal processor. The processor feeds the individual antennas with signals of different relative phases, creating a directed beam of RF signal aimed at the client device. The 802.11n draft amendment proposes this as an optional PHY capability. The technology uses phased-array antenna technology and is often referred to as *smart antenna* technology.

transmit diversity On a multiple-antenna transmitter, a method of transmitting out of the antenna where the last best received signal was heard.

transmit opportunity (TXOP) A limited-duration controlled access phase, providing contention-free transfer of QoS data.

transmit power control (TPC) Part of the 802.11h amendment. TPC is used to regulate the power levels used by 802.11a radio cards. The ERC and the FCC mandate that radio cards operating in the 5 GHz band use TPC to abide by a maximum regulatory transmit power and are able to alleviate transmission power to avoid interference. The TPC service is used to meet the ERC and FCC regulatory requirements.

transmit spectrum mask A mask that defines the frequencies and power levels that a transmission signal and its sidebands must operate within.

transmitter The initial component in the creation of the wireless medium. The computer hands the data off to the transmitter, and it is the transmitter's job to begin the RF communication.

U

unequal modulation In a MIMO system, multiple spatial streams are sent with different (unequal) modulation.

unicast key A dynamically generated encryption key that is generated per session per user is often referred to as the *unicast key*. Unicast keys are used to encrypt and decrypt all unicast 802.11 data frames.

unit of comparison Units of measure that provide comparative measurements, not absolute measurements. Decibel is an example of a unit of comparison.

unit of power Units of measure that provide absolute measurement values, not relative or comparative measurements. Watt is an example of a unit of power.

Unlicensed National Information Infrastructure (UNII) The IEEE 802.11a amendment designated OFDM data transmissions within the frequency space of the 5 GHz UNII bands. The 802.11a amendment defined three groupings, or bands, of UNII frequencies, known as UNII-1 (lower), UNII-2 (middle), and UNII-3 (upper). All three of these bands are 100 MHz wide and each has 4 channels. The IEEE 802.11h amendment introduced the capability for

802.11 radios to transmit in a new frequency band called UNII-2 Extended with 11 more channels. The 802.11h amendment effectively is an extension of the 802.11a amendment. The UNII bands are as follows:

- UNII-1 (lower) is 5.15–5.25 GHz.
- UNII-2 (middle) is 5.25–5.35 GHz.
- UNII-2 Extended is 5.47-5.725 GHz.
- UNII-3 (upper) is 5.725–5.825 GHz.

unscheduled automatic power save delivery (U-APSD) An enhanced power-management method introduced by the IEEE 802.11e amendment. The Wi-Fi Alliance's WMM Power Save (WMM-PS) certification is based on U-APSD.

upfade Increased signal strength caused by multiple RF signal paths arriving at the receiver at the same time, and the signals are in phase or partially out of phase with the primary wave.

user priority (UP) Differentiated access for stations provided by EDCA. User priority uses eight levels. The user priority tags are identical to the 802.1D priority tags.

V

virtual AP Multiple SSIDs configured on a single physical AP, where each SSID is mapped to a unique BSSID.

virtual BSSID The BSSID is typically the MAC address of the access point's radio card and the layer 2 identifier of a basic service set (BSS). Because access points are capable of advertising multiple SSIDs, and because each SSID requires a separate BSSID, the access point will generate virtual BSSID addresses.

virtual carrier-sense A CSMA/CA mechanism used by listening 802.11 stations. When the listening radio hears a frame transmission from another station, it looks at the header of the frame and determines whether the Duration/ID field contains a Duration value or an ID value. If the field contains a Duration value, the listening station will set its NAV timer to this value. The listening station will then use the NAV as a countdown timer, knowing that the RF medium should be busy until the countdown reaches 0.

virtual local area network (VLAN) Virtual local area networks (VLANs) are used to create separate broadcast domains in a layer 2 network and are often used to restrict access to network resources without regard to physical topology of the network. In a WLAN environment, individual SSIDs can be mapped to individual VLANs and users can be segmented by the SSID/VLAN pair, all while communicating through a single access point.

virtual private network (VPN) A private network that is created by the use of encryption, tunneling protocols, and security procedures. VPNs are typically used to provide secure communications when physically connected to an insecure network.

Voice over IP (VoIP) Voice over Internet Protocol. The transmission of voice conversations over a data network using TCP/IP protocols.

Voice over Wi-Fi (VoWiFi) Any software or hardware that uses Voice over IP communications over an 802.11 wireless network is known as VoWiFi. Because of latency concerns, VoWiFi requires QoS mechanisms to function properly in an 802.11 BSS.

voltage standing wave ratio (VSWR) VSWR is a numerical relationship between the measurement of the maximum voltage along the line (what is generated by the transmitter) and the measurement of the minimum voltage along the line (what is received by the antenna). VSWR is therefore a ratio of impedance mismatch, with 1:1 (no impedance) being optimal but unobtainable, and typical values from 1.1:1 to as much as 1.5:1. VSWR military specs are 1.1:1.

VPN wireless routers VPN wireless routers have all of the same features that can be found in a SOHO wireless router, along with providing secure tunneling functionality in addition to 802.11 layer 2–defined security capabilities. Supported VPN protocols may include PPTP, L2TP, IPsec, and SSH2.

W

wardriving Wardriving is the act of looking for wireless networks, usually while in a moving vehicle. Software utilities known as WLAN discovery tools exist for the purpose of finding open WLAN networks. The most common wardriving software tool is a freeware program called NetStumbler.

watt A basic unit of power, named after James Watt, an 18th-century Scottish inventor. One watt is equal to 1 ampere (amp) of current flowing at 1 volt.

wave propagation The way in which the RF waves move—known as *wave propagation*—can vary drastically depending on the materials in the signal's path. Drywall will have a much different effect on an RF signal than metal.

waveform The shape and form of the AC signal, also known as a sine wave.

wavelength The distance between similar points on two back-to-back waves. When measuring a wave, the wavelength is typically measured from the peak of a wave to the peak of the next wave.

Wi-Fi A brand marketing name that is used by the Wi-Fi Alliance to promote 802.11 WLAN technology.

Wi-Fi Alliance The Wi-Fi Alliance is a global, nonprofit industry trade association with more than 300 member companies. The Wi-Fi Alliance is devoted to promoting the growth of wireless LANs (WLANs). One of the Wi-Fi Alliance's primary tasks is to ensure the interoperability of WLAN products by providing certification testing. During the early days of the 802.11 standard, the Wi-Fi Alliance further defined the 802.11 standard and provided a set of guidelines to ensure compatibility among vendors. Products that pass the Wi-Fi certification process receive a Wi-Fi CERTIFIED certificate.

Wi-Fi Multimedia (WMM) The Wi-Fi Alliance maintains the Wi-Fi Multimedia (WMM) certification as a partial mirror of the 802.11e QoS amendment. WMM currently provides for traffic prioritization via four access categories.

Wi-Fi Multimedia Power Save (WMM-PS) WMM-PS is an enhancement over the legacy power-saving mechanisms. The goal of WMM-PS is to have client devices spend more time in a "doze" state and consume less power. WMM-PS also is designed to minimize latency for time-sensitive applications such as voice during the power-management process.

Wi-Fi phishing attack After completing a wireless hijacking attack at a hotspot, a hacker may also use web server software and captive portal software to perform a Wi-Fi phishing attack. After client stations have been hijacked to an evil twin access point, they are redirected to a login web page that looks exactly like a hotspot's login page. The hacker's fake login page will request a credit card number from the hijacked user. Phishing attacks are common on the Internet and are now appearing at Wi-Fi hotspots.

Wi-Fi Protected Access (WPA) Prior to the ratification of the 802.11i amendment, the Wi-Fi Alliance introduced Wi-Fi Protected Access (WPA) certification as a snapshot of the not-yet-released 802.11i amendment, supporting only the TKIP/RC4 dynamic encryption key management. 802.1X/EAP authentication was required in the enterprise, and passphrase authentication was required in a SOHO environment.

Wi-Fi Protected Access 2 (WPA2) WPA2 is based on the security mechanisms that were originally defined in the IEEE 802.11i amendment defining a *robust security network (RSN).* Two versions of WPA2 exist: WPA2-Personal defines security for a small office, home office (SOHO) environment, and WPA2-Enterprise defines stronger security for enterprise corporate networks. Each certified product is required to support WPA2-Personal or WPA2-Enterprise.

Wi-Fi Protected Setup Wi-Fi Protected Setup defines simplified and automatic WPA and WPA2 security configurations for home and small-business users.

Wired Equivalent Privacy (WEP) WEP is a layer 2 encryption method that uses the RC4 streaming cipher. The original 802.11 standard defined 64-bit and 128-bit WEP. WEP encryption has been cracked and is not considered a strong encryption method.

wireless distribution system (WDS) Although the distribution system (DS) typically uses a wired Ethernet backbone, it is possible to use a wireless connection instead. A wireless distribution system (WDS) can connect access points together, using what is referred to as a *wireless backhaul.* WLAN bridges, repeaters and mesh access points all use WDS connectivity.

wireless hijacking Wireless hijacking, also known as the evil twin attack, occurs when a hacker disrupts communications between client stations and a legitimate AP. Client stations lose their connection to the legitimate AP and reconnect to the hacker's access point. The hacker AP hijacks the client stations at layer 1 and layer 2, allowing the hacker to proceed with peer-to-peer attacks.

Wireless Internet Service Provider (WISP) WISPs deliver Internet services via wireless networking. Instead of directly cabling each subscriber, a WISP can provide services via RF communications from central transmitters. WISPs often use wireless technology other than 802.11, allowing them to provide wireless coverage to much greater areas. Service from WISPs is not without its own problems.

wireless intrusion detection system (WIDS) A WIDS is a client/server solution that is used to constantly monitor for 802.11 wireless attacks such as rogue APs, MAC spoofing, layer 2 DoS, and so on. A WIDS usually consists of three components: a server, sensors, and monitoring software. Wireless intrusion detection uses policies and alarms to properly classify attacks and to alert administrators to potential attacks.

wireless intrusion prevention system (WIPS) A WIPS is a wireless intrusion detection system (WIDS) that is capable of mitigating attacks from rogue access points. WIPS use spoofed deauthentication frames, SMNP, and proprietary methods to effectively render a rogue access device useless and protect the network backbone.

wireless LAN bridge A common nonstandard deployment of 802.11 technology. The purpose of bridging is to provide wireless connectivity between two or more wired networks. A bridge generally supports all the same features that a fat access point possesses; however, the purpose is to connect wired networks and not to provide wireless connectivity to client stations. Although bridge links are sometimes used indoors, generally they are used outdoors to connect the wired networks inside two buildings.

wireless local area network (WLAN) The 802.11 standard is defined as a wireless local area network technology. Local area networks provide networking for a building or campus environment. The 802.11 wireless medium is a perfect fit for local area networking simply because of the range and speeds that are defined by the 802.11 standard and its amendments. The majority of 802.11 wireless network deployments are indeed local area networks (LANs) that provide access at businesses and homes.

wireless metropolitan area network (WMAN) A wireless metropolitan area network provides coverage to a metropolitan area such as a city and the surrounding suburbs. The wireless technology that is typically associated with a WMAN is defined by the 802.16 standard. The 802.16 standard defines broadband wireless access and is sometimes referred to as Worldwide Interoperability for Microwave Access (WiMAX).

wireless network management system (WNMS) A central management device originally used to configure and maintain as many as 5,000 autonomous access points. A WNMS can be either a hardware appliance or a software solution. The current WNMS servers are used to manage multiple WLAN controllers from a single vendor and may also be used to manage other vendors' WLAN infrastructure, including autonomous APs.

wireless personal area network (WPAN) A wireless computer network used for communication between computer devices within close proximity of a person. Devices such as laptops, personal digital assistants (PDAs), and telephones can communicate with each other by using a variety of wireless technologies. Wireless personal area networks can be used for communication between devices or as portals to higher-level networks such as a local area network (LAN) and/or the Internet. The most common technologies used in wireless personal area networks are Bluetooth and infrared.

wireless wide area network (WWAN) A wireless computer network that covers broad geographical boundaries but (obviously) uses a wireless medium instead of a wired medium. Wireless wide area networks typically use cellular telephone technologies. Data rates and bandwidth using these technologies are relatively slow when compared to other wireless technologies such as 802.11. However, as cellular technologies improve, so will cellular data transfer rates.

wireless workgroup bridge (WGB) A device that provides wireless connectivity for wired infrastructure devices that do not have radio cards. The radio card inside the wireless workgroup bridge associates with an access point and joins the basic service set (BSS) as a client station. This provides fast and quick wireless connectivity for wired devices through the association the WGB has with the access point. Because the WGB is an associated client of the access point, the WGB does not provide connectivity for other wireless clients. It is also important to understand that only the radio card inside the WGB can contend for the 802.11 wireless medium, and the wired cards behind the WGB cannot contend for the half-duplex RF medium.

Wireless Zero Configuration (WZC) service The most widely used client utility is an integrated operating system client utility, more specifically known as the WZC service utility that is enabled by default in Windows XP.

WLAN array A WLAN controller and multiple access points using sector antennas all combined in a single hardware device.

WLAN controller WLAN controllers are used in a centralized WLAN architecture together with lightweight access points, also known as thin APs. All the intelligence resides in the centralized WLAN controller, and all of the AP configuration settings such as channel and power are distributed to the lightweight APs from the WLAN controller and stored in the RAM of the lightweight AP. The encryption and decryption capabilities might reside in the centralized WLAN controller or may still be handled by the lightweight APs, depending on the vendor. The distribution system service (DSS) and integration service (IS) function within the WLAN controller. Also known as a wireless switch, WLAN controllers provide AP management, user management, RF spectrum planning and management, layer 2 security, layer 3 security, captive portal, VRRP redundancy, WIDS, and VLAN segmentation. Another major advantage of the WLAN controller model is that most of the WLAN controllers support some form of fast secure roaming, which can assist in resolving latency issues often associated with roaming.

WLAN mesh router Wireless mesh routers communicate with each other by using proprietary layer 2 routing protocols, creating a self-forming and self-healing wireless infrastructure (a mesh) over which edge devices can communicate.

WLAN profile A set of configuration parameters that are configured on the WLAN controller. The profile parameters can include the WLAN logical name (SSID), WLAN security settings, VLAN assignment, and QoS parameters.

WLAN switch See *WLAN controller.*

WMM-PS (Power Save) The Wi-Fi Alliance oversees the WMM-PS (Power Save) certification, which uses 802.11e mechanisms to increase the battery life via advanced power-saving mechanisms. The Wi-Fi Alliance's WMM Power Save (WMM-PS) certification is based on automatic power save delivery (APSD).

WMM-SA (Scheduled Access) Wi-Fi Alliance had once proposed a certification called WMM-SA (Scheduled Access), which was to be based on HCCA mechanisms. However, the proposed WMM-SA certification no longer exists. Some VoWiFi vendors were interested in implementing HCCA mechanisms. However, other vendors were not. Support for HCCA mechanisms in the near future is doubtful.

Yagi antenna A type of semidirectional antenna designed to direct a signal in a specific direction. Used for short- to medium-distance communications.

Index

Note to the Reader: Throughout this index **boldfaced** page numbers indicate primary discussions of a topic. *Italicized* page numbers indicate illustrations.

A

B

C

D

E

G

H

I

J

K

L

M

N

O

P

S

T

U

V

W

Y

Z

Wiley Publishing, Inc.
End-User License Agreement

READ THIS. You should carefully read these terms and conditions before opening the software packet(s) included with this book "Book". This is a license agreement "Agreement" between you and Wiley Publishing, Inc. "WPI". By opening the accompanying software packet(s), you acknowledge that you have read and accept the following terms and conditions. If you do not agree and do not want to be bound by such terms and conditions, promptly return the Book and the unopened software packet(s) to the place you obtained them for a full refund.

1. License Grant. WPI grants to you (either an individual or entity) a nonexclusive license to use one copy of the enclosed software program(s) (collectively, the "Software," solely for your own personal or business purposes on a single computer (whether a standard computer or a workstation component of a multi-user network). The Software is in use on a computer when it is loaded into temporary memory (RAM) or installed into permanent memory (hard disk, CD-ROM, or other storage device). WPI reserves all rights not expressly granted herein.

2. Ownership. WPI is the owner of all right, title, and interest, including copyright, in and to the compilation of the Software recorded on the physical packet included with this Book "Software Media". Copyright to the individual programs recorded on the Software Media is owned by the author or other authorized copyright owner of each program. Ownership of the Software and all proprietary rights relating thereto remain with WPI and its licensers.

3. Restrictions On Use and Transfer.

(a) You may only (i) make one copy of the Software for backup or archival purposes, or (ii) transfer the Software to a single hard disk, provided that you keep the original for backup or archival purposes. You may not (i) rent or lease the Software, (ii) copy or reproduce the Software through a LAN or other network system or through any computer subscriber system or bulletin-board system, or (iii) modify, adapt, or create derivative works based on the Software.

(b) You may not reverse engineer, decompile, or disassemble the Software. You may transfer the Software and user documentation on a permanent basis, provided that the transferee agrees to accept the terms and conditions of this Agreement and you retain no copies. If the Software is an update or has been updated, any transfer must include the most recent update and all prior versions.

4. Restrictions on Use of Individual Programs. You must follow the individual requirements and restrictions detailed for each individual program in the About the CD-ROM appendix of this Book or on the Software Media. These limitations are also contained in the individual license agreements recorded on the Software Media. These limitations may include a requirement that after using the program for a specified period of time, the user must pay a registration fee or discontinue use. By opening the Software packet(s), you will be agreeing to abide by the licenses and restrictions for these individual programs that are detailed in the About the CD-ROM appendix and/or on the Software Media. None of the material on this Software Media or listed in this Book may ever be redistributed, in original or modified form, for commercial purposes.

5. Limited Warranty.

(a) WPI warrants that the Software and Software Media are free from defects in materials and workmanship under normal use for a period of sixty (60) days from the date of purchase of this Book. If WPI receives notification within the warranty period of defects in materials or workmanship, WPI will replace the defective Software Media.

(b) WPI AND THE AUTHOR(S) OF THE BOOK DISCLAIM ALL OTHER WARRANTIES, EXPRESS OR IMPLIED, INCLUDING WITHOUT LIMITATION IMPLIED WARRANTIES OF MERCHANTABILITY AND FITNESS FOR A PARTICULAR PURPOSE, WITH RESPECT TO THE SOFTWARE, THE PROGRAMS, THE SOURCE CODE CONTAINED THEREIN, AND/OR THE TECHNIQUES DESCRIBED IN THIS BOOK. WPI DOES NOT WARRANT THAT THE FUNCTIONS CONTAINED IN THE SOFTWARE WILL MEET YOUR REQUIREMENTS OR THAT THE OPERATION OF THE SOFTWARE WILL BE ERROR FREE.

(c) This limited warranty gives you specific legal rights, and you may have other rights that vary from jurisdiction to jurisdiction.

6. Remedies.

(a) WPI's entire liability and your exclusive remedy for defects in materials and workmanship shall be limited to replacement of the Software Media, which may be returned to WPI with a copy of your receipt at the following address: Software Media Fulfillment Department, Attn.: *CWNA: Certified Wireless Network Administrator Official Study Guide*, Wiley Publishing, Inc., 10475 Crosspoint Blvd., Indianapolis, IN 46256, or call 1-800-762-2974. Please allow four to six weeks for delivery. This Limited Warranty is void if failure of the Software Media has resulted from accident, abuse, or misapplication. Any replacement Software Media will be warranted for the remainder of the original warranty period or thirty (30) days, whichever is longer.

(b) In no event shall WPI or the author be liable for any damages whatsoever (including without limitation damages for loss of business profits, business interruption, loss of business information, or any other pecuniary loss) arising from the use of or inability to use the Book or the Software, even if WPI has been advised of the possibility of such damages.

(c) Because some jurisdictions do not allow the exclusion or limitation of liability for consequential or incidental damages, the above limitation or exclusion may not apply to you.

7. U.S. Government Restricted Rights. Use, duplication, or disclosure of the Software for or on behalf of the United States of America, its agencies and/or instrumentalities "U.S. Government" is subject to restrictions as stated in paragraph (c)(1)(ii) of the Rights in Technical Data and Computer Software clause of DFARS 252.227-7013, or subparagraphs (c) (1) and (2) of the Commercial Computer Software - Restricted Rights clause at FAR 52.227-19, and in similar clauses in the NASA FAR supplement, as applicable.

8. General. This Agreement constitutes the entire understanding of the parties and revokes and supersedes all prior agreements, oral or written, between them and may not be modified or amended except in a writing signed by both parties hereto that specifically refers to this Agreement. This Agreement shall take precedence over any other documents that may be in conflict herewith. If any one or more provisions contained in this Agreement are held by any court or tribunal to be invalid, illegal, or otherwise unenforceable, each and every other provision shall remain in full force and effect.

The Best CWNA Book/CD Package on the Market!

Get ready for your Certified Wireless Network Administrator (CWNA) certification with the most comprehensive and challenging sample tests anywhere!

The Sybex Test Engine features

- All the review questions, as covered in each chapter of the book
- Challenging questions representative of those you'll find on the real exam
- Three full-length bonus exams available only on the CD
- An assessment test to narrow your focus to certain objective groups

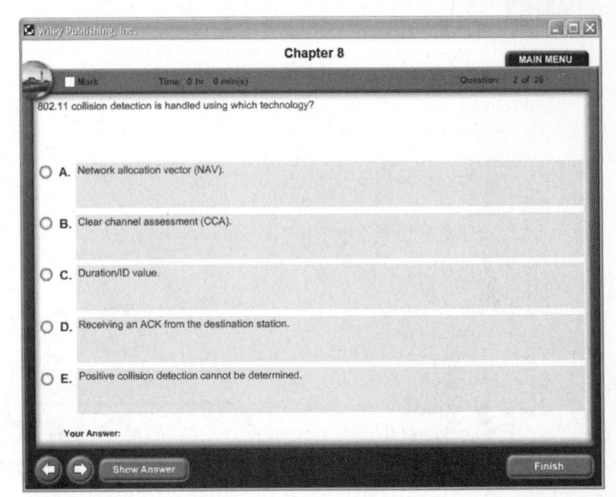

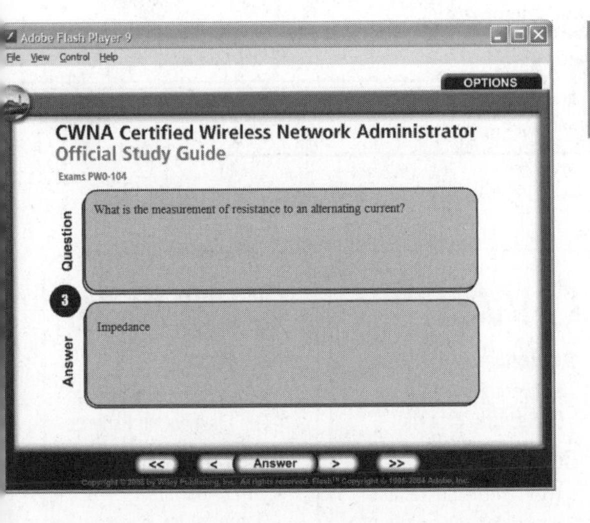

Use the electronic flashcards for PCs or Palm devices to jog your memory and prep last-minute for the exam!

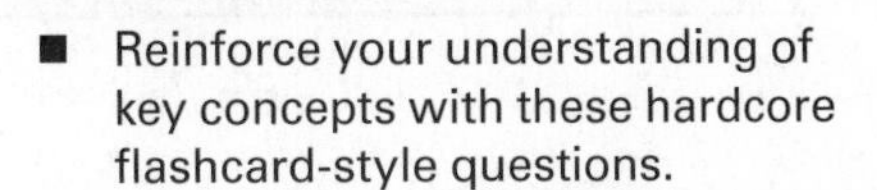

- Reinforce your understanding of key concepts with these hardcore flashcard-style questions.

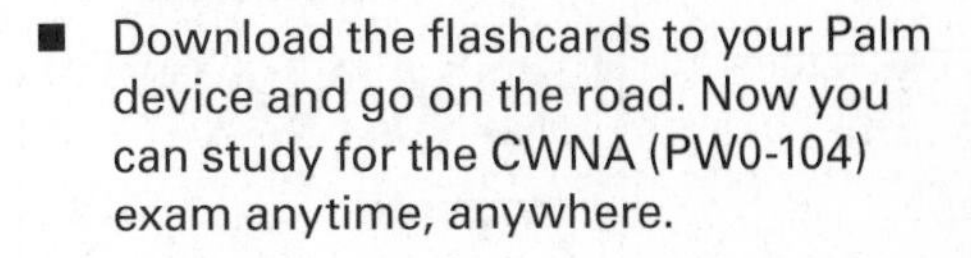

- Download the flashcards to your Palm device and go on the road. Now you can study for the CWNA (PW0-104) exam anytime, anywhere.